SPRINGER HANDBOOK OF
AUDITORY RESEARCH

Series Editors: Richard R. Fay and Arthur N. Popper

Springer

New York
Berlin
Heidelberg
Hong Kong
London
Milan
Paris
Tokyo

SPRINGER HANDBOOK OF AUDITORY RESEARCH

For more information about the series, please visit www.springer-ny.com/shar.

Stephen M. Highstein
Richard R. Fay
Arthur N. Popper
Editors

The Vestibular System

With 135 Illustrations and one color illustration

Springer

Stephen M. Highstein
Department of Otolaryngology
Washington University School
  of Medicine
St. Louis, MO 63110, USA
highstes@medicine.wustl.edu

Richard R. Fay
Department of Psychology and
  Parmly Hearing Institute
Loyola University of Chicago
Chicago, IL 60626, USA
rfay@wpo.it.luc.edu

Arthur N. Popper
Department of Biology and
  Neuroscience and
  Cognitive Science Program
University of Maryland
College Park, MD 20742-4415, USA
apopper@umd.edu

Series Editors: Richard R. Fay and Arthur N. Popper

Cover illustration: The inset figure (in text, Figure 4.1, Panel B) is based on Retzius (1881, 1884) as sketched by Wersall and Bagger-Sjoback (1974).

Library of Congress Cataloging-in-Publication Data
The Vestibular System [edited by] Stephen M. Highstein, Richard R. Fay, Arthur N. Popper.
    p.   cm.—(Springer handbook of auditory research ; v. 19)
    Includes bibliographical references.
    ISBN 0-387-98314-7 (alk. paper)
    1. Vestibular apparatus.   I. Highstein, Stephen M.   II. Fay, Richard R.   III. Popper,
Arthur N.   IV. Series.
    QP471.A58   2003
    612.8′58—dc21   2003050524

ISBN 0-387-98314-7   Printed on acid-free paper.

Printed in the United States of America.

9 8 7 6 5 4 3 2 1   SPIN 10635433

www.springer-ny.com

Springer-Verlag   New York Berlin Heidelberg
A member of BertelsmannSpringer Science+Business Media GmbH

We are pleased to dedicate this volume to Professor Åke Flock whose seminal experiments inspired many, including the editor (SMH) to enter the world of vestibular research. Dr. Flock's contributions have strongly influenced the whole field of inner ear research.

Series Preface

The *Springer Handbook of Auditory Research* presents a series of comprehensive and synthetic reviews of the fundamental topics in modern auditory research. The volumes are aimed at all individuals with interests in hearing research, including advanced graduate students, postdoctoral researchers, and clinical investigators. The volumes are intended to introduce new investigators to important aspects of hearing science and to help established investigators to better understand the fundamental theories and data in fields of hearing that they may not normally follow closely.

Each volume is intended to present a particular topic comprehensively, and each chapter will serve as a synthetic overview and guide to the literature. As such, the chapters present neither exhaustive data reviews nor original research that has not yet appeared in peer-reviewed journals. The volumes focus on topics that have developed a solid data and conceptual foundation rather than on those for which a literature is only beginning to develop. New research areas will be covered on a timely basis in the series as they begin to mature.

Each volume in the series consists of five to eight substantial chapters on a particular topic. In some cases, the topics will be ones of traditional interest for which there is a substantial body of data and theory, such as auditory neuroanatomy (Vol. 1) and neurophysiology (Vol. 2). Other volumes in the series will deal with topics that have begun to mature more recently, such as development, plasticity, and computational models of neural processing. In many cases, the series editors will be joined by a coeditor having special expertise in the topic of the volume.

RICHARD R. FAY, Chicago, Illinois
ARTHUR N. POPPER, College Park, Maryland

Preface

The books in the *Springer Handbook of Auditory Research* generally deal with issues related to the sense of hearing. However, the organ of equilibrium (the vestibular labyrinth) and the organ of hearing (the cochlea in mammals) share some common embryological origins, operational mechanisms, and structural elements that make them highly interrelated. Thus, it is appropriate to include a discussion of this part of the ear in the series.

The vestibular system consists of two sets of inner ear end organs, the semicircular canals and the otolithic organs, that project to the brain with the same cranial nerve as the hearing end organs. These end organs provide vertebrates with sensory information that enables them to move around freely in their environment by supplying information about acceleration and movement of the head and movements and orientations with respect to gravity. Interestingly, as Highstein points out in Chapter 1, the vestibular system may very well be the oldest of the secondary vertebrate sensory systems.

Thus, in including this volume in the *Springer Handbook of Auditory Research series*, we are providing readers of the series with an overview of an important part of the ear and its related central projections. Unlike other SHAR volumes, however, a complete overview of the vestibular system is provided in a single volume, with the goal of providing readers with a broad understanding of the basic biology and the clinical implications of the system.

Following an overview of the vestibular system by Highstein in Chapter 1, Chang and colleagues discuss the molecular biology of development of the ear in Chapter 2. This is followed by a discussion of the morphology and physiology of the vestibular sensory hair cells by Lysakowski and Goldberg in Chapter 3. In Chapter 4, Rabbitt and colleagues consider the biophysics of the semicircular canals, and, in Chapter 5, Steinacker discusses the ionic currents of the sensory hair cells themselves. The function of the vestibular system is detailed in several subsequent chapters. The vestibuloocular reflex is discussed by Cohen and Raphan in Chapter 6, and control of the head and the function of the vestibular nuclei are the subject of Chapter 7 by Balaban and Yates. In Chapter 8, Peterson and Boyle review the neural mechanisms through which multiple sensory cues may trigger compensatory changes in blood pressure and respiratory muscle activity during changes in posture. The molecular biology of the vestibular system is considered by De Zeeuw and colleagues in Chapter 9, and plasticity in the vestibular system is discussed in Chapter 10 by Green and colleagues. Finally, clinical implications of the vestibular system, and its relationship to basic science studies, are discussed in Chapter 11 by Halmagyi and colleagues.

As is often the case, chapters in one SHAR volume are complemented by chapters in other volumes. The anatomy of the vestibular regions of the ear discussed in this volume is complemented by discussions of the auditory part of the ear in chapters by Slepecky in Volume 8 (*The Cochlea*), and biomechanics of vestibular hair cells are complemented by studies of biomechanics of auditory hair cells by Kros and by Holley in the same volume. In addition, the chapter on molecular development of the ear by Chang and colleagues in this volume is paralled by a chapter that highlights the auditory system by Fritzsch and colleagues in Volume 9 (*Development of the Auditory System*).

STEPHEN M. HIGHSTEIN, St. Louis, Missouri
RICHARD R. FAY, Chicago, Illinois
ARTHUR N. POPPER, College Park, Maryland

Contents

Contributors

ARJAN M. VAN ALPHEN
Department of Neuroscience, Erasmus MC, 3000 DR Rotterdam, P.O. Box 1738, The Netherlands

SWEE T. AW
Neurology Department, Royal Prince Albert Hospital, Camperdown, Sydney, NSW 2050, Australia

CAREY D. BALABAN
Departments of Otolaryngology and Neurobiology, University of Pittsburgh, Eye and Ear Institute, Room 107, 203 Lohrop Street, Pittsburgh, PA 15213, USA

RICHARD D. BOYLE
Ames Research Center, National Aeronautics and Space Administration, Mail Stop 239-11, Moffett Field, CA 94035-1000, USA

RAQUEL CANTOS
Laboratory of Molecular Biology, National Institute on Deafness and Other Communication Disorders, 5 Research Court, Room 2B34, Rockville, MD 20850, USA

WEISE CHANG
Laboratory of Molecular Biology, National Institute on Deafness and Other Communication Disorders, 5 Research Court, Room 2B34, Rockville, MD 20850, USA

BERNARD COHEN
Department of Neurology, Box 1135, Mt. Sinai Medical Center, 1 East 100th Street, New York, NY 10029-6574, USA

MICHIEL P.H. COESMANS
Department of Neuroscience, Erasmus MC, 3000 DR Rotterdam, P.O. Box 1738, The Netherlands

IAN S. CURTHOYS
School of Psychology, University of Sydney, Sydney, NSW 2006, Australia

EDWARD R. DAMIANO
Department of Mechanical and Industrial Engineering, University of Illinois at Urbana-Champaign, Urbana, IL 61801, USA

MARCEL T.G. DE JEU
Department of Neuroscience, Erasmus MC, 3000 DR Rotterdam, P.O. Box 1738, The Netherlands

CHRIS I. DE ZEEUW
Department of Neuroscience, Erasmus MC, 3000 DR Rotterdam, P.O. Box 1738, The Netherlands

MAARTEN A. FRENS
Department of Neuroscience, Erasmus MC, 3000 DR Rotterdam, P.O. Box 1738, The Netherlands

HENRIETTA L. GALIANA
Department of Biomedical Engineering, McGill University, Faculty of Medicine, 3775 University Street, Montreal, QC H3A 2B4, Canada

NIELS GALJART
Department of Cell Biology, Erasmus MC, 3000 DR Rotterdam, P.O. Box 1738, The Netherlands

HIERONYMUS H.L.M. GOOSSENS
Department of Neuroscience, Erasmus MC, 3000 DR Rotterdam, P.O. Box 1738, The Netherlands

ANDREA M. GREEN
Department of Anatomy and Neurobiology, Washington University School of Medicine, 4566 Scott Avenue, St. Louis, MO 63110, USA

JAY M. GOLDBERG
Department of Neurobiology, Pharmacology, and Physiology, University of Chicago, 947 E. 58th Street, Chicago, IL 60637, USA

J. WALLACE GRANT
Department Engineering Science and Mechanics, Virginia Polytechnic Institute and State University, Blacksburg, VA 24061, USA

G. MICHAEL HALMAGYI
Hearing and Balance Clinic, Neurology Department, Royal Prince Albert Hospital, Camperdown, Sydney, NSW 2050, Australia

FREEK HOEBEEK
Department of Neuroscience, Erasmus MC, 3000 DR Rotterdam, P.O. Box 1738, The Netherlands

STEPHEN M. HIGHSTEIN
Department of Otolaryngology, Washington University School of Medicine, 4566 Scott Avenue, St. Louis, MO 63110, USA

Y. HIRATA
Department of Otolaryngology, Washington University School of Medicine, 4566 Scott Avenue, St. Louis, MO 63110 USA and Department of Electronic Engineering, Chubu University College of Engineering, 1200 Matsumoto-cho Kasugai Aichi, 487-8501, Japan

DICK JAARSMA
Department of Neuroscience, Erasmus MC, 3000 DR Rotterdam, P.O. Box 1738, The Netherlands

JOANNA JEN
Neurology Department, UCLA School of Medicine, Los Angeles, CA 90095-1769, USA

SEBASTIAAN K.E. KOEKKOEK
Department of Neuroscience, Erasmus MC, 3000 DR Rotterdam, P.O. Box 1738, The Netherlands

CHONGDE LUO
Department of Neuroscience, Erasmus MC, 3000 DR Rotterdam, P.O. Box 1738, The Netherlands

ANNA LYSAKOWSKI
Department of Anatomy and Cell Biology, University of Illinois at Chicago, 808 S. Wood Street, M/C 512, Chicago, IL 60612, USA

GARY PAIGE
Department of Neurology, Box 605, University of Rochester, 601 Elmwood Avenue, Rochester, NY 14642, USA

BARRY PETERSON
Northwestern University Medical School, 303 E. Chicago Avenue, Room 5-095, Chicago, IL 60611, USA

RICHARD D. RABBITT
Department of Bioengineering, 50 S. Central Campus Drive, Room 2480, University of Utah, Salt Lake City, UT 84113, USA

THEODORE RAPHAN
Institute of Neural and Intelligent Systems, Department of Computer & Information Science, Brooklyn College, Brooklyn, NY 11210, USA

MATTHEW T. SCHMOLESKY
Department of Neuroscience, Erasmus MC, 3000 DR Rotterdam, P.O. Box 1738, The Netherlands

JOHANNES VAN DER STEEN
Department of Neuroscience, Erasmus MC, 3000 DR Rotterdam, P.O. Box 1738, The Netherlands

ANTOINETTE STEINACKER
University of Puerto Rico Medical Sciences Campus, Institute of Neurobiology, 201 Boulevard del Valle, San Juan, PR 00901, USA

JOHN SUN
Department of Neuroscience, Erasmus MC, 3000 DR Rotterdam, P.O. Box 1738, The Netherlands

BILL J. YATES
Departments of Otolaryngology and Neuroscience, University of Pittsburgh, Eye and Ear Institute, 200 Lohrop Street, Pittsburgh, PA 15213, USA

DORIS K. WU
Laboratory of Molecular Biology, National Institute on Deafness and Other Communication Disorders, 5 Research Court, Room 2B34, Rockville, MD 20850, USA

1
Anatomy and Physiology of the Central and Peripheral Vestibular System: Overview

STEPHEN M. HIGHSTEIN

The vestibular system is arguably one of the most ancient vertebrate sensory systems. Its evolution became an invaluable acquisition, enabling vertebrates to detect and control their own motion in any environment. The peripheral vestibular apparatus—the semicircular canals and the otolith organs—evolved rapidly to reach a structure and function that are almost completely similar across extant vertebrate phyla. Labyrinthine enteroceptors report the magnitude and direction of angular and linear motion of the head as an animal translates (moves without rotation) (Moore et al. 2001b) and rotates through space. This information is carried to the central nervous system as a frequency code of impulses by the eighth cranial nerve. This information from the periphery is subsequently combined with information from other sensory systems that converge on vestibular nuclear sites and is used to compute a central estimate or vectorial representation of head and body position and motion in space, called the gravitoinertial vector (Gizzi et al. 1994; Cohen et al. 2001; Imai et al. 2001).

The influence of angular and linear forces is pervasive and extends throughout the central nervous system to include brain functions related to sleep, vision, audition, somatosensation, movement, digestion, cognition, and even learning and memory. The diversity of vestibular anatomy is evidenced by the connectivity within the brain stem, thalamus, basal ganglia, hippocampus, cerebellum, and cerebral cortex. Physiological activity originating within the vestibular system modifies the firing of many central nervous system neurons, including intrinsic cerebellar neurons, spinal and brain stem motor and interneurons, and superior collicular and cerebral cortical neurons, where, for example, the orientation of visual receptive fields can be modified by changes in head position. All of this sensory processing is largely unconscious and unrealized. We, as humans, only become conscious of the vestibular system when it malfunctions. Thus, conditions such as vestibular neuronitis (irritation of the nerve due to infection); Meniere's syndrome, or labyrinthine hydrops; benign positional paroxysmal vertigo, or BPPV (due to otoconial particles having been displaced into the semicircular canals from the mass of crystals atop the otolithic organs);

and others have a profound, incapacitating influence on almost every aspect of our lives.

Vestibular and auditory scientists have performed rigorous, sophisticated, and original neurophysiological, anatomical, morphophysiological, developmental, molecular, and genetic studies of the inner ear end organs, hair cells and afferents, and central vestibular neurons. This volume is intended as an overview of such vestibular science for nonvestibular scientists or for those considering entry into the field. As such, it will not be comprehensive in its treatment of the subject but will instead touch on some of the most elegant issues and studies performed to date in the vestibular sciences.

Vestibular research has seen many recent major advances with the advent of new tools, techniques, and ideas. These advances have been realized in studies of both the central and peripheral vestibular systems and in studies performed in both terrestrial and microgravity environments. Advances have been made in the understanding of the contributions of biomechanics, including intralabyrinthine pressure (Yamauchi et al. 2002) and the stiffness and elastic restoring forces of the stereocilia and cupula to the formation of the response dynamics of the semicircular canal nerves. It has been shown that semicircular canal plugging does not completely inactivate the response of the plugged canal but shifts the phase and gain of the response, effectively remapping the response dynamics (Rabbit et al. 1999). The low-frequency responses are attenuated, and the higher-frequency responses have normal gain but are phase-advanced relative to controls. Furthermore, the site of the canal plug along the long and slender portion of the semicircular canal duct has profound effects on the effectiveness of the plugging procedure.

Studies in microgravity have bearing on the tilt–translation hypothesis (that labyrinthine information is sufficient for animals and humans to discriminate between tilting and translation of the head), as humans in microgravity apparently do not perceive translation but still report the sensation of tilt. Centrifugation in microgravity should, in theory, eliminate the gravitational component of linear acceleration and result in a pure translational force vector. However, subjects continue to report tilt (Moore et al. 2001a).

Concepts of vestibular nuclear function have also been completely revised. Previously, it was assumed that these nuclei merely distributed vestibular information throughout the neuraxis without substantial processing or change in the information. Presently, studies show that signals related to head motion expressed in the nuclei when the head is fixed and the animal is passively rotated are not seen when the head is free and the animal makes a voluntary movement (Cullen et al. 1991, 1992, 1993; McCrea and Cullen 1992; Cullen and McCrea 1993). New tools have allowed scientists to begin to examine the brain's frames of reference relative to the gravitoinertial vector in head-free experiments. Finally, clinical testing of peripheral labyrinthine function has also been revolutionized, as it is now possible to test the function of single semicircular canals as well as the integrity of some central pathways using simple head-thrust procedures (Halmagyi et al. 1990a, 1990b).

In Chapter 2 of this volume, Chang and colleagues review the molecular genetics of vestibular end organ development and the roles of various genes on normal and abnormal vestibular organ genesis. Normal development of the peripheral organs of balance and equilibrium depends on the timely and sequential convergence of multiple signals on the developing tissue space. With the discovery of the *Hox* genes, the segmental organization of the vestibular brain stem has come into sharp focus. However, these brain stem genes also play a large role in the development of the labyrinth. Mutant mice have been very important in the analyses of the expression sequences and the roles of specialized genes. The abnormal patterning of these genes leads to recognizable end organ malformations and diseases. The present state of this research is an indication of a major new direction taken in vestibular research—one that promises to further our understanding of the origins and causes of several common clinical conditions.

Broad consideration of labyrinthine structure, such as the structure of the semicircular canals, indicates that there are five broad classes of structure that might potentially influence the formation of the response dynamics recorded from afferent fibers (Highstein et al. 1996). They are: (1) the biomechanics of endolymph flow, including endolymphatic pressure in response to head rotation, the elastic restoring forces of the cupula and stereocilia, for example; (2) variability in the responses (magnitude of gain) of the transduction apparatus atop each sensory hair cell; (3) variation in the voltage-sensitive basolateral currents in the sensory hair cells; (4) variation in the amount of transmitter released for a given receptor potential; and (5) variation of the postsynaptic response to a given quantum of transmitter released by each hair cell. These five broad categories can also be divided into pretransduction and posttransduction contributions to neural responses. Pretransduction and posttransduction contributions and the biomechanics are reviewed in Chapter 4 by Rabbitt and his colleagues. The postsynaptic variation in morphological contacts encompasses the body of the report in Chapter 3 by Lysakowski and Goldberg.

Chapter 3 is thus an overview of the morphology of vestibular sensory hair cells and their contacts with innervated primary afferent and efferent neurons. These contacts play a role in the shaping of neural response dynamics. The intracellular labeling method has enabled the physiological identification of a given nerve fiber and the subsequent correlation of its physiological responses with its morphological innervation pattern within the sensory epithelium (Schessel and Highstein 1981; Schessel et al. 1991). Responses of vestibular afferents differ in their gains to natural stimulation (spikes per second per degree per second of angular or linear velocity or acceleration) and the phase of their peak responses to these same stimuli (Goldberg and Fernandez 1975). Vestibular afferents are generally broadly tuned, in contrast to their auditory counterparts. Broad tuning means that there is no characteristic or best frequency, but instead each fiber generally responds with a similar gain to a broad range of stimuli. However, there is some frequency-dependent variation in gain, and the phase of the response is also generally frequency-dependent. The search for the origins of this diversity in response dynamics has been the goal of a number of research projects during the previous decades. Afferent labeling studies have been carried out in a variety of species, and the results have led to a consensus on the roles of the different patterns of afferent and efferent innervation in the shaping of afferent responses.

Hair cells are the common element in all of the octavolateralis organs, including the vestibular semicircular canals and otolithic organs, the lateral line organs, and the cochlea. The accessory structures into which the hair cell is embedded determine the range of frequencies to which each of these end organs responds. The key to understanding these diverse responses is the study of the biomechanics within each individual end organ. Biomechanics can be divided into macromechanics, micromechanics, and nanomechanics. Macromechanics of the semicircular canals concerns the physical forces acting on the gross structures of the canals, such as the endolymphatic fluid, and the responses of the canal components due to their physical properties. Thus, the physical laws that govern fluid flow loom large in this analysis. In contrast, in the otolithic organs, it is the inertia of the otoconia versus the physical attachment of these particles to the epithelium that determines the responses. Micromechanics of the responses may be due to the attachments of the hair cell sensory cilia to the accessory structures such as the cupula or otolithic membrane and otoconial mass, whereas nanomechanics might be the movements of molecules within the hair bundles themselves, such as gating spring mechanisms or other motions that impart energy into the transduction process. These areas are the purview of Chapter 4. Some of the novel material reviewed in this chapter is the role of endolymphatic pressure in determining endolymph flow as well as the localization of the site of the resultant maximum shear strain within the stereociliary bundles of the crista hair cells.

In Chapter 5, Steinacker considers the ionic currents localized to the basolateral surfaces of hair cells and the influence of these currents on the shaping of the response dynamics of the end organ nerves. The composition of these currents is reviewed with an eye toward explaining how they influence the effects of the changes in membrane potential caused by transduction on transmitter release. Information about both type I and type II hair cells is provided.

There is a diverse set of K^+ channels in hair cells, and the problem in analyzing them and understanding their function is intensified by their variety and kinetics within and between end organs. These differences are sufficiently marked not only in their kinetics but also in the voltage ranges over which they are active, suggesting that they have been configured to transmit specific information probably concerning aspects of head and body movements.

Results concerning the morphophysiological studies reviewed in Chapter 3, the biomechanics in Chapter 4, and the ionic currents in Chapter 5 did not produce the complete understanding or explanation of the origins of afferent response dynamics that was hoped for. Namely, the variation in the morphology, biomechanics, and currents of the population of afferents cannot account for the broad range of variation in responses seen in afferent recording. Although factors such as the size and number of afferent contacts can contribute, they are not in and of themselves sufficient to account for this variation. Furthermore, the biomechanics either alone or in combination with the morphology and currents does not adequately account for the broad range of afferent responses. Other factors still under investigation are necessary to explain the differences in the range of responses recorded in primary afferent nerves of the labyrinth versus those of the pretransduction and posttransduction components up to but not including the synapse. By elimination, evidence to date seems to point to a major role for the synapse in shaping response dynamics.

Cohen and Raphan (Chapter 6) discuss the vestibuloocular reflex and begin with an extensive review of the history of research in this area. The chapter then illuminates the differences between compensatory and orienting vestibular reflex responses. The peripheral vestibular apparatus, and particularly the semicircular canals, define a frame of reference within the head (Dai et al. 1991; Cohen et al. 1999). This useful concept of frames of reference is explored, elucidated, and expanded to include other landmark orienting frames and the physical forces and factors that help to define these reference frames. The major importance of the vestibular system in the orientation of the organism is highlighted, and the role of models in the understanding of orientation is documented.

The vestibuloocular reflex has traditionally been a subject of study because of the simplicity of its neural control structures—namely, the three-neuron arc. Control of the head is supposed to be substantially more complex. Chapter 8, by Peterson and Boyle, is a landmark chapter in this regard because it highlights the novel and emerging view that reexamines the roles of the vestibular nuclei in signal processing of sensory signals. Until recently, the role of these nuclei, defined as the brain stem territory that receives vestibular primary afferent nerve fiber terminals, was merely to relay and distribute incoming information to its targets throughout the central nervous systems. Chapter 8 discusses new results that show that the responses of these vestibular neurons are highly dependent on the behavioral context in which they discharge. It is striking that the discharge of vestibulospinal neurons that receive direct monosynaptic input from the labyrinthine nerves is dynamically modified by the behavioral context in which a movement is made. The central vestibular representation of movement is dynamically controlled by the behavioral context of the head movement. These neurons do not encode volitional head movements. "The head velocity in space signal produced by self-generated, active movements of the head is effectively canceled. As a result, vestibulospinal neurons selectively detect *external* perturbations of the head and translate only those passive components of the overall head movement into control signals to facilitate reflex behaviors and postural stability." (Peterson and Boyle, Chapter 8). Thus, even for very large head saccades, the vestibulospinal neuron's firing rate is largely unaffected. "Signals created by voluntary head movement are canceled from the cell's firing, leaving the cell able to detect unexpected passive movement signals. The functional significance of this finding is clear: individuals can actively explore the environs using rapid, orienting head (and gaze) movements and maintain head and body stability in the event of an unexpected and externally induced head perturbation." (Peterson and Boyle, Chapter 8). It is profoundly important to emphasize that this complexity of signal processing is carried out within the vestibular nuclei at neurons that are the recipients of primary labyrinthine afferents. Such sophisticated signal processing is usually touted to the cerebral cortex. Yet Chapter 8 provides an example of signal processing at the incoming level of the central nervous system. These findings effectively require the field to reevaluate the function of the nuclei and are a major step forward in understanding the processing of sensory vestibular information in particular, and they suggest that such processing of sensory information in general may need to be reevaluated.

The regulation of blood flow, blood pressure, and the distribution of bodily fluids are also under the control of the vestibular system. There are challenges imposed on the organism by gravitoinertial acceleration that must be met. Evidence suggests that somatic and visceral receptors in combination with the vestibular system participate in detecting the position of the body in space. This information helps to coordinate position-dependent autonomic and motor activity. In Chapter 7, Balaban and Yates review recent experimental evidence concerning the neural mechanisms through which multiple sensory cues may trigger compensatory changes in blood pressure and respiratory muscle activity during changes in posture.

Molecular science has made major contributions to our understanding of neurobiology; vestibular science has also greatly benefited from the development of these novel techniques. Chapter 9, by De Zeeuw et al., epitomizes the molecular approach to the study of the brain. In highly detailed work, the responses and learning abilities of specific mutant mice are reviewed. This chapter highlights the advantages and disadvantages of the molecular approach employing a highly studied neural system, the vestibuloocular reflex. The input and output connections of the flocculus of the vestibulocerebellum are topographically organized in line with the organization of the semicircular canals. The simple spike and complex spike activities of its Purkinje cells are optimally modulated following optokinetic and/or vestibular stimulation about the canal axes. The role of the flocculus in the control of compensatory eye movements affects both the optokinetic response and the vestibuloocular reflex, but the effects on the optokinetic response are more prominent than those on the vestibuloocular reflex. Usually, the flocculus exerts a gain-enhancing and a phase-leading effect on these reflexes. However, recent genetic lesion studies in mouse mutants indicate that the optokinetic response and vestibuloocular reflex performances can be separately altered and that their gain and phase parameters can also be partially separately influenced. In general, the gain values can be influenced by merely changing the expression of genes that may influence the synaptic activity of neurons but that do not necessarily affect the cytoarchitecture of the vestibulocerebellum; phase changes, on the other hand, usually require more robust genetic effects that do affect the cytoarchitecture and the hardware wiring.

The vestibuloocular reflex is a compensatory, plastic behavior. The gain of this reflex, defined as eye velocity divided by head velocity, is usually 1 in the light. Thus, for every head movement there is an equal, but opposite, eye movement. These reflex eye movements result in the eye motion resembling that of a gyroscope that is stable in space and ensure that the viewed target is usually located on the fovea of the retina, ensuring clear vision. However, the gain of this reflex is plastic and can change throughout life as the head size and the interrelationships of the eyes and ears change. This plasticity has also been extremely well-studied in the laboratory during the past 30 or more years (Lisberger and Fuchs 1978a, 1978b; Ito 1985, 1987, 1989, 1993a, 1993b; Miles et al. 1985; Nagao and Ito 1991; Nagao et al. 1991; Shojaku et al. 1993; Watanabe et al. 1993; Hirata and Highstein 2001). The fact that the circuitry that moves the eyes and drives the reflex is intimately connected to the cerebellum results in the formation of multiple interconnected and recursive neural loops (Hirata and Highstein 2001; Hirata et al. 2002). Therefore, microelectrode recording of neural responses at a given node within the circuit that may accompany a gain change is not always indicative that the site of the changes originates at the recorded site. However neural models can be highly useful and instructive in ferreting out the sites and changes of neuronal behavior. Mathematical models of the nervous system have been routinely employed since the advent of neurophysiology. In the case of motor learning of the vestibuloocular reflex, they are necessary for complete understanding. Chapter 10 by Green et al. begins with simple, static models and simple equations to introduce the reader to this subject. The chapter then goes on to more complex dynamic models and finally gives an example of the system identification approach to understanding neural structure and performance. This is a very important and timely subject.

Finally, Halmagyi et al. in Chapter 11 provide examples of how the study of basic science can apply to clinical practice. Interestingly, they also provide an example of how clinical medicine can actually lead to basic science discoveries. The vestibular systems is perhaps the primary system that exemplifies the melding of basic and applied science. Borrowing a page from the laboratory, the authors illustrate how clinical tests of vestibular function can elucidate abnormalities in the responses of a single semicircular canal. This is timely and important information because as clinical acumen improves, so does the ability to pinpoint the site and cause of a particular vestibular malfunction. For example, deciding whether a particular vestibular deficit is bilateral or unilateral, and if unilateral whether the defect is on the left or right, has long plagued neurologists and otolaryngologists. Further, it has often been difficult to ascribe a particular finding to the periphery or brain stem. The type of sophisticated clinical testing detailed here can solve these common dilemmas for the practitioner in most cases.

Chapter 11 also provides an excellent explanation of the physiology of the caloric test, a commonly applied technique in neurology and otolaryngology. The results of reorienting the head during this test are reviewed and the determinants of the response documented. The vestibular-evoked myogenic potential is also similarly detailed and explained. Finally, the genetic causes of some common vestibular conditions are reviewed and updated. In general, this chapter exemplifies the close relationship between the laboratory and the bedside that permits otolaryngology and neurology to be among the most integrated of medical and surgical specialties.

References

Cohen B, Wearne S, Dai M, Raphan T (1999) Spatial orientation of the angular vestibulo-ocular reflex. J Vestib Res 9:163–172.

Cohen B, Maruta J, Raphan T (2001) Orientation of the eyes to gravitoinertial acceleration. Ann N Y Acad Sci 942:241–258.

Cullen KE, McCrea RA (1993) Firing behavior of brain stem neurons during voluntary cancellation of the horizontal vestibuloocular reflex. I. Secondary vestibular neurons. J Neurophysiol 70:828–843.

Cullen KE, Belton T, McCrea RA (1991) A non-visual mechanism for voluntary cancellation of the vestibulo-ocular reflex. Exp Brain Res 83:237–252.

Cullen KE, Chen-Huang C, McCrea RA (1992) Participation of secondary vestibular neurons in nonvisual mechanisms of vestibuloocular reflex cancellation. Ann N Y Acad Sci 656:920–923.

Cullen KE, Chen-Huang C, McCrea RA (1993) Firing behavior of brain stem neurons during voluntary cancellation of the horizontal vestibuloocular reflex. II. Eye movement related neurons. J Neurophysiol 70:844–856.

Dai MJ, Raphan T, Cohen B (1991) Spatial orientation of the vestibular system: dependence of optokinetic after-nystagmus on gravity. J Neurophysiol 66: 1422–1439.

Gizzi M, Raphan T, Rudolph S, Cohen B (1994) Orientation of human optokinetic nystagmus to gravity: a model-based approach. Exp Brain Res 99:347–360.

Goldberg JM, Fernández C (1975) Vestibular mechanisms. Annu Rev Physiol 37: 129–162.

Halmagyi GM, Curthoys IS, Gremer PD, Henderson CJ, Staples M (1990a) Head impulses after unilateral vestibular deafferentation validate Ewald's second law. J Vestib Res 1:187–197.

Halmagyi GM, Curthoys IS, Cremer PD, Henderson CJ, Todd MJ, Staples MJ, D'Cruz DM (1990b) The human horizontal vestibulo-ocular reflex in response to high-acceleration stimulation before and after unilateral vestibular neurectomy. Exp Brain Res 81:479–490.

Highstein SM, Rabbitt RD, Boyle R (1996) Determinants of semicircular canal afferent response dynamics in the toadfish, *Opsanus tau*. J Neurophysiol 75: 575–596.

Hirata Y, Highstein SM (2001) Acute adaptation of the vestibuloocular reflex: signal processing by floccular and ventral parafloccular Purkinje cells. J Neurophysiol 85:2267–2288.

Hirata Y, Lockard JM, Highstein SM (2002) Capacity of vertical VOR adaptation in squirrel monkey. J Neurophysiol 88:3194–3207.

Imai T, Moore ST, Raphan T, Cohen B (2001) Interaction of the body, head, and eyes during walking and turning. Exp Brian Res 136:1–18.

Ito M (1985) Synaptic plasticity in the cerebellar cortex that may underlie the vestibulo-ocular adaptation. Rev Oculomot Res 1:213–221.

Ito M (1987) Cerebellar adaptive function in altered vestibular and visual environments. Physiologist 30:S81.

Ito M (1989) Long-term depression. Annu Rev Neurosci 12:85–102.

Ito M (1993a) Cerebellar flocculus hypothesis. Nature 363:24–25.

Ito M (1993b) Neurophysiology of the nodulofloccular system. Rev Neurol (Paris) 149:692–697.

Lisberger SG, Fuchs AF (1978a) Role of primate flocculus during rapid behavioral modification of vestibuloocular reflex. I. Purkinje cell activity during visually guided horizontal smooth-pursuit eye movements and passive head rotation. J Neurophysiol 41:733–763.

Lisberger SG, Fuchs AF (1978b) Role of primate flocculus during rapid behavioral modification of vestibuloocular reflex. II. Mossy fiber firming patterns during horizontal head rotation and eye movement. J Neurophysiol 41:764–777.

McCrea RA, Cullen KE (1992) Responses of vestibular and prepositus neurons to head movements during voluntary suppression of the vestibuloocular reflex. Ann N Y Acad Sci 656:369–395.

Miles FA, Optican LM, Lisberger SG (1985) An adaptive equalizer model of the primate vestibulo-ocular reflex. Rev Oculomot Res 1:313–326.

Moore ST, Clement G, Raphan T, Cohen B (2001a) Ocular counterrolling induced by centrifugation during orbital space flight. Exp Brain Res 137:323–335.

Moore ST, Hirasaki E, Raphan T, Cohen B (2001b) The human vestibulo-ocular reflex during linear locomotion. Ann N Y Acad Sci 942:139–147.

Nagao S, Ito M (1991) Subdural application of hemoglobin to the cerebellum blocks vestibuloocular reflex adaptation. Neuroreport 2:193–196.

Nagao S, Yoshioka N, Hensch T, Hasegawa I, Nakamura N, Nagao Y, Ito M (1991) The role of cerebellar flocculus in adaptive gain control of ocular reflexes. Acta Otolaryngol Suppl (Stockh) 481:234–236.

Rabbitt RD, Boyle R, Highstein SM (1999) Influence of surgical plugging on horizontal semicircular canal mechanics and afferent response dynamics. J Neurophysiol 82:1033–1053.

Schessel DA, Highstein SM (1981) Is transmission between the vestibular type I hair cell and its primary afferent chemical? Ann N Y Acad Sci 374:210–214.

Schessel DA, Ginzberg R, Highstein SM (1991) Morphophysiology of synaptic transmission between type I hair cells and vestibular primary afferents. An intracellular study employing horseradish peroxidase in the lizard, *Calotes versicolor*. Brain Res 544:1–16.

Shojaku H, Watanabe Y, Ito M, Mizukoshi K, Yajima K, Sekiguchi C (1993) Effect of transdermally administered scopolamine on the vestibular system in humans. Acta Otolaryngol Suppl (Stockh) 504:41–45.

Watanabe Y, Ohmura A, Ito M, Shojaku H, Mizukoshi K (1993) Quantitative evaluation of the function of looking at visual target in optokinetic nystagmus. Acta Otolaryngol Suppl (Stockh) 504:21–25.

Yamauchi A, Rabbitt RD, Boyle R, Highstein SM (2002) Relationship between inner-ear fluid pressure and semicircular canal afferent nerve discharge. J Assoc Res Otolaryngol 3:26–44.

2
Molecular Genetics of Vestibular Organ Development

Weise Chang, Laura Cole, Raquel Cantos, and Doris K. Wu

1. Introduction

Normal development of the vertebrate inner ear depends on signals emanating from multiple surrounding tissues, including the hindbrain, neural crest, mesenchyme, and notochord (for reviews, see Fritzsch et al. 1998; Torres and Giraldez 1998; Fekete 1999; Kiernan et al. 2002). Primarily through the analyses of mutant mice with spontaneous mutations or targeted deletions (knockouts), several genes involved in the patterning of the inner ear have been identified. Analyses of the phenotypes resulting from mutations within some of these genes, as well as analyses of their spatial and temporal expression patterns, indicate that they play specific, and sometimes multiple, roles in the patterning of the vestibular and auditory components of the inner ear (Table 2.1). Here, we summarize our current knowledge of the molecular mechanisms governing the development of the inner ear and the roles played by a variety of genes, focusing on the vestibular apparatuses of the chicken and mouse.

2. Gross Development of the Vestibular Apparatus

The membranous portion of the vertebrate inner ear originates from a thickening of the ectoderm adjacent to the hindbrain (Fig. 2.1). This thickened epithelium, known as the otic placode, invaginates to form the otic cup, which closes to form the otic vesicle/otocyst. A subpopulation of epithelial cells in the anteroventral lateral region of the otic cup and otic vesicle delaminate and coalesce to form the eighth (vestibulocochlear) ganglion. The otic vesicle proper undergoes a series of elaborate morphogenetic changes to give rise to an intricate, mature inner ear.

Figure 2.2 illustrates the gross development of the mouse inner ear from a late stage of otic vesicle formation through maturity, a period covering the complete development of the vestibular apparatus (Morsli et al. 1998). The vestibular component of the inner ear develops largely from the dorsal

TABLE 2.1. Genes affecting vestibular patterning.

Gene	Human disease	Type of protein	Distribution in the inner Ear and surrounding structures	Mutant or knockout phenotype	Ref.
Bmp4	—	secreted factor	three presumptive cristae, Hensen's and Claudius' regions of the cochlea	Bmp4 +/−: absence of lateral canal	Morsli et al. 1998; Teng et al. 2000
Brn4/Pou3f4	DFN3	POU domain transcription factor	periotic mesenchyme	Brn4 −/−, Slf: defects in fibroblasts of spiral ligament; shortened cochlea; constricted superior canal	de Kok et al. 1995; Phippard et al. 1998, 1999; Minowa et al. 1999
Dlx 5	—	homeobox transcription factor	dorsal posterior region of otic vesicle; semicircular canals and endolymphatic duct; sensory epithelium	no anterior or posterior canal; reduced lateral canal; poorly formed cristae; reduced maculae; abnormal endolymphatic duct and cochlea	Acampora et al. 1999b; Depew et al. 1999; Merlo et al. 2002
Eya 1	Branchial-oto-renal syndrome	transcription coactivator	ventrolateral otic vesicle; eighth ganglion; neurogenic and sensory regions	no eighth ganglion; amorphic inner ear	Abdelhak et al. 1997; Xu et al. 1997, 1999; Kalatzis et al. 1998;
Fgf3	—	growth factor	r5 and r6; prospective otic placode region; neurogenic and sensory regions	no endolymphatic duct or sac; reduced spiral ganglion; enlarged membranous labyrinth	Mansour et al. 1993; Mansour 1994; McKay et al. 1996
Fgf10		growth factor	neurogenic area; all prospective sensory patches; vestibular and spiral ganglia	lacks all three canals and the posterior crista; malformed anterior crista	Pauley et al. 2003
Fgfr2 (IIIb)	—	growth factor receptor	otic placode; dorsal and medial wall of otic vesicle; nonsensory regions of the inner ear	rudimentary inner ear with no sensory organs; loss of eighth ganglion; 50% of mutants lack endolymphatic duct	Pirvola et al. 2000

Gene	Disorder	Protein	Expression	Phenotype	References
Fidgetin	—	AAA protein	epithelial cells in canal outpocket; cochlear duct	*fidget*: missing lateral canal; malformed anterior and posterior canals	Cox et al. 2000
GATA3	HDR syndrome	zinc-finger transcription factor	regions of periotic mesenchmye around prospective canal region; hindbrain; vestibular sensory components except the saccule; cochlear duct	rudimentary inner ear with a poorly developed endolymphatic duct; misrouted efferent projections	Karis et al. 2001
Gli3	—	zinc-finger transcription factor	periotic mesenchyme	*Extratoes*: truncated anterior canal; no lateral canal, but lateral crista is present	Schimmang et al. 1992; Hui et al. 1994
Hmx2 (Nkx5.2)	—	homeobox transcription factor	anterodorsal region of otic vesicle; canals and ampullae; utricle, saccule, and endolymphatic duct; stria vascularis of the cochlea	absence of three canals and cristae, fusion of the utricle and saccule	Wang et al. 2001
Hmx3 (Nkx5.1)	— —	homeobox transcription factor	otic placode; canal outpocket; semicircular canals	reduced anterior canal, missing posterior and lateral canals, loss of lateral crista (Bober's group); missing lateral crista and ampulla, fusion of utricle and saccule (Lufkin's group)	Hadrys et al. 1998; Wang et al. 1998
Hoxa1	—	homeobox transcription factor	8 dpc: r3/4 boundary to spinal cord	no endolymphatic duct or sac; amorphic inner ear; no organ of Corti; reduced eighth ganglion	Gavalas et al. 1998
Hoxa1/ Hoxb1	—	homeobox transcription factors	*Hoxb1*: 8 dpc: r3/4 boundary to spinal cord; 9 dpc: expression up-regulated in r4	amorphic inner ear—more severe than *Hoxa1* alone	Rijli et al. 1993; Maconochie et al. 1996

TABLE 2.1. Continued

Gene	Human disease	Type of protein	Distribution in the inner Ear and surrounding structures	Mutant or knockout phenotype	Ref.
Hoxa2	—	homeobox transcription factor	r1/2 boundary to spinal cord; expression up-regulated in r3 and r5	enlarged membranous labyrinth; scala vestibuli lacking or collapsed	Deol 1964; Cordes and Barsh 1994; McKay et al. 1996
Jagged1/ Serrate1	Allagille syndrome	Transmembrane protein, Notch ligand	all prospective sensory organs; later restricts to supporting cells; subpopulation of endolymphatic duct cells	Htu, Slm: small or missing one or both anterior and posterior ampullae and canals; decreased outer hair cell number and increased inner hair cell number	Adam et al. 1998; Kiernan et al. 2001; Tsai et al. 2001
Kreisler	—	bZIP transcription factor	r5 and r6	misplaced otocyst; inner ear usually cyst-like	Deol 1964; Cordes and Barsh 1994; McKay et al. 1996; Ma et al. 1998, 2000
Lmx1a	—	LIM homeodomain protein	dorsal and lateral regions of otic vesicle	Dreher: distended endolymphatic duct and sac; constricted canals; poorly coiled cochlea	Giraldez 1998; Millonig et al. 2000
Netrin 1	—	secreted protein, related to laminin	central region of canal outpocket; semicircular canals; nonsensory region of utricle, saccule, and cochlea	defect in fusion plate formation; reduced anterior canal; no posterior or lateral canal	Salminen et al. 2000
NeuroD	—	HLH transcription factor	neurogenic area and eighth ganglion; all sensory regions	severe reduction in eighth ganglion; shortened cochlear duct	Liu et al. 2000; Kim et al. 2001

Gene	Human syndrome	Protein class	Expression domain	Mutant phenotype	References
Ngn1	—	bHLH transcription factor	anteroventrolateral otic vesicle	no eighth ganglion; fusion of utricle and saccule; reduced utricle and saccule; shortened cochlea	Ma et al. 2000
Nor-1	—	nuclear receptor transcription factor	central region of canal outpocket; semicircular canals	thin semicircular canals and flattened ampullae	Ponnio et al. 2002
Otx1	—	transcription factor	lateral wall of otic vesicle; lateral canal and ampulla; lateral wall of saccule and cochlea	no lateral canal or ampulla; no lateral crista; incomplete separation of utricle and saccule; misshapen saccule and cochlea	Acampora et al. 1996; Morsli et al. 1999
Otx2	—	transcription factor	ventral tip of otic vesicle; lateral wall of saccule and cochlea	$Otx1^{-/-}$, $Otx2^{+/-}$: incomplete separation of utricle and saccule; misshapen saccule and cochlea	Morsli et al. 1999; Cantos et al. 2000
Pax3	Waardenberg syndrome type I	paired box transcription factor	dorsal half of neural tube	Splotch: aberrant endolymphatic duct; misshapen cochlear and vestibular components	Deol 1966; Epstein et al. 1991; Goulding et al. 1991, 1993
Prx1/Prx2	—	paired-related homeobox transcription factor	Prx1—periotic mesenchyme; Prx2—otic epithelium; periotic mesenchyme	$Prx1^{-/-}$, $Prx2^{-/-}$: no lateral canal; reduced anterior and posterior canals; smaller otic capsule	ten Berge et al. 1998
Shh	Holoprosencephaly	secreted factor	notochord and floor plate of neural tube, otic epithelium and ganglion	absence of the lateral canal and crista; no discernible ampulla, utricle, saccule, cochlea, and endolymphatic duct; reduced cochleovestibular ganglion	Liu et al. 2002; Riccomagno et al. 2002

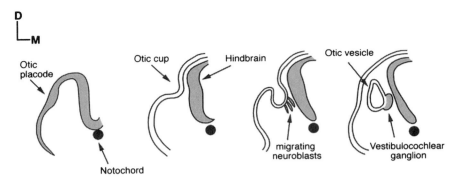

FIGURE 2.1. A schematic diagram summarizing the stages of inner ear development from an otic placode to an otic vesicle. These stages span approximately 8.5–9.5 dpc (days postcoitium) in mice, and embryonic days 1.5–2.5 (Hamburger and Hamilton stages 9–17) in chickens. Orientations: D, dorsal; M, medial. (Adapted from Wu and Choo 2003.)

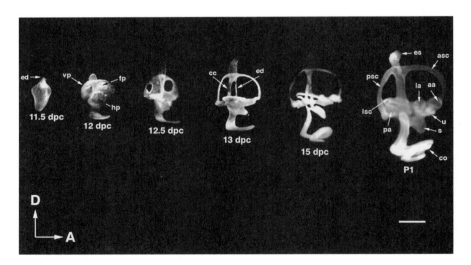

FIGURE 2.2. Lateral views of paint-filled membranous labyrinths of mice from 11.5 dpc to postnatal day 1. Specimens were fixed in paraformaldehyde, dehydrated in ethanol, and cleared in methyl salicylate. The gross anatomy of the developing inner ears was revealed by microinjecting a 0.1% white latex paint solution in methyl salicylate to the lumen of the membranous labyrinths. Abbreviations: aa, anterior ampulla; asc, anterior semicircular canal; cc, common crus; co, cochlea; dpc, days postcoitium; ed, endolymphatic duct; es, endolymphatic sac; fp, fusion plate; hp, horizontal canal plate; la, lateral ampulla; lsc, lateral semicircular canal; pa, posterior ampulla; psc, posterior semicircular canal; s, saccule; u, utricle; vp, vertical canal plate. Orientations: D, dorsal; A, anterior. Scale bar = 30 μm.

region of the otic vesicle, and it consists of the utricle, saccule, and three semicircular canals (anterior, lateral, and posterior) and their associated ampullae. At one end of each semicircular canal is an enlarged structure known as the ampulla that contains a sensory organ, the crista ampullaris. Together, the three cristae sense angular acceleration. Two additional sensory organs, the maculae of the utricle and saccule, are located in their respective chambers. The macula of the utricle detects gravity, and the macula of the saccule detects linear acceleration. The total number of vestibular sensory organs varies among different vertebrate species. For example, there are seven vestibular sensory organs in the chicken (three cristae, two maculae, the lagena, and macula neglecta) and only five major ones in the mouse. The number varies even more among anamniotes (Wersäll and Bagger-Sjöbäck 1974). However, the five vestibular sensory organs in the mouse (three cristae and two maculae) are consistently found among all species of amniotes, including humans.

The anterior and posterior semicircular canals develop from the vertical canal plate, and the lateral semicircular canal develops from the horizontal plate (vp, hp in Fig. 2.2). Over time, the opposing epithelia in the central region of each presumptive canal merge to form a fusion plate (fp), which is eventually resorbed, leaving behind a tube-shaped canal. In mice, this process is completed by 13 days postcoitium (dpc). After the canals and ampullae are formed, they continue to increase in size at least until birth. During this same developmental period, the auditory component of the inner ear, the cochlea, develops from the ventral portion of the otocyst and assumes its characteristic coiled structure (Cantos et al. 2000). The development of the chicken inner ear closely parallels that of the mouse except that the cochlear duct in the chicken is a relatively straight tube rather than a coiled structure (Bissonnette and Fekete 1996).

Although not generally considered part of the vestibular apparatus, the endolymphatic duct is the first structure that forms on the medial side of the otic vesicle. Fate mapping studies of the rim of the chicken otic cup using lipophilic dye have shown that the endolymphatic duct derives from the dorsal rim of the otic cup. Three lineage-restricted boundaries appear to specify the position of the endolymphatic duct: anterior and posterior boundaries at the dorsal pole of the otic cup that bisect the endolymphatic duct into anterior and posterior halves, and a lateral boundary that defines the lateral edge of the duct. It has been proposed that signaling across compartment boundaries may play a role in duct specification (Brigande et al. 2000a, 2000b). Thus, failure in the formation of these boundaries would result in the absence or improper specification of the endolymphatic duct and may have other deleterious effects on inner ear development. Consistent with this hypothesis, malformed inner ears that lack an endolymphatic duct are often associated with other abnormalities of the inner ear (see below). As the endolymphatic duct and sac mature, they become essential for maintaining the fluid homeostasis of the endolymph that fills the membranous labyrinth. Abnormal fluid homeostasis also leads to functional deficits in vestibular and auditory systems (see below). Molecular mechanisms regulating the proper development and function of the vestibular apparatus involve signals that originate from several different tissues, including the hindbrain, periotic mesenchyme, and otic epithelium itself. In the following discussion, we address the roles played by each of these tissues, beginning with the hindbrain.

3. Genes Expressed in the Hindbrain

Experimental manipulations have established a critical role of the hindbrain in the development of the inner ear (for reviews, see Fritzsch et al. 1998; Torres and Giraldez 1998; Fekete 1999; Anagnostopoulos 2002). Based on analyses of mutant and knockout mice, several genes expressed in the hindbrain have been shown to be required for normal development of the inner ear, including the vestibular system. *HoxA1*, *HoxA2*, *Kreisler*, and *Raldh2* are all expressed in the developing hindbrain. Loss of function of these gene products affects the development of the hindbrain—in particular, rhombomeres 4, 5, and 6, regions that are closest to the developing inner ear (for a review, see Kiernan et al. 2002). Inner ears of all of these mutant mice often fail to form endolymphatic ducts and remain cystlike, suggesting that rhombomeric regions 4 to 6 of the hindbrain, in particular rhombomere 5, are required for the formation of vestibular and auditory structures.

The expression of the *Fibroblast growth factor 3* (*Fgf3*) in rhombomeres 5 and 6 is also thought to be important for inner ear development. In both *Kreisler* and *HoxA1* mutant mice, *Fgf3* expression in the hindbrain is downregulated (Carpenter et al. 1993; Mark et al. 1993; McKay et al. 1996). This down-regulation of *Fgf3* expression has been proposed to contribute to the *Kreisler* and *HoxA1* phenotypes. This hypothesis is supported by the fact that inner ears of *Fgf3* knockout mice also lack endolymphatic ducts. Furthermore, morphogeneses of the mutant inner ears are often incomplete, and the spiral ganglia are reduced in size (Mansour et al. 1993; Mansour 1994; McKay et al. 1996). It is interesting that the knockout of one of the FGF3 receptors, *Fgfr-2* (*IIIb*), that is expressed in the otic epithelium results in severe dysmorphogenesis of the inner ear, including the absence of the endolymphatic duct and sac (Pirvola et al. 2000). Part of the phenotype observed in *Fgfr-2* (*IIIb*) knockout mice might be attributable to the inability of the otic epithelium to respond to FGF3 signals produced in the hindbrain (Pirvola et al. 2000).

Analysis of the role of hindbrain-derived FGF3 in the development of vestibular structures has been compounded by the observation that *Fgf3* is expressed not only in the hindbrain but also within the inner ear itself. Early in development, *Fgf3* is expressed in the head ectoderm, including the otic placode region. It is also expressed in the presumptive neurogenic region of the otocyst as well as in individual sensory organs of the inner ear before birth (Wilkinson et al. 1989; Mansour 1994; McKay et al. 1996; Pirvola et al. 2000). Whereas the endolymphatic duct phenotype is thought to be mediated by hindbrain-derived FGF3, *Fgf3* expression in the neurogenic region is thought to be important for the proper formation of the spiral ganglion that is reduced in the *Fgf3* knockout mice (Mansour et al. 1993; Mansour 1994; McKay et al. 1996). Although *Fgf3* is presumably expressed in the sensory regions, no obvious sensory phenotypes were associated with the *Fgf3* knockout (Mansour et al. 1993; Mansour 1994). Because *Fgf10* is expressed in the sensory regions as well, there could be overlapping functions among *Fgf*s in these regions (Pirvola et al. 2000). Therefore, in the case of genes such as *Fgf3* that have a dynamic spatial and temporal expression pattern in the hindbrain as well as in the otic epithelium, it is important to decipher its specific function in each expression domain.

A more recently identified dominant mouse mutant, *Wheels*, may also serve as a model for studying effects of the hindbrain on inner ear development (Alavizadeh et al. 2001). *Wheels* homozygotes are embryonic lethal and have an abnormal hindbrain with an extended rhombomere 4 that could affect inner ear development. Although the hindbrain segmentation in heterozygotes appears normal, these mice have a truncated lateral canal and small or absent posterior canal, suggesting that the otic epithelium itself and/or tissues other than the hindbrain are involved. Identification of the mutated gene and determination of its normal expression pattern will help to discern the role of this gene in inner ear patterning.

All of the hindbrain genes that have been discussed thus far most likely function to ensure correct positioning of the developing inner ear along the anterior/posterior axis of the body. The hindbrain could also function to specify the dorsal/ventral axis of the inner ear. Mutations in genes such as (*Sonic Hedgehog*) (*Shh*), *Pax3*, and *Lmx1* that are known to perturb the dorsal/ventral patterning of the neural tube also affect inner ear development. Because these genes may be expressed in both the inner ear and hindbrain, it is often difficult to determine the relative contributions played by signals produced by the hindbrain or inner ear. Nonetheless, due to the severe inner ear phenotypes observed in mice with mutant alleles of these neural tube specifying genes, it is clear that these genes are also essential for proper inner ear development.

Inner ears of *Shh* knockout mice have no discernible ventral structures, including the utricle, saccule, and cochlea. The delamination of neuroblasts from the anteroventral region of the otic cup or otocyst is also affected in these mutant ears. Even though it has been postulated that SHH released from the ventral midline patterns the inner ear (Riccomagno et al. 2002), the presence of low levels of *Shh* within the otic epithelium has been reported (Liu et al. 2002). Although the source of SHH for patterning the inner ear remains an open question, it is clear that the otic epithelium responds directly to SHH as indicated by the presence of *Patched* (receptor for *Shh*) and *Gli1* (a downstream target of *Shh*) mRNA transcripts within the epithelium (Liu et al. 2002; Riccomagno et al. 2002).

Furthermore, two additional mouse models, *Splotch* and *Dreher*, have disrupted neural tubes along the dorsal/ventral axis as well as malformed inner ears. *Splotch* mutants have an open neural tube and inner ear defects that include vestibular and auditory components (Deol 1966; Epstein et al. 1991; Goulding et al. 1991; Rinkwitz et al. 2001). Consistent with the phenotype, *Pax3*, which is mutated in *Splotch*, is expressed in the dorsal one-third of the neural tube. A detailed study of *Pax3* expression in the inner ear has not been reported, although *Pax3* does not appear to be expressed during early stages of inner ear development (Goulding et al. 1991).

In *Dreher*, the roof plate of the neural tube fails to form, and defects in the inner ear involve both vestibular and cochlear components (Deol 1983). In addition, the endolymphatic duct and sac are greatly distended. The gene responsible for this mutant is *Lmx1a*, a LIM homeodomain transcription factor (Manzanares et al. 2000; Millonig et al. 2000). The expression of *Lmx1* or *Lmx1a* has been described in both chickens and mice, respectively (Giraldez 1998; Failli et al. 2002). This gene is expressed in the roof plate of the neural tube as well as the dorsal and lateral regions of the otocyst. Its expression domain in the otic placode is altered as a result of neural tube ablation, suggesting that the otic expression of this gene, at least in the chicken, is regulated by hindbrain signals (Giraldez 1998).

4. Genes Expressed in the Mesenchyme

In addition to signals produced by the hindbrain, the development of the inner ear is also influenced by mesenchyme-derived signals. In fact, the epithelium of the otic placode/otocyst and the surrounding periotic mesenchyme are thought to exert reciprocal influences on each other during normal inner ear development. Results from explant cultures show that morphogenesis of the inner ear does not proceed when the majority of the periotic mesenchyme is removed (Van de Water et al. 1980). Similarly, chondrogenesis *in vitro* requires growth factors that are thought to be released by the otic epithelia such as bone morphogenetic proteins (BMP), transforming growth factor-β (TGF-β), and FGF2 (Frenz et al. 1992, 1994, 1996). Recently, ectopic expression studies in the chicken using avian retroviruses encoding dominant-negative or a constitutive active form of bone morphogenetic protein receptor IB (BMPRIB) show that BMPs are indeed important for otic chondrogenesis *in vivo*. BMPs for some regions of the otic capsule, such as areas around the canals, are thought to emanate from the otic epithelium (Chang et al. 2002).

Analyses of genetically altered mice indicate that three transcription factors, *Prx1*, *Prx2*, and *Brn4*, regulate genes important for mesenchymal–epithelial signaling. *Prx1* and *Prx2* are paired-related homeobox genes. *Prx1* is expressed in the periotic mesenchyme, and *Prx2* is expressed in the otic epithelium as well as the periotic mesenchyme. A knockout of *Prx1* results in a reduction in the size of the otic capsule, whereas a knockout of *Prx2* has no apparent phenotype in the inner ear (ten Berge et al. 1998). *Prx1* and *Prx2* share redundant functions in other tissues. Therefore, it is not surprising that the double knockout of both *Prx1* and *Prx2* results in a more severe inner ear phenotype. In addition to the reduction in the size of the otic capsule observed in the knockout of *Prx1*, in the double knockout, the lateral semicircular canal does not form, and there is a reduction in the size of both the anterior and posterior canals (ten Berge et al. 1998). These results suggest that the coexpression of *Prx1* and *Prx2* in the periotic mesenchyme is important for mediating mesenchymal–epithelial signaling in the vestibular apparatus.

Brn4 (*Pou3f4*), a transcription factor belonging to the POU-domain gene family, is expressed in the periotic mesenchyme (Phippard et al. 1998). Knockout mice of *Brn4* are deaf, and vestibular phenotypes such as head bobbing have been reported in one of the two knockout lines (Minowa et al. 1999; Phippard et al. 1999). The primary cell type affected in the *Brn4* knockout mice appears to be the fibrocytes of the spiral ligament that have been postulated to be important in maintaining the endocochlear potential (Minowa et al. 1999; Phippard et al. 1999). Interestingly, in one of the *Brn4* knockout lines, patterning defects in the cochlea were reported (Phippard et al. 1999). The number of cochlear turns in this mutant line is often affected, and the anterior semicircular canal is constricted. The constriction of the anterior semicircular canal is thought to be the cause of the vestibular deficits (Phippard et al. 1999). The reason for the phenotypic variation observed between the two knockout lines is not clear because the gene-targeted region and the genetic background of the mutant mice are similar. However, *sex-linked fidget* (*slf*) mice have an inversion on the X chromosome that eliminates expression of *Brn4* in the developing inner ear but not the neural tube. These mice, like one of the *Brn4* knockout lines, display both cochlear and vestibular deficits (Phippard et al. 2000). These results provide the first evidence that a gene, expressed primarily in the periotic mesenchyme, mediates otic epithelial morphogenesis. Identifying possible upstream signaling molecules and downstream targets for this transcriptional factor, whether they are epithelium- or mesenchyme-derived, will be important. It is interesting that, in the *Shh* mutants, both *Brn4* and *Tbx1* are down-regulated in the otic mesenchyme (Riccomagno et al. 2002). The otic capsule is reduced in *Shh* mutants, indicating that other molecular pathways that mediate otic chondrogenesis are not perturbed by the loss of *Shh*. However, the cochlear defects observed in *Brn4* knockout mice suggest that

Shh could mediate its effects on inner ear patterning through activating *Brn4* as well as *Tbx1* in the mesenchyme.

5. Genes Expressed within the Otic Epithelium

It is not surprising that genes expressed in the otic epithelium itself are important for the development of the vestibular apparatus (Table 2.1). Some of these genes, when knocked out, result in a rudimentary inner ear with poorly developed vestibular as well as cochlear components. These inner ears often lack endolymphatic ducts as well as the vestibular and spiral ganglia. *Fgfr-2* (*IIIb*), *GATA-3*, and *Eyes absent* (*Eya1*) are good examples of genes in this category (Xu et al. 1999; Pirvola et al. 2000; Karis et al. 2001). All three genes are activated early in development and are broadly expressed in the inner ear, particularly during the otic cup and otocyst stages (Xu et al. 1997; Pirvola et al. 2000; Karis et al. 2001). As described above, the severe phenotype of the *Fgfr-2* (*IIIb*) knockout could be a result of its inability to respond to growth factor signals produced by the hindbrain as well as by sensory regions of the otic epithelium.

GATA-3 is a member of a zinc-finger transcription factor family that recognizes a specific GATA consensus sequence in promoter regions. Genes in this family are important for differentiation of multiple tissues during embryogenesis, including the brain and hematopoietic system (Simon 1995). In the otocyst, *GATA-3* is broadly expressed within the otic epithelium, and, as differentiation progresses, *GATA-3* is expressed in all of the vestibular sensory organs except the saccule. The vestibular ganglion is also devoid of *GATA-3* expression (Karis et al. 2001). Within the auditory structures of the inner ear, both the cochlear duct and spiral ganglion are positive for *GATA-3*. Interestingly, the repression of *GATA-3* expression is correlated spatially and temporally with hair cell differentiation, which proceeds in a gradient from the base to the apical region of the cochlea (Rivolta and Holley 1998). *GATA-3* null mutants die between 11 and 12 dpc and have rudimentary inner ears (Karis et al. 2001). Correlating phenotypes with expression domains will be a challenge for this gene because *GATA-3* is expressed not only in the inner ear but also in the hindbrain and periotic mesenchyme (Nardelli et al. 1999).

Eya-1 is a homolog of the *Drosophila eyes absent* gene. In the *Drosophila* eye imaginal disk, *eya* functions as a transcription coactivator that interacts with other transcription factors but does not bind DNA directly (Chen et al. 1997; Pignoni et al. 1997). Mutations in this gene in humans cause branchiootorenal syndrome, which is associated with defects in the kidney as well as the external, middle, and inner ear (Abdelhak et al. 1997). Expression of *Eya-1* in the inner ear is extensive at the otocyst stage, and *Eya-1* null mutants have rudimentary inner ears that lack the eighth ganglion (Xu et al. 1999). A hypomorphic allele of *Eya-1* has also been identified. In this case, the vestibular portion of the inner ear appears intact but the cochlear duct is truncated, suggesting that *Eya-1* is particularly essential for cochlear development (Johnson et al. 1999).

Because knockouts of genes such as *Fgfr-2* (*IIIb*), *GATA-3*, and *Eya-1* have such deleterious effects on inner ear development in general, it is often difficult to discern their specific effects on individual inner ear components. On the other hand, knockouts of transcription factor genes such as *Otx1*, *Hmx2* and *Hmx3* (*Nkx5.2* and *Nkx5.1*), and *Dlx5* affect the development of specific components of the inner ear (Hadrys et al. 1998; Wang et al. 1998; Acampora et al. 1999b; Depew et al. 1999). More detailed descriptions of the functions of these and other genes in the development of individual vestibular components are given below.

5.1. Development of the Sensory Organs

The origin and the lineage relationships among the vestibular sensory organs within the inner ear are not known. However, early in inner ear development, prior to any discernible histological differentiation, the presumptive cristae of the semicircular canals can be molecularly distinguished from the presumptive maculae of the utricle and saccule. Based on the different morphologies of the cristae and maculae at maturity, it is not surprising that multiple genes are differentially expressed in these sensory organs during the course of their development. Therefore, it is important to identify those essential for the specification and differentiation of each type of sensory organ.

Thus far, genes that are expressed in the sensory tissues can be divided into two groups: those that do and do not act in the Notch-signaling pathway (Fig. 2.3). The Notch signaling pathway is used in a variety of tissues to generate cell type diversity during development (for reviews, see Artavanis-Tsakonas and Simpson 1991; Artavanis-Tsakonas et al. 1999). Originally delineated by studies of neurogenesis in invertebrate systems, the Notch signaling pathway relies on local cell interactions to control the differential specification of otherwise equivalent cells. For example, in the case of invertebrate neurogenesis, Notch signaling mediates the decision of whether ectodermal cells become neuroblasts or epidermal cells. Several molecules acting in the Notch pathway have been identified and include the Notch receptors and several membrane-associated Notch ligands such as Delta and Serrate. During fruit fly (*Drosophila*) central nervous system development, clusters of neural precursor cells develop within the ectodermal epithelium via the expression of proneural genes, encoded by the achaete-scute complex. Then, one cell from each cluster will become committed to the neural fate, and others will cease to express achaete-scute genes and switch to the epidermal fate. This process is mediated by the Notch pathway. Notch ligands displayed on the committed neural cell acti-

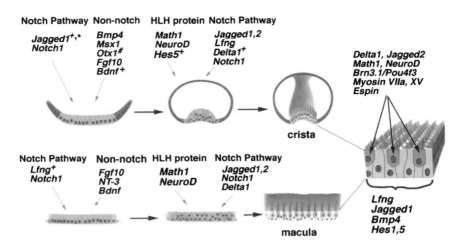

FIGURE 2.3. A schematic diagram outlining genes expressed in different stages of the crista and macula development. For simplification, sensory organ development is divided into three stages corresponding to 9.5–11, 12–14, and 15–18 dpc in mice. Readers should refer to cited references for specific timing of individual gene activation. (+) represents initiation of gene expression in the indicated prospective sensory organ before others. (#) represents expression only in the lateral crista and not the anterior or posterior cristae. (*) represents expression data from chickens. *Hes1* is expressed in supporting cells of the rat utricle at 17.5 dpc, and it is not clear whether it is expressed in other vestibular sensory tissues as well. In mice, *Bmp4* is only expressed in supporting cells of cristae and not in maculae.

vate Notch receptors in its neighboring cells and thus activate an alternate developmental pathway, an epidermal fate in this case.

The development of the sensory patches in the vertebrate inner ear has been compared with that of the mechanoreceptor organs in fruit flies (*Drosophila*) (Adam et al. 1998; Eddison et al. 2000; Fritzsch et al. 2000; Caldwell and Eberl 2002). Based on expression studies of Notch signaling molecules, it has been proposed that the expression of Notch ligands, Delta and Jagged/Serrate, on the surface of presumptive sensory hair cells activated Notch receptors present on neighboring cells (Adam et al. 1998; Lewis et al. 1998). This activation of Notch receptors in the neighboring cells induced them to develop into supporting cells. Consistent with this model, mutation of genes in the Notch signaling pathway usually results in changes in the number of hair cells and presumably supporting cells in the sensory organs. For example, knockout of a Notch ligand, *Jagged2*, results in an increase in the number of inner and outer hair cells in the cochlea (Lanford et al. 1999). In the zebrafish (*Brachydanio rerio*) *mind bomb* mutant, in which the *Delta–Notch* signaling pathway is thought to be affected, the inner ear contains only hair cells and no supporting cells

(Haddon et al. 1998). In addition, treatment of rat cochlear cultures with antisense oligonucleotides of *Jagged1* and *Notch1* result in supernumerary hair cells (Zine et al. 2000).

5.1.1. Development of the Crista Ampullaris

5.1.1.1 Notch Signaling Pathway

In the mouse, both *Notch1* and *Serrate1/Jagged1* are expressed in all presumptive sensory organs of the inner ear. During later stages of development, the expression domains of both of these genes are restricted to nonsensory cells within each sensory patch (Lewis et al. 1998; Morrison et al. 1999; Shailam et al. 1999). Recent data suggest that the function of the Notch signaling pathway is not restricted to hair cell/supporting cell determination in the inner ear but is also required for the patterning of the ampulla and canal. Two mouse mutants, *Headturner* and *Slalom*, with missense mutations in the Notch ligand, *Jagged1*, have recently been characterized. Homozygotes of both mutants die at early embryonic stages due to vasculature defects, and heterozygotes have an aberrant number of hair cells in the cochlea (Kiernan et al. 2001; Tsai et al. 2001). Interestingly, *Headturner* and *Slalom* are missing one or both of the anterior and posterior ampullae. The ampulla phenotype is accompanied by truncation of its corresponding canal. Despite the phenotype in the anterior and posterior canals, the lateral canal and ampulla appear to be intact in these mutants. It is not clear why the anterior and posterior ampullae are preferentially affected because *Jagged1* is expressed in all prospective sensory organs (Morrison et al. 1999). Coincidentally, in the chicken, *Jagged1/Serrate1* is expressed in the presumptive anterior and posterior cristae earlier than in other sensory organs (Myat et al. 1996; Cole et al. 2000). Therefore, if the expression pattern of *Jagged1* in mice is similar to that of the chicken, the patterning phenotype observed in *Slalom* and *Headturner* might be due to the requirement of *Jagged1* function prior to hair cell/supporting cell determination.

Some genes in the Notch signaling pathway, such as *Jagged2* and *Hes5*, however, are activated slightly later during sensory organ development and are correlated with the period of hair cell and supporting cell commitment (Fig. 2.3). *Jagged2* is expressed in presumptive hair cells of each sensory patch (Lanford et al. 1999; Shailam et al. 1999). *Hes5*, a basic-helix-loop-helix (bHLH) transcription factor, is a homolog of the *Drosophila hairy* and *enhancer-of-split*. It is one of the downstream genes activated by Notch. *Hes5* is preferentially expressed in the presumptive cristae at 12.5 dpc and is later expressed in supporting cells of the cristae and striolar region of the utricle (Shailam et al. 1999; Zheng et al. 2000). In other systems, members of the bHLH family of transcription factors have been shown to be both upstream mediators and downstream targets of the Notch signaling pathway (for a review, see Anderson and Jan 1997). In addition to

Hes5, other examples of downstream targets of the Notch signaling pathway that are expressed in the inner ear include *Hes1*, *Hes6*, *Hey1*, and *Hey2* (Leimeister et al. 1999; Pissarra et al. 2000; Zheng et al. 2000). Detailed expression studies and the consequences of loss of some of these encoded proteins during inner ear development have not been reported. However, *Math1*, a bHLH transcription factor, might be an upstream mediator of the Notch pathway in the inner ear. *Math1* is a homolog of the fruit fly (*Drosophila*) proneural gene *atonal*, which is important for the formation of chordotonal organs (mechanoreceptor organ) in flies. In mice, *Math1* –/– inner ears have no sensory hair cells even though the gross anatomy of the sensory organs appears normal (Bermingham et al. 1999). In addition, ectopic expression of *Math1* in rat cochlear cultures resulted in an ectopic appearance of sensory hair cells in nonsensory regions (Zheng and Gao 2000). The onset of *Math1* expression in individual sensory organs appears to precede that of *Jagged2*, consistent with its postulated role as a proneural gene (Shailam et al. 1999; Liu et al. 2000). However, more recent studies suggest that *Math1* functions in hair cell determination rather than specification of the sensory primordium (Chen et al. 2002). The important role of *Math1* in sensory development will undoubtedly be revealed with further experiments.

NeuroD belongs to a subfamily of bHLH proteins that are widely expressed in the nervous system of vertebrates and are potent neuronal differentiation factors (Lee et al. 1995). *NeuroD* is expressed in the presumptive cristae, but the cristae of *NeuroD* knockout mice appeared normal, even though the number of sensory hair cells in the cochlea is aberrant (Liu et al. 2000; Kim et al. 2001). In addition, *NeuroD* is important for the development of the eighth ganglion (see below).

5.1.1.2. Non-Notch Pathway

Examples of genes that are expressed in the presumptive cristae but are not components of the Notch-signaling pathway include *Bmp4* and *Msx1* (Fig. 2.3). *Bmp4* belongs to the *TGF*-β gene family and plays an important role in the development of multiple tissues (for a review, see Hogan 1996). In the mouse inner ear, *Bmp4* is expressed at the rim of the invaginating otic cup (Morsli et al. 1998). After the otic cup closes to form the otic vesicle, *Bmp4* expression is restricted to two domains, an anterior streak and a posterior focus (as, pf in Fig. 2.4A,B; Morsli et al. 1998). The posterior focus corresponds to the position of the future posterior crista. The posterior expression domain later splits to form the dorsal posterior crista and a ventral streak that corresponds to Hensen's and Claudius' regions of the cochlea in mice (pc and lco in Fig. 2.5A). The anterior streak also splits to form the anterior and lateral cristae at a later time of development (Figs. 2.4A, 2.5A; Morsli et al. 1998). The early expression of *Bmp4* in the otic cup and otocyst stages is conserved in the chicken, frog, and zebrafish, but the

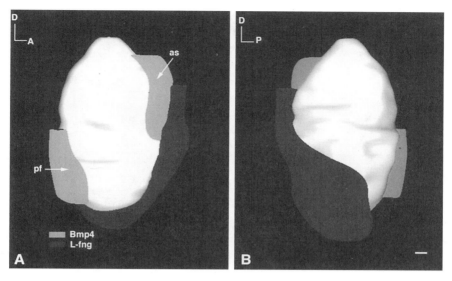

FIGURE 2.4. A three-dimensional reconstruction of *Bmp4* and *L-fng* expressions in the mouse inner ear at 10.5 dpc. The lateral (**A**) and medial (**B**) views of the right inner ear are shown. The emerging endolymphatic duct on the medial side at this stage is not drawn. *Bmp4* positive regions are displayed in light gray and the *L-fng* positive area in dark gray. Alternate 12 μm serial sections were probed for *Bmp4* or *L-fng* mRNA and reconstructed using ROSS software (Biocomputation Center, Ames Research Center, NASA). Data for the reconstruction were obtained from Morsli et al. (1998). The anterior streak (as) of the *Bmp4* hybridization signal later splits to form the anterior and lateral cristae (see Fig. 2.5A). The posterior focus (pf) encompasses the presumptive posterior crista. *L-fng* is broadly expressed at this stage with an expression domain that spans from the anterolateral region to the ventromedial region of the otocyst. *L-fng* and *Bmp4* expression domains are largely nonoverlapping. Orientation: D, dorsal; A, anterior; P, posterior. Scale bar = 30 μm.

role of *Bmp4* in formation of the crista or other parts of the inner ear is not clear because *Bmp4* null mice die before sufficient inner ear development (Hemmati-Brivanlou and Thomsen 1995; Mowbray et al. 2001; Wu and Oh 1996). However, some *Bmp4* heterozygotes have a malformed lateral canal, indicating that BMP4 is essential for proper inner ear development (Teng et al. 2000). Because the receptors for *Bmp4* are ubiquitously expressed in the otic epithelium and adjacent mesenchyme, *Bmp4* could function both autonomously within the presumptive cristae and through effects on the adjacent nonsensory otic epithelium and periotic mesenchyme (Dewulf et al. 1995).

In the chicken, the early expression of *Brain-derived nerve growth factor* (*Bdnf*) has an expression pattern similar to that of *Bmp4* (Hallbook et al.

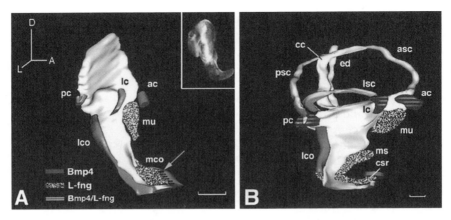

FIGURE 2.5. A three-dimensional reconstruction of *Bmp4* and *L-fng* expression domains in the mouse inner ear at 12 (**A**) and 13 (**B**) dpc. *Bmp4*-positive areas are in light gray, and *L-fng*-positive areas are spotted. The arrow in A is pointing to black stripes that represent a region of *Bmp4* and *L-fng* coexpression in the distal tip of the growing cochlea. The insert in A is a 12 dpc paint-filled inner ear shown in a view similar to the reconstructed image. By 13 dpc (B), the cristae are positive for both *Bmp4* and *L-fng*, highlighted in light and dark gray stripes. Data analysis and three-dimensional reconstructions were carried out as described in the legend to Figure 2.4. Abbreviations: ac, anterior crista; asc, anterior semicircular canal; cc, common crus; csr, cochlear sensory region; ed, endolymphatic duct; lc, lateral crista; lco, lateral cochlear hybridization signal; lsc, lateral semicircular canal; mco, medial cochlear hybridization signal; ms, macula sacculi; mu, macula utriculi; pc, posterior crista; psc, posterior semicircular canal. Orientation: A, anterior; D, dorsal; L, lateral. Scale bar = 100 µm. (Adapted from Morsli et al. 1998.)

1993). In mice, the early *Bdnf* expression pattern is also thought to overlap with that of *Bmp4* (Fritzsch et al. 1999). BDNF is required for proper innervation of the cristae by the vestibular ganglion (Fritzsch et al. 1999). Later in development, *Bdnf* is also expressed in the maculae.

Msx1 and *Msx2* are orthologs of the *Drosophila msh* (muscle segment homeobox) gene and are important for mediating epithelial–mesenchymal interactions in several tissues during embryogenesis (Satokata and Maas 1994; Chen et al. 1996). The role of *Msx1* in crista formation is not clear, but it is expressed in the presumptive cristae and not the maculae (Dewulf et al. 1995; Wu and Oh 1996; Alavizadeh et al. 2001). *Msx1* knockout mice have no apparent phenotype in the inner ear (Satokata and Maas 1994). However, *Msx1* may share redundant functions with *Msx2*. Inner ear analyses of mice with double knockouts of *Msx1* and *Msx2* have not been reported.

Fgf10 is expressed in the vestibulocochlear ganglion as well as each of the prospective sensory organs (Pirvola et al. 2000). Knockout of *Fgf10* results in absence of all three semicircular canals and the posterior crista. The anterior crista is malformed and misaligned relative to the utricle (Pauley et al. 2003).

5.1.1.3. Genes Expressed in Specific Cristae

The anterior and posterior cristae are anatomically indistinguishable from each other except for their positions within the inner ear, whereas the lateral crista is different in appearance and resembles half of an anterior or posterior crista (Landolt et al. 1975). Furthermore, the lateral canal and ampulla are the last among the three canals and ampullae to have arisen during vertebrate evolution and are absent in Agnatha (jawless vertebrates; for a review, see Wersäll and Bagger-Sjöbäck 1974). So far, no genes have been demonstrated to be exclusively expressed in either anterior or posterior cristae even though some genetic mutations differentially affect the two cristae (see below).

On the other hand, *Otx1* is expressed in the presumptive lateral crista and canal but not in the anterior or posterior cristae or their canals (Morsli et al. 1999). *Otx1* and *Otx2* are both vertebrate orthologs of *Drosophila orthodenticle*, which is important for sense organ and head development (Acampora et al. 1995; Hirth et al. 1995; Royet and Finkelstein 1995; Acampora et al. 1996; Ang et al. 1996). In *Otx1* knockout mice, the lateral crista and canal fail to develop (Acampora et al. 1996; Morsli et al. 1999). However, *Bmp4* expression in the *Otx1* mutant inner ears is normal at the early otic vesicle stage, suggesting that the specification of the lateral crista may be normal initially and that *Otx1* may be important for the subsequent differentiation of the sensory organ (Morsli et al. 1999). More recently, an ectopic sensory patch located on the medial side of the mutant inner ear by the endolymphatic duct was reported in *Otx1* mutant inner ears (Fritzsch et al. 2001). It is not clear whether this sensory patch is a mispositioned lateral crista or the result of an aberrant segregation of sensory patches. Nevertheless, the function of *Otx1* in lateral canal and ampulla formation is indispensable and not compensated by replacing a human *Otx2* cDNA in the disrupted *Otx1* locus despite the sequence homology between the two genes and the ability of human *Otx2* to rescue the brain phenotype observed in *Otx1* mutant mice (Acampora et al. 1999a; Morsli et al. 1999).

5.1.2. Development of the Maculae

5.1.2.1. Notch Signaling Pathway

The positions of the two presumptive maculae are marked by the expression of *Lunatic fringe* (*L-fng*). *L-fng* is an ortholog of the *Drosophila fringe* gene that acts in the Notch signaling pathway to establish boundaries during the development of both flies and vertebrates (Laufer et al. 1997; Panin et al. 1997; Evrard et al. 1998; Papayannopoulos et al. 1998; Zhang and Gridley 1998). Recent data show that Fringe mediates its effect by forming complexes with Notch receptors and modulating their ligand preferences (Hicks et al. 2000; Ju et al. 2000). In the inner ear, *L-fng* is expressed in an anterolateral domain of the otic cup that later expands medially (Morsli et al. 1998; Fig. 2.4A,B). The *L-fng* positive domain encompasses three presumptive sensory organs: the maculae of the utricle and saccule and the sensory tissue of the cochlea. In addition, based on its location, the lateral region of the *L-fng* positive area most likely encompasses the cells that are delaminating at this stage to form the eighth cranial ganglion even though *L-fng* transcripts were not detected in the migrating neuroblasts (Morsli et al. 1998). Note that the *L-fng* expression domain is ventral to and largely nonoverlapping with the *Bmp4* positive region. By 12 dpc, the *L-fng* expression domain splits into a dorsal and a ventral region. The dorsal region is destined to become the macula of the utricle (mu, Fig. 2.5A). The ventral region (mco in Fig. 2.5A) encompasses the future macula of the saccule and the cochlear sensory region, which are distinguishable from each other by 13 dpc (ms and csr, Fig. 2.5B). By 13 dpc, the three cristae also coexpress *Bmp4* and *L-fng* (dark and light gray stripes, Fig. 2.5B). Given the role of *L-fng* in the Notch signaling pathway and its role in boundary formation in other tissues, it was suggested that this gene might play a role in hair cell and supporting cell determination as well as in the positioning of sensory organs within the inner ear (Morsli et al. 1998). So far, there is no obvious gross anatomical defect in *L-fng* knockout mice, suggesting that L-fng is not essential for positioning of sensory organs (Zhang et al. 2000; Johnson and Wu, unpublished results). However, lack of *L-fng* suppresses the increase in the number of inner hair cells in *Jagged2* knockout mice but has no effect on the increase in the number of outer hair cells (Zhang et al. 2000). These results, although not straightforward to interpret, suggest that *L-fng* plays a role in modulating the ligand preference for Notch similar to its role in other systems.

Math1 and *NeuroD* are also expressed in the presumptive maculae, and loss of *Math1* results in the absence of macular sensory hair cells, similar to the phenotype observed in the cristae (see above). In addition, ectopic expression of *Math1* in rat utricule cultures induces the conversion of supporting cells into hair cells (Zheng and Gao 2000). Two downstream targets of Notch are expressed in supporting cells of the macula of the utricle, Hes1 and Hes5 (Zheng et al. 2000). Knockout of *Hes1* leads to the formation of supernumerary hair cells in the utricle. It is not clear whether *Hes1* is expressed in the cristae as well (Zheng et al. 2000). *Neurogenin1* (*Ngn*), another bHLH transcription factor, when knocked out affects the development of the utricle, saccule, cochlea, and formation of the eighth ganglion (Ma et al. 2000, see below). However, the expression of this gene in prospective sensory organs has not been reported.

5.1.2.2. Non-Notch Pathway

Comparing published results, the *L-fng* positive domain at the otocyst stage appears to be also positive for *Neurotrophin3* (*NT-3*), and later on *NT-3* is expressed in the presumptive maculae and cochlea (Fritzsch et al. 1999). In addition to *NT-3*, *Bdnf* is also expressed in the presumptive maculae. Both NT-3 and BDNF are required for the survival of sensory ganglia neurons that innervate the two maculae. Given the anatomical differences between the maculae and cristae, it is surprising that, besides *NT-3*, no other genes have been reported to be differentially expressed in the maculae and not the cristae. Even otoconin-95, a major component of the otoconia, is not restricted to the utricle and saccule but rather broadly expressed in the non-sensory regions of the inner ear (Verpy et al. 1999). However, several genes, although not exclusively expressed in the utricle or saccule, such as *Otx1*, *Otx2*, *Hmx2* and *Hmx3*, and *Ngn1*, when knocked out resulted in an incomplete separation of the utricle and saccule that often affected the development of the two maculae (Wang et al. 1998; Cantos et al. 2000; Ma et al. 2000). Furthermore, even though *Otx2* null mutants die too early, before sufficient inner ear development, analysis of mutant mice with *Otx1* cDNA inserted into the disrupted *Otx2* locus suggests that the role of *Otx2* in the development of the saccule and cochlea is not compensated by *Otx1* (Cantos et al. 2000).

5.1.3. Summary

For simplification, the discussion in the section above was organized into genes that do and do not act in the Notch signaling pathway. However, it is important to note that there may be substantial interplay among the pathways. For example, genes in the Non-Notch category could interact with proneural genes upstream of Notch as well as interact with genes within the Notch signaling pathway. Although such interactions have not been demonstrated during sensory organ formation in the inner ear, in the fruit fly (*Drosophila*), a wingless signaling pathway component, Dishevelled, has been shown to bind the carboxy-terminal of the Notch receptor and block *Notch* signaling (Axelrod et al. 1996).

Multiple lines of research indicate that the *Notch* signaling pathway in inner ear development is more complicated than the simple paradigm presented at the beginning of this section. Although *Notch* appears to be ubiquitously expressed in the developing inner ear, the ligands for Notch are not. For example, in the chicken otic cup, *Jagged1* expression is concentrated in the medial-posterior region, whereas *Delta* is expressed in the anterior, neurogenic region, suggesting that these ligands have different functions (Myat et al. 1996; Adam et al. 1998). However, in later stages of inner ear development, Notch ligands and their modulator, L-fng, tend to be coexpressed in the prosensory domains. The temporal sequence of how different Notch ligands interact to achieve cell type diversity is not clear.

Experiments designed to block the *Notch* signaling pathway in the developing chicken inner ears show that *Jagged1* expression was down-regulated in the sensory regions rather than up-regulated as the conventional model might have predicted (Haddon et al. 1998; Eddison et al. 2000). This result suggests that not all Notch ligands respond in a similar manner to changes in *Notch* signaling. The complex phenotypes observed in *Headturner*, *Slalom*, and *Jagged2* and *L-fng* double knockouts also lend support to the complexity of the *Notch* signaling pathway in inner ear development. Furthermore, there are other existing vertebrate, Notch ligands and receptors whose expression patterns and possible functions in the inner ear have not been explored.

Besides the sensory patches, both *Jagged1* and *Delta1* have restricted patterns of expression in a subpopulation of cells within the endolymphatic sac (Morrison et al. 1999). Thus, most likely, the Notch signaling pathway also plays a role in cell type determination in the endolymphatic sac.

5.2. Development of the Eighth Cranial Ganglion

No vestibular sensory organs can function properly without appropriate innervations from the sensory ganglion. Based on analyses of knockout mice, the development of the eighth ganglion (vestibulocochlear ganglion) can also be divided into several phases (for a review, see Fritzsch et al. 1999). First, cells in the anteroventral lateral region of the otic cup or otocyst delaminate from the otic epithelium. Then, these neuroblasts migrate away and undergo further proliferation before coalescing to form a ganglion that later divides to form the vestibular and spiral ganglia (Carney and Couve 1989). The *Notch* signaling pathway is important for the neuroblast determination, as indicated by the expression of *Delta1*, *Jagged1*, and *L-fng* in the neurogenic domain of the otic cup and otocyst (Adam et al. 1998; Lewis et al. 1998; Morsli et al. 1998). In addition, the number of vestibulocochlear neurons is increased in the zebrafish (*B.rerio*) *mind bomb* mutant in which the *Notch* signaling pathway is postulated to be affected (Haddon et al. 1998). Based on gene expression patterns, the neurogenic region appears to overlap with some prospective sensory domains; however, whether neuroblasts share a common lineage with hair cells and supporting cells within these domains remains to be determined (for a review, see Fekete and Wu 2002).

Two HLH transcription factors, *Ngn1* and *NeuroD*, have been shown to be important for the early phases of ganglion development (Liu et al. 2000; Ma et al. 2000; Kim et al. 2001). *NeuroD* knockout mice show defects in neuroblast delamination from the otic epithelium and subsequent neuronal differentiation (Liu et al. 2000). As a result, sensory organs are poorly innervated in *NeuroD* mutants. In *Ngn1* knockout mice, inner ear sensory neurons are completely absent (Ma et al. 2000). Presumably, *Ngn1* is acting upstream of *NeuroD* and functions in a pathway similar to *NeuroD* in the development of sensory neurons (Ma et al. 1998). Gene expression analyses of *Shh* knockout mice as well as a transgenic line that ectopically expresses *Shh* in the otic vesicle (*ShhP1*) suggest that *Shh* may act upstream of *Ngn1* (Riccomagno et al. 2002). In *Shh* knockout mice, *Ngn1* and *NeuroD* are down-regulated and the cochleovestibular ganglia are greatly reduced in size. In contrast, both *Ngn1* and *NeuroD* are up-regulated in *ShhP1* mice, which have enlarged ganglia.

Brn3.a/Brn3.0, a POU-domain transcription factor, is expressed in the neuroblasts shortly after they delaminate from the otic epithelium. Loss of *Brn3.a* affects the differentiation of the sensory neurons, expression of downstream genes such as *TrkB* and *TrkC*, normal projections, and target innervations (Huang et al. 2001). The expressions of the neurotrophin receptors *TrkA*, *TrkB*, and *TrkC* in the differentiating neurons mark a later phase of ganglionic development. The survival of these neurons becomes dependent on neurotrophins such as BDNF and NT-3 synthesized in the differentiating sensory tissues (Fritzsch et al. 1999). Knockout of *Bdnf* or its high-affinity receptor, *TrkB*, results in no innervation of the three cristae and poor innervation of the two maculae (Fritzsch et al. 1995; Schimmang et al. 1995; Bianchi et al. 1996). Despite the fact that *NT-3* is expressed in the maculae, knockout of *NT-3* or its receptor, *TrkC*, results in only a limited loss of saccular and utricular innervations (Fritzsch et al. 1995; Fritzsch et al. 1997). In contrast to the ganglion cell dependency on sensory tissues for neuronal survival, the development, differentiation, and survival of sensory hair cells appear independent of afferent and efferent innervations (Fritzsch et al. 1997; Silos-Santiago et al. 1997; Liu et al. 2000; Kim et al. 2001).

5.3. Development of the Semicircular Canals

Semicircular canal development can be divided into four phases: outgrowth and patterning of the epithelial outpocket, fusion plate formation, resorption, and continued growth of the canal after its formation. The patterning process is most evident by examining the formation of the prospective posterior canal in a series of frontal views of paint-filled chicken inner ears (Fig. 2.6). In chickens, as in mice, the anterior and posterior canals arise from the same vertical outpouch initially, and between embryonic day 4.5 (E4.5) and 5.5, the presumptive posterior canal forms at approximately a right angle to the presumptive anterior canal, possibly via differential growth (Fig. 2.6). By E5.5, the alignment of the anterior and posterior canals is established, but the resorption process for the posterior canal is just beginning and is quite evident by E6. In the chicken, programmed cell death seems to be the main mechanism for the resorption process (Fekete et al. 1997). Ectopic expression of *Bcl2* that inhibits normal programmed cell death in the chicken resulted in the blockage of canal fusion (Fekete et al. 1997). However, in mice, retraction of cells to the inner margin of the future

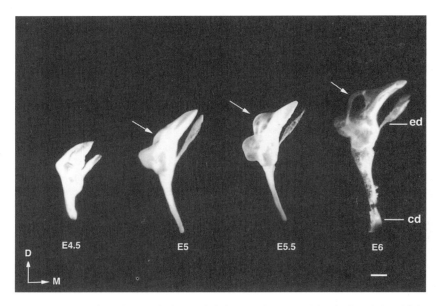

FIGURE 2.6. A series of frontal views of right membranous labyrinths of the chicken from E4.5 to E6. Various steps in the process of posterior canal formation, including outgrowth of the epithelial outpocket (E4.5 to E5), fusion plate formation (E5.5), and resorption (E5.5 to E6), are shown. Arrows point to the developing posterior canal. Abbreviations: ed, endolymphatic duct; cd, cochlear duct. Orientations: D, dorsal; M, medial. Scale bar = 30 μm.

canal has been proposed to be the main mechanism for the elimination of cells from the center of the canal pouch. Surrounding periotic mesenchyme has also been proposed to be a driving force in the formation of the fusion plate (Salminen et al. 2000; see below).

Thus far, previously identified genes expressed during semicircular canal formation can be roughly divided into two groups: those expressed in the early canal outpocket stage and those expressed slightly later in development (Fig. 2.7). The first group of genes are transcription factors, such as *Hmx2*, *Hmx3*, and *Dlx5*, that are activated early at the otic placode stage or shortly after placode formation. These genes are expressed in the epithelium of the canal outpockets and later primarily in the semicircular canals and ampullae. Knockouts of these genes affect the normal development of ampullae and canals (Hadrys et al. 1998; Wang et al. 1998; Acampora et al. 1999b; Depew et al. 1999). *Hmx2* and *Hmx3* are members of a homeobox-containing family of transcription factors that are distinct from *Hox* and other homeobox-containing genes. Similar to *Hox* genes, *Hmx* are evolutionarily conserved from fruit flies (*Drosophila*) to humans. There are three

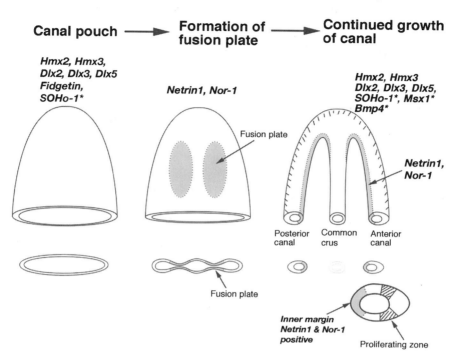

FIGURE 2.7. A schematic diagram summarizing genes expressed during development of the semicircular canals. The lower panel is a cross-sectional view of the upper panel. An enlarged cross-sectional view of a canal is shown on the lower right. Genes such as *Dlx5* and *Hmx3* are expressed in the canal outpocket, whereas *Netrin 1* and *Nor-1* are expressed in the central region of the outpocket that is destined to form the fusion plate. Once the canals are formed, *Netrin 1* and *Nor-1* are expressed in the inner margin of the canals, and other genes such as *Hmx3* are broadly expressed in the canal epithelia. Asterisks represent gene expression patterns reported in the chicken. Refer to the legend of Figure 2.2 for abbreviations and orientations.

members in the mammalian genome: *Hmx1*, *Hmx2*, and *Hmx3*. Both *Hmx2* and *Hmx3* are expressed in the developing mouse inner ear, with *Hmx3* having a slightly earlier onset of expression than *Hmx2* starting at the otic placode stage (Rinkwitz-Brandt et al. 1995, 1996; Wang et al. 2001). Targeted deletions of *Hmx3* have been reported by two independent laboratories. Bober's group reported a reduction in the size of the anterior canal, missing posterior and lateral canals, and the absence of a lateral crista in their *Nkx5.1/Hmx3* knockout mice (Hadrys et al. 1998). Lufkin's group observed a much milder canal phenotype in their *Hmx3* mutants: only the lateral crista and ampulla were missing. In addition, the two maculae were fused (Wang et al. 1998). However, they reported a much more severe inner ear phenotype for the *Hmx2* knockout: loss of all three canals and their associated cristae, as well as a fused utriculosaccular chamber (Wang et al. 2001). In fact, the phenotypes of the *Hmx2* knockout closely resembled the phenotypes observed in the *Hmx3* knockout mice generated in the Bober laboratory. A negative effect of the inserted *Hmx3-Pkneo* allele on the closely linked *Hmx2* gene in Bober's *Hmx3* knockout line was put forth as a plausible explanation for these paradoxical results (Wang et al. 2001). Nevertheless, these combined results suggest that *Hmx2* and *Hmx3* both have unique and overlapping functions in vestibular development.

Dlx5 belongs to a family of homeobox-containing genes that is related to the *Distal-less* (*Dll*) gene of the fruit fly (*Drosophila*). In *Drosophila*, *Dll* is required for correct development of the distal portion of the legs, antennae, and mouth parts (Cohen et al. 1989; O'Hara et al. 1993). In mice, there are at least six *Dlx* genes, four of which are expressed in the developing inner ear (Robinson and Mahon 1994; Simeone et al. 1994; Acampora et al. 1999b; Depew et al. 1999). So far, only a knockout of *Dlx5* has been reported to result in malformations of the inner ear, including a smaller lateral canal and missing anterior and posterior canals (Acampora et al. 1999b; Depew et al. 1999). The three cristae are malformed, and the two maculae are also reduced in size (Merlo et al. 2002).

In addition to transcription factors, Fidgetin, a chaperone protein that is a member of the AAA (ATPase associated with different cellular activities) family of proteins, was also identified to be important for proper canal formation. AAA proteins are a group of ATPases that share common sequence features in addition to an ATP-binding motif. These proteins participate in a variety of cellular functions such as cell-cycle regulation, proteolysis, and membrane fusion (Patel and Latterich 1998). Using a positional cloning approach, *Fidgetin* was identified as the gene causing the inner ear and retinal phenotypes in the spontaneous mouse mutant *fidget* (Cox et al. 2000). In the inner ear, *Fidgetin* is expressed in the canal outpocket and the cochlear duct (Cox et al. 2000). *Fidget* mice are missing the lateral canal and crista and have malformed anterior and posterior canals (Truslove 1956). The function of Fidgetin in mediating canal development remains unclear. It has a unique N-terminal domain compared with other members of its family and, unlike other members of the family, is not predicted to have ATPase activity.

The expression of a second group of genes is initiated slightly later during canal formation. These genes include *Netrin 1* and *Nor-1*, which are expressed in the central region of the canal outpocket that is destined to form the fusion plate (Fig. 2.7). Netrin 1 is a laminin-like, secreted molecule that functions as an axonal guidance molecule in the brain (Livesey 1999). In the inner ear, *Netrin 1* knockout mice fail to form a fusion plate and, as a result, no resorption takes place in the prospective canals. It was proposed that the lack of proliferation in the surrounding mesenchyme fails to drive the opposing otic epithelia of the outpocket to come together to form the fusion plate (Salminen et al. 2000).

Nor-1 is a member of the nuclear receptor family of transcription factors. Members of this subclass of nuclear receptors are thought to function as constitutively active transcription factors (Maruyama et al. 1998). A ligand for Nor-1, if one exists, has not been identified. Although the expression patterns of *Netrin 1* and *Nor-1* in the inner ear are similar (highest in the fusion plate region), loss of *Nor-1* function does not affect canal resorption. In *Nor-1* knockout mice, the canals and ampullae are smaller than in wild-type mice (Ponnio et al. 2002). Cell proliferation is initially widespread in the prospective canal region, but after canal formation it becomes restricted to two regions of the canal (Fig. 2.7; Chang et al. 1999; Ponnio et al. 2002). The loss of *Nor-1* affected the proliferation and continual growth of all three canals and ampullae. Molecularly, it is not clear how *Nor-1* regulates cell proliferation in canals because *Nor-1* does not appear to be expressed in the proliferative zones.

Furthermore, in contrast to the expression of *Netrin 1* and *Nor-1* in the inner margin of the canals, several genes are asymmetrically distributed in the outer margins, such as *SOHo-1* (sensory organ homeobox), *Msx1*, and *Bmp4* (Kiernan et al. 1997; Chang et al. 1999). Together, these results indicate that the semicircular canals are molecularly more complex than their simple tube-shaped structures imply.

In addition to the two groups of genes mentioned above, *Otx1* and *Shh* are specifically important for the development of the lateral canal. In addition, *Gli3*, a negative regulator of *Shh* functions, also plays a role in canal development. In mouse mutant *Extratoes*, in which the *Gli3* gene is mutated, the lateral canal is missing and the anterior canal is truncated (Johnson 1967). Detailed expression of *Gli3* in the inner ear has not been reported, but its expression in the periotic mesenchyme has been demonstrated (Hui et al. 1994). Therefore, *Gli3* is another candidate gene that may influence canal development via a mesenchymal–epithelial signaling mechanism.

The anterior and posterior semicircular canals are connected to the common crus at one end. It is not clear whether the formation of the common crus is governed by common crus-specific molecules or is the consequence of resorption in the surrounding tissues. So far, there is no report of any gene that is specifically expressed in the common crus and not in the canals. However, two lines of evidence suggest that the common crus development is regulated differently from that of the canals. First, there has been a report of a patient with Goldenhar syndrome who has no common crus but has intact anterior and posterior canals (Manfre et al. 1997). Second, by implanting beads soaked with retinoic acid in the developing chicken otocyst, it has been shown that formation of the semicircular canal is sensitive to retinoic acid treatment in a dose-dependent manner (Choo et al. 1998). In the most severe cases, where none of the semicircular canals formed properly, the common crus was still intact, suggesting that genes regulating common crus development are insensitive to retinoic acid treatment and thus might be different from those governing canal formation.

A given gene could function in multiple phases of canal formation. For example, BMPs are important for multiple stages of canal development in the chicken. Noggin, an antagonist of BMPs, in particular BMP2 and BMP4, was delivered to the developing chicken otocyst using either Noggin-producing cells, beads soaked with Noggin protein, or a replication-competent avian retrovirus encoding the *Noggin* cDNA (Chang et al. 1999; Gerlach et al. 2000). These treatments consistently result in truncations of the canals and sometimes involve malformations of the ampullae. The defect in the canal formation is evident at the canal outpocket phase. Interestingly, even after the canals are formed at E7, implantation of beads soaked with Noggin protein leads to canal truncation 2 days later, indicating that the continual presence of BMPs is important for canal development. More recent data suggest that Noggin mediates its effect on canal development by blocking the action of BMP2 (Chang et al. 2002).

5.4. Relationship of Sensory and Nonsensory Tissue Development

Even though distinct molecular mechanisms govern the differentiation of sensory versus nonsensory components of the inner ear, the two pathways are most likely coordinated during early developmental stages to ensure a functional end product. One way that this can be accomplished molecularly is to activate genes that can initiate different developmental pathways in different tissues simultaneously. For example, *Otx1* is activated in both the prospective lateral ampulla and canal at the same time of development and may serve to synchronize their development. Another way to mediate the coordinated development of sensory and nonsensory tissues is through signaling molecules such as growth factors released by either tissue that couple the two developmental programs. Under these models, one would predict that most morphogenetic mutants would have both sensory and nonsensory defects. Indeed, most mutants, both in mice and zebrafish (*B.rerio*), that lack a sensory component such as a crista also show defects in the corresponding canal (Malicki et al. 1996; Whitfield et al. 1996). However, the reverse is not true. There are mutants that have defective canals but intact cristae, such as *eselsohr* in zebrafish (*B.rerio*) and *Rotating* and *Extratoes* in mice (Deol 1983; Whitfield et al. 1996). The existence of such mutants suggests that sensory tissues may play a dominant role in coordinating inner ear development by specifying nonsensory tissue formation. Axial rotation experiments performed in the chicken are consistent with this idea and suggest that the specification of sensory structures precedes specification of nonsensory structures (Wu et al. 1998). By reversing the anteroposterior (A/P) axis of the otocyst relative to the body axis, these studies indicate that the A/P axes of the sensory organs are fixed during a development period when nonsensory components of the inner ear remain unspecified.

Identification of signaling molecules that coordinate the two developmental pathways is essential to understanding the development of this complex organ. However, it is not always straightforward to extrapolate the function of a given gene based on mutation analyses or expression patterns alone. For example, *Bmp4* is expressed in both sensory and nonsensory components of the inner ear during development. Furthermore, the ubiquitous expression of its receptors suggests that BMP4 could affect multiple target tissues. A more revealing expression pattern might be that of *Fgf10* and *EphB2*. *Fgf10* is predominantly expressed in the sensory tissues, whereas its receptor, *Fgfr-2 (IIIb)* is exclusively expressed in the surrounding nonsensory component of the inner ear (Pirvola et al. 2000). So far, a majority of the phenotypes reported for *Fgf10* knockout mice are consistent with *Fgf10*'s postulated role in mediating nonsensory tissue development. However, the associated sensory phenotypes observed in *Fgf10* knockout mice suggest that other FGF receptors besides *FGFR-2 (IIIb)* are responsible for mediating this development (Pauley et al. 2003).

Eph and its ligand ephrin participate in bidirectional signaling cascades that operate in both receptor- and ligand-expressing cells. These molecules are important for multiple cell–cell communication processes, including axonal guidance, boundary formation in the brain, and vascular development (Flanagan and Vanderhaeghen 1998; Frisen et al. 1999). In the inner ear, *EphB2*, a tyrosine kinase receptor, is expressed in the nonsensory, vestibular dark cells bordering the sensory tissues of the cristae and maculae. In contrast, its putative ligand, ephrinB2, is expressed in the supporting cells of the sensory organs. *EphB2* knockout mice show a defect in fluid homeostasis in the endolymph, and their vestibular dark cells are disorganized (Cowan et al. 2000; see below). A possible role of *ephrinB2*-expressing cells in the development or differentiation of *EphB2*-expressing cells warrants further investigation.

Although it appears that sensory tissue induction precedes nonsensory tissue induction, it is possible that once nonsensory tissues are specified, genes expressed in these tissues feed back on sensory tissue and affect its development. The best supporting evidence for this comes from the analysis of the *Otx1* knockout mice. *Otx1* is not expressed in the presumptive maculae of the utricle and saccule; however, its expression domain abuts the lateral region of both maculae. The absence of *Otx1* results in incomplete separation of the maculae of the utricle and saccule, which could result from abnormal morphogenesis of the surrounding nonsensory tissues. Alternatively, Otx1 produced by nonsensory tissue may lead to the activation of factors that in turn affect sensory development (Morsli et al. 1999; Fritzsch et al. 2001).

5.5. Genes that Affect Fluid Homeostasis

Apart from genes that are important for patterning of the vestibular apparatus, there are genes that regulate fluid homeostasis of the endolymph. Absence of these gene products can also lead to changes in the shape of the membranous labyrinth and deficits in vestibular system function (Table 2.2).

The endolymph that fills the membranous labyrinth has an unusually high potassium ion concentration, which is important for proper signal transduction in sensory hair cells. It has been proposed that, in the cochlea, potassium ions enter the hair cells during the process of mechanotransduction and are subsequently taken up by the supporting cells and recycled back into the endolymph via the stria vascularis in the lateral wall of the cochlear duct (Kikuchi et al. 1995; Spicer and Schulte 1998). Similar mechanisms may be involved in the vestibular apparatus; light and dark cells with secretory and resorption functions are located in close proximity to each of the vestibular sensory organs (Dohlman 1961).

TABLE 2.2. Genes affecting fluid homeostasis of the inner ear.

Gene	Type of protein	Distribution in the inner ear	Functional deficits	
			Vestibular	Cochlear
KCNE1/ isk	protein that coassembles with K$^+$ channel subunits	stria vascularis	+	+
Ephb2	tyrosine kinase receptor	stria vascularis, dark cells of vestibule	+	–
Kvlqt1/ KCNQ1	K$^+$ channel	stria vascularis	+	+
*KCNQ4**	K$^+$ channel	outer hair cells of the cochlea; hair cells of vestibular organs	?	+*
Pendrin	anion transporter	endolymphatic sac and duct; between macula utriculi and anterior and lateral cristae; nonsensory region of the saccule; external sulcus region of the cochlea	+	+
Slc12a2	Na$^+$–K$^+$–Cl^{2-} transporter	marginal cells of stria vascularis; spiral ligament; dark cells of vestibule	+	+
Slc12a7	K–Cl$^-$ cotransporter	supporting cells for inner and outer hair cells	–	+

* No animal model available yet; the functional deficits are based on data from humans.

KCNQ4 encodes a potassium channel and is primarily expressed in the outer hair cells of the cochlea and type I hair cells of the vestibular organs (Kharkovets et al. 2000). Immunostaining studies localized this protein to the basolateral membrane of the sensory hair cells, supporting its postulated role in recycling potassium ions from hair cells back to the endolymph. Mutations in *KCNQ4* cause dominant, progressive deafness in humans. However, no animal models for this gene are yet available (Kubisch et al. 1999). More recently, a K–Cl cotransporter, Kcc4, that is expressed in the Deiters' and phalangeal cells has been postulated to participate in the recycling of potassium ions that have exited hair cells into supporting cells (Boettger et al. 2002). Mice lacking Kcc4 function are deaf and display renal tubular acidosis.

So far, three genes expressed in the stria vascularis region are believed to be important for recycling potassium ions into the endolymph. *Kcnq1*(*Kvlqt1*) or *isk* (*KCNE1*) are both expressed in the marginal cells of the stria (Sakagami et al. 1991; Wangemann et al. 1995; Neyroud et al. 1997). *Kcnq1* encodes a potassium channel subunit in the same family as *Kcnq4*. *Isk* encodes a transmembrane protein that assembles with potassium channel subunits including *Kcnq1*. Mutations in both *KCNQ1* and *KCNE1* cause Jervell and Lange–Nielsen syndrome in humans (Neyroud et al. 1997; Schulze-Bahr et al. 1997), a syndrome associated with ventricular tachyarrhythmias of the heart and deafness. Knockout mouse models for both genes show a collapsed membranous labyrinth indicative of endolymph secretion failure and disruption of fluid homeostasis in the inner ear (Vetter et al. 1996; Lee et al. 2000; Casimiro et al. 2001). A spontaneous mouse mutant, *Punk Rocker*, with a nonsense mutation in *Kcne1* that results in a truncated protein, also shows an inner ear phenotype similar to the knockout mice (Letts et al. 2000).

Slc12a2, which encodes a K–Na–Cl cotransporter, is also postulated to participate in recycling potassium ions back into the endolymph. This protein is expressed in the basolateral membrane of the marginal cells in the stria vascularis, fibrocytes in the spiral ligament, and dark cells of the vestibule (Crouch et al. 1997; Goto et al. 1997; Mizuta et al. 1997). Three mouse models are available for *Slc12a2*: a targeted deletion mutant; a radiation-induced mutant (*Shaker-with-syndactylism* (*sy*)) with a deletion that includes the *Slc12a2* locus; and an allele of *sy*, *sy*^{ns} (*Shaker with no syndactylism*), that has a frame-shift mutation in *Slc12a2* (Delpire et al. 1999; Dixon et al. 1999). All three mutant lines are deaf, with waltzer/shaker behavior indicative of vestibular deficits. In addition, their membranous labyrinths are collapsed, indicating a problem with endolymph secretion (Delpire et al. 1999; Dixon et al. 1999).

As indicated earlier, lack of *EphB2* also causes reduction of endolymph production. *EphB2* is postulated to regulate fluid homeostasis by interacting indirectly with anion exchangers and aquaporins (Cowan et al. 2000). Interestingly, despite the expression of *EphB2* in the nonsensory components of both the vestibule and cochlea, *EphB2* knockout mice display vestibular dysfunction but are not deaf. Furthermore, their cochlear ducts appear normal, suggesting that fluid homeostasis in the cochlea is not affected. Because the membranous labyrinth of the mouse is largely two separate compartments by 16.5 dpc, genes that affect fluid homoestasis may not necessarily affect both auditory and vestibular functions, depending on the expression domain and mode of action of a given gene (Cantos et al. 2000).

Another example of a gene that regulates fluid homeostasis is *Pendrin* (*Pds*), which is responsible for causing Pendred syndrome as well as a non-syndromic form of deafness in humans (Everett et al. 1997; Li et al. 1998). Patients with Pendred syndrome have sensorineural deafness and goiter. Widened vestibular aqueducts are commonly found in the inner ears of these patients. In addition, cochleae of Mondini phenotype characterized by incomplete coiling have also been described (Johnsen et al. 1986; Cremers et al. 1998).

In the mouse, *Pds* mRNA is found in the inner ear, thyroid, and kidney (Everett et al. 1997; Everett et al. 1999). Within the inner ear, *Pds* is highly expressed in the endolymphatic sac and duct. It is also expressed in non-sensory regions of the utricle and saccule and the external sulcus region (adjacent to the stria vascularis) of the cochlea (Everett et al. 1999). The expression of *Pds* is first activated in the endolymphatic sac and duct around 13 dpc. *Pds* knockout mice are deaf and show a variable spectrum of vestibular problems such as circling, head tilting, and bobbing behaviors (Everett et al. 2001). Unlike other knockout mice that have defects in fluid homeostasis, *Pds*–/– mutants show swelling of the membranous labyrinth instead of shrinkage. The endolymphatic duct and sac are the first structures to swell, starting at 15 dpc (Fig. 2.8A,B, arrows). The swelling later spreads into the vestibular and cochlear regions. The deafness and balancing problems in these mice are most likely due to sensory hair-cell degeneration resulting from an ionic imbalance within the endolymph (Everett et al. 2001). Functional studies in frog (*Xenopus*) oocytes suggest that PENDRIN is a chloride and iodide transporter (Scott et al. 1999). However, whether chloride and/or possibly other anions are being transported by PENDRIN within the inner ear remains to be directly determined.

In the mouse inner ear, as morphogenesis proceeds, the connection between the utricle and saccule becomes restricted such that, by 16.5 dpc, the endolymphatic sac and duct, as well as the saccule and cochlea, are one continuous chamber, and the utricle and three canals and their ampullae are joined in another chamber (Cantos et al. 2000). Figure 2.8C illustrates a paint-filled inner ear that has been injected in the endolymphatic sac at P1. Only the saccule and cochlea, but not the utricle or the rest of the labyrinth, were filled with paint from such an injection. Despite the prenatal malformations and swelling of the membranous labyrinth of the *Pds* knockout mice, a similar paint-fill pattern was observed in *Pds* mutants,

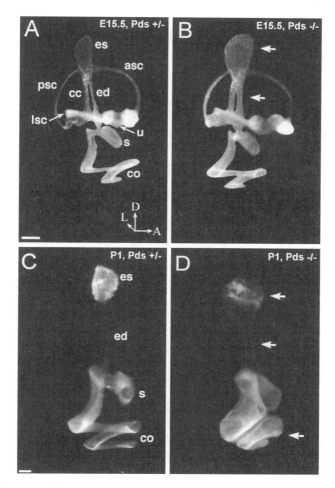

FIGURE 2.8. Paint-filled mouse membranous labyrinths of the wild type (**A, C**) and *Pendrin* −/− mutant (**B, D**). Swelling of the membranous labyrinth of *Pnd* mutants is first apparent in the endolymphatic duct and sac at 15.5 dpc (arrows in B). Latex paint solution is injected only into the endolymphatic sac in wild-type (C) and mutant (D) inner ears at P1. Injection into the endolymphatic sac only fills the sac and its duct, the saccule, and cochlea (C). Despite the enlarged membranous labyrinth in *Pnd* null mutants, injection of latex paint to the endolymphatic sac shows a pattern similar to the wild type, indicating that the utricle and saccule are in separate compartments (D). For abbreviations, see Figure 2.2. Scale bar = 100 μm.

indicating that the utricle and saccule still separated into individual compartments (Fig. 2.8D). This is in contrast to the morphogenetic mutants such as *Hmx2*, *Hmx3*, *Ngn1*, *Otx1*, and *Otx2* knockouts, where the utricle and saccule fail to separate from each other (Wang et al. 1998; Morsli et al. 1999; Ma et al. 2000).

Furthermore, because of the unique ionic composition and high resting potential of the endolymph, the epithelial cells of the membranous labyrinth might require specialized intercellular communication networks and proper "sealing" from their surrounding tissues. Consistent with this hypothesis, mutations in genes encoding for gap junction proteins such as connexin 26 and 31 and tight junction proteins such as claudin 14 have been implicated in causing human deafness (Wilcox et al. 2001; for a review, see Steel et al. 2002). The etiologies of these human syndromes will be apparent as more animal models become available.

6. Conclusion

Two areas of inner ear development have not been discussed thus far: otic induction and differentiation of sensory hair cells. *Fgf19* and *Wnt-8c* are implicated in otic induction in the chicken (Ladher et al. 2000; Vendrell et al. 2000); *Fgf3* and *Fgf8* are implicated in otic induction in zebrafish (*B. rerio*) (Phillips et al. 2001; Leger and Brand 2002; Maroon et al. 2002, whereas *Fgf3* and *Fgf10* are important for otic induction in mice (Wright and Mansour 2003). Recent reviews on otic induction and related topics can be found in a special issue of Journal of Neurobiology (Kil and Collazo 2002; Noramly and Grainger 2002; Whitfield 2002). Furthermore, many genes have been identified to be essential for hair cell development/differentiation, such as *Pou4f3* (*Brn3.1*), *myosinVIIa*, *Espin*, and *Cadherins*. Mutations of these genes lead to vestibular and auditory deficits in both humans and mice. Readers are referred to recent reviews on these topics (Steel and Kros 2001; Caldwell and Eberl 2002; Steel et al. 2002). For additional readings on genes associated with morphogenesis of the inner ear, readers are referred to two excellent reviews by Anagnostopoulos (Anagnostopoulos 2002) and Kiernan et al. (Kiernan et al. 2002).

Correlating a specific gene's knockout phenotype with its expression pattern is essential to understanding its role in inner ear development. However, multiple examples given here show that a gene's expression pattern does not necessarily predict the phenotype that results from loss of the gene product. *Dlx5* and *Netrin1*, for example, are both equivalently expressed in each of the three presumptive canals; however, knockouts of these genes show different degrees of phenotypic severity among the three canals. Also, although loss of *Math1* affected hair cell formation in all inner ear sensory organs, *Jagged1* and *Jagged2* seem to have differential effects on hair cell formation in different sensory organs.

Such disparities may be explained by differential control and functional redundancy. Despite the apparent morphological similarities in the formation of the canals and the arrangement of hair cells and supporting cells in different sensory organs, the molecular mechanisms underlying each of these processes are most likely regulated differently. Furthermore, the developmental pathways for inner ear structures are likely to be influenced by a variety of genes whose expression patterns and actions within the individual inner ear structures have thus far not been assessed. Finally, the differential expression and/or efficacy of functionally redundant genes in the different inner ear structures may determine the extent to which the knockout of any given gene affects a particular structure. For example, four out of the six *Dlx* genes are expressed in the inner ear; one or more of these genes could share a redundant function with *Dlx5* in the formation of the lateral canal.

The creation of multiple and conditional knockouts in mice will continue to be a powerful tool for molecularly unraveling the organogenesis of this complex organ. With the aid of the mouse genome project, the identification of genes responsible for existing and upcoming mutants will be expedited. Contributions from other genetic models such as zebrafish (*B. rerio*) and models that are ideal for misexpression studies and embryonic manipulations, such as the chicken and frog (*Xenopus*) will also be indispensable. An in-depth molecular understanding of this complex organ during development will pave the way for better strategies to alleviate vestibular and auditory deficits associated with this sense organ.

Acknowledgments. The authors wish to thank Quianna Burton, Jenny Bai, and Michael Mulheisen for figure preparation and three-dimensional reconstructions and Dr. Susan Sullivan for critical reading of the manuscript and discussions. The three-dimensional reconstruction software was provided by the Biocomputation Center at Ames Research Center, NASA. The authors also wish to thank Drs. Bernd Fritzsch, MengQing Xiang, Amy Kiernan, and Suzanne Mansour for preprints prior to publication. Data provided in Figure 2.8 are done in collaboration with Lorraine Everett and Eric Green in NHGRI, NIH.

References

Abdelhak S, Kalatzis V, Heilig R, Compain S, et al. (1997) A human homologue of the *Drosophila* eyes absent gene underlies branchio-oto-renal (BOR) syndrome and identifies a novel gene family. Nat Genet 15:157–164.

Acampora D, Mazan S, Lallemand Y, Avantaggiato V, et al. (1995) Forebrain and midbrain regions are deleted in Otx2–/– mutants due to a defective anterior neuroectoderm specification during gastrulation. Development 121:3279–3290.

Acampora D, Mazan S, Avantaggiato V, Barone P, et al. (1996) Epilepsy and brain abnormalities in mice lacking the Otx1 gene. Nat Genet 14:218–222.

Acampora D, Avantaggiato V, Tuorto F, Barone P, et al. (1999a) Differential transcriptional control as the major molecular event in generating Otx1–/– and Otx2–/– divergent phenotypes. Development 126:1417–1426.

Acampora D, Merlo GR, Paleari L, Zerega B, et al. (1999b) Craniofacial, vestibular, and bone defects in mice lacking the Distal-less-related gene Dlx5. Development 126:3795–3809.

Adam J, Myat A, Le Roux I, Eddison M, et al. (1998) Cell fate choices and the expression of Notch, Delta and Serrate homologues in the chick inner ear: parallels with *Drosophila* sense-organ development. Development 125:4645–4654.

Alavizadeh A, Kiernan AE, Nolan P, Lo C, et al. (2001) The Wheels mutation in the mouse causes vascular, hindbrain, and inner ear defects. Dev Biol 234:244–260.

Anagnostopoulos A (2002) A compendium of mouse knockouts with inner ear defects. Trends Genet 18:S21–S38.

Anderson DJ, Jan YN (1997) The determination of the neuronal phenotype. In: Cowan WM, Jessell TM, Zipursky SL (eds) Molecular and Cellular Approaches to Neural Development. New York: Oxford University Press, pp. 26–63.

Ang SL, Jin O, Rhinn M, Daigle N, et al. (1996) A targeted mouse Otx2 mutation leads to severe defects in gastrulation and formation of axial mesoderm and to deletion of rostral brain. Development 122:243–252.

Artavanis-Tsakonas S, Simpson P (1991) Choosing a cell fate: a view from the Notch locus. Trends Genet 7:403–408.

Artavanis-Tsakonas S, Rand MD, Lake RJ (1999) Notch signaling: cell fate control and signal integration in development. Science 284:770–776.

Axelrod JD, Matsuno K, Artavanis-Tsakonas S, Perrimon N (1996) Interaction between Wingless and Notch signaling pathways mediated by dishevelled. Science 271:1826–1832.

Bermingham NA, Hassan BA, Price SD, Vollrath MA, et al. (1999) Math1: an essential gene for the generation of inner ear hair cells. Science 284:1837–1841.

Bianchi LM, Conover JC, Fritzsch B, DeChiara T, et al. (1996) Degeneration of vestibular neurons in late embryogenesis of both heterozygous and homozygous BDNF null mutant mice. Development 122:1965–1973.

Bissonnette JP, Fekete DM (1996) Standard atlas of the gross anatomy of the developing inner ear of the chicken. J Comp Neurol 368:620–630.

Boettger T, Hubner CA, Maier H, Rust MB, et al. (2002) Deafness and renal tubular acidosis in mice lacking the K–Cl co-transporter Kcc4. Nature 416:874–878.

Brigande JV, Iten LE, Fekete DM (2000a) A fate map of chick otic cup closure reveals lineage boundaries in the dorsal otocyst. Dev Biol 227:256–270.

Brigande JV, Kiernan AE, Gao X, Iten LE, et al. (2000b) Molecular genetics of pattern formation in the inner ear: do compartment boundaries play a role? Proc Natl Acad Sci USA 97:11700–11706.

Caldwell JC, Eberl DF (2002) Towards a molecular understanding of *Drosophila* hearing. J Neurobiol 53:172–189.

Cantos R, Cole LK, Acampora D, Simeone A, et al. (2000) Patterning of the mammalian cochlea. Proc Natl Acad Sci USA 97:11707–11713.

Carney PR, Couve E (1989) Cell polarity changes and migration during early development of the avian peripheral auditory system. Anat Rec 225:156–164.

Carpenter EM, Goddard JM, Chisaka O, Manley NR, et al. (1993) Loss of Hox-A1 (Hox-1.6) function results in the reorganization of the murine hindbrain. Development 118:1063–1075.

Casimiro MC, Knollmann BC, Ebert SN, Vary JC Jr, et al. (2001) Targeted disruption of the Kcnq1 gene produces a mouse model of Jervell and Lange–Nielsen syndrome. Proc Natl Acad Sci USA 98:2526–2531.

Chang W, Nunes FD, De Jesus-Escobar JM, Harland R, et al. (1999) Ectopic noggin blocks sensory and nonsensory organ morphogenesis in the chicken inner ear. Dev Biol 216:369–381.

Chang W, ten Dijke P, Wu DK (2002) BMP pathways are involved in otic capsule formation and epithelial–mesenchymal signaling in the developing chicken inner ear. Dev Biol 251:380–394.

Chen P, Johnson JE, Zoghbi HY, Segil N (2002) The role of Math1 in inner ear development: uncoupling the establishment of the sensory primordium from hair cell fate determination. Development 129:2495–2505.

Chen R, Amoui M, Zhang Z, Mardon G (1997) Dachshund and eyes absent proteins form a complex and function synergistically to induce ectopic eye development in *Drosophila*. Cell 91:893–903.

Chen Y, Bei M, Woo I, Satokata I, et al. (1996) Msx1 controls inductive signaling in mammalian tooth morphogenesis. Development 122:3035–3044.

Choo D, Sanne JL, Wu DK (1998) The differential sensitivities of inner ear structures to retinoic acid during development. Dev Biol 204:136–150.

Cohen SM, Bronner G, Kuttner F, Jurgens G, et al. (1989) Distal-less encodes a homoeodomain protein required for limb development in *Drosophila*. Nature 338:432–434.

Cole LK, Le Roux I, Nunes F, Laufer E, et al. (2000) Sensory organ generation in the chicken inner ear: contributions of bone morphogenetic protein 4, serrate 1, and lunatic fringe. J Comp Neurol 424:509–520.

Cordes SP, Barsh GS (1994) The mouse segmentation gene *kr* encodes a novel basic domain-leucine zipper transcription factor. Cell 79:1025–1034.

Cowan CA, Yokoyama N, Bianchi LM, Henkemeyer M, et al. (2000) EphB2 guides axons at the midline and is necessary for normal vestibular function. Neuron 26:417–430.

Cox GA, Mahaffey CL, Nystuen A, Letts VA, et al. (2000) The mouse fidgetin gene defines a new role for AAA family proteins in mammalian development. Nat Genet 26:198–202.

Cremers CW, Admiraal RJ, Huygen PL, Bolder C, et al. (1998) Progressive hearing loss, hypoplasia of the cochlea and widened vestibular aqueducts are very common features in Pendred's syndrome. Int J Pediatr Otorhinolaryngol 45:113–123.

Crouch JJ, Sakaguchi N, Lytle C, Schulte BA (1997) Immunohistochemical localization of the Na–K–Cl co-transporter (NKCC1) in the gerbil inner ear. J Histochem Cytochem 45:773–778.

de Kok YJ, van der Maarel SM, Bitner-Glindzicz M, Huber I, et al. (1995) Association between X-linked mixed deafness and mutations in the POU domain gene POU3F4. Science 267:685–688.

Delpire E, Lu J, England R, Dull C, et al. (1999) Deafness and imbalance associated with inactivation of the secretory Na–K–2Cl co-transporter. Nat Genet 22:192–195.

Deol MS (1964) The abnormalities of the inner ear in kreisler mice. J Embryol Exp Morphol 12:475–490.

Deol M (1966) Influence of the neural tube on the differentiation of the inner ear in the mammalian embryo. Nature 209:219–220.

Deol MS (1983) Development of auditory and vestibular systems in mutant mice. In: Romand R (ed) Development of Auditory and Vestibular Systems. New York: Academic Press, pp. 309–333.

Depew MJ, Liu JK, Long JE, Presley R, et al. (1999) Dlx5 regulates regional development of the branchial arches and sensory capsules. Development 126: 3831–3846.

Dewulf N, Verschueren K, Lonnoy O, Moren A, et al. (1995) Distinct spatial and temporal expression patterns of two type I receptors for bone morphogenetic proteins during mouse embryogenesis. Endocrinology 136:2652–2663.

Dixon MJ, Gazzard J, Chaudhry SS, Sampson N, et al. (1999) Mutation of the Na–K–Cl co-transporter gene Slc12a2 results in deafness in mice. Hum Mol Genet 8:1579–1584.

Dohlman G (1961) Excretion and absorption of endolymph in the vestibular apparatus. In: de Reuck AVS, Knight H (eds) Ciba Foundation Symposium in Motatic, Kinesthetic and Vestibular Mechanisms. London: Churchill, pp. 138–143.

Eddison M, Le Roux I, Lewis J (2000) Notch signaling in the development of the inner ear: lessons from *Drosophila*. Proc Natl Acad Sci USA 97:11692–11699.

Epstein DJ, Vekemans M, Gros P (1991) Splotch (Sp2H), a mutation affecting development of the mouse neural tube, shows a deletion within the paired homeodomain of Pax-3. Cell 67:767–774.

Everett LA, Glaser B, Beck JC, Idol JR, et al. (1997) Pendred syndrome is caused by mutations in a putative sulphate transporter gene (PDS). Nat Genet 17: 411–422.

Everett LA, Morsli H, Wu DK, Green ED (1999) Expression pattern of the mouse ortholog of the Pendred's syndrome gene (Pds) suggests a key role for pendrin in the inner ear. Proc Natl Acad Sci USA 96:9727–9732.

Everett LA, Belyantseva IA, Noben-Trauth K, Cantos R, et al. (2001) Targeted disruption of mouse Pds provides key insight about the inner-ear defects encountered in Pendred syndrome. Hum Mol Genet 10:153–161.

Evrard YA, Lun Y, Aulehla A, Gan L, et al. (1998) Lunatic fringe is an essential mediator of somite segmentation and patterning. Nature 394:377–381.

Failli V, Bachy I, Retaux S (2002) Expression of the LIM-homeodomain gene Lmx1a (dreher) during development of the mouse nervous system. Mech Dev 118: 225–228.

Fekete DM (1999) Development of the vertebrate ear: insights from knockouts and mutants. Trends Neurosci 22:263–269.

Fekete DM, Wu DK (2002) Revisiting cell fate specification in the inner ear. Curr Opin Neurobiol 12:35–42.

Fekete DM, Homburger SA, Waring MT, Riedl AE, et al. (1997) Involvement of programmed cell death in morphogenesis of the vertebrate inner ear. Development 124:2451–2461.

Flanagan JG, Vanderhaeghen P (1998) The ephrins and Eph receptors in neural development. Annu Rev Neurosci 21:309–345.

Frenz DA, Galinovic-Schwartz V, Liu W, Flanders KC, et al. (1992) Transforming growth factor beta 1 is an epithelial-derived signal peptide that influences otic capsule formation. Dev Biol 153:324–336.

Frenz DA, Liu W, Williams JD, Hatcher V, et al. (1994) Induction of chondrogenesis: requirement for synergistic interaction of basic fibroblast growth factor and transforming growth factor-beta. Development 120:415–424.

Frenz DA, Liu W, Capparelli M (1996) Role of BMP-2a in otic capsule chondrogenesis. Ann NY Acad Sci 785:256–258.

Frisen J, Holmberg J, Barbacid M (1999) Ephrins and their Eph receptors: multitalented directors of embryonic development. EMBO J 18:5159–5165.

Fritzsch B, Silos-Santiago I, Smeyne R, Fagan M, et al. (1995) Reduction and loss of inner ear innervation in trkB and trkC receptor knockout mice: a whole mount DiI and scanning electron microscopic analysis. Aud Neurosci 1:401–417.

Fritzsch B, Silos-Santiago I, Bianchi LM, Farinas I (1997) Effects of neurotrophin and neurotrophin receptor disruption on the afferent inner ear innervation. Semin Cell Dev Biol 8:277–284.

Fritzsch B, Barald KF, Lomax MI (1998) Early embryology of the vertebrate ear. In: Rubel EW, Popper AN, Fay RR (eds) Development of the Auditory System. Springer Handbook of Auditory Research, Volume 9. New York: Springer, pp. 80–145.

Fritzsch B, Pirvola U, Ylikoski J (1999) Making and breaking the innervation of the ear: neurotrophic support during ear development and its clinical implications. Cell Tissue Res 295:369–382.

Fritzsch B, Beisel KW, Bermingham NA (2000) Developmental evolutionary biology of the vertebrate ear: conserving mechanoelectric transduction and developmental pathways in diverging morphologies. Neuroreport 11:R35–44.

Fritzsch B, Signore M, Simeone A (2001) Otx1 null mutants show partial segregation of sensory epithelial comparable to lamprey ears. Dev Genes Evol 211:388–396.

Gavalas A, Studer M, Lumsden A, Rijli FM, et al. (1998) Hoxa1 and Hoxb1 synergize in patterning the hindbrain, cranial nerves and second pharyngeal arch. Development 125:1123–1136.

Gerlach LM, Hutson MR, Germiller JA, Nguyen-Luu D, et al. (2000) Addition of the BMP4 antagonist, noggin, disrupts avian inner ear development. Development 127:45–54.

Giraldez F (1998) Regionalized organizing activity of the neural tube revealed by the regulation of lmx1 in the otic vesicle. Dev Biol 203:189–200.

Goto S, Oshima T, Ikeda K, Ueda N, et al. (1997) Expression and localization of the Na–K–2Cl cotransporter in the rat cochlea. Brain Res 765:324–326.

Goulding M, Sterrer S, Fleming J, Balling R, et al. (1993) Analysis of the Pax-3 gene in the mouse mutant splotch. Genomics 17:355–363.

Goulding MD, Chalepakis G, Deutsch U, Erselius JR, et al. (1991) Pax-3, a novel murine DNA binding protein expressed during early neurogenesis. EMBO J 10:1135–1147.

Haddon C, Jiang YJ, Smithers L, Lewis J (1998) Delta-Notch signalling and the patterning of sensory cell differentiation in the zebrafish ear: evidence from the mind bomb mutant. Development 125:4637–4644.

Hadrys T, Braun T, Rinkwitz-Brandt S, Arnold HH, et al. (1998) Nkx5-1 controls semicircular canal formation in the mouse inner ear. Development 125:33–39.

Hallbook F, Ibanez CF, Ebendal T, Persson H (1993) Cellular localization of brain-derived neurotrophic factor and neurotrophin-3 mRNA expression in the early chicken embryo. Eur J Neurosci 5:1–14.

Hemmati-Brivanlou A, Thomsen GH (1995) Ventral mesodermal patterning in *Xenopus* embryos: expression patterns and activities of BMP-2 and BMP-4. Dev Genet 17:78–89.

Hicks C, Johnston SH, diSibio G, Collazo A, et al. (2000) Fringe differentially modulates Jagged1 and Delta1 signalling through Notch1 and Notch2. Nat Cell Biol 2:515–520.

Hirth F, Therianos S, Loop T, Gehring WJ, et al. (1995) Developmental defects in brain segmentation caused by mutations of the homeobox genes orthodenticle and empty spiracles in *Drosophila*. Neuron 15:769–778.

Hogan BL (1996) Bone morphogenetic proteins: multifunctional regulators of vertebrate development. Genes Dev 10:1580–1594.

Huang EJ, Liu W, Fritzsch B, Bianchi LM, et al. (2001) Brn-3a is a transcriptional regulator of soma size, target field innervation, and axon pathfinding of Inner ear sensory neurons. Development 126:2869–2882.

Hui CC, Slusarski D, Platt KA, Holmgren R, et al. (1994) Expression of three mouse homologs of the *Drosophila* segment polarity gene cubitus interruptus, Gli, Gli-2 and Gli-3, in ectoderm- and mesoderm-derived tissues suggests multiple roles during postimplantation development. Dev Biol 162:402–413.

Johnsen T, Jorgensen MB, Johnsen S (1986) Mondini cochlea in Pendred's syndrome. A histological study. Acta Otolaryngol 102:239–247.

Johnson DR (1967) Extra-toes: a new mutant gene causing multiple abnormalities in the mouse. J Embryol Exp Morphol 17:543–581.

Johnson KR, Cook SA, Erway LC, Matthews AN, et al. (1999) Inner ear and kidney anomalies caused by IAP insertion in an intron of the Eya1 gene in a mouse model of BOR syndrome. Hum Mol Genet 8:645–653.

Ju BG, Jeong S, Bae E, Hyun S, et al. (2000) Fringe forms a complex with Notch. Nature 405:191–195.

Kalatzis V, Sahly I, El-Amraoui A, Petit C (1998) Eya1 expression in the developing ear and kidney: towards the understanding of the pathogenesis of Branchio-Oto-Renal (BOR) syndrome. Dev Dyn 213:486–499.

Karis A, Pata I, van Doorninck JH, Grosveld F, et al. (2001) Transcription factor GATA-3 alters pathway selection of olivocochlear neurons and affects morphogenesis of the ear. J Comp Neurol 429:615–630.

Kharkovets T, Hardelin JP, Safieddine S, Schweizer M, et al. (2000) KCNQ4, a K$^+$ channel mutated in a form of dominant deafness, is expressed in the inner ear and the central auditory pathway. Proc Natl Acad Sci USA 97:4333–4338.

Kiernan AE, Nunes F, Wu DK, Fekete DM (1997) The expression domain of two related homeobox genes defines a compartment in the chicken inner ear that may be involved in semicircular canal formation. Dev Biol 191:215–229.

Kiernan AE, Ahituv N, Fuchs H, Balling R, et al. (2001) The Notch ligand Jagged1 is required for inner ear sensory development. Proc Natl Acad Sci USA 98:3873–3878.

Kiernan AE, Steel KP, Fekete DM (2002) Development of the mouse inner ear. In: Rossant JT, Tam PPL (eds) Mouse Development: Patterning, Morphogenesis, and Organogenesis. Orlando, FL: Academic Press, pp. 539–566.

Kikuchi T, Kimura RS, Paul DL, Adams JC (1995) Gap junctions in the rat cochlea: immunohistochemical and ultrastructural analysis. Anat Embryol (Berl) 191: 101–118.

Kil SH, Collazo A (2002) A review of inner ear fate maps and cell lineage studies. J Neurobiol 53:129–142.

Kim WY, Fritzsch B, Serls A, Bakel LA, et al. (2001) NeuroD-null mice are deaf due to a severe loss of the inner ear sensory neurons during development. Development 128:417–426.

Kubisch C, Schroeder BC, Friedrich T, Lutjohann B, et al. (1999) KCNQ4, a novel potassium channel expressed in sensory outer hair cells, is mutated in dominant deafness. Cell 96:437–446.

Ladher RK, Anakwe KU, Gurney AL, Schoenwolf GC, et al. (2000) Identification of synergistic signals initiating inner ear development. Science 290:1965–1967.

Landolt JP, Correia MJ, Young ER, Cardin RP, et al. (1975) A scanning electron microscopic study of the morphology and geometry of neural surfaces and structures associated with the vestibular apparatus of the pigeon. J Comp Neurol 159:257–287.

Lanford PJ, Lan Y, Jiang R, Lindsell C, et al. (1999) Notch signalling pathway mediates hair cell development in mammalian cochlea. Nat Genet 21:289–292.

Laufer E, Dahn R, Orozco OE, Yeo CY, et al. (1997) Expression of Radical fringe in limb-bud ectoderm regulates apical ectodermal ridge formation [see comments] Nature 386:366–373. [published erratum appears in Nature 388:400 (1997)].

Lee JE, Hollenberg SM, Snider L, Turner DL, et al. (1995) Conversion of *Xenopus* ectoderm into neurons by NeuroD, a basic helix- loop-helix protein. Science 268:836–844.

Lee MP, Ravenel JD, Hu RJ, Lustig LR, et al. (2000) Targeted disruption of the Kvlqt1 gene causes deafness and gastric hyperplasia in mice. J Clin Invest 106:1447–1455.

Leger S, Brand M (2002) Fgf8 and Fgf3 are required for zebrafish ear placode induction, maintenance and inner ear patterning. Mech Dev 119:91.

Leimeister C, Externbrink A, Klamt B, Gessler M (1999) Hey genes: a novel subfamily of hairy- and Enhancer of split related genes specifically expressed during mouse embryogenesis. Mech Dev 85:173–177.

Letts VA, Valenzuela A, Dunbar C, Zheng QY, et al. (2000) A new spontaneous mouse mutation in the Kcne1 gene. Mamm Genome 11:831–835.

Lewis AK, Frantz GD, Carpenter DA, de Sauvage FJ, et al. (1998) Distinct expression patterns of notch family receptors and ligands during development of the mammalian inner ear. Mech Dev 78:159–163.

Li XC, Everett LA, Lalwani AK, Desmukh D, et al. (1998) A mutation in PDS causes non-syndromic recessive deafness. Nat Genet 18:215–217.

Liu M, Pereira FA, Price SD, Chu M, et al. (2000) Essential role of BETA2/NeuroD1 in development of the vestibular and auditory systems. Genes Dev 14:2839–2854.

Liu W, Li G, Chien JS, Raft S, et al. (2002) Sonic hedgehog regulates otic capsule chondrogenesis and inner ear development in the mouse embryo. Dev Biol 248:240–250.

Livesey FJ (1999) Netrins and netrin receptors. Cell Mol Life Sci 56:62–68.

Ma Q, Chen Z, Barrantes I, de la Pompa JL, et al. (1998) Neurogenin 1 is essential for the determination of neuronal precursors for proximal cranial sensory ganglia. Neuron 20:469–482.

Ma Q, Anderson DJ, Fritzsch B (2000) Neurogenin 1 null mutant ears develop fewer, morphologically normal hair cells in smaller sensory epithelia devoid of innervation. J Assoc Res Otolaryngol 1:129–143.

Maconochie M, Nonchev S, Morrison A, Krumlauf R (1996) Paralogous Hox genes: function and regulation. Annu Rev Genet 30:529–556.

Malicki J, Schier AF, Solnica-Krezel L, Stemple DL, et al. (1996) Mutations affecting development of the zebrafish ear. Development 123:275–283.

Manfre L, Genuardi P, Tortorici M, Lagalla R (1997) Absence of the common crus in Goldenhar syndrome. Am J Neuroradiol 18:773–775.

Mansour SL (1994) Targeted disruption of int-2 (fgf-3) causes developmental defects in the tail and inner ear. Mol Reprod Dev 39:62–68.

Mansour SL, Goddard JM, Capecchi MR (1993) Mice homozygous for a targeted disruption of the proto-oncogene int-2 have developmental defects in the tail and inner ear. Development 117:13–28.

Manzanares M, Trainor PA, Ariza-McNaughton L, Nonchev S, et al. (2000) Dorsal patterning defects in the hindbrain, roof plate and skeleton in the dreher (dr(J)) mouse mutant. Mech Dev 94:147–156.

Mark M, Lufkin T, Vonesch JL, Ruberte E, et al. (1993) Two rhombomeres are altered in Hoxa-1 mutant mice. Development 119:319–338.

Maroon H, Walshe J, Mahmood R, Kiefer P, et al. (2002) Fgf3 and Fgf8 are required together for formation of the otic placode and vesicle. Development 129:2099–2108.

Maruyama K, Tsukada T, Ohkura N, Bandoh S, et al. (1998) The NGFI-B subfamily of the nuclear receptor superfamily (review). Int J Oncol 12:1237–1243.

McKay IJ, Lewis J, Lumsden A (1996) The role of FGF-3 in early inner ear development: an analysis in normal and kreisler mutant mice. Dev Biol 174:370–378.

Merlo GR, Paleari L, Mantero S, Zerega B, et al. (2002) The Dlx5 homeobox gene is essential for vestibular morphogenesis in the mouse embryo through a BMP4-mediated pathway. Dev Biol 248:157–169.

Millonig JH, Millen KJ, Hatten ME (2000) The mouse Dreher gene Lmx1a controls formation of the roof plate in the vertebrate CNS. Nature 403:764–769.

Minowa O, Ikeda K, Sugitani Y, Oshima T, et al. (1999) Altered cochlear fibrocytes in a mouse model of DFN3 nonsyndromic deafness [see comments]. Science 285:1408–1411.

Mizuta K, Adachi M, Iwasa KH (1997) Ultrastructural localization of the Na–K–Cl cotransporter in the lateral wall of the rabbit cochlear duct. Hear Res 106:154–162.

Morrison A, Hodgetts C, Gossler A, Hrabe de Angelis M, et al. (1999) Expression of Delta1 and Serrate1 (Jagged1) in the mouse inner ear. Mech Dev 84:169–172.

Morsli H, Choo D, Ryan A, Johnson R, et al. (1998) Development of the mouse inner ear and origin of its sensory organs. J Neurosci 18:3327–3335.

Morsli H, Tuorto F, Choo D, Postiglione MP, et al. (1999) Otx1 and Otx2 activities are required for the normal development of the mouse inner ear. Development 126:2335–2343.

Mowbray C, Hammerschmidt M, Whitfield TT (2001) Expression of BMP signaling pathway member in the developing zebrafish inner ear and lateral line. MOD 108:179–184.

Myat A, Henrique D, Ish-Horowicz D, Lewis J (1996) A chick homologue of Serrate and its relationship with Notch and Delta homologues during central neurogenesis. Dev Biol 174:233–247.

Nardelli J, Thiesson D, Fujiwara Y, Tsai FY, et al. (1999) Expression and genetic interaction of transcription factors GATA-2 and GATA-3 during development of the mouse central nervous system. Dev Biol 210:305–321.

Neyroud N, Tesson F, Denjoy I, Leibovici M, et al. (1997) A novel mutation in the potassium channel gene KVLQT1 causes the Jervell and Lange–Nielsen cardio-auditory syndrome. Nat Genet 15:186–189.

Noramly S, Grainger RM (2002) Determination of the embryonic inner ear. J Neurobiol 53:100–128.

O'Hara E, Cohen B, Cohen SM, McGinnis W (1993) Distal-less is a downstream gene of Deformed required for ventral maxillary identity. Development 117: 847–856.

Panin VM, Papayannopoulos V, Wilson R, Irvine KD (1997) Fringe modulates Notch–ligand interactions. Nature 387:908–912.

Papayannopoulos V, Tomlinson A, Panin VM, Rauskolb C, et al. (1998) Dorsal–ventral signaling in the *Drosophila* eye. Science 281:2031–2034.

Patel S, Latterich M (1998) The AAA team: related ATPases with diverse functions. Trends Cell Biol 8:65–71.

Pauley S, Wright T, Pirvola U, Ornitz DM, Beisel KW, et al. (2003) Expression and function of FGF-10 in mammalian inner ear development. Dev Dynamics 227:203–215.

Phillips BT, Bolding K, Riley BB (2001) Zebrafish Fgf3 and Fgf8 encode redundant functions required for otic placode induction. Dev Biol 235:351–365.

Phippard D, Heydemann A, Lechner M, Lu L, et al. (1998) Changes in the subcellular localization of the Brn4 gene product precede mesenchymal remodeling of the otic capsule. Hear Res 120:77–85.

Phippard D, Lu L, Lee D, Saunders JC, et al. (1999) Targeted mutagenesis of the POU-domain gene Brn4/Pou3f4 causes developmental defects in the inner ear. J Neurosci 19:5980–5989.

Phippard D, Boyd Y, Reed V, Fisher G, et al. (2000) The sex-linked fidget mutation abolishes Brn4/Pou3f4 gene expression in the embryonic inner ear. Hum Mol Genet 9:79–85.

Pignoni F, Hu B, Zavitz KH, Xiao J, et al. (1997) The eye-specification proteins So and Eya form a complex and regulate multiple steps in *Drosophila* eye development. Cell 91:881–891.

Pirvola U, Spencer-Dene B, Xing-Qun L, Kettunen P, et al. (2000) Fgf/Fgfr-2 (IIIb) signaling is essential for inner ear morphogenesis. J Neurosci 20:6125–6134.

Pissarra L, Henrique D, Duarte A (2000) Expression of hes6, a new member of the Hairy/Enhancer-of-split family, in mouse development. Mech Dev 95:275–278.

Ponnio T, Burton Q, Pereira FA, Wu DK, et al. (2002) The nuclear receptor Nor-1 is essential for proliferation of the semicircular canals of the mouse inner ear. Mol Cell Biol 22:935–945.

Riccomagno MM, Martinu L, Mulheisen M, Wu DK, et al. (2002) Specification of the mammalian cochlea is dependent to Sonic hedgehog. Genes Dev 16:2365–2378.

Rijli FM, Mark M, Lakkaraju S, Dierich A, et al. (1993) A homeotic transformation is generated in the rostral branchial region of the head by disruption of Hoxa-2, which acts as a selector gene. Cell 75:1333–1349.

Rinkwitz S, Bober E, Baker R (2001) Development of the vertebrate inner ear. Ann NY Acad Sci 942:1–14.

Rinkwitz-Brandt S, Justus M, Oldenettel I, Arnold HH, et al. (1995) Distinct temporal expression of mouse Nkx-5.1 and Nkx-5.2 homeobox genes during brain and ear development. Mech Dev 52:371–381.

Rinkwitz-Brandt S, Arnold HH, Bober E (1996) Regionalized expression of Nkx5-1, Nkx5-2, Pax2 and sek genes during mouse inner ear development. Hear Res 99:129–138.

Rivolta MN, Holley MC (1998) GATA3 is downregulated during hair cell differentiation in the mouse cochlea. J Neurocytol 27:637–647.

Robinson GW, Mahon KA (1994) Differential and overlapping expression domains of Dlx-2 and Dlx-3 suggest distinct roles for Distal-less homeobox genes in craniofacial development. Mech Dev 48:199–215.

Royet J, Finkelstein R (1995) Pattern formation in *Drosophila* head development: the role of the orthodenticle homeobox gene. Development 121:3561–3572.

Sakagami M, Fukazawa K, Matsunaga T, Fujita H, et al. (1991) Cellular localization of rat Isk protein in the stria vascularis by immunohistochemical observation. Hear Res 56:168–172.

Salminen M, Meyer BI, Bober E, Gruss P (2000) Netrin 1 is required for semicircular canal formation in the mouse inner ear. Development 127:13–22.

Satokata I, Maas R (1994) Msx1 deficient mice exhibit cleft palate and abnormalities of craniofacial and tooth development. Nat Genet 6:348–356.

Schimmang T, Lemaistre M, Vortkamp A, Ruther U (1992) Expression of the zinc finger gene Gli3 is affected in the morphogenetic mouse mutant extra-toes (Xt). Development 116:799–804.

Schimmang T, Minichiello L, Vazquez E, San Jose I, et al. (1995) Developing inner ear sensory neurons require TrkB and TrkC receptors for innervation of their peripheral targets. Development 121:3381–3391.

Schulze-Bahr E, Wang Q, Wedeking H, Haverkamp W, et al. (1997) KCNE1 mutations cause Jervell and Lange–Nielsen syndrome. Nat Genet 17:267–268.

Scott DA, Wang R, Kreman TM, Sheffield VC, et al. (1999) The Pendred syndrome gene encodes a chloride-iodide transport protein. Nat Genet 21:440–443.

Shailam R, Lanford PJ, Dolinsky CM, Norton CR, et al. (1999) Expression of proneural and neurogenic genes in the embryonic mammalian vestibular system. J Neurocytol 28:809–819.

Silos-Santiago I, Fagan AM, Garber M, Fritzsch B, et al. (1997) Severe sensory deficits but normal CNS development in newborn mice lacking TrkB and TrkC tyrosine protein kinase receptors. Eur J Neurosci 9:2045–2056.

Simeone A, Acampora D, Pannese M, D'Esposito M, et al. (1994) Cloning and characterization of two members of the vertebrate Dlx gene family. Proc Natl Acad Sci USA 91:2250–2254.

Simon MC (1995) Gotta have GATA. Nat Genet 11:9–11.

Spicer SS, Schulte BA (1998) Evidence for a medial K$^+$ recycling pathway from inner hair cells. Hear Res 118:1–12.

Steel KP, Kros CJ (2001) A genetic approach to understanding auditory function. Nat Genet 27:143–149.

Steel KP, Erven A, Kiernan AE (2002) Mice as models for human hereditary deafness. In: Keats BJB, Popper AN, Fay Rr (eds) Genetics and Auditory Disorders. Springer Handbook of Auditory Research, Volume 14. New York: Springer-Verlag, pp. 247–296.

ten Berge D, Brouwer A, Korving J, Martin JF, et al. (1998) Prx1 and Prx2 in skeletogenesis: roles in the craniofacial region, inner ear and limbs. Development 125:3831–3842.

Teng X, Ahn K, Bove M, Frenz D, et al. (2000) Malformations of the lateral semicircular canal occur in heterozygous Bmp4 mice. Assoc Res Otolaryngol Abstr 181:51.

Torres M, Giraldez F (1998) The development of the vertebrate inner ear. Mech Dev 71:5–21.

Truslove GM (1956) The anatomy and development of the *Fidget* mouse. J Genet 54:64–86.

Tsai H, Hardisty RE, Rhodes C, Kiernan AE, et al. (2001) The mouse slalom mutant demonstrates a role for Jagged1 in neuroepithelial patterning in the organ of Corti. Hum Mol Genet 10:507–512.

Van de Water TR, Li CW, Ruben RJ, Shea CA (1980) Ontogenic aspects of mammalian inner ear development. Birth Defects 16:5–45.

Vendrell V, Carnicero E, Giraldez F, Alonso MT, et al. (2000) Induction of inner ear fate by FGF3. Development 127:2011–2019.

Verpy E, Leibovici M, Petit c (1999) Characterization of Otoconin-95, the major protein of murine otoconia, provides insights into the formation of these inner ear biominerals. Proc Natl Acad Sci USA 96:529–534.

Vetter DE, Mann JR, Wangemann P, Liu J, et al. (1996) Inner ear defects induced by null mutation of the isk gene. Neuron 17:1251–1264.

Wang W, Van de Water T, Lufkin T (1998) Inner ear and maternal reproductive defects in mice lacking the Hmx3 homeobox gene. Development 125:621–634.

Wang W, Chan EK, Baron S, Van de Water T, et al. (2001) Hmx2 homeobox gene control of murine vestibular morphogenesis. Development 128:5017–5029.

Wangemann P, Liu J, Marcus DC (1995) Ion transport mechanisms responsible for K^+ secretion and the transepithelial voltage across marginal cells of stria vascularis in vitro. Hear Res 84:19–29.

Wersäll J, Bagger-Sjöbäck D (1974) Morphology of the vestibular sense organ. In: Autrum H, Jung R, Loenstein WR, Mackay DM (eds) Handbook of Sensory Physiology: Vestibular System, Part I. New York: Springer-Verlag, pp. 124–170.

Whitfield TT (2002) Zebrafish as a model for hearing and deafness. J Neurobiol 53:157–171.

Whitfield TT, Granato M, van Eeden FJ, Schach U, et al. (1996) Mutations affecting development of the zebrafish inner ear and lateral line. Development 123:241–254.

Wilcox ER, Burton QL, Naz S, Riazuddin S, et al. (2001) Mutations in the gene encoding tight junction claudin-14 cause autosomal recessive deafness DFNB29. Cell 104:165–172.

Wilkinson DG, Bhatt S, McMahon AP (1989) Expression pattern of the FGF-related proto-oncogene int-2 suggests multiple roles in fetal development. Development 105:131–136.

Winnier G, Blessing M, Labosky PA, Hogan BLM (1995) Bone morphogenetic protein-4 is required for mesoderm formation and patterning in the mouse. Genes Dev 9:2105–2116.

Wright TJ, Mansour SL (2003) Fgf3 and Fgf10 are required for mouse otic placode induction. Development, 130:3379–3390.

Wu DK, Oh SH (1996) Sensory organ generation in the chick inner ear. J Neurosci 16:6454–6462.

Wu DK, Choo DI (2003) Development of the ear. In: Snow JB Jr (eds) Ballengers, Manual of Otorhinolaryngology Head and Neck Surgery. Hamilton, Ontario, Canada: BC Decker, Inc., pp. 25–37.

Wu DK, Nunes FD, Choo D (1998) Axial specification for sensory organs versus non-sensory structures of the chicken inner ear. Development 125:11–20.

Xu PX, Woo I, Her H, Beier DR, et al. (1997) Mouse Eya homologues of the *Drosophila* eyes absent gene require Pax6 for expression in lens and nasal placode. Development 124:219–231.

Xu PX, Adams J, Peters H, Brown MC, et al. (1999) Eya1-deficient mice lack ears and kidneys and show abnormal apoptosis of organ primordia. Nat Genet 23: 113–117.

Zhang N, Gridley T (1998) Defects in somite formation in Lunatic fringe-deficient mice. Nature 394:374–377.

Zhang N, Martin GV, Kelley MW, Gridley T (2000) A mutation in the Lunatic fringe gene suppresses the effects of a Jagged2 mutation on inner hair cell development in the cochlea. Curr Biol 10:659–662.

Zheng JL, Gao WQ (2000) Overexpression of Math1 induces robust production of extra hair cells in postnatal rat inner ears. Nat Neurosci 3:580–586.

Zheng JL, Shou J, Guillemot F, Kageyama R, et al. (2000) Hes1 is a negative regulator of inner ear hair cell differentiation. Development 127:4551–4560.

Zine A, Van de Water TR, de Ribaupierre F (2000) Notch signaling regulates the pattern of auditory hair cell differentiation in mammals. Development 127: 3373–3383.

3
Morphophysiology of the Vestibular Periphery

Anna Lysakowski and Jay M. Goldberg

1. Introduction

With the development of intraaxonal labeling methods, it has become possible to relate the discharge properties of a vestibular afferent with its peripheral innervation patterns. In this chapter, we review the results of such morphophysiological studies. To provide a context for the review, we first consider the morphology of the vestibular organs from their gross anatomy to their ultrastructure. Because the species used in the morphophysiological studies have ranged from fish to mammals, we adopt a comparative approach to the morphology. A second perspective is provided by considering general features of afferent physiology. We next summarize results for each of the four species in which intraaxonal labeling has been used. In a final section, we describe general comparative trends that have emerged, consider the strengths and weaknesses of a morphophysiological approach, and speculate about the relation between the diversity of afferent physiology and the several stages of vestibular transduction.

2. Morphology

2.1. Evolution and Gross Morphology

The peripheral vestibular apparatus is similar in its gross anatomy from fish to mammals (Fig. 3.1) (Retzius 1881; Baird 1974; Lewis et al. 1985; Ramprashad et al. 1986). Jawless fishes provide some exceptional features.[1] Jawed vertebrates have three semicircular canals, each consisting

---

[1] Jawless fishes (agnatha) differ from all other (jawed, Gnathastome) craniates in having one or two vertical semicircular canals and two vertical ampullae but no horizontal canal and no horizontal ampulla (Fig. 3.1A,B) (Stensiö 1927; Mazan et al. 2000). In addition, they have a utricular sac but neither an utriculosaccular foramen or a separate sacculus. Extant forms of jawless fishes include lampreys and hagfishes. These animals share these features and have only a single otolithic macula (Lowenstein et al. 1968; Lowenstein and Thornhill 1970).

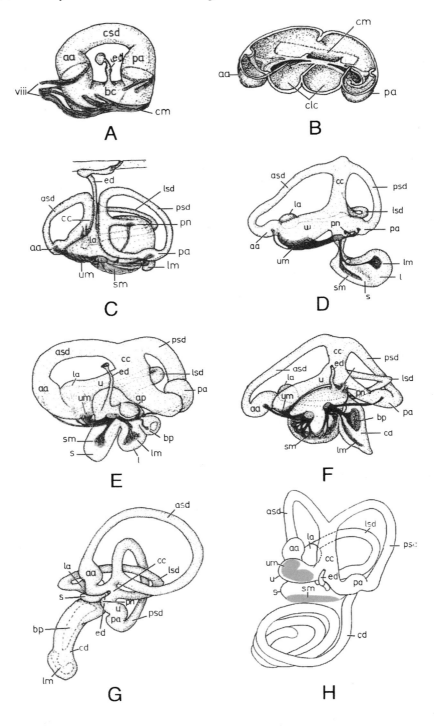

of a canal duct and an enlarged ampulla, the latter containing the crista ampullaris (Fig. 3.1C–H). There are three otolith organs: the utricular, saccular, and lagenar maculae. Two of these, the utricular and saccular maculae, are found in all jawed vertebrates. A lagena is absent only in eutherian (placental) mammals (Fig. 3.1H). Organs not including otoconia and not associated with a semicircular canal are usually referred to as *papillae*. One of these, the papilla neglecta, is present in many, but not all, vertebrates. The basilar papilla functions as a hearing organ in amphibians, reptiles, birds, and mammals (Fig. 3.1E–H) and is already present in coelacanths (*Latimeria*) (Fritzsch 1987). It is represented in placental mammals by the cochlea. A third, or amphibian, papilla is found in urodeles (salamanders) and anurans (frogs, toads) (Fig. 3.1E) but not in apodans (wormlike amphibians).

The presumed original function of the inner ear was to monitor rotational and linear movements of the head as well as the orientation of the head relative to the Earth vertical gravitational vector. As such, the inner ear functions as a proprioceptor. At several points during the ear's evolutionary history, one or another organ has taken on an exteroceptive function, the detection of sound and/or substrate-borne vibrations. Such transformations have usually involved the saccular or lagenar maculae or the newly evolved basilar or amphibian papillae. Most of the organs that have changed to or incorporated an exteroceptive function are in the sacculus or in one of its recesses. This has led to the notion that the ear has two separate divisions (Lowenstein 1936; Lewis et al. 1985). According to this view, the superior division, consisting of the three semicircular canals with its cristae and the utriculus with its macula, have retained their vestibular or proprioceptive functions. The inferior division, in contrast, is involved in exteroception. Although there is some merit to this dichotomy, it is far from

---

FIGURE 3.1. Inner ear structures from several vertebrate classes. (**A**) Primitive vertebrate, hagfish (*Myxine*). (**B**) Primitive vertebrate, lamprey (*Lampetra*). (**C**) Elasmobranch fish, ray (*Raja*). (**D**) Teleost fish, catfish (*Malapterurus*). (**E**) Anuran amphibian, toad (*Bufo*). (**F**) Reptile, turtle (*Trionychid*). (**G**) Bird, duck (*Anser*). (**H**) Mammal, gerbil (*Meriones*). Abbreviations: aa, anterior ampulla; ap, amphibian papilla (E only); asd, anterior semicircular duct; bc, basal chamber (A only); bp, basilar papilla; cc, common crus; cd, cochlear duct; clc, ciliated chamber (B only); cm, common macula (A,B only); csd, common semicircular duct (A only); ed, endolymphatic duct; l, lagena; la, lateral ampulla; lm, lagenar macula; lsd, lateral semicircular duct; pa, posterior ampulla; pn, papilla neglecta; psd, posterior semicircular duct; s, saccule; sm, saccular macula; u, utricle; um, utricular macular; usd, utriculosaccular duct; viii, eighth nerve. (A–G. Modified with permission from Baird 1974. H. Modified with permission from Wersäll and Bagger-Sjöback 1974, Copyright 1974, Springer-Verlag.)

absolute. For example, although the saccular macula functions as a hearing organ in many fish (Furukawa and Ishii 1967) and as a vibratory organ in frogs (Koyama et al. 1982; Lewis et al. 1982; Christensen-Dalsgaard and Narins 1993), it is a vestibular organ in mammals (Fernández et al. 1972; Tomko et al. 1981a). Similarly, the utricular macula may function as a hearing organ in herring (*Clupeus*) and related fish (Denton and Gray 1979; Popper and Platt 1979). As a final example, the papilla neglecta is part of the superior division. It may function as an auditory or vibratory organ in elasmobranchs (Lowenstein and Roberts 1951; Fay et al. 1975; Corwin 1981) but is a detector of head rotations in turtles and probably many other vertebrates (Brichta and Goldberg 1998).

Mention should also be made of the relation between vestibular and lateral line organs. In fish and aquatic amphibians (including larval forms), the two sets of organs develop from adjacent placodes, the otic and six lateral line placodes (Northcutt 1997; Baker and Bronner-Fraser 2001), have similar transduction machinery (Flock 1971; Hudspeth 1989), and can share their efferent innervation (Claas et al. 1981; Bleckmann et al. 1991). From these similarities, it is apparent that the two sensory systems are closely related, so much so that they are collectively referred to as the *octavolateralis* or *acousticolateralis* system. Because of their presumably simpler structure, it has been supposed that the lateral lines are phylogenetically more primitive and may have given rise to the vestibular organs (van Bergeijk 1966; Baird 1974). Another equally plausible conjecture is that the two sets of organs arose independently, possibly from the same primordium (Denison 1966).

2.2. Hair Cells and Afferents

Each vestibular organ has a neuroepithelium composed of hair cells and supporting cells (Wersäll and Bagger-Sjöbäck 1974; Hunter-Duvar and Hinojosa 1984; Lewis et al. 1985). Hair cells are innervated by afferent nerve fibers projecting to the brain and efferent nerve fibers coming from the brain. Sensory hair bundles arise from the apical surfaces of the hair cells and insert into a gelatinous accessory structure, termed a cupula in each crista ampullaris, an otolithic or otoconial membrane in each otolith organ, and a tectorial membrane or cupula in each papilla. The hair bundles consist of a single, eccentrically located kinocilium and several stereocilia arranged in rows, which increase in height as they approach the kinocilium (Figs. 3.2A, 3.3A,B).

A morphological polarization vector (Fig. 3.3) is defined as the axis of bilateral symmetry of the hair bundle and points toward the kinocilium (Fig. 3.3) (Flock 1964; Spoendlin 1965; Lindeman 1969). As we shall see, this morphological polarization determines the directional properties of the hair cells.

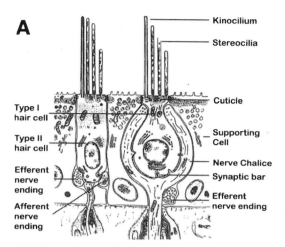

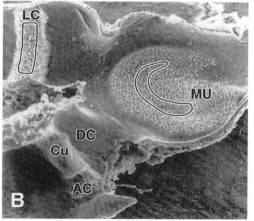

FIGURE 3.2. Hair cell ultrastructure. (**A**) Two types of hair cells found in the vestibular periphery are distinguished by their afferent and efferent innervations. (**B**) Scanning electron micrograph illustrating the anterior and lateral cristae and the utricular macula. Superimposed on the lateral crista (LC) and the utricular macula (MU) are lines showing the boundaries of the central zone and striola, respectively. Overlying the anterior crista (AC) is a remnant of the cupula (Cu). DC indicates dark cells. (**C**) Scanning electron micrograph of the saccular macula, with a line showing the boundary of the striolar region (st). TE indicates transitional epithelium. (A. Modified with permission from Wersäll 1956. Copyright 1956, Taylor and Francis, Inc. B,C. Modified with permission from Hunter-Duvar and Hinojosa 1984, Copyright 1984, Elsevier Science.)

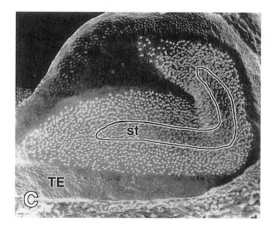

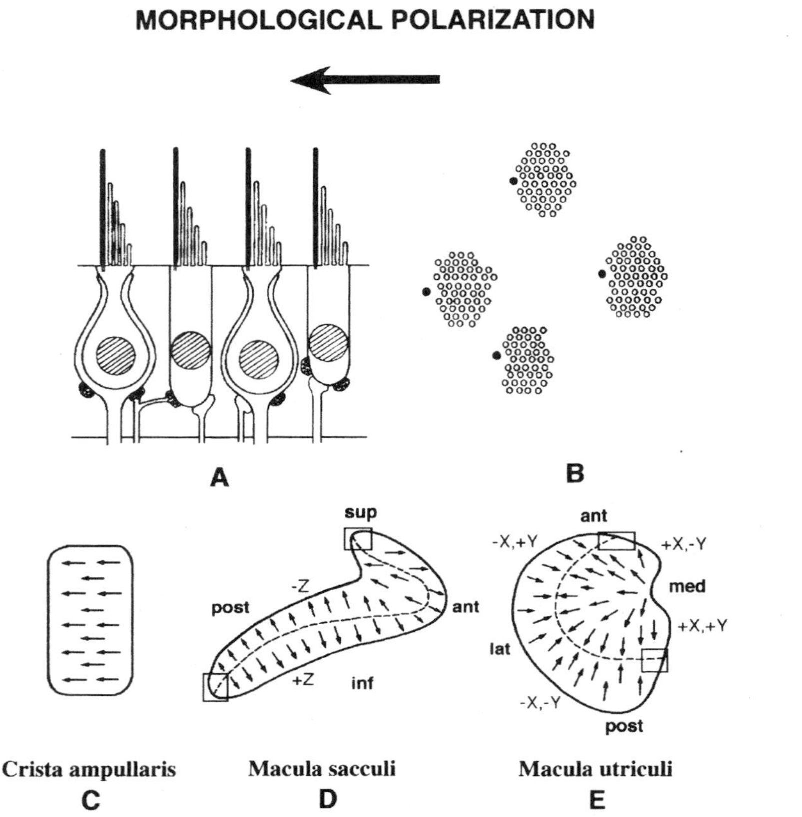

MORPHOLOGICAL POLARIZATION

FIGURE 3.3. Morphological polarization of hair cells. (**A**, **B**) Each hair cell has a hair bundle consisting of a single, eccentrically placed kinocilium (dark rod) and several stereocilia (light rods) arranged in rows of increasing length. When the hair bundle is displaced in the direction of the large arrow (toward the kinocilium), the hair cell is depolarized, leading to excitation of its afferents. (**C–E**) The small arrows indicate the morphological polarizations of hair cells (i.e., the direction of excitatory hair bundle displacement in a crista ampullaris (**C**), macula sacculi (**D**), and macula utriculi (**E**). Note that, in the saccular macula, hair cells are polarized away from the reversal (dashed) line, whereas, in the utricular macula, hair cells are polarized toward the reversal line. (Modified with permission from Lindeman 1969, Copyright 1969, Springer-Verlag.)

Accessory structures and the apical surfaces of the hair cells, including their hair bundles, are bathed in endolymph, an extracellular fluid unusual in being rich in potassium. Perilymph, a more typical extracellular fluid, bathes the basolateral surfaces of the hair cells and supporting cells as well as the nerve fibers. This section focuses on hair cells, afferent and efferent nerve fibers, and their peripheral terminations. Transitional cells and dark cells, which border the neuroepithelium at its base, will be briefly described.

2.2.1. Hair Cell Types

Wersäll (1956), using electron microscopy in the guinea pig crista ampullaris, described two types of hair cells (Fig. 3.2A). The first, which he called type I, was enclosed in a calyceal ending, a nerve terminal that had been described earlier (Retzius 1881; Cajal 1911; Lorente de Nó 1926; Poljak 1927). The second kind of hair cell, which Wersäll called type II, was contacted on its basolateral surface by bouton endings. From a study of surface preparations of the cristae, Lindeman (1969) observed that there were regional differences in the sizes and spacing of both types of hair cells and in their hair bundle morphology. Based on these differences, the crista can be divided into three concentric zones of approximately equal areas. The boundary of the centralmost zone is depicted in the upper left in Figure 3.2B. Hair cells are larger and more widely spaced in the central zone than in the intermediate and peripheral zones. Hair bundles are shorter and thicker centrally. Consideration of these and other regional differences led to a reexamination of the ultrastructure in the crista ampullaris (Lysakowski and Goldberg 1997), which revealed other regionally based differences (Fig. 3.4).

Lindeman (1969) made similar observations in the maculae, which as originally described by Werner (1933) can be divided into a narrow striola bordered on each side by a broader extrastriola (Fig. 3.2B,C). Hair cells in the striola are larger and more widely spaced than in the extrastriola. Hair bundles are shorter and thicker in the striola (Lindeman 1969; Lapeyre et al. 1992). A reversal line running within the striola separates hair bundles of opposite morphological polarization (Fig. 3.3D,E) (Flock 1964; Spoendlin 1965; Lindeman 1969). In the saccular macula, each hair bundle has its kinocilium pointing away from the reversal line (Fig. 3.3D), whereas the opposite is true in the utricular macula (Fig. 3.3E). Crystals in the otoconial layer are smallest above the striola, and there are characteristic regional differences in the thickness of the otoconial layers in both maculae (Werner 1933; Lindeman 1969).

Type I hair cells are present in reptiles, birds, and mammals, but not in fish or amphibians (Wersäll and Bagger-Sjöbäck 1974; Lysakowski 1996). In mammals, both type I and type II hair cells are found throughout the cristae and maculae. In reptiles and birds, the same is true for type II hair cells, but type I hair cells have a more restricted distribution to the central zones of the cristae and the striola of the utricular macula (Rosenhall 1970; Jørgensen and Anderson 1973; Jørgensen 1974, 1975; Brichta and Peterson 1994; Lysakowski 1996). When compared with the utricular macula, the distribution of type I hair cells is more extensive in the saccular macula of reptiles (Jørgensen 1974, 1975) and birds (Rosenhall 1970; Jørgensen and

A   **Central Zone**

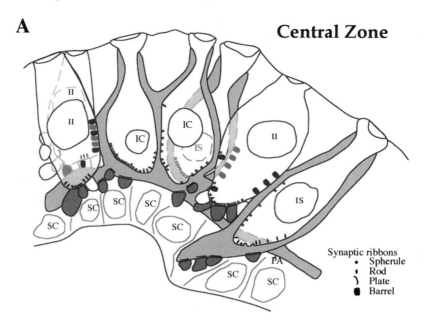

B   **Peripheral Zone**

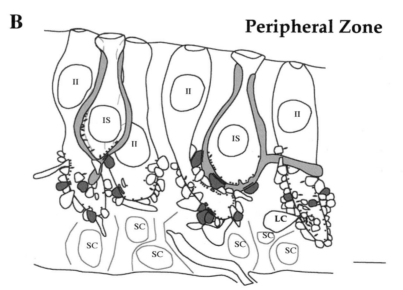

Anderson 1973). In reptiles, calyx endings do not extend past the constricted neck of the type I hair cell (Baird and Lowman 1978; Jørgensen 1988), whereas in mammals and birds, the ending extends beyond the neck, almost reaching the junctional complex near the apical surface of the neuroepithelium (Wersäll and Bagger-Sjöbäck 1974; Correia et al. 1985).

Type II hair cells, which are defined by their lack of calyx endings, are present in all vertebrate species (Wersäll and Bagger-Sjöbäck 1974; Lysakowski 1996). Despite their being given a common designation, type II hair cells are far from homogeneous in their morphology. As already noted, type II hair cells in mammals show regional differences in their size, spacing, and hair bundle morphology (Lindeman 1969; Lapeyre et al. 1992; Lysakowski and Goldberg 1997). In describing the morphological diversity of type II hair cells in nonmammalian vertebrates, three classification criteria have been used: (1) hair bundle morphology (Lewis and Li 1975; Baird and Lewis 1986; Myers and Lewis 1990); (2) hair cell shape (Guth et al. 1994; Gioglio et al. 1995; Lanford et al. 2000); and (3) ultrastructural morphology (Chang et al. 1992; Popper et al. 1993; Saidel et al. 1995; Lanford and Popper 1996). The first two criteria have been most extensively used in the frog, and their consideration is best done when we review morphophysiological studies in this species. Concerning the third criterion, Chang et al. (1992) used ultrastructural features to define "striolar" and "extrastriolar" hair cell types. In lamprey (*Lampetra fluviatalis* and *Entosephanus japonicus*) macular organs, hair cell shape (Lowenstein and Osborne 1964; Hoshino 1975) and in the thorn back ray (*Raja clavata*), stereociliar diam-

---

FIGURE 3.4. Regional variations in cellular architecture and synaptic innervation in a chinchilla superior crista based on serial ultrastructural reconstructions of the central (A) and peripheral (B) zones. (**A**) This portion of the central zone contains three type II hair cells (II) and four type I hair cells, including two in a complex calyx (IC) and two others, each in a simple calyx (IS). Supporting cell (SC) nuclei are seen at the bottom. Efferent boutons are shaded dark gray, whereas afferent boutons are unshaded. Calyx endings are shaded light gray, as is the parent axon (PA) running from right to left and giving rise to the complex calyx. Hair cells are wider and calyces are thicker than in the peripheral zone. There are fewer afferent boutons per type II hair cell in the central zone, but each central bouton makes multiple ribbon synapses. Type II hair cells in the central zone also synapse with the outer faces of neighboring calyx endings. Synaptic ribbons occur in different shapes (see key). Efferent synapses are made with type II hair cells, with calyx endings, and with other afferent processes. (**B**) Four type II hair cells (II) and two type I hair cells (IS) are shown from the peripheral zone. Hair cells and calyces are thinner, and there are many more afferent boutons per type II hair cell. Most afferent boutons contact one ribbon synapse. Complex calyx endings and outer-face ribbons are rare. The efferent innervation appears similar in the two regions. (Modified with permission from Lysakowski and Goldberg 1997, Copyright 1997, Wiley-Liss, Inc.)

eter (Lowenstein et al. 1964) have been used in defining two varieties of type II hair cells.

2.2.2. Afferent Fiber Types

More than seventy years ago, morphologists recognized in silver-stained material that vestibular nerve fibers innervating the central region of the mammalian crista were thicker than those innervating more peripheral parts of the crista and that the thick, central fibers ended in calyx endings (Retzius 1881; Cajal 1911; Lorente de Nó 1926; Poljak 1927). In contrast, thin peripheral fibers were observed to enter an intraepithelial plexus and to end in boutons (Retzius 1881; Poljak 1927). Later ultrastructural studies showed that the thin fibers innervated type II hair cells (Wersäll 1956; Engström 1958). Although Wersäll claimed that type I and type II hair cells were separately innervated, the conclusion seemed contrary to earlier descriptions of medium-sized fibers that gave rise to calyx endings, as well as noncalyceal collaterals (Retzius 1881; Cajal 1911; Lorente de Nó 1926; Poljak 1927), and to later ultrastructural observations that some axons provide a mixed innervation, including calyx endings to type I hair cells and bouton endings to type II hair cells (Ades and Engström 1965). Such afferents have been termed *dimorphic* (Schessel 1982).

With the advent of extracellular labeling techniques, it was confirmed that afferent fibers in mammals could be placed into three classes based on their peripheral terminations (Fernández et al. 1988, 1995). Calyx fibers exclusively terminate on type I cells and bouton fibers, because they do not have calyx endings, presumably only innervate type II cells. Dimorphic fibers contact both kinds of hair cells. In addition, extracellular labeling showed that dimorphic fibers make up most of the afferent innervation of the chinchilla (*Chinchilla laniger*) and squirrel monkey (*Saimiri sciureus*) cristae (Fig. 3.12) (Fernández et al. 1988, 1995) and the chinchilla utricular macula (Fig. 3.16) (Fernández et al. 1990). That the innervation of type II hair cells in the utricular macula comes from dimorphic fibers was emphasized by Ross (Ross 1985; Ross et al. 1986), who was unable to find bouton afferents in ultrastructural studies of the rat maculae. In fact, although there are bouton fibers in the chinchilla utricula macula, they could easily be missed because they make up such a small proportion of the afferent innervation (Fernández et al. 1990). As will be described below, the three kinds of afferents differ in their regional distribution in both the cristae and the utricular macula (Fernández et al. 1988, 1990, 1995). Recent immunohistochemical studies in mammals have found specific markers for calyx fibers and possibly for bouton fibers. Calretinin, a calcium-binding protein, is present, among afferents, only in calyx units (Desmadryl and Dechesne 1992; Desai et al. 2000). Peripherin, an intermediate-filament protein (Lysakowski et al. 1999), and possibly Substance P (Usami et al.

1993, 1995), may be selective markers for bouton units. Dimorphic fibers are neither calretinin- nor peripherin-immunoreactive. It would be important to find a selective marker for the latter fibers.

Calyx, dimorphic, and bouton fibers have been observed in the cristae of reptiles (Schessel 1982; Schessel et al. 1991; Brichta and Peterson 1994; Brichta and Goldberg 2000a) and in the utricular (Si et al. 2003) and saccular (Mridhar et al. 2001; Zakir et al. 2003) maculae and horizontal crista (Haque and Dickman 2001) of birds. As expected, calyx and dimorphic units are only found in the central zone of the cristae and in the macular striolae, the only regions of the neuroepithelium having type I hair cells in these animals.

In mammals, bouton endings, including those on bouton or dimorphic fibers, are round or spheroidal enlargements along the course and at the ends of thin collaterals; the latter are typically less than 0.5 μm in diameter (Fernández et al. 1988, 1990, 1995). Calyx endings, whether on calyx or dimorphic fibers, are seen at the ends of thick processes, either the parent axon or one of its thick branches. Similar arrangements are seen in the turtle (*Pseudemys* (*Trachemys*) *scripta*) posterior crista (Brichta and Peterson 1994) and in the pigeon (*Columba livia*) utricular macula (Si et al. 2003).

Because the hair cells of fish and amphibians are all classified as type II, this should not obscure the fact that afferent fibers can differ in their branching patterns (Lewis et al. 1982; Baird and Lewis 1986; Honrubia et al. 1989; Myers and Lewis 1990; Boyle et al. 1991; Baird and Schuff 1994) or types of endings (Lowenstein et al. 1968; Boyle et al. 1991; Baird and Schuff 1994; Lanford and Popper 1996). Four types of endings have been described: bouton endings, whether *en passant* or *terminaux*, on thin collaterals (Boyle et al. 1991; Baird and Schuff 1994); club endings, consisting of a large, round, or blunt terminal attached to a thick branch (Boyle et al. 1991; Baird and Schuff 1994); cup-shaped or claw endings, which can embrace the bottom of one to several hair cells (Honrubia et al. 1989; Baird and Schuff 1994; Lanford and Popper 1996); and candelabra endings, which have only been described in the lamprey (*Lampetra fluviatilis*) (Lowenstein et al. 1968).

2.2.3. Afferent and Efferent Synapses

Synaptic transmission between hair cells and their afferents is mediated by ribbon synapses (Smith and Sjöstrand 1961; Iurato et al. 1972; Liberman et al. 1990). Similar structures are found in the vertebrate retina (Sjöstrand 1958; Dowling and Boycott 1966; Rao-Mirotznik et al. 1995), where a major structural protein, termed *ribeye*, has recently been purified (Schmitz et al. 2000). Presynaptically, there is a dense body or ribbon surrounded by a halo of tethered, clear vesicles. The ribbon is attached by pedicles to a presynaptic density, which lies opposite a postsynaptic density. Because they occur in cells that continually transduce, it has been supposed that ribbon synapses are specialized for the rapid and continual release of neurotransmitter (Parsons et al. 1994; von Gersdorff and Matthews 1999; von Gersdorff 2001).

Ribbon morphology in vestibular hair cells varies across the vertebrate scale (Wersäll and Bagger-Sjöbäck 1974; Lysakowski 1996). Synaptic ribbons are spherical in all vertebrate classes up to and including reptiles. Spherical ribbons are particularly large (200–500nm in diameter) in frogs and turtles. Smaller ribbons are found in birds and mammals, and some of them are elongated. In amniotes, synaptic ribbons are found in both type I and type II hair cells. Besides contacting their own bouton afferents, synaptic ribbons in type II hair cells can contact the outer faces of calyx endings. The ribbons found in the mammalian crista vary with cell type and region (Lysakowski and Goldberg 1997). Most of the ribbons in central type I cells are spherical. Those in peripheral hair cells of either type can be spherical or elongate, with the latter usually being small rods. Ribbons are especially heterogeneous in central type II cells and include spheres, rods, barrels, and plates (see the legend to Fig. 3.4). There are typically 10–14 ribbons in each mammalian hair cell. The number of attached vesicles varies with ribbon size, ranging from 20 for small spherules to 100–300 for large barrels and plates. Large spherical ribbons in the frog sacculus can have more than 500 tethered vesicles (Lenzi et al. 1999).

Efferent boutons are highly vesiculated and make synapses with type II hair cells and with afferent processes, including calyces, boutons, and dendrites (Fig. 3.4 and Smith and Rasmussen 1968; Wersäll 1968; Iurato et al. 1972). The innervation of hair cells has been termed *presynaptic*; that of afferent processes, *postsynaptic* (Flock 1971). In frogs, efferents only contact hair cells (Lysakowski 1996). The contacts on type II hair cells are marked by a postsynaptic cistern (Figs. 3.2A, 3.4). There is an accumulation of vesicles, although this need not occur directly opposite the cistern (Smith and Rasmussen 1968). Efferent synapses with calyces and other afferent processes are characterized by a dense accumulation of vesicles and by slight presynaptic and postsynaptic thickenings (Figs. 3.2A, 3.4). In addition to the accumulation of vesicles, efferent boutons and their processes are distinctive in having a more uniform size, smaller mitochondria, and a darker, more filamentous cytoplasm than afferent boutons (Lysakowski and Goldberg 1997). The latter contain a small number of vesicles, but these are less homogeneous in size than their efferent counterparts. Individual efferent boutons can make contacts with both afferent processes and type II hair cells (Lysakowski and Goldberg 1997).

2.2.4. Supporting, Transitional, and Dark Cells

Supporting cells have a homogeneous appearance. Their nuclei are located at the basal end of the sensory epithelium, just above the basement membrane (Figs. 3.2A, 3.4). The cytoplasm of supporting cells contains intermediate filaments, secretory granules, a conspicuous rough endoplasmic reticulum, and a prominent Golgi apparatus. This appearance is consistent with the conclusion that supporting cells make and secrete the mucopolysaccharides and collagen fibrils of the cupula and otolithic membranes (Lim 1973, 1984; Kachar et al. 1990; Silver et al. 1998). Cells of the transitional epithelium, which border the neuroepithelium (Fig. 3.2C), resemble supporting cells in many respects but differ, among other ways, in the size and position of their nuclei (Hunter-Duvar and Hinojosa 1984).

Dark cells are positioned at the margins of the transitional epithelium (Fig. 3.2B). Pigment cells are located directly underneath the dark cells. The basal membranes of the dark cells are deeply invaginated and contain numerous mitochondria, an appearance consistent with the cells having a secretory function. An array of ion channels, ion transporters, and ion pumps are found in the apical and basal membranes of dark cells and are thought to be responsible for producing the unique ionic composition of endolymph (Vetter et al. 1996; Wangemann et al. 1996; see Wangemann 1995 for a review).

2.3. Efferent Vestibular System

2.3.1. Anatomical Organization in Mammals

In mammals, efferents arise bilaterally in the brain stem from three collections of neurons. The first, and by far the largest, of these is a slender column of medium-sized multipolar neurons that extends about 1mm in length rostrocaudally (Goldberg and Fernández 1980, 1984; Marco et al. 1993), between the abducens and superior vestibular nucleus, and lies just dorsolateral to the facial genu (Gacek and Lyon 1974; Warr 1975; Goldberg and Fernández 1980; Strutz 1982b; Dechesne et al. 1984; Schwarz et al. 1986; Perachio and Kevetter 1989; Ohno et al. 1991; Marco et al. 1993). The group has been referred to as group *e* (Fig. 3.5; Goldberg and Fernández 1980; Goldberg et al. 2000). A second, more or less compact group of somewhat smaller fusiform neurons lies dorsomedial to the facial genu (Goldberg and Fernández 1980; Strutz 1982b; Schwarz et al. 1986; Ohno et al. 1991; Marco et al. 1993). A third or ventral group of neurons with large dendritic trees is scattered in the caudal pontine reticular formation (Strutz 1982b; Schwarz et al. 1986; Perachio and Kevetter 1989; Ohno et al. 1991; Marco et al. 1993).

Roughly equal numbers of group *e* neurons project to the ipsilateral and contralateral labyrinths in rats (Schwarz et al. 1986), guinea pigs (Strutz 1982b), cats (Gacek and Lyon 1974; Warr 1975; Dechesne et al. 1984), and monkeys (Goldberg and Fernández 1980; Carpenter et al. 1987), whereas in chinchillas (Marco et al. 1993; Lysakowski, unpublished data) and gerbils (Perachio and Kevetter 1989), two-thirds of the neurons project contralaterally. It is likely that there are also bilaterally projecting neurons, but the relative numbers of unilaterally and bilaterally projecting neurons remain

Lamprey (Fritzsch et al., 1989)

**Gymnophion, amphibia
(Fritzsch & Crapon de Crapona, 1984)   Salamander (Fritzsch, 1981)**

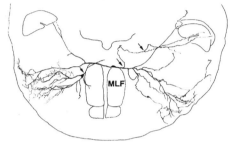

Toadfish (Highstein and Baker, 1986)

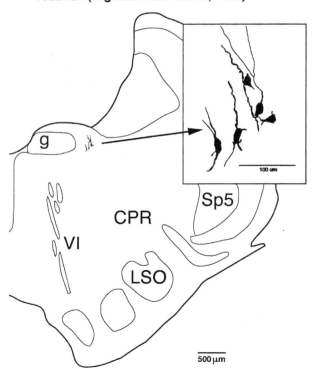

Chinchilla (Lysakowski and Singer, 2000)

controversial (Dechesne et al. 1984; Perachio and Kevetter 1989). Other details of vestibular efferent innervation in mammals have been reviewed recently (Goldberg et al. 2000).

2.3.2. Anatomical Organization in Nonmammalian Species

In this section, we emphasize general trends that distinguish lower vertebrates from each other and from mammals. As can be seen in Figure 3.5, the spread of efferent dendrites, both relatively and absolutely, is smallest in mammals. In fish, vestibular efferent neurons are part of a group of about 20–40 octavolateralis efferent neurons (OEN) that includes auditory, lateral line, and electrosensory efferents (Strutz et al. 1980; Bell 1981; Koester 1983; Meredith and Roberts 1987a, 1987b; Fritzsch et al. 1989; Koyama et al. 1989; Bleckmann et al. 1991; Gonzalez and Anadon 1994). The toadfish (*Opsanus tau*) has a considerably larger number of efferent neurons than has been reported in other fish (Highstein and Baker 1986), which may be correlated with the high density of efferent terminals in toadfish (*O. tau*) sensory organs (Sans and Highstein 1984). As summarized by Meredith (1988), 80–95% of efferent neurons are located ipsilaterally in bony fish (Bell 1981; Highstein and Baker 1986; Meredith and Roberts 1987b), whereas the distribution is more bilateral in the dogfish, a cartilaginous fish (Meredith and Roberts 1987a). Individual efferent axons in fish can branch to innervate two or more vestibular organs or vestibular and lateral line organs (Claas et al. 1981; Bleckmann et al. 1991).

Among vertebrates, anurans and apodan amphibians are unusual in having an exclusively ipsilateral efferent projection (Strutz et al. 1981; Will 1982; Fritzsch and Crapon de Caprona 1984; Pellegrini et al. 1985). In contrast, salamanders (Fritzsch 1981; Fritzsch and Crapon de Caprona 1984) resemble reptiles (Strutz 1981, 1982a; Fayyazuddin et al. 1991) and many fish (Meredith 1988) in that their efferent neurons, although predominantly

◄─────────────────────────────────────

FIGURE 3.5. Locations of efferent cell bodies, their dendrites, and axons (small arrows) in several vertebrates. All sections are drawn to the same scale, except for the inset, which shows efferent neurons at higher magnification in the chinchilla group *e*. In the lamprey (**A**), amphibians (**B,C**), and toadfish (*O. tau*) (**D**), efferent dendrites sample from a much larger portion of the brain stem than in mammals (**E**). Abbreviations: CPR, caudal pontine reticular formation; g, genu of the facial nerve; LSO, lateral superior olive nucleus; MLF, medial longitudinal fasciculus; Sp5, spinal nucleus of the trigeminal nerve; VI, abducens nerve. (Modified with permission from Elsevier Science, Fritzsch 1981 (salamander, *Salamandra salamandra*); Fritzsch and Crapon de Crapona 1984 (gymnophion); Fritzsch et al. 1989 (lamprey, *L. fluviatilis*); and with permission from Wiley-Liss, Inc. from Highstein and Baker 1986 (toadfish, *O. tau*); Lysakowski and Singer 2000 (chinchilla, *C. laniger*, Copyright Wiley-Liss, Inc.)

ipsilateral, are also found on the contralateral side. Electrophysiological studies in the frog demonstate that efferents can branch to more than one organ (Rossi et al. 1980; Prigioni et al. 1983; Valli et al. 1986; Sugai et al. 1991). A differential distribution of efferents destined for auditory and vestibular organs is seen in the caiman (*Caiman crocodilus*) (Strutz 1981) but not in the turtle (*Terrapene ornata*) (Strutz 1982a).

In birds, efferent neurons are found bilaterally, either equally distributed on the two sides or with an ipsilateral predominance (Schwarz et al. 1978, 1981; Whitehead and Morest 1981; Eden and Correia 1982; Strutz and Schmidt 1982), with efferent vestibular neurons being more dorsally located than their auditory counterparts. Double-labeling experiments indicate that efferent axons in the pigeon (*Columba livia*) branch to two or more vestibular organs on one side or, less commonly, on both sides (Schwarz et al. 1981).

The efferent nuclei, which include the fish nucleus motorius tegmenti or OEN, the amphibian nucleus reticularis medius, and the avian caudal pontine reticular nuclei, have all been considered by Strutz (1982b) to be homologous to the caudal pontine reticular nucleus of mammals. In fish, efferent neurons are found in close association with facial motoneurons, which has led to the suggestion that the efferents are branchiomotor neurons that innervate neuroepithelial tissue rather than musculature (Meredith 1988). Even in chicks and mice, where there is a separation of efferent neurons and facial motoneurons in the adult brain, the two groups of neurons have an overlapping location early in their postmitotic development (Fritzsch 1996; Bruce et al. 1997). Contralateral efferent neurons in chicks result not from axonal outgrowth of neurons settled on the contralateral side but rather from the cellular migration across the midline of efferent neurons already sending axons to the periphery (Simon and Lumsden 1993).

3. Afferent Physiology

3.1. General Features

3.1.1. Resting Discharge

Even in the absence of rotations, semicircular canal afferents continue to discharge. A resting discharge was observed in the earliest recordings of vestibular nerve activity. Ross (1936) thought that the resting discharge might be artifactual. In contrast, Lowenstein and Sand (1936) suggested that this was a normal feature offering two advantages. First, a resting discharge allowed each fiber to respond bidirectionally to vestibular stimulation. So, for example, a fiber innervating the horizontal semicircular canal increases its discharge when the animal is rotated toward the ipsilateral side and decreases firing during contralateral rotations. Second, a background discharge could provide a source of postural tone. Later, Lowenstein (1956) also pointed out that such activity could eliminate the existence of a sensory threshold. Since Lowenstein and Sand's (1936) original observations, the presence of a resting discharge in semicircular canal afferents has been confirmed in a wide variety of preparations (Lowenstein and Sand 1940a, 1940b; Ledoux 1949; Goldberg and Fernández 1971a; Blanks and Precht 1976; Fernández and Goldberg 1976a; Hartmann and Klinke 1980; Honrubia et al. 1989; Boyle and Highstein 1990a).

Because semicircular canal afferents respond to rotational forces, the resting discharge can be measured simply by keeping the head stationary or having it move at a constant velocity with respect to an inertial frame. Otolith afferents respond to linear gravitoinertial forces. Because gravity is ever-present, an absence of stimulation is not possible on Earth. Here use is made of the fact that each otolith afferent can be characterized by a polarization vector, which summarizes its directional properties when the head is tilted in various directions with respect to the vertical (Fernández et al. 1972; Loe et al. 1973). The resting (zero-force) discharge is obtained when the polarization and gravity vectors are orthogonal. This will occur at two points separated by 180° as the animal is tilted through a great circle about any horizontal axis. The two points are recognized as having the same discharge rate.

Resting discharges depend on species, discharge regularity, organ, and type of preparation. Discharge regularity will be considered in the next section. For now, we note that afferents can have a regular or an irregular spacing of action potentials (Fig. 3.6). Background rates among regularly discharging canal afferents in anesthetized mammals range from 60 to 120 spikes/s, being somewhat higher in monkeys (Goldberg and Fernández 1971b; Lysakowski et al. 1995) than in cats (Estes et al. 1975; Anderson et al. 1978), gerbils (Schneider and Anderson 1976), or chinchillas (Baird et al. 1988). Rates are lower in afferents innervating otolith organs (Fernández et al. 1972; Fernández and Goldberg 1976a; Perachio and Correia 1983; Goldberg et al. 1990a) and in irregularly discharging, as compared to regularly discharging, afferents innervating either canal (Goldberg and Fernández 1971b; Estes et al. 1975; Tomko et al. 1981b; Perachio and Correia 1983; Lysakowski et al. 1995) or otolith organs (Fernández and Goldberg 1976a; Tomko et al. 1981a; Perachio and Correia 1983; Goldberg et al. 1990a). Canal afferents in barbiturate-anesthetized pigeons have resting rates near 90 spikes/s, much lower than those recorded in unanesthetized pigeons (Anastasio et al. 1985). Only small effects of anesthesia have been seen in mammals (Keller 1976; Louie and Kimm 1976; Blanks and Precht 1978; Perachio and Correia 1983).

One might expect that testing rates would be lower in cold-blooded animals than in warm-blooded ones. This is true in turtles (Brichta and Goldberg 2000a) and frogs (Honrubia et al. 1989; Myers and Lewis 1990, 1991), where rates are 10–30 spikes/s among regularly discharging fibers and are even lower among irregularly discharging fibers. But regularly dis-

Regular

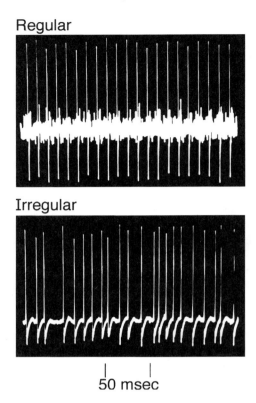

Irregular

50 msec

FIGURE 3.6. Discharge regularity in vestibular nerve afferents. Spike trains are shown during the resting discharge for two afferents, each innervating the superior semicircular canal in a squirrel monkey (*S. sciureus*). Although both afferents have a similar discharge rate, just under 100 spikes/s, they differ in the spacing of their action potentials, which is regular in the top afferent and irregular in the bottom afferent. These two discharge patterns can each be correlated with several other physiological and morphological features that distinguish the two classes of afferents (Table 3.1). (Modified with permission from Goldberg and Fernández 1971a, Copyright 1971, The American Physiological Society.)

charging fibers in fish can have rates of 50–120 spikes/s, similar to those in mammals; once again, irregularly discharging fibers have lower rates (Hartmann and Klinke 1980; Boyle and Highstein 1990a; Boyle et al. 1991).

3.1.2. Discharge Regularity

Fibers differ not only in their discharge rates but also in their discharge regularity. Afferents can have a regular or an irregular spacing of action potentials (Fig. 3.6). Discharge regularity is characteristic of each afferent and can be summarized by a cv^*, the coefficient of variation (cv) at a standard mean interval (Goldberg and Fernández 1971b; Goldberg 2000). Interest in this discharge property stems in part from the fact that fibers classified as regularly or irregularly discharging differ in several other respects as well. Table 3.1 summarizes some of the differences for mammalian afferents, including those innervating the cristae and the maculae. In many respects, similar differences in axon diameter, response dynamics, and vestibular sensitivity are seen in many lower vertebrates (Honrubia et al. 1989; Myers and Lewis 1990; Brichta and Goldberg 2000a, 2000b). In this respect, the toadfish (*O. tau*) may be an exception (Boyle and Highstein 1990a; Boyle et al. 1991).

Which of the differences listed in Table 3.1 is causally related to discharge regularity? To answer this question requires an understanding of the cellular mechanisms determining the spacing of action potentials. Because the topic has been recently reviewed (Goldberg 2000), we will only state the conclusions of the analysis. A spike encoder, located in the afferent terminal, converts postsynaptic depolarizations into trains of action potentials. There is a causal relation between discharge regularity and the sensitivity of the encoder to synaptic inputs and to external currents. Of the several

TABLE 3.1. Characteristics of regularly and irregularly discharging afferents in the mammalian vestibular nerve.

Irregularly discharging	Regularly discharging
[1]Thick and medium-sized axons ending as calyx and dimorphic terminals in the central (striolar) zone.	Medium-sized and thin axons ending as dimorphic and bouton terminals in the peripheral (peripheral extrastriolar) zone.
[2]Phasic-tonic response dynamics, including a sensitivity to the velocity of cupular (otolith) displacement.	Tonic response dynamics, resembling those expected of end organ macromechanics.
[2]High sensitivity to angular or linear forces acting on the head. (Calyx units innervating the cristae have an irregular discharge and low sensitivities.)	Low sensitivity to angular or linear forces.
[3]Large responses to electrical stimulation of efferent fibers.	Small responses to electrical stimulation of efferent fibers.
[4]Low thresholds to short shocks and large responses to constant galvanic currents, both delivered via the perilymphatic space.	High thresholds and small responses to the same galvanic stimuli.

[1] Goldberg and Fernández 1977; Baird et al. 1988; Goldberg et al. 1990a; Lysakowski et al. 1995.
[2] Goldberg and Fernández 1971b; Fernández and Goldberg 1976b; Yagi et al. 1977; Baird et al. 1988; Goldberg et al. 1990a; Lysakowski et al. 1995.
[3] Goldberg and Fernández 1980; McCue and Guinan 1994.
[4] Goldberg et al. 1982, 1984, 1987; Ezure et al. 1983; Brontë-Stewart and Lisberger 1994.

differences listed in Table 3.1, irregular fibers would be expected to have a greater sensitivity to sensory inputs, to efferent activation, and to externally applied galvanic currents. Fiber size, although it is known to affect electrical excitability (Rushton 1951), has a much smaller effect on galvanic sensitivity than does discharge regularity (Goldberg et al. 1984). The only difference listed in the table that is not causally related to discharge regularity involves response dynamics. Confirmation of the latter conclusion was obtained from a comparison of the response dynamics to sinusoidal galvanic currents and head rotations (Goldberg et al. 1982; Ezure et al. 1983; Highstein et al. 1996). Results are consistent with the conclusion that differences in response dynamics arise at an earlier stage of hair cell transduction than do differences in discharge regularity. In addition, the conclusion illustrates that two discharge properties—in this case, discharge regularity and response dynamics—can be highly correlated without being causally related.

3.1.3. Directional Properties of Hair Cells and Afferents

There is a relation between the directional properties of hair cells and their morphological polarization. As first deduced by Lowenstein and Wersäll (1959), deflections of the hair bundle toward the kinocilium are excitatory. This rule holds for all hair cell organs and is consistent with the "gating spring" model of hair cell transduction (Hudspeth 1989; Pickles and Corey 1992).

3.1.4. Responses of Afferents to Electrical Stimulation of Efferent Pathways

Repetitive electrical stimulation of efferents results in inhibition of afferent activity in auditory (Fex 1962, 1967; Wiederhold and Kiang 1970; Furukawa 1981; Art et al. 1984), vibratory (Ashmore and Russell 1982; Sugai et al. 1991), and lateral line receptors (Russell 1968; Flock and Russell 1973, 1976; Sewell and Starr 1991). Effects are largely due to a hyperpolarization of hair cells (Flock and Russell 1976; Art et al. 1984; Fuchs and Murrow 1992a, 1992b). The efferent neurotransmitter is acetylcholine (ACh), and the $\alpha9$ nicotinic receptor is likely to be involved (Elgoyhen et al. 1994; Sridhar et al. 1997). When the ligand-gated nicotinic channel is opened, it becomes permeable to monovalent cations and to Ca^{2+}. As Ca^{2+} enters the hair cell, it triggers a small-conductance (sK) Ca-activated K^+ channel, which produces a hyperpolarizing inhibitory postsynaptic potential (IPSP) (Fuchs and Murrow 1992a, 1992b).

Quite a different situation occurs in vestibular organs. In mammals, efferent activation excites afferents, as evidenced by an increase in background discharge (Goldberg and Fernández 1980; McCue and Guinan 1994; Marlinsky et al. 2000). As can be seen in Figure 3.7, the excitation consists of fast and slow components with kinetics of 10–100 ms and 5–20 s, respec-

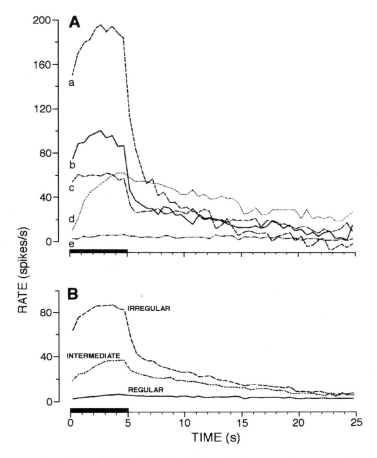

FIGURE 3.7. The effect of electrical stimulation of efferent pathways on afferent discharge in the squirrel monkey (*S. sciureus*). (**A**) Increase in discharge rate during and after shock trains of 5s duration, 333 shocks/s. Responses for four irregularly discharging (a–d) and one regularly discharging (e) unit. (**B**) Responses averaged for 14 irregular, 10 intermediate, and 10 regular vestibular nerve afferents obtained from the same (ipsilateral) stimulating locus in one animal. Same shock-train parameters as in A. (Modified with permission from Goldberg and Fernández 1980, Copyright 1980, The American Physiological Society.)

tively. Responses differ depending on discharge regularity. Irregular afferents have large responses consisting of both fast and slow components. Responses in regular afferents are small and almost entirely slow. The cellular basis of the response has not been studied in mammals. Excitation is also seen in calyx and dimorphic fibers in turtles. Here excitation is mediated by a monosynaptic excitatory postsynatptic potential (EPSP) on the afferent (Holt et al. 2001).

Similar to results in mammals, efferent responses are predominantly excitatory in the toadfish (*O. tau*) (Boyle and Highstein 1990b; Boyle et al. 1991). Responses are more heterogeneous in the posterior crista of frogs (Rossi et al. 1980; Bernard et al. 1985; Rossi and Martini 1991) and turtles (Brichta and Goldberg 2000b): some afferents are excited, whereas others are inhibited or show a mixed inhibitory–excitatory response. Because such heterogeneous responses are seen in afferents innervating different regions of the crista, this topic will be pursued in the next section. Confirming the distinction in the efferent responses from vestibular and vibratory receptors, both excitation and inhibition are seen in all vestibular organs in the toad, including the three semicircular canals, the utricular macula, and the lagena (Sugai et al. 1991). The one exception is the saccular macula, which functions as a vibratory organ in anurans (Koyama et al. 1982; Christensen-Dalsgaard and Narins 1993). In this organ, almost all fibers are inhibited.

The pharmacology of vestibular efferent neurotransmission has been recently reviewed (Guth et al. 1998; Goldberg et al. 2000). It is possible that the inhibition seen in frog and turtle vestibular organs involves the same receptor mechanisms as described above for the inhibition of auditory and vibratory organs (Holt et al. 2001). Excitation in frogs involves a modulation of neurotransmitter release from the hair cell to the afferent (Rossi et al. 1980; Bernard et al. 1985; Rossi and Martini 1991). Nicotinic receptors (Bernard et al. 1985) or purinergic receptors (Rossi et al. 1994) may be involved. In any case, a nicotinic action not coupled to a potassium channel should cause excitation (Goldberg et al. 2000). Slow excitation, which has also been seen in lateral lines (Flock and Russell 1973, 1976; Sewell and Starr 1991), may involve slow nicotinic actions (Sridhar et al. 1995), muscarinic receptors (Bernard et al. 1985; Guth et al. 1998), or peptidergic transmission (Sewell and Starr 1991).

3.2. Physiology of Semicircular Canals

Canal afferents respond to angular head rotations. When there is a density gradient within the endolymphatic ring, the canal can also respond to linear forces. Linear sensitivity is the basis of the convective component of the caloric response (Coats and Smith 1967; Paige 1985; Minor and Goldberg 1990) and may provide a basis for various forms of positional nystagmus and positional vertigo (Money and Myles 1974). Under normal conditions, however, canal afferents are exclusively sensors of rotational forces (Correia et al. 1992).

3.2.1. Directional Properties

Were the fluid circuit comprising each canal independent of the two other canals, it would be expected that endolymph displacement would be proportional to the cosine of the angle between the plane of the canal duct and the plane of motion. The fact that all three canals have two openings into the utriculus complicates the physics. According to Rabbitt's (1999) analysis, the cosine law is obeyed, but the optimal plane may deviate from the geometric canal plane. Experimentally, the deviation is less than 10° (Estes et al. 1975; Reisine et al. 1988; Dickman 1996). What cannot be determined from physical principles are the directions of excitatory and inhibitory rotations. These were deduced by Ewald (Ewald 1892) and summarized in his so-called First Law (Camis 1930). Ewald's conclusions were later verified by afferent recordings (Lowenstein and Sand 1940a). All afferents innervating a given canal have the same directional properties. Deflections of hair bundles toward the utriculus are excitatory for the horizontal canal and inhibitory for the vertical canals. This correlates with hair bundle morphological polarization, which is uniform in each crista (Fig. 3.3C) but oppositely directed in the horizontal and vertical cristae (Lowenstein and Wersäll 1959; Lindeman 1969).

Detailed measurements of canal planes are available for several species (Blanks et al. 1972, 1985; Curthoys et al. 1975; Reisine et al. 1988; Dickman 1996). From these and the cosine law, one can infer the directional properties of afferents innervating each of the three canals. The canals are arranged in coplanar pairs with the two horizontal canals forming a pair, as do the anterior canal on one side and the contralateral posterior canal. Any head rotation causing an excitatory or inhibitory response from a canal will result in an oppositely directed response from the contralateral coplanar canal. It is the difference in discharge between coplanar canals that is interpreted by the brain as a head rotation.

3.2.2. Response Dynamics

In the hair cell literature, the bulk motion of the accessory structure relative to the apical surface of the neuroepithelium is sometimes referred to as macromechanics to distinguish it from micromechanics, which reflects the coupling of the hair bundles to the accessory structure, as well as local deformations of the latter. As summarized in Chapter 4 of this volume by Rabbitt and colleagues, the expected response dynamics contributed by macromechanics can be studied by solving the so-called torsion pendulum equation for sinusoidal inputs. This is done in Figure 3.8A for frequencies ranging from 0.0125 to 8 Hz. In a midband frequency range, which may extend from 0.025 to 30 Hz, the canal should act as an angular velocity transducer. It becomes a displacement transducer above this range and an acceleration transducer below it.

The question arises as to whether the response dynamics are modified by any of the several stages of vestibular transduction interposed between macromechanics and afferent nerve discharge. The situation is summarized in Figure 3.8A based on data obtained in the squirrel monkey (*S. Sciureus*) (Fernández and Goldberg 1971; Goldberg and Fernández 1971b). Note that

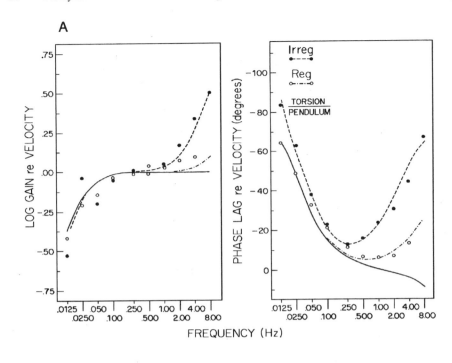

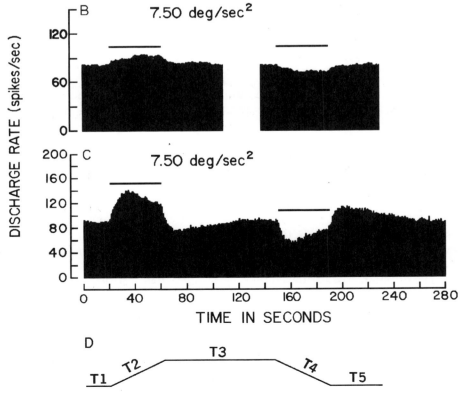

the discharge of vestibular afferents is sufficiently linear that response dynamics can be characterized by gains and phases as a function of sinusoidal frequency. Irregularly discharging afferents have response dynamics that deviate both at low and at high frequencies from predicted macromechanics.

Consider the high-frequency deviation (Fig. 3.8A). As frequency is increased above 0.5 Hz, there is a progressive phase lead and gain enhancement. In the monkey, these high-frequency effects can be interpreted as indicating that afferent response, $r(t)$, is proportional to a weighted sum of cupular displacement, $\xi(t)$, and the velocity of cupular displacement, $d\xi/dt$, with the weights being independent of frequency. That is, afferent response, $r(t) \propto \xi(t) + \tau_v d\xi/dt$, where t_v is the fixed weight. But in this frequency range, cupular displacement, $\xi(t)$, and cupular velocity, $d\xi/dt$, are proportional to head angular velocity, $\omega(t)$, and angular acceleration, $\alpha(t) = d\omega/dt$, respectively, in which case, $r(t) \propto \omega(t) + \tau_v \alpha(t)$. Because the acceleration of a fixed-velocity sinusoidal head rotation is proportional to frequency, the acceleration term increases with frequency, whereas the velocity term does not. As a result, the overall response shows a phase lead with regard to velocity of 45° and a gain enhancement with regard to velocity of $\sqrt{2}$ at the so-called corner frequency, $f_c = 1/2\pi\tau_v$, which for irregular units is near 2 Hz.

The high-frequency deviation is smaller in regularly discharging afferents. Remarkably, though, the same formalism can describe the deviation in the latter units, but the weight, t_v, is smaller and the corner frequency is higher, between 10 and 15 Hz. This is an example of what might be termed *parameterization*, which is to say that the differences in response dynamics of afferents can be described by the variation of one or a few parameters in a transfer function. A practical consequence of parameterization is that

---

FIGURE 3.8. Response dynamics of semicircular canal afferents in the squirrel monkey (*S. sciureus*). (**A**) Because their responses are nearly linear, the afferents can be characterized by their gains (left) and phases (right) to sinusoidal head rotations. Such Bode plots are compared for an irregular unit, a regular unit, and the torsion-pendulum model. (Modified with permission from Goldberg and Fernández 1971b, Copyright 1971, The American Physiological Society.) (**B, C**) Some afferents show a low-frequency adapatation, which is best demonstrated with velocity trapezoids, consisting of two 60 s velocity ramps separated by a 60 s (B) or a 90 s (C) velocity plateau. Velocity profile indicated in (**D**). An irregular unit (C) shows adaptation, which consists of a per-stimulus response decline and poststimulus secondary responses. A regular unit (B) shows little adaptation as its response increases exponentially during velocity ramps and decreases exponentially during the velocity plateau. There is a 30 s break in B so that the inhibitory ramps for the two units are in register. (Modified with permission from Goldberg and Fernández 1971a, Copyright 1971, The American Physiological Society.)

differences between units can be characterized by determining their gains and phases at a single frequency. The simplification holds for afferents recorded in other mammals, even though the precise form of the cupular velocity sensitivity may involve fractional operators (Schneider and Anderson 1976; Tomko et al. 1981b; Baird et al. 1988). In lower vertebrates, differences in high-frequency response dynamics may be too complicated to be adequately described in terms of one or a few parameters (Boyle and Highstein 1990a; Brichta and Goldberg 2000a). Even in these situations, it is still possible to find a single frequency whose gain and phase can be used, possibly along with discharge regularity, to discriminate between various afferent classes.

We now consider the low-frequency deviation. This is best seen in the responses to long-duration angular velocity trapezoids, which are shown for a regular unit (Fig. 3.8B) and an irregular unit (Fig. 3.8C). Except for asymmetries in its excitatory and inhibitory responses, the regular unit conforms to expectations from macromechanics. During long-duration constant angular accelerations and decelerations, excitatory and inhibitory responses build up exponentially with time constants near 5 s. The return to the resting discharge from either kind of response is also exponential. In contrast, the discharge of the irregular unit shows per-acceleratory response declines and postacceleratory secondary responses, including an undershoot in rate following excitation and an overshoot following inhibition. These deviations are a form of adaptation and may reflect, if only partly, processes in the sensory axon (Taglietti et al. 1977; Goldberg et al. 1982; Ezure et al. 1983). Adaptation is also reflected as phase leads and gain decreases in the response to very-low-frequency (≤ 0.01 Hz) sinusoidal head rotations (Fernández and Goldberg 1971).

3.3. Physiology of the Papilla Neglecta

In most animals where it exists, the papilla neglecta is a small, ovoidal patch of hair cells surmounted by a cupula extending only a short distance into the endolymphatic space. The organ is situated in the ventrolateral wall of the utricular sac near the utriculosaccular duct, ventral to the entrance of the crus commune (Fig. 3.1). Although it is innervated by a branch of the posterior ampullary nerve, the papilla in many animals is not associated with a specific semicircular canal.

The papilla is considered here because recent studies in the turtle show that it responds to angular head rotations, albeit with unique directional properties and response dynamics (Brichta and Goldberg 1998). Concerning the directional properties of the papilla, its response plane does not coincide with the plane of any semicircular canal. Rather, afferents innervating the papilla in the turtle respond maximally to pitch rotations and minimally to roll and yaw rotations. Upward pitches excite and downward pitches inhibit. These directional properties can be explained if the organ were excited by the caudal flow of endolymph out of the anterior ampulla and into the posterior ampulla. Such a proposition is consistent with the position of the organ in the turtle, intermediate between the two ampullae and also consistent with the polarization vectors of its hair cells, which point in a caudal direction.

The response dynamics of papilla afferents are also distinctive. When stimulated by midband and high-frequency rotations, semicircular canal afferents in the turtle respond from near angular velocity to near angular acceleration. Papilla afferents, in contrast, respond halfway between angular acceleration and angular jerk. Differences in response dynamics between canal and papilla afferents can be explained by the coupling of their respective cupulae to endolymph flow in vertical canal ducts. Because the cupula of each vertical canal occludes the endolymphatic space of its ampulla, cupular displacement will be proportional to endolymph displacement. The papilla cupula, because it extends only a short distance into the endolymph, will be coupled to endolymph flow by viscous forces, in which case its displacement will be proportional to endolymph velocity. In terms of head motion, the effective stimulus for the papilla is the first time derivative of the effective stimulus for the vertical canals.

A papilla neglecta, with a structure similar to that seen in turtles, is found in many vertebrates (Lewis et al. 1985). Three groups of animals are exceptional in this regard. In some elasmobranchs, the organ presumed homologous to the papilla has a location different from that described above (Fig. 3.1C) (Baird 1974; Corwin 1978) and can be quite large (Corwin 1978). Physiological studies indicate that the papilla neglecta in elasmobranchs is sensitive to vibratory and auditory stimuli (Lowenstein and Roberts 1951; Fay et al. 1975; Corwin 1981), but the physical basis for the sensitivity remains controversial (Kalmijn 1988). Within amphibians, a papilla is present in apodans but not in anurans or urodeles (Baird 1974; Lombard and Bolt 1979; Fritzsch and Wake 1988). It has been suggested that the papilla in the latter animals has been transformed into the amphibian papilla, an auditory organ peculiar to amphibians (Fritzsch and Wake 1988). Finally, a papilla neglecta is found in some mammals but is lacking in others (Lewis et al. 1985). The organ has been reported in a small fraction of human temporal bones (Montandon et al. 1970; Okano et al. 1978), but reservations have been expressed whether the structure being described is homologous to the papilla as seen in other species (for a discussion, see Brichta and Goldberg 1998).

3.4. Physiology of Otolith Organs

As was mentioned in Section 2.1, the saccular macula can serve as a hearing organ or as a detector of substrate-borne vibrations. In mammals, however, both the utricular and saccular maculae are sensors of linear forces, and it is only in this context that they will be discussed.

Before turning to the ultricular and saccular maculae, we briefly consider the lagenar macula, which is an otolith organ present in most jawed vertebrates, with the exception of placental mammals (Baird 1974; Lewis et al. 1985). In fish and amphibians, the macula is located in the posterior part of the sacculus, usually in its own recess (Fig. 3.1C–E). In reptiles and birds, it is located at the apical end of the cochlear duct (Fig. 3.1F,G). Lagenar afferents in the thornback ray (*Raja Clavata*) are sensitive to head tilts and other linear forces (Lowenstein 1950) but not particularly sensitive to vibrations (Lowenstein and Roberts 1951). In the goldfish (*Carassius auratus*), lagenar fibers are primarily sensitive to vibrations, with all fibers having a best frequency near 200 Hz (Furukawa and Ishii 1967). In frogs, many fibers respond to linear forces, whereas others are vibration sensors (Caston et al. 1977; Baird and Lewis 1986; Cortopassi and Lewis 1998).

3.4.1. Directional Properties

The directional properties of utricular and saccular afferents in mammals have been studied by determining the discharge rate of afferents as the animal is tilted into various static positions (Fernández et al. 1972; Loe et al. 1973; Fernández and Goldberg 1976a; Tomko et al. 1981a). Discharge rate is found to be an approximately sinusoidal function of the tilt angle (θ) about a horizontal axis (Fig. 3.9). To understand this, we note that the effective force (F) acting on the hair cells innervated by an afferent is $F = |\mathbf{P}||\mathbf{G}| = g\cos\theta$, where θ is the angle between a polarization vector, $\mathbf{P}$, and the vertically oriented gravity vector, $\mathbf{G}$. Presumably, the polarization vector is averaged over the ensemble of innervated hair cells. By convention, the magnitude of the polarization vector is normalized to unity and the

---

FIGURE 3.9. Discharge rates in several otolith afferents in the squirrel monkey (*S. sciureus*) are plotted as functions of tilt angle about pitch and roll axes. (**A–H**) Each graph represents data from one unit recorded from the superior vestibular nerve (SN) and presumably innervating the utricular macula. (**I**) Two units recorded from the inferior vestibular nerve (IN) and presumably innervating the saccular macula. (**J**) Static-tilt data, such as shown in A–I, were used to calculated a polarization vector for each unit. Vectors for SN (shaded bars) and IN units (unshaded bars) differ in the angle they make with the horizontal plane. SN units lie near the horizontal (utricular) plane, whereas IN units lie near the vertical (saccular) plane. (**K**) Head coordinate system used in this chapter, including A–I. (**L**) Locations of afferents can be inferred by comparing polarization vectors calculated from tilt data with morphological polarization maps (Fig. 3.3). Here the presumed locations of utricular afferents (A–H) are indicated on the macular diagram. (A–J. Reproduced with permission from Fernández and Goldberg 1976a, Copyright 1976, Elsevier Science. K. Reproduced with permission from Goldberg 1979, Copyright 1979, The American Physiological Society.)

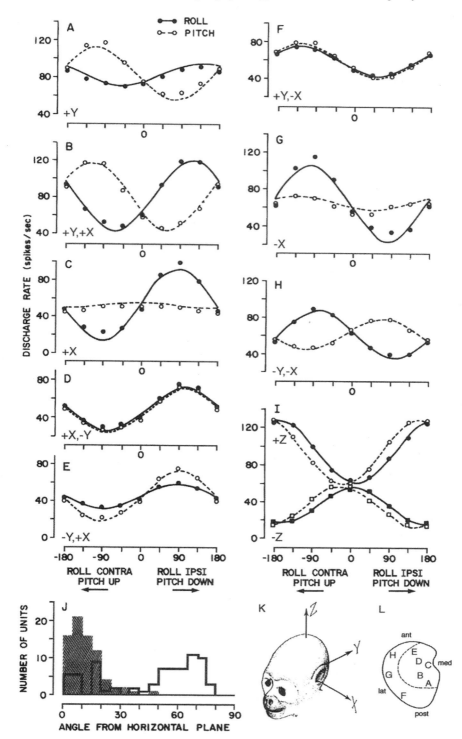

magnitude of the gravity vector = $1\,g$ ($980\,cm/s^2$). The results are consistent with the observation, made on isolated hair cells, that the mechanoelectric transducer current arising from hair bundle displacement is proportional to $\cos\theta$, where θ is the angle between the polarization and displacement vectors (Shotwell et al. 1981). Afferent response, measured as the difference between the discharge rate (d) at a particular tilt angle and its resting or zero-force value (d_0), should be a function of F. The function is approximately linear for small forces. Under these circumstances, $d = sF + d_0$, where s is a sensitivity factor in spikes·s^{-1}/g.

By fitting tilt data, such as is illustrated in Figure 3.9, we can estimate the coordinates of the polarization vector, $\{X, Y, Z\}$, as well as s and d_0. It is of particular interest to compare the responses to static tilts of large populations of saccular and utricular units (Fernández et al. 1972; Fernández and Goldberg 1976a; Tomko et al. 1981a).[2] Utricular afferents respond in opposite directions to ipsilateral and contralateral rolls or to upward and downward pitches (Fig. 3.9A–H). In some units, responses to rolls are larger than those to pitches (Fig. 3.9C,G,H); in others, the reverse is true (Fig. 3.9A,E); and in still others, pitch and roll responses are of comparable size (Fig. 3.9B,D,F). Given the coordinate system depicted in Figure 3.9K, units excited by ipsilateral or contralateral rolls are assigned $+X$ and $-X$ vector components, respectively. Similarly, excitation by upward or downward pitches is associated with $+Y$ and $-Y$ components, respectively. From the polarization map (Fig. 3.3E), the four vector combinations should correspond to units located in the anterolateral ($-X,+Y$), posterolateral ($-X,-Y$), anteromedial ($+X,-Y$), and posteromedial ($+X,+Y$) quadrants. Utricular units show similar discharge rates in the prone ($0°$) and supine ($180°$) positions, which is consistent with their having small Z components. Vectors for utricular afferents are broadly distributed in a horizontal plane. Saccular afferents have quite different directional properties (Fig. 3.9I). Discharge rates are similar for $90°$ tilts in any direction from the prone position, implying that X and Y components are small. Maximum excitation occurs either in the prone or supine positions. From Figure 3.3D, the former units should be located in the inferior ($-Z$) part of the saccular macula; the latter units, in the superior ($+Z$) part.

Because of the broad distribution of their vectors in a horizontal plane, utricular units provide a two-dimensional picture of linear forces acting in a horizontal plane. Saccular units, with their vertically oriented vectors,

---

[2] In the studies summarized here, units were divided into those traveling in the superior (SN) or inferior (IN) vestibular nerves. This was based on the semicircular canal afferents recorded in the same puncture (see Fernández and Goldberg 1976a for details). In the present summary, we have taken the liberty of referring to SN units as "utricular" and IN units as "saccular." There is one potential confusion. Afferents innervating the anterior part of the sacculus actually travel in the SN and then reach their destination by way of Voit's anastomosis.

provide the third dimension. Results are consistent with the assumption that otolith organs are only responsive to shearing forces (i.e., to forces in the plane of the macula). This assumption was tested (Fernández and Goldberg 1976b). As predicted, compressional forces are ineffective in and of themselves and do not influence the response to simultaneously applied shearing forces. This raises a question: Do utricular afferents respond to forces directed in more than one plane? In particular, although much of the utricular macula lies in a horizontal plane, the anterolateral part of the organ curves sharply upward (Fig. 3.15). Anterolateral (AL) afferents, which can be recognized by the directional properties of their tilt responses, do not have distinctively large, downwardly pointing ($-Z$) components. Although it would be good to verify this negative conclusion, it suggests that, under the influence of linear forces, the utricular otoconial membrane is not locally deformable but rather moves as a rigid body (Kachar et al. 1990).

3.4.2. Response Dynamics

It has been suggested that the otoconial membrane of either the utricular or saccular macula can be approximated by a damped second-order system with a lower corner frequency much higher than the frequency spectrum of linear forces exerted on the head during everyday tasks (de Vries 1950; Grant et al. 1994; Rabbitt et al., Chapter 4). The responses of regularly discharging otolith afferents in mammals are approximately consistent with this suggestion (Fernández and Goldberg 1976b; Goldberg et al. 1990a). Once again, it is important to note that responses are sufficiently linear that those at a given sinusoidal frequency can be characterized by a gain and a phase, in this case taken with respect to linear force or the negative of linear acceleration. When this is done for regular units (Fig. 3.10A), gain is almost constant in a frequency bandwidth ranging from dc to 2 Hz. Responses are almost in phase with linear force, small phase leads at low frequencies being replaced by slightly larger phase lags at higher frequencies. The low-frequency phase lead may reflect an adaptive process, at least partly located in the afferent terminal (Goldberg et al. 1982; Ezure et al. 1983). The high-frequency phase lag presumably reflects the macromechanics of otolith motion. Responses of irregular afferents are more phasic, being characterized by phase leads of 20–40° between 0.01 and 2 Hz (Fig. 3.10B) and an approximately fivefold frequency-dependent gain enhancement (Fig. 3.10B).

Based on the responses to controlled transitions from one force level to another, mammalian otolith afferents can be classified as tonic or phasic-tonic (Fernández and Goldberg 1976c). Tonic units show a response that is proportional to the instantaneous force (Fig. 3.11A,B), whereas phasic-tonic units show an additional sensitivity to the velocity of the transition (Fig. 3.11C,D). As predicted from their responses to sinusoidal linear forces,

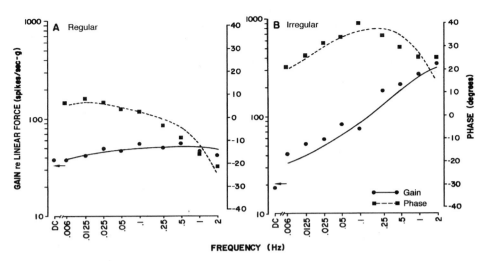

FIGURE 3.10. Response dynamics for two otolith afferents in the squirrel monkey (*S. sciureus*). Gains (left ordinates) and phases (right ordinates) with regard to linear force versus sinusoidal frequency for a regular unit (**A**) and an irregular unit (**B**). The irregular unit has more phasic response dynamics, as indicated by its larger phases and its larger high-frequency gain enhancements. Curves are from empirical transfer functions, with arrows indicating predicted static (DC) gains. (Modified with permission from Fernández and Goldberg 1976c, Copyright 1976, The American Physiological Society.)

regular units are tonic, whereas irregular units are phasic-tonic. Tonic (Fig. 3.11E) and phasic-tonic responses (Fig. 3.11F) to controlled head tilts have also been seen in otolith afferents in nonmammalian vertebrates (Macadar et al. 1975; Blanks and Precht 1976; Baird and Lewis 1986). In addition, some units in the latter preparations are almost entirely phasic, responding only during transitions between tilt positions (Fig. 3.11G). Phasic units are common in the frog (Blanks and Precht 1976; Baird and Lewis 1986). Based on a comparison of the responses to tilts and to galvanic polarizations, it has been suggested that the differences between tonic, phasic-tonic, and phasic units reflect transduction mechanisms preceding the postsynaptic spike encoder (Macadar and Budelli 1984).

4. Morphophysiological Relations in Vestibular Organs

In this section, we summarize information concerning the relation between the physiology of vestibular afferents, their branching patterns and terminal endings, and their location in the neuroepithelium. The most straightforward way to correlate the physiology and terminal morphology of an

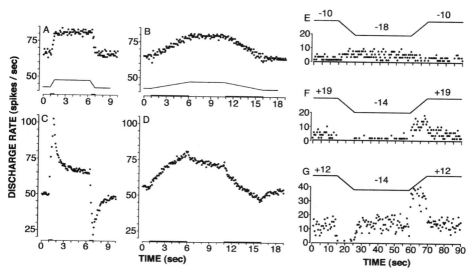

FIGURE 3.11. Responses to variations in linear force for a regular unit and an irregular unit. The regular unit (**A**, **B**) responds to force magnitude but not to the rate of force application. Two components are seen in the irregular unit (**C**, **D**), one proportional to the rate of force application, the other to force magnitude. (Reproduced with permission from Fernández and Goldberg 1976b, Copyright 1976, The American Physiological Society.) (**E–G**) Responses of three utricular afferents in the frog to dynamic tilts. (E) Response of the tonic unit is determined by instantaneous head position during changes in head position and during maintained head position. (F) The phasic-tonic unit responds both to maintained head position and to the velocity of head movement. (G) The phasic unit only responds during head movements. (Modified with permission from Blanks and Precht 1976, Copyright 1976, Springer-Verlag.)

axon is to impale it, characterize its physiology, labeling it with an intracellular marker, and then recover both the parent axon and the peripheral terminal. Because the method is technically difficult, the sample of physiologically characterized and adequately labeled axons is usually small. In addition, it remains difficult to penetrate the thinnest axons, so intraaxonal samples are usually biased toward large-diameter fibers. In most studies, the intraaxonally recorded sample has been supplemented by a large sample of extracellularly recorded units, so the representative nature of labeled afferents in terms of their discharge properties can be evaluated. What about the terminal morphology of intraaxonally labeled fibers? This is where purely anatomical studies are of enormous help. Because the latter only require the extracellular deposit of a tracer, large samples of labeled afferents can be obtained and there is less of a size-related bias. Unfortunately, such a dual labeling approach has been confined to the cristae and utricular macula of mammals (Baird et al. 1988; Fernández et al. 1988, 1990; Goldberg et al. 1990b), the posterior crista of turtles (Brichta and Peterson 1994; Brichta and Goldberg 2000a), and the utricular macula of frogs (Baird and Lewis 1986; Baird and Schuff 1994). The results of extracellular labeling studies can be used to predict details of synaptic innervation, which can then be explored at ultrastructural levels. Although there have been several ultrastructural investigations of hair cells and their innervation, only in mammals has synaptic ultrastructure been correlated with afferent labeling studies (Goldberg et al. 1990c, 1992; Lysakowski and Goldberg 1997).

4.1. Mammalian Cristae

4.1.1. Structural Considerations

As already noted, the cristae can be divided into central, intermediate, and peripheral zones based on the size and spacing of hair cells and the configuration of hair bundles (Lindeman 1969; Fernández et al. 1988, 1995; Lysakowski and Goldberg 1997). There are species differences in the relative numbers of type I and type II hair cells. In the guinea pig (Lindeman 1969) and the chinchilla (Fernández et al. 1988), the two kinds of hair cells occur with nearly equal frequency throughout the neuroepithelium. In the squirrel monkey (*S. sciureus*), type I hair cells outnumber type II hair cells by a ratio, averaged over the entire crista, of 3:1 (Fernández et al. 1995). The ratio varies from nearly 5:1 in the central zone to less than 2:1 in the peripheral zone. Counts in humans resemble those in the squirrel monkey (*S. sciureus*) in that the type I to type II ratio approaches 3:1. At the same time, there is less variation in the ratio for central and peripheral zones (Merchant et al. 2000).

Modern extracellular labeling studies have confirmed the conclusions of earlier workers (Lorente de Nó 1926; Poljak 1927) that there are regional variations in afferent innervation. In both the chinchilla (Fig. 3.12) (Fernández et al. 1988) and squirrel monkey (*S. sciureus*) (Fernández et al. 1995), calyx units are confined to the central zone, while bouton units are restricted to the peripheral zone. Dimorphic units are found throughout all three zones and can vary from having 1–4 calyces and 1–100 bouton endings. There is no correlation within dimorphic units between the number of calyx and bouton endings. Complex calyx endings, each of which enclose 2–4 type I hair cells, are most common in the central zone and are much more likely to innervate calyx, rather than dimorphic, units. It might be expected that innervation patterns would be matched to the relative numbers of type I and type II hair cells in different species. This was the case when the squirrel monkey (*S. sciureus*) was compared to the chinchilla (Fernández et al. 1995).

To check whether an extracellular sample is representative, one can calculate a so-called afferent reconstruction (Fernández et al. 1988, 1995). In

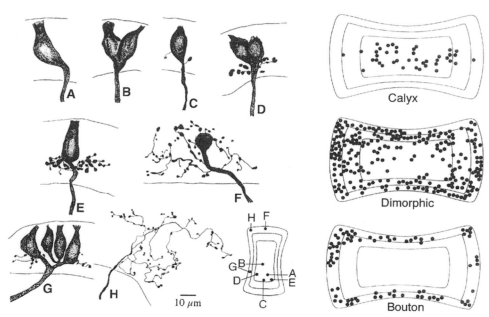

FIGURE 3.12. Branching patterns of individual semicircular canal afferents labeled by the extracellular deposit of horseradish peroxidase in the chinchilla vestibular nerve (**A–H**). Locations of the afferents are indicated on a flatterned map of the crista (lower middle). Right column: Distribution of calyx, dimorphic, and bouton units on a flattened reconstruction of the crista; in all maps, the crista is divided into central, intermediate, and peripheral zones of equal areas. (Modified with permission from Fernández et al. 1988, Copyright 1988, The American Physiological Society.)

each zone, hair cell counts are compared to the results from extracellularly labeled material, including the proportion of calyx, dimorphic, and bouton fibers, as well as the average number of type I hair cells innervated and the average number of bouton endings for each of the three afferent types. The calculations are straightforward, involving simple algebraic manipulations of empirical data. There are two ways to check the accuracy of the calculations: (1) the total number of afferents estimated from the reconstruction can be compared with actual fiber counts; and (2) the reconstruction provides estimates of the number of boutons per type II cell in each zone, which can be checked by quantitative electron microscopy. Reconstructions have been done for the squirrel monkey (*S. sciureus*) and the chinchilla cristae. The calculated afferent counts compare well with empirical counts in the squirrel monkey (*S. sciureus*) (Honrubia et al. 1987; Lysakowski, unpublished data) and in the chinchilla (Carney et al. 1990; Lysakowski, unpublished data). One prediction of the reconstructions is that there should be many more boutons per type II hair cell in the peripheral, as compared to the central, zone. The prediction has been confirmed in both species (Fig. 3.4) (Lysakowski and Goldberg 1993; Fernández et al. 1995; Lysakowski and Goldberg 1997).

There are a large number of efferent synapses in the neuroepithelium as compared with the number of parent efferent axons innervating a crista. This has led to the conjecture that individual efferent axons provide a highly divergent innervation. The supposition has been confirmed by Purcell and Perachio (1997), who used anterograde tracer techniques in the gerbil to describe the peripheral innervation patterns of efferent neurons arising from ipsilateral or contralateral efferent groups. On average, the number of bouton endings provided by each efferent axon ranges from 80 to 90 in the central zone to 160–170 and 340–360 in the intermediate and peripheral zones on the crista slopes and near the planum, respectively. These numbers are about ten times larger than the number of bouton endings found on dimorphic or bouton afferents in the same zones (Fernández et al. 1988, 1995). Purcell and Perachio (1997) made two other important findings. First, individual efferent axons respect zonal boundaries by innervating either the central or peripheral–intermediate zones. Second, their results suggest a laterality in the projections. Efferent neurons projecting to the contralateral labyrinth ended almost exclusively in intermediate and peripheral zones. The central zone was innervated only from efferent neurons on the ipsilateral side. Other ipsilateral efferents innervated the intermediate and peripheral zones.

4.1.2. Discharge Properties

Resting discharges in regular afferents range from 60 to 120 spikes/s in anesthetized mammals. Gain and phase vary with discharge regularity. Due to the parameterization of response dynamics, it is only necessary to determine the response to sinusoidal rotations at a single frequency. Because response dynamics of regular and irregular units diverge at 2 Hz in mammals (Fig. 3.8), this is a convenient testing frequency. In Figure 3.13, gain and phase are plotted versus cv^* for several units in the chinchilla. The gain curve provides evidence for two populations. Units in the first population ($\bigcirc$) range from regular to irregular and, for these units, gain increases linearly with cv^*. The largest gains are 2–3 spikes·s^{-1}/deg·s^{-1}, some 10–20 times smaller than seen in some lower vertebrates (see below). The second population ($\bullet$) consists of irregular units with gains about five times lower than units of the first population with comparably irregular discharge. Remarkably, for the units of the two populations, there is only a single relation between phase and cv^* (Fig. 3.13B).

Two factors contribute to the gain versus cv^* relation for the first group. By far the most important factor is the sensitivity of the postsynaptic spike

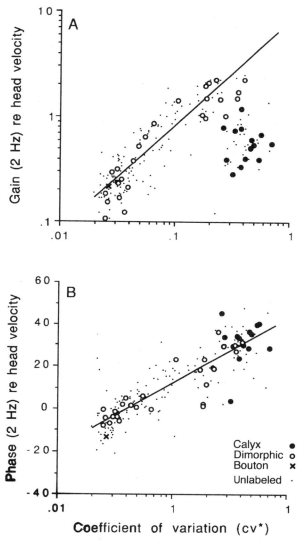

FIGURE 3.13. Responses of labeled and unlabeled semicircular canal afferents in the chinchilla to sinusoidal head rotations. Each point represents one afferent unit. (**A**) Sinusoidal gain at 2 Hz versus normalized coefficient of variation (cv^*), a measure of discharge regularity. Straight line is best-fitting power-law relation between gain and cv^* for the dimorphic and the bouton afferents. (**B**) Sinusoidal phase at 2 Hz versus cv^* for the same units. Straight line is the best-fitting semilogarithmic relation between phase and cv^* for all afferents. See the legend for symbols. (Modified with permission from Baird et al. 1988, Copyright 1988, The American Physiological Society.)

encoder, which can be measured by the sensitivity of individual afferents to galvanic currents delivered via the perilymphatic space (Goldberg et al. 1984; Smith and Goldberg 1986). The other factor is the high-frequency gain enhancement associated with the more phasic response dynamics of irregular units. When the influence of response dynamics is eliminated, the gain curve of the first group parallels the relation between galvanic sensitivity to perilymphatic currents and cv^*. The parallel between the two relations implies that the synaptic input to the encoder is constant for units of the first group. By the same reasoning, synaptic input is about five times lower in units of the second group. These relations, first established in the chinchilla (Baird et al. 1988), were later confirmed in the squirrel monkey (*S. sciureus*) (Lysakowski et al. 1995).

4.1.3. Relation Between Afferent Morphology and Physiology

The first attempts to relate afferent diversity with morphology compared physiological estimates of fiber size with discharge regularity and other discharge properties. Fiber size was estimated by electrically stimulating afferent fibers and measuring the conduction time from the stimulating to the recording electrodes. Orthodromic conduction times were measured in the cat (Yagi et al. 1977) and antidromic conduction times in the squirrel monkey (*S. sciureus*) (Goldberg and Fernández 1977; Lysakowski et al. 1995). Results are illustrated for the monkey (Fig. 3.14). Fibers with long conduction times and, by inference, slow conduction velocities and small diameters, are regularly discharging. By the same reasoning, the largest fibers are irregularly discharging. Medium-sized fibers can be regularly or irregularly discharging. This last finding demonstrates that fiber size, except near the extremes of the size distribution, is not a reliable marker of discharge properties.

A more direct approach, done in the chinchilla, involves the intraaxonal labeling of physiologically characterized fibers (Baird et al. 1988). Unlabeled (small symbols) and labeled fibers (large symbols) are included in Figure 3.13. As expected, labeled calyx units terminate in the central zone. These fibers are irregularly discharging units with relatively small gains and large phases (●). In short, calyx units are the second group of irregularly discharging units identified by extracellular recording. Dimorphic units belong to the first group (○). The physiology of dimorphic units depends on their location in the crista. Those terminating in the central zone are irregularly discharging with large gains and phases. Of the two centrally located groups, calyx units are only slightly more irregular in their discharge and have only slightly larger phases. The two groups can be distinguished by their gains, which are much smaller in calyx fibers. Peripheral dimorphic fibers are regularly discharging and have small gains and phases. Only one intraaxonally labeled bouton unit could be traced to its neuroepithelial termination. As expected, it ended in the peripheral zone and, like peripheral

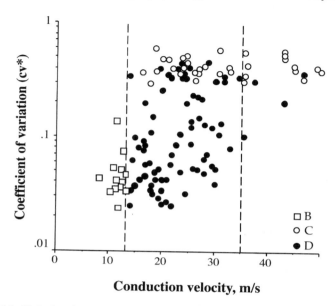

Conduction velocity, m/s

FIGURE 3.14. Relation between conduction velocity and normalized coefficient of variation (*cv**), a measure of discharge regularity. Units were typed as calyx (C) units by a combination of their irregular discharge and low head-rotation gains, as bouton (B) units by their low conduction velocities, and otherwise as dimorphic (D) units. (Reproduced with permission from Lysakowski et al. 1995, Copyright 1995, The American Physiological Society.)

dimorphic units, was regularly discharging and had similarly small gains and phases.

The paucity of labeled bouton fibers presumably reflects their small diameter, which makes them difficult to impale. Fortunately, their small diameter also makes it possible to identify bouton units in extracellular recordings by their small conduction velocities. This was done in the squirrel monkey (*S. sciureus*) (Fig. 3.14) (Lysakowski et al. 1995). As might be expected, the presumed bouton units are regularly discharging with small gains and phases.

4.2. Mammalian Utricular Macula

4.2.1. Structural Considerations

The macula is a kidney-shaped neuroepithelium situated at the anterior end of the utricular sac and measuring about 1 mm in both its anteroposterior and mediolateral dimensions. It can be divided into a flattened posterior portion lying in a horizontal plane and a curved anterior portion (Fig. 3.15). Fascicles of utricular axons pass through channels in the bony meatus,

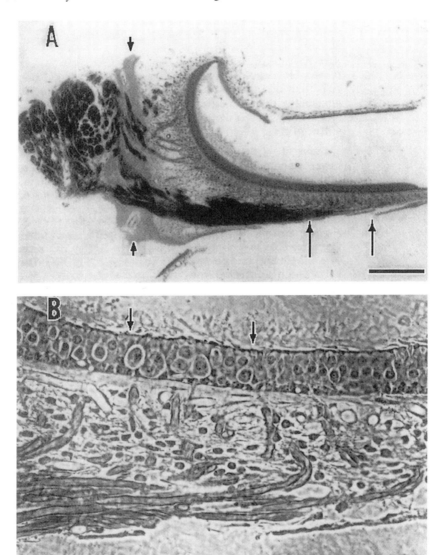

FIGURE 3.15. Section passing through the anterior–posterior axis of the utricular macula of the chinchilla. (**A**) Axons pass from the main trunk of the superior vestibular nerve on the left as fascicles through channels in the bony labyrinth (short arrows) to innervate the anterior, curved portion of the macula. Fibers innervating the posterior, flattened part of the macula first travel in a fiber layer at the bottom of the stroma. Long arrows point to the enlarged region in B. (**B**) On reaching their destination, individual fibers bend sharply upward and run directly to the neuro-epithelium. Arrows point to the border of the striola. Bars: 250 μm (A) and 100 μm (B). (Modified with permission from Fernández et al. 1990, Copyright 1990, The American Physiological Society.)

immediately anterior to the macula. Axons destined for the anterior part run directly to it. Those innervating the posterior part first enter a fiber layer at the base of the connective tissue stroma. Individual fibers of the layer, when they reach their destination, turn sharply upward to reach the neuroepithelium (Fig. 3.15).

The striola is an ≈100-μm-wide ribbon-shaped zone that runs throughout much of the length of the macula and divides it into lateral and medial extrastriolae (Fig. 3.16). As already noted, the striola can be distinguished by its histological appearance from either extrastriola (Fig. 3.15, lower

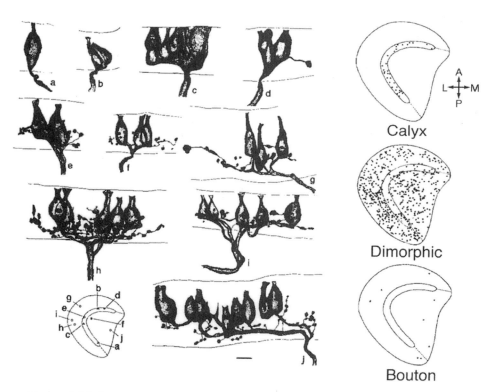

FIGURE 3.16. Branching patterns of individual utricular afferents (**A–J**) labeled by the extracellular deposit of horseradish peroxidase in the vestibular nerve. Locations of the units are indicated on a flattened map of the macula (lower left). Right column: Distribution of calyx, dimorphic, and bouton units on flattened reconstructions of the utricular macula; in all maps of the macula, the central curved area is the striola and the compass indicates anatomical directions. The striola and its continuation (dashed lines) divide the macula into medial and lateral portions. (Modified with permission from Fernández et al. 1990, Copyright 1990, The American Physiological Society.)

panel). There are differences not only in the size and spacing of hair cells but also in the proportion of type I and type II hair cells and of simple and complex calyces. About two-thirds of the striolar hair cells are type I, whereas type I and type II hair cells occur in approximately equal numbers in the extrastriola (Lindeman 1969). Almost half of the calyx endings in the striola are complex, innervating from 2 to 4 type I hair cells (Fernández et al. 1990; Desai et al. 2000). In contrast, simple calyx endings predominate in the extrastriola. In the squirrel monkey (*S. sciureus*), there is a 60:40 ratio between the areas of the medial and lateral extrastriolae (Fernández et al. 1972), whereas in the chinchilla, the comparable ratio is 45:55 (Fernández et al. 1990). Much larger ratios favoring the medial extrastriola are found in lower vertebrates (Flock 1964; Rosenhall 1970; Jørgensen 1974; Baird and Schuff 1994; Si et al. 2003; see Lewis et al. 1985 for a review).

As was the case for the crista, the afferent innervation of the chinchilla utricular macula was studied by extracellular labeling and was found to consist of calyx, dimorphic, and bouton fibers (Fig. 3.16 and Fernández et al. 1990). Percentages of the three fiber types were estimated from an afferent reconstruction. Calyx fibers make up 2–3% of the innervation and are confined to the striola. Dimorphic fibers are found throughout the neuroepithelium and are the predominant innervation, constituting nearly 75% of the afferent innervation of the striola and almost 90% of the extrastriolar innervation. Bouton fibers make up 10% of the afferent fibers, and their terminal fields are found in the extrastriola at some distance from the striola. The absence of bouton afferents helps to define the juxtastriola, a region surrounding the striola, which is also distinguished from the peripheral extrastriola in having dimorphic afferents with more compact terminal trees and fewer bouton endings.

The afferent innervation respects zonal boundaries. Striolar afferents do not cross the middle of the striola to innervate hair cells of opposite polarity; nor do they typically send processes into the juxtastriola. In a similar way, the juxtastriola and the lateral and medial extrastriolae are each provided with a largely independent innervation. Calyx, dimorphic, and bouton fibers have thick, medium-sized, and thin axons, respectively (Fernández et al. 1990).

There have been no published afferent labeling studies of the utricular macula in mammals other than the chinchilla or of the saccular macula in any mammal.

4.2.2. Discharge Properties

Zero-force (d_0) discharge rates are similar for regular and irregular utricular units (Fernández and Goldberg 1976a; Goldberg et al. 1990b). Static or dc gains are somewhat larger in regular afferents. In almost all units, discharge is seldom abolished by static tilts even to head positions leading to minimal discharge. In the squirrel monkey (*S. sciureus*), units excited by ipsilateral roll tilts outnumbered those excited by contralateral roll tilts by a 75:25 ratio (Fernández et al. 1972; Fernández and Goldberg 1976a). A similar preponderance of ipsilaterally excited units was observed in the cat (Loe et al. 1973). In contrast, in the chinchilla, ipsilaterally excited units were only in a slight majority (55:45) (Goldberg et al. 1990a). The ratios based on tilt responses can be compared to the ratio of macular areas with laterally and medially directed hair bundles summarized above. In both the squirrel monkey (*S. sciureus*) and chinchilla, the proportion of ipsilaterally excited units is slightly larger than would be predicted on an areal basis.

As already noted, regular and irregular units differ in their response dynamics (Fig. 3.10). This is reflected by the fact that both the 2 Hz gain ($g_{2\,Hz}$) and phase ($\phi_{2\,Hz}$) increase with cv^* (Fig. 3.17). Similar trends are seen in canal units (Fig. 3.13). Nevertheless, there are obvious differences between the two sets of organs. As was the case for canal units, there is a single, semilogarithmic relation between phase and cv^* (Fig. 3.17B). But unlike the case for canal units, $g_{2\,Hz}$ for irregular utricular units do not fall into two discrete clusters (Fig. 3.17A). One other difference may be noted. In both kinds of organs, regular and irregular units have dissimilar response dynamics. Such differences fall into distinct low- and high-frequency ranges for canal units in mammals (Fig. 3.8) but are broadly distributed across the frequency range for mammalian utricular units (Fig. 3.10).

4.2.3. Relation Between Afferent Morphology and Physiology

Intraaxonal labeling studies have been done on utricular afferents in the chinchilla (Goldberg et al. 1990a). None of the labeled units were of the bouton variety. Their absence from the sample is hardly surprising considering their small axons and that they make up only ≈10% of the total innervation. For the labeled afferents, polarization vectors were determined by static tilts. The locations and vectors of the labeled units were compared with published morphological polarization maps (Spoendlin 1965; Lindeman 1969). Most (50/52) of the labeled units had vectors consistent with the maps. The simplest interpretation of the two aberrant units is that the wrong afferent was labeled. This would suggest an error rate of 5–10% in the labeling of physiologically characterized units. Because of this potential error, undue weight should not be given to exceptional units.

As expected, intraaxonally labeled calyx units were exclusively found in the striola, whereas dimorphic units were found in the striola, juxtastriola, and extrastriola. Calyx units were the most irregularly discharging units in the sample. The discharge regularity of dimorphic units depended on their location. Most of those located in the striola were irregular, whereas most peripheral extrastriolar dimorphs were regular. Juxtastriolar dimorphs were never as regular as the most regularly discharging units in the peripheral extrastriola or as irregularly discharging as some striolar dimorphic or calyx units.

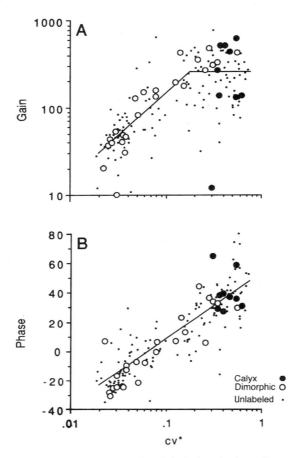

FIGURE 3.17. Responses of labeled and unlabeled utricular afferents in the chinchilla to sinusoidal head rotations. Each point represents one afferent unit. The key indicates labeled calyx and dimorphic afferent, as well as unlabeled afferents. (**A**) Sinusoidal gain at 2 Hz versus normalized coefficient of variation (cv^*), a measure of discharge regularity. Diagonal line is the best-fitting power-law relation between gain and cv^* for unlabeled afferents with $cv^* \leq 0.2$. Horizontal line implies that gain does not continue to increase for $cv^* > 0.2$. Note that unlike canal afferents (Fig. 3.13), there is no clear difference in gains between irregular calyx and dimorphic afferents. (**B**) Sinusoidal phase at 2 Hz versus cv^* for the same units. Straight line is the best-fitting semilogarithmic relation between phase and cv^* for the entire population of unlabeled afferents. As in canal afferents (Fig. 3.13), all units conform to the same relation. (Modified with permission from Goldberg et al. 1990b, Copyright 1990, The American Physiological Society.)

The relation among intraaxonally labeled fibers between $g_{2\,Hz}$ and cv^* (Fig. 3.17A) and that between $\phi_{2\,Hz}$ and cv^* (Fig. 3.17B) resembles those for the extracellular sample (Fig. 3.17A,B, small dots). There is a single, semi-logarithmic relation between phase and cv^* for the intraaxonal sample, including dimorphic and calyx units. In addition, a strong power-law relation exists in the intraaxonal sample between $g_{2\,Hz}$ and cv^* for regular dimorphic units. Many irregular units have relatively high gains, but none of the gains are as high as would be suggested from an extrapolation of the power law for regular units. In addition, there is no clear separation in the values of $g_{2\,Hz}$ for calyx and irregular dimorphic units, a feature distinguishing otolith data from canal data. In an attempt to reconcile the two sets of data, it has been suggested that the comparable 2 Hz gains of utricular calyx and irregular dimorphs can be explained by the former units having a somewhat more irregular discharge and more phasic response dynamics (for details, see Goldberg et al. 1990b).

4.2.4. Efferent Responses of Inferred Afferent Classes and Locations

As stated earlier, efferent responses are similar in canal and otolith organs, so both sets of organs can be considered together. Units characterized by their efferent responses have not been labeled. Based on morphophysiological studies of the relation between discharge regularity and neuro-epithelial location, we can conclude that units with large efferent responses, including both fast and slow components, are located in central (striolar) regions, whereas units with small, slow efferent responses are found in peripheral (extrastriolar) regions. The original study of efferent responses (Goldberg and Fernández 1980) was done before the physiological differences between calyx and irregular dimorphic crista units were appreciated. Recently, the two categories of irregular afferents have been distinguished by their rotational gains in the chinchilla crista (Marlinsky et al. 2000), and both were found to have large excitatory responses, including fast and slow components. No attempt has been made to distinguish the efferent responses of striolar calyx and dimorphic units in the maculae.

The anatomical study of Purcell and Perachio (1997) in the gerbil suggests that electrical stimulation of contralateral group *e* would only activate efferent neurons supplying the peripheral and intermediate zones and hence would not affect central dimorphic and calyx units, both of which are irregular. Physiological studies in the chinchilla do not confirm this suggestion (Marlinsky et al. 2000). Irregular afferents showed similarly large responses when efferent pathways were stimulated on the ipsilateral or contralateral sides or in the midline. Whether the responses from contralateral or midline stimulation were due to direct or transsynaptic activation of efferent fibers is not entirely clear (Marlinsky et al. 2000). Because the anatomical and physiological studies were done in different species, it would be good to do both kinds of studies in a single species.

4.3. Turtle Posterior Crista

4.3.1. Structural Considerations

The turtle (*Pseudemys scripta*) posterior crista consists of two triangular-shaped hemicristae (Fig. 3.18). Each hemicrista extends from the planum semilunatum to the torus, a nonsensory region equivalent to the eminentia cruciatum in other vertebrates, and consists of a central zone (CZ) and a surrounding peripheral zone (PZ). Type I hair cells are confined to the CZ, which also contains a smaller number of type II hair cells (Jørgensen 1974; Brichta and Peterson 1994; Lysakowski 1996). The afferent innervation was described in an extracellular labeling study (Brichta and Peterson 1994). In the central zone, the type I hair cells are innervated by calyx and dimorphic fibers; the type II hair cells by dimorphic and bouton fibers (Fig. 3.18). Only type II hair cells and bouton afferents are found in the peripheral zone. Two classes of bouton afferents, alpha and beta, were recognized by Brichta and Peterson (1994). Alpha fibers have thin axons, sparse terminal trees with thin dendritic branches, and relatively few bouton endings when compared with beta fibers. The alpha fibers were found throughout the neuroepithelium, whereas the more robust beta fibers were concentrated near the torus. There was a longitudinal gradient in the morphology of the alpha fibers, with those nearer the torus having thicker axons and larger terminal trees than those closer to the planum. Including the beta fibers accentuated these trends. Calyx endings can enclose 1–5 type I hair cells. Dimorphic units contact fewer type I hair cells than calyx units and have many fewer bouton endings than do bouton units. Calyx-bearing units have relatively thick axons, similar in size to those of beta units and larger than those of alpha units.

Type I and type II cells differ in cell shape. Most type I cells have a constricted neck, whereas type II cells are cylindrical or club-shaped (Jørgensen 1974; Lysakowski 1996). The two kinds of cells also differ in their hair bundles (Peterson et al. 1996). Compared with PZ type II cells, type I hair cells have more stereocilia, and individual stereocilia are thicker, with the result that type I hair bundles are wider and occupy a larger fraction of the hair cell's apical surface. The tapering of successive ranks of stereocilia is more gradual in the type I hair cells. In many respects, hair bundles of CZ type II cells have characteristics intermediate between those found on type I and PZ type II cells.

4.3.2. Discharge Properties

Turtle posterior crista fibers differ in their discharge regularity and in the gains and phases of their responses to sinusoidal head rotations (Brichta and Goldberg 2000a). Resting discharges range from 0 to 40 spikes/s. Discharge regularity, measured as a cv^* normalized to a mean interval of 50 ms, ranges from 0.1 to 1.0. Firing rates of regular units are lower in the

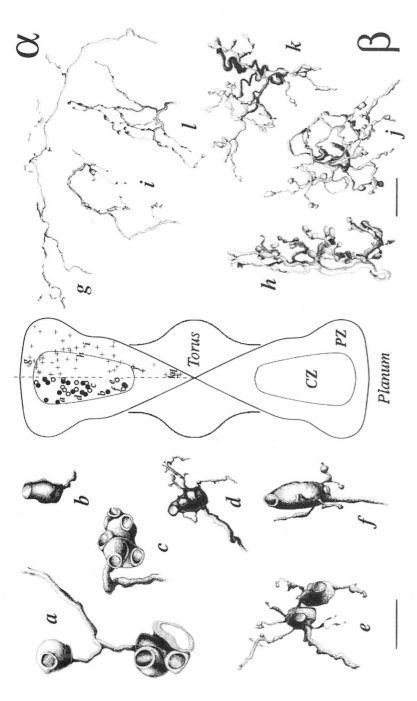

FIGURE 3.18. Turtle posterior crista ampullaris. Within each hemicrista is a central zone (CZ), which contains a mixture of type I and type II hair cells, surrounded by a peripheral zone (PZ) containing only type II chair cells. Calyx afferents (**A–C**, filled circles) and dimorphic afferents (**D–F**, unfilled circles) are found in the CZ. Bouton afferents (**G–L**, pluses) are found in both the CZ and PZ. Individual calyx and dimorphic units are illustrated on the left. Individual bouton units are seen on the right. Bouton fibers have been distinguished into α and β varieties. The former have sparse dendritic trees and thin axons (G, I, L); the latter, robust trees and thicker axons (H, J, K). (Modified with permission from Brichta and Peterson 1994, Copyright 1994, Wiley-Liss, Inc.)

turtle than in mammals. Despite the difficulty of comparing units with little overlap in rates, it would appear that the discharge in the turtle is never as regular as the most regularly discharging mammalian afferents (Goldberg 2000).

Regularly discharging afferents have relatively low gains, and their response dynamics conform to the torsion-pendulum model with a first-order time constant near 3 s (Fig. 3.19A,E). The one discrepancy from the model occurs at low frequencies, where phase leads are somewhat larger than expected. In other species, a phase lead of this sort has been interpreted in terms of a low-frequency adaptation (Fernández and Goldberg 1971). Irregular units come in two varieties. Some irregular units have very large gains and phase leads. Taking values at 0.3 Hz, a frequency where differences in the phase of different unit groups are largest, gains of these

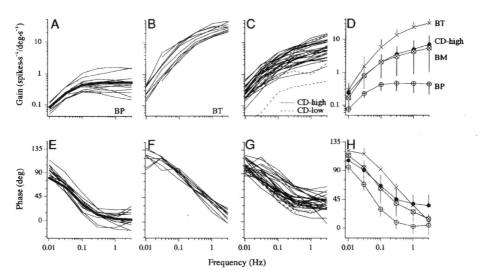

FIGURE 3.19. Gains (**A–D**) and phases (**E–H**) versus sinusoidal frequency for different groups of unlabeled units from the turtle posterior crista. By comparing the physiological properties of the unlabeled units with those of labeled units, the former could be distinguished into bouton and calyx-bearing groups and their longitudinal positions estimated (see Fig. 3.20 and the text for details). BP, BT, and BM units are classified as bouton units presumed to be located near the planum (P), near the torus (T), or in midportions (M) of the hemicrista, respectively. CD units are classified as calyx-bearing (calyx or dimorphic) and are distinguished into those with high (CD-high, >5 spikes/s) and low (CD-Low, 0–5 spikes/s) resting discharges. Thin lines in A–C and E–G are for individual units. The thick lines in A and E are the best-fitting torsion-pendulum model for regular units. Group means and standard deviations are found in D and H. (Modified with permission from Brichta and Goldberg 2000a, Copyright 2000, The American Physiological Society.)

irregular units (Fig. 3.19B,D) are almost 50 times those in regular units (Fig. 3.19A,D), and response leads head velocity by 45° (Fig. 3.19F,H) as compared to the near-zero phases of regular units (Fig. 3.19E,H). The large phase leads of these irregular units diminish at higher frequencies, amounting to only 5–10° at 3 Hz (Fig. 3.19F,H). Other irregular units have intermediate gains (Fig. 3.19C,D) and phase leads (Fig. 3.19G,H) at 0.3 Hz. Furthermore, the phase leads of the latter irregular units, rather than declining at high frequencies, reach an asymptote of 40–50° (Fig. 3.19G,H). Another feature of the latter group concerns their resting discharge. Regular units and irregular units of the first group have typical background rates of 5–35 spikes/s. This is also the case for many irregular units of the second group, but other irregular units of this group have rates less than 5 spikes/s, and some of them are even silent at rest. The silent units have lower gains and phase leads than do units of the second group having an appreciable background discharge.

4.3.3. Relation Between Afferent Morphology and Discharge Properties

Intraaxonal labeling was used to identify the various groups of units distinguished by their extracellular discharge properties (Brichta and Goldberg 2000a). Testing was done at 0.3 Hz because differences between the phases of the groups were nearly maximal at this frequency (Fig. 3.19D,H). It was found that the regular units, as well as irregular units with the highest gains and phase leads, were bouton (B) afferents. In fact, for the entire population of B afferents, there was a power-law relation between gain and cv^* (Fig. 3.20A) and a semilogarithmic relation between phase and cv^* (Fig. 3.20B). The large differences in cv^*, gain, and phase were correlated with the longitudinal position of the units. In describing the differences, it is convenient to use different abbreviations for B units with terminal fields near the planum (BP), near the torus (BT), or in midportions of the crista (BM).

BP units are regularly discharging and have low gains and small phase leads, whereas BT units are irregularly discharging and have high gains and large phase leads. BM units, including bouton units in the central zone, are intermediate in their discharge properties. Calyx (C) and dimorphic (D) units cannot be distinguished in their physiology and are placed in a single calyx-bearing (CD) group. They are the irregular units with lower gains and phases than BT units and almost all of the units with low background rates. Based on whether their resting discharge was greater or less than 5 spikes/s, the CD units are placed into CD-high or CD-low categories. Several morphological features were measured for each B and CD unit, and a multiple regression was run to identify possible morphological correlates of physiological properties. Longitudinal position was the only morphological feature that was correlated with the cv^*, gains, and phases of B units. The same was true for CD units except that even longitudinal position was not

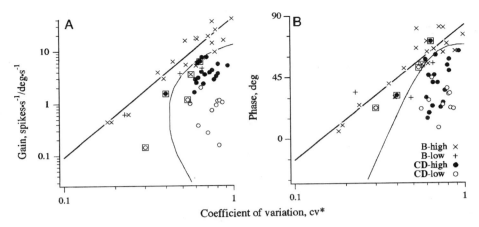

FIGURE 3.20. Distinguishing bouton (B) and calyx-bearing (CD, calyx and dimor-phic) afferents in the turtle posterior crista. Responses of labeled posterior canal afferents in the turtle to 0.3 Hz sinusoidal head rotations. Each point represents one afferent unit. Units are divided into low and high categories based on their background discharge rate. (**A**) Sinusoidal gains versus normalized coefficient of variation (cv^*), a measure of discharge regularity. Straight line is the best-fitting power-law relation between gain and cv^* for B units. (**B**) Sinusoidal phase versus cv^* for B and CD units. Straight line is the best-fitting semilogarithmic relation between phase and cv^* for B units. A discriminant function (curved line) separates B and CD units. Of the 54 afferents, five (<10%, squares) were misclassified. (Modified with permission from Brichta and Goldberg 2000a, Copyright 2000, The American Physiological Society.)

significantly related to cv^*. There was a differential distribution of CD-high and CD-low units in the central zone, with CD-low units being concentrated nearer the planum and CD-high units nearer the torus. There was no obvious relation between the resting discharge rates of B units and their longitudinal position.

CD units have an irregular discharge, but their gains and phases are lower than those of BT units with comparable discharge regularity (Brichta and Goldberg 2000a). Stating the difference another way, CD units have a more irregular discharge than B units with comparable gains and phases. The dif-ference was used to calculate quadratic discriminant functions, which sorted the two populations with an error rate of <10% (Fig. 3.20). In addition, multiple-regression equations were developed that predicted the longitu-dinal position of B and CD units with rms errors of <10% of the hemicrista length. The discriminant and multiple-regression equations were then used to infer the type (B vs CD) and longitudinal position of extracellularly recorded units. One use of the sorting procedure was to determine the response dynamics of large samples of extracellularly recorded BP, BM, BT, and CD-high units. In fact, Figure 3.19 was sorted in this way. A second use was to study response intensity functions of the various unit groups. On this basis, it was concluded that BT units linearly encode head rotations only for angular velocities <10°/s; larger velocities saturate discharge. Given this and their very high sensitivities, BT units are suited to monitor the small head movements involved in postural control. In contrast, BP and CD units have lower sensitivities and extended linear ranges, making them more suitable for monitoring large, volitional movements.

4.3.4. Efferent Responses of Inferred Afferent Classes and Locations

A third use of the sorting procedure was to characterize the responses to electrical stimulation of efferent fibers for various groups of extracellularly recorded units (Brichta and Goldberg 2000b). BT units show inhibitory responses that greatly outlast the shock train (Fig. 3.21A). BM units show mixed responses consisting of an early inhibition and a late excitation (Fig. 3.21B). Similar inhibitory–excitatory responses have been observed in lateral lines (Russell 1971; Sewell and Starr 1991) and frog cristae (Rossi and Martini 1991; Sugai et al. 1991), where the posttrain excitation has been interpreted as a release from inhibition. This interpretation is probably incorrect in the present situation. One reason for believing so is shown by data from another BM unit (Fig. 3.21F,G). Lowering shock frequency and lengthening the train results in a brief inhibition followed by a small excitation. Even though inhibition had ceased several seconds earlier, the end of the train is marked by a large excitation. Small excitatory responses are observed in BP units (Fig. 3.21C) and large excitatory responses in CD-high units (Fig. 3.21D). In the latter units, prolonging the shock train could result in a slow posttrain response lasting several tens of seconds (Fig. 3.21E). Responses could be small or absent in CD-low units even when their background rates were raised by rotations.

Recent intraaxonal recordings near the crista show that the efferent responses of B units reflect a modulation in the rate of mEPSPs ascribable to quantal neurotransmitter release from the hair cell (Holt et al. 2001). In contrast to this presynaptic efferent action, efferent actions in CD units include a postsynaptic action because a large, relatively slow EPSP can be recorded from the latter units during efferent stimulation.

4.4. Lizard Horizontal Crista

In the lizard (*Calotes versicolor*), Schessel et al. (1991) found that discharge rates range from 0 to 160 spikes/s and *cv* ranges from 0.12 to 0.92. Intraaxonal labeling was done in an attempt to relate afferent morphology and physiology. Many calyx fibers, but relatively few dimorphic and bouton fibers, were labeled. Calyx fibers were irregularly discharging, leading to the suggestion that any unit having a *cv* > 0.4 belonged to this category. Fibers so classified were found to have mEPSPs that could be modulated by caloric

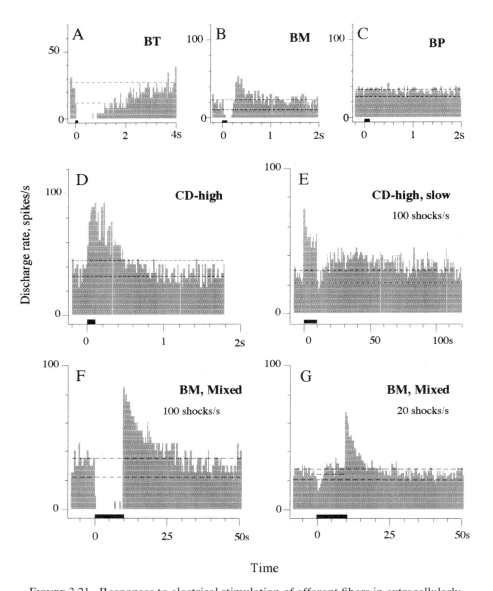

FIGURE 3.21. Responses to electrical stimulation of efferent fibers in extracellularly recorded afferents from the turtle posterior crista. Unit types and locations were determined by comparing their discharge properties with those of labeled afferents. Except where noted, the shock rate was 200/s. (**A**) Inhibitory response of a bouton (BT) unit. (**B**) Mixed (inhibitory–excitatory) response of a bouton (BM) unit. (**C**) Small excitatory response of a bouton (BP) unit estimated to be located near the planum. (**D**) Large excitatory response of a calyx-bearing (CD-high) unit. (**E**) A 10s shock train evokes in a CD-high unit a per-stimulus response and, after a delay, a poststimulus response lasting >100s. (**F, G**) A 10s shock train evokes in a BM unit a mixed inhibitory-excitatory response when the shock rate is 100 shocks/s. When the shock rate is lowered to 20 shocks/s, the initial inhibitory response is shortened and is followed by a small per-stimulus excitatory response. Although inhibition had terminated 8s previously, there is still a large posttrain excitation, indicating that the latter is not merely a release from inhibition. (Modified with permission from Brichta and Goldberg 2000b, Copyright 2000, The American Physiological Society.)

stimulation, apparently contradicting the notion that there is no conventional synaptic transmission between type I hair cells and their calyx endings (Bennett 1972; Gulley and Bagger-Sjöbäck 1979; Yamashita and Ohmori 1990). Unfortunately, the classification criterion used by Schessel et al. (1991) may be questioned. In the turtle, there are bouton (BT) fibers with high *cv*. Such fibers might be expected to exist in the lizard also.

4.5. Frog Cristae

4.5.1. Structural Considerations

The vertical cristae are dumbbell-shaped and can be divided into an isthmus region (zone III), bordered on either side by an intermediate region (zone II) that interconnects the isthmus with a region abutting the planum (zone I) (Fig. 3.22A) (Hillman 1976; Myers and Lewis 1990). The three zones differ in the size of the fibers innervating them (Honrubia et al. 1989), in the density (Myers and Lewis 1990) and shapes of their hair cells (Guth et al. 1994; Gioglio et al. 1995), and in the configurations of their hair bundles (Flock and Orman 1983; Myers and Lewis 1990). It would appear that the horizontal crista is analogous to half of a vertical crista, consisting of a planar (zone I) region that is attached to the isthmus (zone III) by a single zone II (Fig. 3.23B) (Hillman 1976; Myers and Lewis 1990). Unless otherwise noted, the following descriptions pertain to the vertical cristae.

Fiber diameter has been studied in the anterior crista by Honrubia et al. (1989). In most specimens, the ampullary nerve splits into two branches, each innervating one or the other longitudinal half crista. Just below the crista, each branch splits into three bundles (A, B, and C) that reach zones III, II, and I, respectively. Fibers range in external diameter from <1 μm to 15 μm or more. There is a heavier concentration of large-diameter fibers in bundle A than in bundle C. Fibers >7 μm make up almost one-third of the fibers in bundle A but less than 5% of those in bundle C. At the same time, the thinnest fibers (<3 μm) make up a considerable fraction of the fibers in all bundles, including bundle A.

There may be a regional difference in hair cell shape (Guth et al. 1994; Gioglio et al. 1995). Hair cells have been divided into three categories: cylindrical or cigar-shaped, club-shaped, and pear-shaped. Cylindrical and club-shaped hair cells are both elongated, 30–40 μm tall. The two types differ in that the club-shaped cells have a basal expansion, whereas the cylindrical cells do not. Pear-shaped cells are short and squat, with lengths as small as 15 μm. Guth and his colleagues (Guth et al. 1994) report in the bullfrog (*Rana catesbiana*) that all three types are found throughout the crista, with cigar- and club-shaped cells being more heavily concentrated near the planum and pear-shaped cells near the center. Gioglio et al. (1995) find a somewhat different distribution in the European green frog (*Rana esculenta*): club-like cells are found only near the planum, cylindrical cells

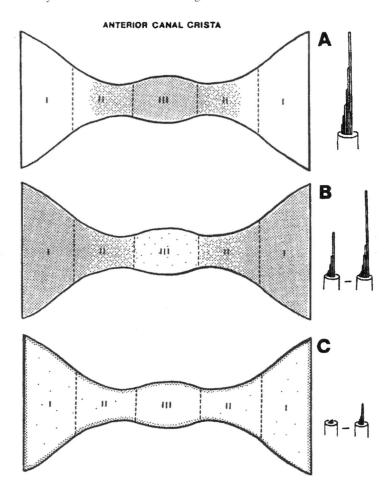

FIGURE 3.22. The frog vertical cristae have each been divided into three zones (I, II, and III). Three classes of hair bundles are illustrated at the right, and their distributions are indicated in the corresponding maps. (**A**) Hair cells with stereocilia almost as long as the tall, eccentrically placed kinocilium are found in the isthmus portion (zone III) and a part of zone II. (**B**) Hair cells with short and medium-sized kinocilia. The tallest stereocilia reach only half the height of the kinocilium. Hair bundles of this variety are found in the planum region (zone I) and in zone II. (**C**) Around the margins of the sensory epithelium and occasionally within its borders is found the third type of hair bundle, which has a very short kinocilium and even shorter stereocilia. (Modified with permission from Myers and Lewis 1990, Copyright 1990, Elsevier Science.)

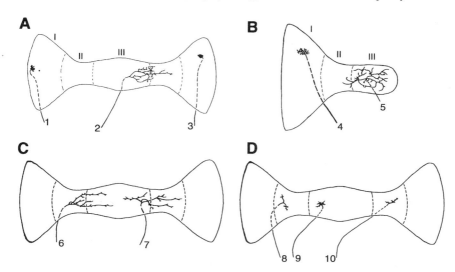

FIGURE 3.23. Afferent innervation patterns in the frog crista ampullaris as revealed by intraaxonal labeling. Maps of vertical cristae (**A, C–D**) and horizontal crista (**B**). Units in zone I (I) have restricted terminal fields (fibers 1, 3, 4). Those in zones II (II) and III (III) can have large fields (LF, fibers 2, 5–7) or small fields (SF, fibers 8–10). (Modified with permission from Myers and Lewis 1990, Copyright 1990, Elsevier Science.)

throughout the crista, and pear-shaped cells in the intermediate (zone II) region. It is unclear whether differences are species-related. Hair cell density is higher in the planum region than in the isthmus (Hillman 1976; Myers and Lewis 1990; Gioglio et al. 1995).

As was described by Hillman (1976) and by Flock and Orman (1983), different types of hair bundles can be recognized (Fig. 3.22). The latter authors referred to hair bundle types as A, B, or C (Fig. 3.22). According to Flock and Orman (1983), A bundles are located in the isthmus region, B bundles throughout the crista, and C bundles along the perimeter of the crista. Myers and Lewis (1990) confirm the presence of the three bundle types. The distribution of bundle types is similar to that described by Flock and Orman (1983) except that there are relatively few B types in zone III. It is also suggested that the C type belongs to immature hair cells.

A few terminal fields are described by Honrubia et al. (1989), but a thorough description of innervation patterns based on extracellular labeling is lacking. An important point, as seen in intracellularly labeled fibers, is that some have small terminal fields confined to a single zone (Fig. 3.23A1,A3,B4,D8–10), whereas others have axons running longitudinally and innervating more than one zone (Fig. 3.23A2,B5,C6–7) (Myers and Lewis 1990).

4.5.2. Discharge Properties

Although the frog crista has been used since the pioneering studies of Ross (Ross 1936; Ledoux 1949; Precht et al. 1971; Blanks and Precht 1976), only recently has it become apparent that there is a diversity in the discharge properties of afferents innervating a particular crista. That the diversity had not been recognized sooner may have been due to a bias in earlier studies toward recording from the very large fibers in the frog ampullary nerves. The most complete study of discharge properties is that by Honrubia et al. (1989) of fibers innervating the anterior crista in an *in vivo*, anesthetized preparation of *R. catesbiana*. Many fibers have a background discharge of 10–30 spikes/s. These fibers show a range of discharge regularities with *cv* from 0.1 to 1.0. Still other, presumably irregular, fibers have lower rates, and nearly 10% of the sample are silent at rest. No attempt was made to normalize the *cv* to a standard mean interval. Response dynamics were measured with sinusoidal head rotations whose effective magnitude was held constant at 22.2 deg/s in a frequency range from 0.0125 to 0.4 Hz. Phases were related to discharge regularity. The discharge of regular units at 0.4 Hz was in phase with head velocity, and response dynamics conformed to that expected of a torsion pendulum with a slow time constant approaching 5 s. For irregular units, discharge led head velocity by as much as 50°. Gains were also related to *cv*. Gain at 0.4 Hz increased 20-fold, from 0.4 to 8–10 spikes·s^{-1}/deg·s^{-1}, as *cv* increased from 0.1 to 0.4. Gains of more irregularly discharging units remained relatively constant as *cv* increased from 0.4 to 1.0.

Two points can be made about these results. First, the diversity in gains and phases is much larger than that seen in mammals. A similarly large diversity is seen in other nonmammalian species studied, including the turtle (*P. scripta*) (Brichta and Goldberg 2000a) and the toadfish (*O. tau*) (Boyle and Highstein 1990a). Second, the rotations used by Honrubia and colleagues (Honrubia et al. 1989) may have been too large in magnitude to stay in the linear range of the most sensitive units (Brichta and Goldberg 2000b). Using much smaller rotation amplitudes, Myers and Lewis (1991) have observed gains as high as 40 spikes·s^{-1}/deg·s^{-1} in a limited sample of irregularly discharging units. High-gain units show a skew distortion to sinusoidal stimuli beyond their linear range, which would lead to an overestimation of their phase leads (Segal and Outerbridge 1982).

4.5.3. Relation Between the Morphology and Physiology of Afferents

The issue has been approached in two ways. By labeling individual axons, Honrubia et al. (1989) were able to show that there was a positive correlation between discharge and fiber diameter. It had already been established that there was a trend for the thickest fibers to be heading toward the isthmus and for the thinnest fibers to be heading toward the planar zone. In addition, there was a correlation between the discharge regularity, gain, and phase of a fiber. Based on these trends, it was suggested that thin, regularly discharging afferents with small gains and phase leads were destined for planar regions, whereas thick, irregularly discharging afferents with larger gains and phase leads supplied the isthmus.

A more direct approach was used by Myers and Lewis (1990, 1991). Individual axons were characterized by their discharge regularity. The fibers were then intraaxonally labeled and traced to their peripheral terminations (Fig. 3.23). Regularly discharging fibers ($cv \approx 0.2$) were traced to the planar region (Fig. 3.23A1,A3,B4), whereas irregularly discharging fibers went to zones II and III (Fig. 3.23A2,B5,C6–7,D8–10). Terminal fields in zone I are the most compact of any in the crista. Afferents in zones II and III can have large (LF, Fig. 3.23A2,B5,C6–7) or small fields (SF, Fig. 3.23D8–10). Most LF fibers innervate both zones II and III. Presumably reflecting a sampling bias toward large axons, none of the labeled axons had calculated external diameters less than 3.0 μm, even though such fibers make up more than half the fibers destined for the isthmus and almost two-thirds of those supplying the planar region (Honrubia et al. 1989). The same bias may have led to the result that there was no relation between fiber diameter and either peripheral destination or terminal-field size. Small-amplitude rotations were used to characterize some of the labeled irregular fibers going to the isthmus. Resting discharge rates ranged from 5 to 20 spikes/s for regular fibers and were somewhat lower for irregular fibers. There was a suggestion that LF fibers had higher rotational gains than SF fibers.

4.5.4. Efferent Responses of Inferred Afferent Classes and Locations

There have been several studies of the response of posterior crista afferents to electrical stimulation of efferent fibers (Fig. 3.24) (Rossi et al. 1980; Bernard et al. 1985; Rossi and Martini 1991). Some afferents are inhibited (Fig. 3.24A), whereas others are excited (Fig. 3.24B). Both the excitation and the inhibition are mediated by presynaptic actions on hair cells. No attempt has been made to label afferents whose efferent responses have been characterized. Sugai et al. (1991), working in a toad (*Bufo vulgaris japonicus*) found that electrical stimulation of efferents did not produce detectable responses in regular afferents but did so in irregular afferents. Some irregular fibers were excited by efferent activation, whereas others were inhibited. Based on the evidence reviewed above, the affected fibers should be confined to zones II and III. Whether excited and inhibited fibers innervate separate zones of the anuran crista remains to be determined.

4.6. Frog Otoconial Organs

As in most vertebrates, there are three otoconial organs in the anuran ear: the utricular, lagenar, and saccular maculae (Hillman 1976; Lewis et al. 1985). The saccular macula is sensitive to substrate-borne vibrations in a

A   B

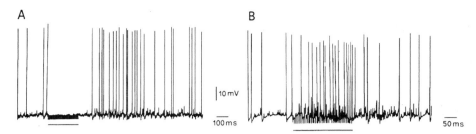

FIGURE 3.24. Responses to electrical stimulation of efferent fibers in two intra-axonally recorded units from the frog posterior crista. Efferents were electrically stimulated by an axon reflex from the anterior ampulla. Duration of the shock train is indicated by a line below each trace; shock frequency, 200 Hz. (**A**) Spikes are inhibited during the train and excited thereafter. Note elimination of hair cell mEPSPs (small, subthreshold events) during the train and an increase in their frequency during posttrain excitation. (**B**) In another unit, spikes are excited both during and immediately after the shock train. mEPSP frequency is increased during both periods. The two afferents both exhibit irregular resting discharge patterns. (Reproduced with permission from Rossi and Martini 1991, Copyright 1991, Elsevier Science.)

frequency range of 10–100 Hz (Koyama et al. 1982; Lewis et al. 1982; Christensen-Dalsgaard and Narins 1993). In contrast, the utricular macula is a vestibular organ, sensitive to head tilts relative to the gravity vector and to other linear forces acting on the head (Lewis et al. 1982; Baird and Lewis 1986). The lagena may be a mixed organ because some lagenar afferents respond to head tilts, others to substrate-borne vibrations, and still others to both kinds of stimuli (Caston et al. 1977; Lewis et al. 1982; Baird and Lewis 1986; Cortopassi and Lewis 1998). In the following, we concentrate on the utricular and lagenar maculae.

4.6.1. Structural Considerations

The utricular macula is a horizontally oriented, crescent-shaped neuroepithelium with a medially positioned hilus (Fig. 3.25G), where the parent nerve reaches the organ (Hillman 1976; Lewis et al. 1985). A striola is found near the lateral margin of the macula and is bordered by a large medial extrastriola and a much smaller lateral extrastriola. The posterior part of the macula lies in a horizontal plane, whereas the anterior part curves upward (Baird and Schuff 1994). Hair cells in the striola tend to be larger and more widely spaced than those in either the lateral extrastriola or medial extrastriola (Lewis et al. 1985; Baird and Lewis 1986; Baird and Schuff 1994). An imaginary reversal line running through the middle of the striola separates hair bundles of opposite morphological polarization. Both

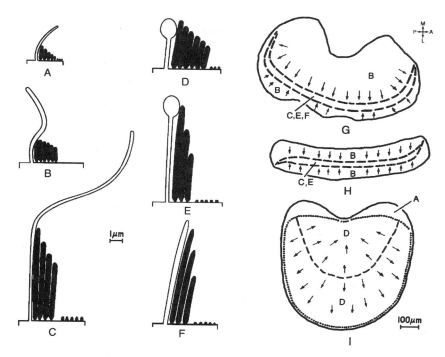

FIGURE 3.25. Hair bundle (**A–F**) types and their distribution in the frog utricular (**G**), lagenar (**H**), and saccular macula (**I**). Types A–C have stereocilia, which are shorter than the kinocilium, and various lengths of kinocilium. In types D–F, the tallest stereocilia are almost as long as the kinocilium, which ends as a bulb in types D and E. The utricular and lagenar maculae have a striola (dashed lines). There is no striola in the saccular macula, but a reversal line (dashed line) separates hair cells of opposite polarization. Compass indicates anatomical directions in G. (Modified with permission from Lewis and Li 1975, Copyright 1975, Elsevier Science.)

lateral and medial to the reversal line, hair bundles point toward the dividing line. Morphological polarization vectors run perpendicular to the curved striola and, hence, diverge like a fan (Fig. 3.25G). Hair bundles in the utricular macula have diverse shapes (Fig. 3.25A–F) (Lewis and Li 1975). Throughout the medial and lateral extrastriolae, most hair bundles are of the B type, consisting of several rows of stereocilia much shorter than the kinocilium. At the margins of the striola, type C bundles predominate and are replaced by types F and E as one moves inward from the extra-striola border. Type E is most numerous on either side of the reversal line. The lagenar macula resembles the utricular macula in many of its internal features (Hillman 1976; Baird and Lewis 1986) but unlike the latter is vertically oriented, more elongated, its striola is more centrally located within the organ, and the hair bundles in the internal striolar rows do not include type F. Morphological polarization vectors, although similar to those in the utricular macula in pointing toward the striola, are not arranged fanlike but point along a single axis.

Extracellular labeling has been used to characterize the innervation patterns of utricular afferents (Fig. 3.26) (Baird and Schuff 1994). Striolar afferents (Fig. 3.26A, e–h) have thicker axons, larger terminal fields, and more synaptic endings than do extrastriolar afferents (Fig. 3.26A, a–c). A juxtastriolar region interposed between the striola and the medial extrastriola has innervation patterns that are in some respects intermediate between the striola and extrastriola (Fig. 3.26A, i, j). As compared to the fibers innervating the medial extrastriola (Fig. 3.26A, a–c), the relatively few fibers innervating the lateral extrastriola have larger axons, larger terminal fields, larger numbers of terminal endings, and contact many more hair cells (Fig. 3.26A, d). Sizes and shapes of terminal fields are illustrated in Figure 3.26B. This figure illustrates the difference in terminal-field size of the various zones. Three additional points can be made. First, fibers seldom cross boundaries from one region to another. Second, most striolar afferents predominantly innervate one or the other side of the reversal line (see also Fig. 3.26A, g, h). Third, individual afferents, even those with elongated terminal fields, innervate hair cells with similar morphological polarization (Fig. 3.26B,C). A comparable study of the innervation of the lagenar macula has not been published.

FIGURE 3.26. Afferent innervation patterns in the frog utricular macula. (**A**) Drawings of individual afferents, whose locations are indicated on a standard map (inset). In the map, the macula is divided into medial and lateral extrastriolae by the striola (shaded); a juxtastriola (thin line) is found immediately medial to the striola and its border with the medial extrastriola. Afferents in the medial extrastriola (a–c) can be characterized by relatively fewer branches and large boutons. A cuplike ending is seen in part a (asterisk). Lateral extrastriolar afferents (d) have many branches and small terminal endings. Striolar afferents are the most varied, having either the appearance of lateral extrastriolar afferents (e, f) or else short, thick branches with thick terminal endings (g) or very thin elongated fibers with short terminal branches (h). Terminal fields of striolar afferents are restricted to one side of the reversal line (dotted line). Juxtastriolar afferents (i, j) all had a similar appearance: long, relatively unbranched processes with a terminal cluster of many branches near the striola. Shapes of terminal fields (**B**) and morphological polarizations (**C**) are plotted on the standard map; the various regions are as described in A (inset). Scale bar in A, 10 µm, holds for individual afferents; bar in C, 100 µm, is used for all standard maps. (Modified with permission from Baird and Schuff 1994, Copyright 1994, Wiley-Liss, Inc.)

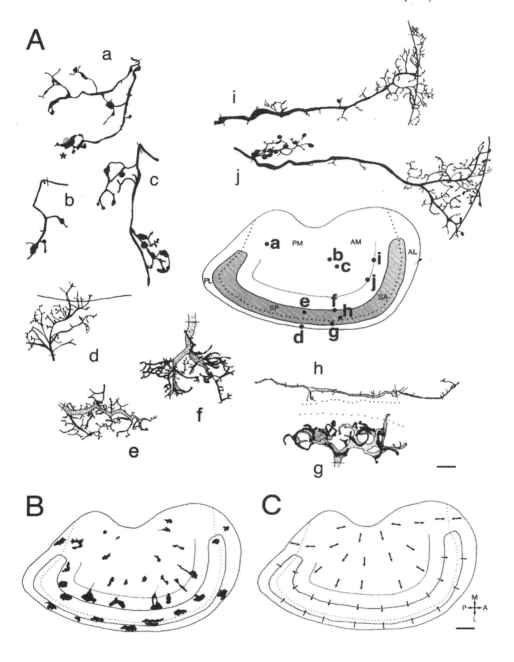

4.6.2. Discharge Properties

Most vestibular nerve afferents innervating frog otolith organs have a background discharge in a horizontal position of 0–30 spikes/s (Blanks and Precht 1976; Caston et al. 1977; Baird and Lewis 1986). Discharge can be regular or irregular (Baird and Lewis 1986; Sugai et al. 1991). In their response dynamics, tilt-sensitive units can be classified as tonic, phasic-tonic, or phasic (Fig. 3.11E–G) (Blanks and Precht 1976; Caston et al. 1977; Baird and Lewis 1986). There may be a relation between response dynamics and discharge regularity (Baird and Lewis 1986). Tonic afferents have a relatively regular discharge, whereas phasic-tonic units are irregular. Remarkably, some phasic afferents have a regular discharge. Although three-dimensional polarization vectors have not been determined, most tilt-sensitive neurons show oppositely directed responses when the animal is tilted in opposite directions from a horizontal position (Blanks and Precht 1976; Caston et al. 1977; Baird and Lewis 1986). In phasic-tonic units, both phasic and tonic response components are affected in the same way by any particular tilt maneuver.

4.6.3. Relation Between Afferent Morphology and Discharge Properties

Baird and Lewis (1986) labeled physiologically characterized afferents innervating the utricular or lagenar maculae. Units were characterized as tilt-sensitive, vibration-sensitive, or both. Tilt-sensitive units were further characterized by the pitch direction (nose up or down) leading to excitation and by the dynamics of the tilt responses (tonic, phasic-tonic, and phasic). Directional selectivity was consistent with the polarization maps for the two organs. For example, units innervating the lateral posterior or medial anterior sectors of the utricular macula were excited by nose-down tilts, whereas those supplying medial posterior or lateral anterior sectors were excited by nose-up tilts (see Fig. 3.25G). As expected, units in the middle of the utricular macula were more sensitive to rolls than to pitches.

None of the utricular afferents were vibration-sensitive, but this may reflect a lack of labeled afferents in the middle of the striola. Tonic units were located in the extrastriola of both the utricular and lagenar maculae. Much of the extrastriola, particularly in the medial part of the utricular macula, was not represented in the intraaxonal sample. Based on trends present in the data, units located in the medial extrastriola are likely to be tonic. Extracellular labeling studies indicate that these extrastriolar afferents can be quite thin (Baird and Schuff 1994), suggesting that tonic afferents may also be underrepresented in extracellular studies reporting a dearth of such units (Blanks and Precht 1976). Phasic-tonic and phasic units are located within striolar regions of both the lagenar and utricular maculae. Some lagenar striolar units showed vibration sensitivity either alone or in combination with tilt sensitivity (Baird and Schuff 1994).

The physiology of afferents can be correlated with hair bundle morphology (Baird and Schuff 1994): tonic and phasic-tonic afferents innervate hair cells with type B and type C bundles, respectively. Phasic afferents innervate hair cells with type F and type C bundle morphologies. Vibration sensitivity of lagenar afferents is associated with bulbed kinocilia (type E). Hair bundles with bulbed kinocilia are also characteristic of the saccular macula and the two auditory organs in anurans (the amphibian and basilar papillae) (Lewis and Li 1975). In addition to location and bundle morphology, there was some suggestion of a correlation between response dynamics and fiber diameter, with tonic units being thin, phasic-tonic units being thick, and phasic units being medium-sized or large (Baird and Schuff 1994). Lagenar vibratory units, including those with tilt sensitivity, were small to medium-sized. Remarkably, there was no correlation between an afferent's physiology and its branching patterns and number of hair cells contacted.

4.6.4. Efferent Responses of Inferred Afferent Classes and Locations

As is the case for the cristae, all otoconial organs are affected by repetitive electrical stimulation of efferents (Sugai et al. 1991). Responses are evident in irregularly discharging, but not in regularly discharging, fibers. Irregular fibers innervating the lagenar or utricular maculae can be excited or inhibited by efferent stimulation. Consistent with the role of the sacculus as a vibratory organ, its afferents are invariably inhibited. No attempt has been made to label afferents characterized by their efferent responses. Nevertheless, the irregular units showing large efferent responses presumably innervate the utricular and lagenar striolae.

Efferents presumably branch to innervate more than one organ, so the efferents going to an organ can be electrically stimulated by an axon reflex arising in the branch going to another organ (Prigioni et al. 1983). Sugai et al. (1991) found that stimulation of the saccular branch could either excite or inhibit afferents innervating other organs. Saccular efferents inhibit all saccular afferents yet can excite afferents arising elsewhere. This implies that efferent actions are not specific to the presynaptic efferent fiber but rather reflect postsynaptic receptor mechanisms.

4.7. Toadfish Horizontal Crista

4.7.1. Structural Considerations

There is a relatively flat neuroepithelium at the apex of the toadfish (*O. tau*) horizontal crista; the slopes of the crista lack hair cells (Boyle and Highstein 1990a). The expanded distal or planum semilunatum end of the neuroepithelium is firmly attached to the ampullary wall, whereas the proximal end is freestanding. The length of the neuroepithelium is approximately 1 mm, and its width varies from 130 μm at the isthmus (Fig. 3.27C,E)

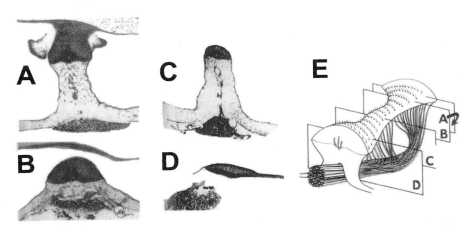

FIGURE 3.27. Toadfish (*O. tau*) horizontal crista ampullaris. The neuroepithelium is the darkly stained region on top of the crista (**A–D**). A and D are near the distal and proximal ends of the crista, respectively. C is at the isthmus in the center of the hemicrista. E shows a three-dimensional representation of the crista ampullaris with the levels of sections in A–D indicated with the horizontal ampullary nerve underneath. (Modified with permission from Boyle et al. 1991, Copyright 1991, The American Physiological Society.)

to about 400 μm at the planum (Fig. 3.27A,D–E). About 350 nerve fibers make up the horizontal crista nerve and range in diameter from 1 to 12 μm. Larger fibers (>7 μm) constitute about 20% of the population and are surrounded by smaller fibers (<4 μm) as the nerve courses underneath the crista. Individual fibers turn sharply to run to the neuroepithelium. In doing so, the fibers enter a channel in the middle of the acellular, cartilaginous stroma. Afferent endings are considered to be of three forms: club, bouton, and cup-shaped (Boyle and Highstein 1990a).

The toadfish (*O. tau*) cupula consists of a multipartite central "pillar" situated over the isthmus region, with more peripherally located "wings" extending to the ends of the neuroepithelium (Silver et al. 1998). Hair bundles insert into a fibrous connective tissue matrix embedded in an isotropic, mucopolysaccharide gel consisting of N-acetylneuroaminic acid and β-N-acetyl glucosaminic acid. A so-called "subcupular space," observed in earlier studies on fixed tissue, was not seen in vital preparations and was thus considered to be a fixation artifact. The absence of a subcupular space, together with the relatively long stereocilia, suggests that there is a tight coupling between the hair bundles and the cupula.

4.7.2. Discharge Properties

Based on their resting discharge characteristics and their responses to sinusoidal head rotations, afferents innervating the toadfish (*O. tau*) horizontal crista have been divided into three possibly overlapping groups (Boyle and Highstein 1990a). Low-gain afferents have a relatively high (40–100 spikes/s) and regular resting discharge ($cv < 0.1$) (Fig. 3.28A). Some high-gain and acceleration afferents also have a high and regular resting discharge, but in other such units the background discharge may be lower in rate and/or characterized by a higher cv. Low-gain, high-gain, and acceleration afferents can also be distinguished by their phase leads with regard to angular head velocity (Fig. 3.28B). Low-gain units have gains of <1 spike·s^{-1}/deg·s^{-1} (Fig. 3.28C). Gains are similar for high-gain and acceleration afferents, 1–10 spikes·s^{-1}/deg·s^{-1}. Figure 3.29 summarizes response dynamics for the three groups in terms of gains and phases versus sinusoidal frequency. Response dynamics for low-gain units conform to a torsion pendulum with a time constant of 15–20 s, three to six times longer than seen in other animals (Fig. 3.29A). As compared with low-gain units, acceleration units have much larger phases and much larger gain enhancements (Fig. 3.29C); high-gain afferents are intermediate in both respects (Fig. 3.29B). Because of the presence of high-gain and acceleration afferents with a regular spacing of action potentials, discharge regularity has not proved a reliable marker of other discharge properties in the toadfish (*O. tau*).

Diversity has also been described in afferents innervating the horizontal semicircular canal of the goldfish (*C. auratus*); in the frequency range from 0.1 to 1 Hz, phase can vary almost 90° and gain almost 20-fold (Hartmann and Klinke 1980).

4.7.3. Relation Between Afferent Morphology and Physiology

Several impaled afferents in the toadfish (*O. tau*) were first characterized by their responses to sinusoidal head rotations and efferent activation and then labeled (Boyle et al. 1991). The location and extent of terminal arbors are indicated in Figure 3.30. Low-gain afferents are located near the distal end of the crista, adjacent to the planum semilunatum. Terminal fields are relatively small. High-gain and acceleration afferents are more centrally located and can branch more extensively within the crista. As illustrated in Figure 3.31, there is a correlation between the longitudinal position of a unit and the gain and phase of its response to 1 Hz sinusoidal rotations. On the other hand, there was only a weak correlation between axon diameter and physiological properties, which may reflect that none of the labeled fibers had an internal diameter less than 3 µm. For reasons that are not entirely clear, many more fibers were labeled in the distal, as compared with the proximal, half of the neuroepithelium. In particular, no low-gain afferents were labeled at the proximal end of the organ. Finally, a few very thick afferents were found to branch extensively within the crista yet were silent at rest and were difficult to stimulate either with rotations or with efferent activation.

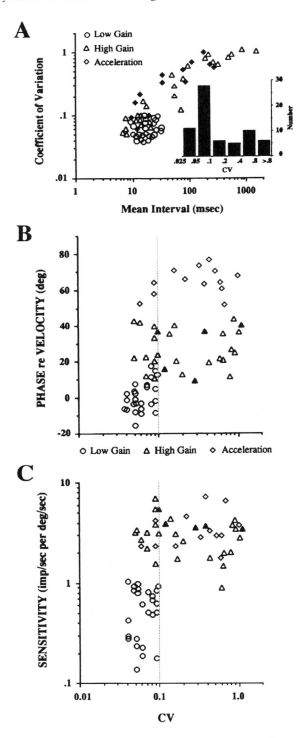

4.7.4. Efferent Responses of Labeled Afferents

Responses to electrical stimulation of efferent fibers have been studied in the toadfish (*O. tau*) (Boyle and Highstein 1990b). Similar results were obtained when efferent fibers were activated by behavioral arousal (Highstein and Baker 1985; Boyle and Highstein 1990b). Efferent activation results in an excitation of almost all afferents, including an increase in backround firing and a decrease in afferent gain. Effects are larger in high-gain (Fig. 3.32C,D) and acceleration units (Fig. 3.32E,F) than in low-gain units (Fig. 3.32A,B). In many units, there is a slow response following the fast, perstimulus response.

These results have been confirmed in labeled afferents (Boyle et al. 1991). The different locations of the three afferent groups imply that efferent effects are targeted for the center of the crista. It is unclear whether the efferent responses are the result of efferent synapses on hair cells or on afferent processes because both kinds of contacts have been observed (Sans and Highstein 1984).

5. Synthesis

5.1. The Cristae: A Historical Narrative

"When a biologist tries to answer a question about a unique occurrence such as "Why are there no hummingbirds in the Old World?" or "Where did the species *Homo sapiens* originate?" he cannot rely on universal laws. The biologist has to study all the known facts relating to the particular problem, infer all sorts of consequences from the reconstructed constellations of factors, and then attempt to construct a scenario that would explain the observed facts of this particular case. In other words, he constructs a historical narrative." Ernst Mayr, *This is Biology. The Science of the Living World* (Mayr 1997).

Morphophysiological studies have now been done in four widely separated species of vertebrates. In this section, we attempt to construct a narrative summarizing the similarities and differences across species in the relation

---

FIGURE 3.28. Discharge properties of horizontal semicircular canal afferents in the toadfish (*O. tau*). (**A**) Coefficient of variation versus mean interval for three groups of afferents distinguished by their rotational response dynamics. Note that all three groups can have a high resting rate and a regular discharge. Units with lower rates and a more irregular discharge have relatively high gains and relatively phasic response dynamics. (**B, C**) Phase and gain responses, respectively, to 0.3 Hz sinusoidal head rotations versus coefficient of variation. Regular units can have low or high gains and small, intermediate, or large phase with regard to head velocity. (Modified with permission from Boyle and Highstein 1990a, Copyright 1990, The Society for Neuroscience.)

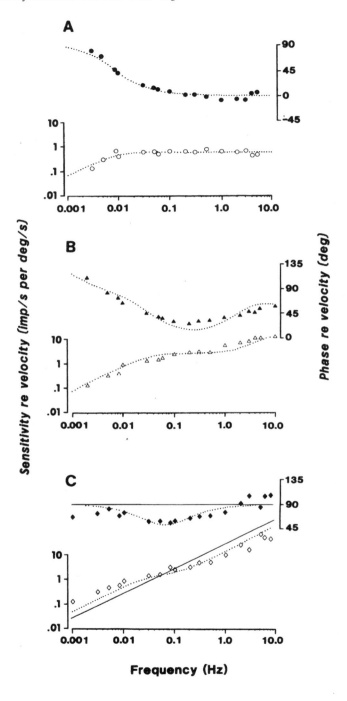

between structure and function in the vertebrate crista. We exclude the maculae from consideration, in part because morphophysiological studies have been done in only two species, but also because in some nonmammalian species, one or another macula has become transformed to monitor substrate-borne or media-borne vibrations. In contrast, the cristae, it would seem, have had only one function: to monitor head rotations. At first glance, this would seem a simple task. So, if we were only interested in statements of general function, there would be no need for a historical narrative. We could simply state that monitoring rotational movements of the head is an important biological function, which was successfully solved near the dawn of vertebrate evolution and has not changed since. But morphophysiological studies indicate that in any particular species there is a diversity of afferents, differing in their locations within the cristae and in their discharge properties. Some of this diversity would seem to share common themes across widely divergent species. At the same time, there are marked differences, say, between fish and mammals.

The purpose of a scenario is to account for both similarities and differences. Yet, we should recognize that any scenario must, of necessity, remain incomplete. There are at least three reasons for this. First, our information comes from only a few species. Second, because we are dealing with soft tissues, information from extant species cannot be supplemented by the fossil record. Hence, the narrative must remain conjectural: one has to use judgment in deciding whether a similarity between species represents inheritance from a common ancestor or is an example of parallel evolution. Third, changes in the periphery are likely to be adaptive responses to the needs of central processing. Yet, we know very little about the relation of different afferent contingents and signal processing in the brain (Goldberg 2000).

Let us first emphasize the similarities across vertebrates. The neuro-epithelium is pseudostratified in all vertebrates, and the ultrastructure of the hair cells and supporting cells is similar in many respects. Hair cells have in common a morphologically polarized hair bundle. Remarkably, the directions of the polarization vectors and the corresponding excitatory cupular deflections remain the same in all gnathostomes: utriculopetal for the horizontal canal and utriculofugal for the vertical canals. Hair bundles are

---

FIGURE 3.29. Bode plots for horizontal canal afferents in the toadfish (*O. tau*), including mean gains with regard to head velocity (open symbols) and mean phases with regard to head velocity (solid symbols) versus sinusoidal frequency for low-gain (**A**), high-gain (**B**), and acceleration afferents (**C**). Dotted lines are empirical transfer functions. Solid lines in C indicate gain and phase of a pure accelerometer. (Modified with permission from Boyle and Highstein 1990a, Copyright 1990, The Society for Neuroscience.)

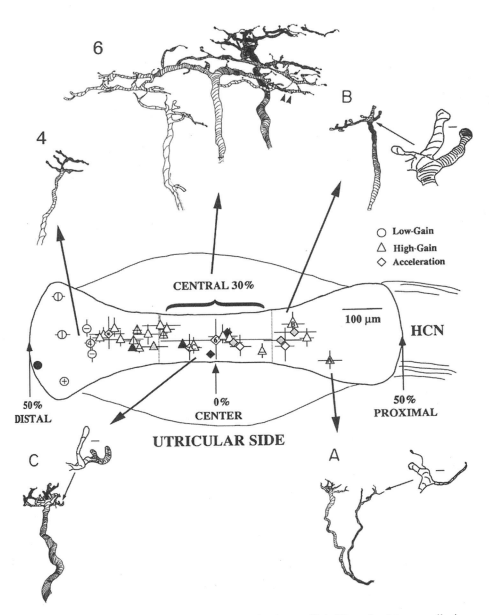

FIGURE 3.30. Afferent innervation patterns in the toadfish (*O. tau*) crista ampullaris. Low-gain afferents (circles) tend to be in the planum region (peripheral 40%), whereas acceleration afferents (diamonds) tend to be in the central 50% of the sensory epithelium. High-gain afferents (triangles) are found within the central 60% of the sensory epithelium. The locations of selected labeled afferents (4, 6, A, B, C) are indicated. Crosses behind each symbol indicate the transverse and longitudinal extents of the dendritic tree for each labeled afferent. Afferent **4** is a low-gain afferent with a thin axon and a limited dendritic tree. Afferent **6** is an acceleration afferent with a thick parent axon that branches below the sensory epithelium into three large-diameter branches, each with a broad dendritic tree. Afferents **A–C** are high-gain afferents. (Modified with permission from Boyle et al. 1991, Copyright 1991, The American Physiological Society.)

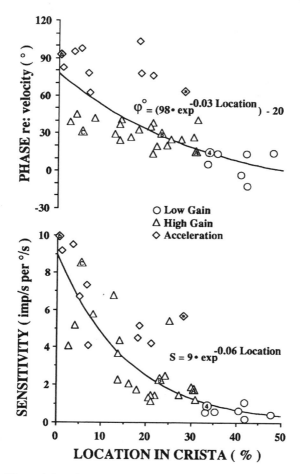

$$\varphi^{\circ} = (98 \cdot \exp^{-0.03\, \text{Location}}) - 20$$

$$S = 9 \cdot \exp^{-0.06\, \text{Location}}$$

○ Low Gain
△ High Gain
◇ Acceleration

FIGURE 3.31. Phase (above) and gain (below) versus longitudinal location in the crista for labeled horizontal canal afferents in the toadfish (*O. tau*). Location of 0% is the center of the crista and 50% is the planar (extreme) edge. Labeled afferents (4, 6, A, B, C) from Figure 3.30 are indicated by alphanumerics within symbols. Data are fit by empirical equations as indicated. (Reproduced with permission from Boyle et al. 1991, Copyright 1991, The American Physiological Society.)

embedded in a gelatinous cupula, which has a similar structure in all vertebrates studied. At the basal pole of the hair cell are found ribbon synapses, which are qualitatively similar even if there are enormous differences in the sizes and shapes of their central, electron-opaque cores and in the number of vesicles surrounding the cores. Efferent fibers are found in all vertebrates and make morphologically similar synapses with hair cells and, in some species, with afferent processes. We have reviewed the evidence that effer-

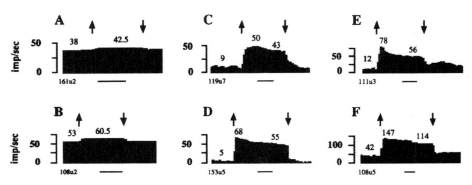

FIGURE 3.32. Responses of horizontal canal afferents in the toadfish (*O. tau*) to electrical stimulation of central afferent pathways. Arrows indicate start (upward) and end (downward) of efferent shock trains; bars are 10 s. Each panel indicates a separate afferent. (**A, B**) Low-gain afferents. (**C, D**) High-gain afferents. (**E, F**) Acceleration afferents. Responses are considerably smaller for low-gain afferents than for high-gain and acceleration afferents. For the latter two groups, per-stimulus responses are followed by post-stimulus responses. (Reproduced by permission from Boyle and Highstein 1990b, Copyright 1990, The Society for Neuroscience.)

ent neurons are a variety of branchiomotor neurons in all vertebrates, yet it might be supposed that the signals carried by efferents and their regulatory roles would differ between fish and mammals. The reason for the suspicion is that vestibular, auditory, and lateral lines receive their efferent innervation from a common set of neurons in fish, whereas the efferents to auditory and vestibular organs are entirely separate in mammals (Goldberg et al. 2000).

We now consider afferent diversity. When variations in afferent discharge properties were first described in mammals, it was suggested that these were related to the distinctive functions of type I and type II hair cells (Goldberg and Fernández 1971b). The suggestion became less attractive when similar differences were found in fish having only type II hair cells (O'Leary et al. 1974; O'Leary and Dunn 1976). Morphophysiological studies have clarified the relation between the organization of the cristae and the physiological diversity of afferents in different species. In both the toadfish (*O. tau*) (Boyle et al. 1991) and the frog (Honrubia et al. 1989; Myers and Lewis 1990, 1991), there may be a single longitudinal gradient in afferent physiology running from the neuroepithelial zone bordering the planum to the isthmus of the organ. Units near the planum are regularly discharging and have low gains and low phases with regard to angular head velocity. As the isthmus is approached, gains and phases become larger. In the frog, as in the other vertebrates studied, there is a correlation between discharge regularity and these other properties; units located near the isthmus have a more irregular discharge than those near the planum. In the toadfish (*O. tau*), discharge regularity is not well-correlated with gain, phase, or longitudinal position. In both frogs and toadfish (*O. tau*), there is a longitudinal gradient in terminal morphology as well as in afferent physiology. Units located near the planum have circumscribed fields. Nearer the longitudinal center of the organ, fields become larger. In the frog, central fields (zones II and III) can be large or small, but even the smallest fields are not as compact as those near the planum (zone I) (Myers and Lewis 1990, 1991).

With the advent of type I cells in reptiles, a new organization emerges (Jørgensen 1974, 1975; Brichta and Peterson 1994). A similar organization is seen in birds (Masetto and Correia 1997; Kevetter et al. 2000; Lysakowski and Dickman, unpublished observations). There is now a central zone consisting of type I and some type II hair cells, surrounded by a peripheral zone only containing type II cells. We can view the hemicrista of reptiles and birds as having a peripheral zone with a longitudinal organization on which is superimposed a central zone.

Morphophysiological studies in the turtle (Brichta and Goldberg 2000a) are consistent with this conclusion. As in the frog, regularly discharging bouton afferents with small gains and small phases are found near the planum, whereas irregularly discharging bouton units near the torus have large gains and large phases. Bouton units within the central zone or in the peripheral zone at the same longitudinal position have intermediate properties. There are corresponding longitudinal gradients in the morphology of bouton terminal fields (Brichta and Peterson 1994). Calyx-bearing units, all of which are located in the central zone, are as irregular as bouton units located in the peripheral zone near the torus but have lower gains and phases (Brichta and Goldberg 2000a).

Three differences are found between the mammalian crista, on the one hand, and the reptilian and bird cristae, on the other. First, type I hair cells, rather than being confined to the central zone, are found in all parts of the crista and are intermixed with type II hair cells. This is so in a variety of mammals, possibly including the echidna (*Tachyglossus aculeatus*) (Jørgensen and Locket 1995). The presence of this feature in the echidna (*T. aculeatus*) would be remarkable because in many respects, including the presence of an uncoiled basilar papilla and the presence of a lagena, the inner ear of this species resembles that of reptiles. Second, based on morphophysiological studies in the chinchilla (Baird et al. 1988), the units with the highest gains in mammals are centrally located dimorphic afferents. These may be compared to units with high gains and high phases in fish, frogs, and turtles. The latter are bouton units located near the longitudinal center of the crista. When compared with high-gain, high-phase bouton units in other species, all mammalian units have gains that are at least 5–20 times lower and phases that are at least 20–40° lower. Some version of high-gain, high-phase units is found in all nonmammalian species studied, so their absence in mammals represents a potentially important difference. Third, the mammalian crista, unlike those of other vertebrates, has a concentric organization. By this is meant that there are similar morphological (Lindeman 1969; Fernández et al. 1988, 1995; Lysakowski and Goldberg 1997) and physiological gradients (Baird et al. 1988) as one proceeds from the geometric center of the organ longitudinally toward the planum, transversely toward the base of the crista, or in any intermediate direction. Strong evidence for the conclusion comes from a consideration of dimorphic and bouton units in the mammalian peripheral zone. Gains and phases are similar and discharge is regular throughout the concentrically shaped peripheral zone, including the region bordering the planum and that at the base of the crista near the center of the organ. The concentric organization may be related, with the neuroepithelium in mammals occupying almost the entire outside of the saddle-shaped crista, whereas in nonmammalian vertebrates, the neuroepithelium runs only a short distance down the crista's slopes.

There is a diversity in afferent physiology in all species studied. The similarity in the constellation of discharge properties defining different afferent groups is striking. Table 3.1, which is based on mammals, provides an approximate description of the afferent physiology in most vertebrates. Afferents with a regular discharge have low gains and phases with regard to angular head velocity and relatively small efferent responses. Irregularly discharging afferents have large gains and phases with regard to velocity and large efferent responses. Even in the toadfish (*O. tau*), the one species where discharge regularity has not proved a good predictor of other discharge properties, units with the lowest gains and phases are regularly discharging. If our scenario has merit, there have been two important events in the transformation from the fish or frog neuroepithelium to that of mammals. The first was the appearance of type I hair cells in reptiles (Jørgensen 1974, 1975; Brichta and Peterson 1994) and birds (Rosenhall 1970; Wersäll and Bagger-Sjöbäck 1974; Masetto and Correia 1997) and the second was the spread of type I hair cells into all regions of the neuroepithelium (Lindeman 1969; Wersäll and Bagger-Sjöbäck 1974; Fernández et al. 1995). Afferent diversity, as such, would appear to be a common feature of the vertebrate crista. Moreover, discharge regularity is but one of several potential makers of this diversity. We shall return to this subject in Section 5.3.1 below.

One advantage of a scenario is to highlight those facts that do not fit its outline. In this spirit, we consider three facts that may not support the proposed scheme. First, it will be noted that we have not considered efferent responses in the context of the scenario. This is because there is no obvious pattern that holds across species. In the toadfish (*O. tau*) (Boyle and Highstein 1990b; Boyle et al. 1991) and in mammals (Goldberg and Fernández 1980; McCue and Guinan 1994; Marlinsky et al. 2000), electrical stimulation of efferents invariably results in excitation. Both excitation and inhibition are seen in anurans (Rossi et al. 1980; Bernard et al. 1985; Rossi and Martini 1991; Sugai et al. 1991) and in the turtle (Brichta and Goldberg 2000b). Of the latter two animals, a morphological basis and a relation to afferent discharge properties are only available for the turtle. Second, according to the scenario, the crista should have a single longitudinal organization in fish and frogs. But there are difficulties with this proposition, which can be exemplified by considering the frog crista. One difficulty concerns the presence of both large and small terminal fields in zones II and III, which would suggest an organization more complicated than that of a single longitudinal gradient (Myers and Lewis 1990, 1991). Another difficulty concerns the cell types found in various regions. The middle region (zone II) interconnecting the isthmus and planar regions has a large concentration of short, so-called pear-shaped cells not found in the isthmus region and only infrequently found in the planar region (Gioglio et al. 1995). Furthermore, the pear-shaped cells of zone II have a distinctive set of ionic currents (Prigioni et al. 1996). The question arises: Could the intermediate zone in the frog form an incipient central zone? Third, in reptiles, birds, and some mammals, the vertical, but not the horizontal, cristae are partly or completely divided by a nonsensory structure that extends both toward and away from the utriculus. Among mammals, this feature, whose function is unknown, is present in some bats, the cat, mouse, and rat, but not in the guinea pig, chinchilla, monkey or human (Lewis et al. 1985). The structure is referred to as an *eminentia cruciatum* or *cruciate septum* in most animals in which it is present but has been called a *torus* in reptiles. The eminentia divides each vertical crista into two hemicristae in reptiles and birds, which does not alter our narrative. In mammals, there is a potential complication that depends on the parts of the neuroepithelium bordering the nonsensory region. If the borders resemble the central zone, then the *eminentia* would merely bisect a single concentric organization. Were the bordering regions more like the peripheral zone, then the vertical cristae should be viewed as two hemicristae, similar to those in reptiles and birds. Even then, the mammalian cristae would be distinguishable because of the widespread distribution of type I hair cells and the lack of units with very high gains and phases. Unfortunately, little is known about the bordering region. The fact that efferent fibers that are otherwise restricted to the peripheral zone travel along the bordering region would seem to favor the first alternative (Purcell and Perachio 1997). At the same time, calretinin-positive endings in the rat and gerbil, which are a marker for calyx fibers in the central zone in other species, abut the eminentia, suggesting the second alternative (A. Lysakowski, unpublished observations). Fourth, in the toadfish (*O. tau*) and in frogs, thin fibers innervate cylindrical cells at the edges of the neuroepithelium (Lysakowski 2001). Although physiological evidence is lacking, the edges may constitute a peripheral zone continuous with the zone bordering the planum, in which case a concentric organization would be a common vertebrate feature.

5.2. Morphophysiological Studies: A Correlative Approach

"A statement (a theory, a conjecture) has the status of belonging to the empirical sciences if and only if it is falsifiable." Karl R. Popper, *Realism and the Aim of Science* (Popper 1993)

"I shall be content simply to exclude [the 'principle of causality'] as 'metaphysical' from the sphere of science. I shall, however, propose a simple rule that we are not to abandon the search for universal laws and for a coherent theoretical system, nor ever give up our attempts to explain causally any kind of event we can describe." Karl R. Popper, *The Logic of Scientific Discovery* (Popper 1968)

Morphophysiological studies correlate afferent physiology with terminal morphology. Correlative strategies are weak paradigms when it comes to defining causal relations. We can cite two examples. First, there is a strong correlation between discharge regularity and sinusoidal response phase. Yet, we know that the two variables are not causally related. Discharge regularity largely reflects the ionic currents in the postsynaptic spike encoder responsible for postspike afterhyperpolarizations (Smith and Goldberg 1986), whereas response dynamics may largely reflect earlier stages of transduction (Baird 1994; Highstein et al. 1996). As a second example, we can consider a suggestion of Boyle et al. (1991). These authors have correlated the transverse dendritic spread of an afferent with its phase. Specifically, units with the largest phases have the largest transverse spreads. Yet, no quantitative models have been developed to explain the proposed relation, and one suspects once again that there is no causal relation. Two lines of evidence support this suspicion. In the turtle, there was no correlation between terminal spread and response phase (Brichta and Goldberg 2000a). In mammals, including the cristae and utricular macula, units with the largest spread are found in the peripheral (extrastriolar) zone (Fernández et al. 1988, 1990, 1995) and have the smallest phases (Baird et al. 1988; Goldberg et al. 1990b). What these examples exemplify is that a correlation between morphology and physiology can provide hints about causality, but some of the hints may be misleading.

At the same time, the lack of a correlation can be used to eliminate certain morphological variables as potential contributors to afferent physiology. Work in the turtle posterior crista provides an example. There are three differences between bouton afferents supplying the peripheral zone near the torus and near the planum: an almost 100-fold variation in gain, a nearly 45° difference in phase, and a large difference in discharge regularity (Brichta and Goldberg 2000a). Yet, a multiple correlation indicates that the only morphological variable that correlates with the three physiological variables is the longitudinal position of the terminal field within the hemicrista. The other morphological variables, which were statistically unrelated to the physiological variables, included axon diameter, number of boutons, number of terminal branches, and terminal-field diameter.

Location, as such, cannot determine physiological properties. Presumably, it is one or more cellular properties that vary with location that is important. This is an important conclusion because we can begin to relate the properties of hair cells, afferents, and efferents with their location. In one attempt to do this, basolateral currents were characterized in turtle posterior crista hair cells selectively harvested from the peripheral zone near the torus and near the planum. It was found that the differences in hair cell currents were too small to account for the large differences in gain and phase of the afferents (Brichta et al. 2002; Goldberg and Brichta 2002). Nevertheless, a strategy of relating cellular properties with location should prove helpful in identifying mechanisms important in determining afferent physiology.

We have referred to morphophysiology as a weak strategy because it is correlative and, hence, can be misleading in drawing causal inferences. One way to challenge the strategy is to adopt a comparative approach. This is so because it is unlikely that diverse species would show the same covariance among a multitude of morphological and physiological variables. We have already used the strategy in discussing the possible role of dendritic spread in determining response phase. As a second example, we take advantage of the difference in functional organization of the cristae in fish and anurans, where the organization is predominantly longitudinal, and in mammals, where the organization is concentric. It has been speculated that differences in afferent response dynamics may be related to longitudinal gradients in cupular mechanics (Boyle et al. 1991). Although the suggestion remains reasonable for species with longitudinally organized cristae, it cannot hold in mammals where peripheral zone afferents have almost identical physiological properties whether they are located near the planum or near the longitudinal center of the crista. Once again, morphophysiology can be used to disprove a proposition, in this case that the longitudinal location of an afferent is an important determinant of its physiology, at least in mammals. But it cannot be used to prove, for example, that regional variations in cupular mechanics are important.

5.3. Morphophysiology and Sensory Transduction

As summarized in the preceding section, there is a pattern to the diversity of afferent discharge properties that is common to most vertebrates. In this section, we review possible transduction mechanisms underlying this diversity. Our goal is to see whether our understanding of the cellular basis of vestibular transduction is consistent with the morphophysiological results summarized above.

5.3.1. Discharge Regularity

Discharge regularity has proved to be an important marker helping to distinguish different kinds of afferents in individual species. As has been observed directly (Highstein and Politoff 1978; Schessel et al. 1991; Xue et al. 2000), regularly discharging afferents have a more prominent after-hyperpolarization (AHP) than do irregular fibers. The conclusion has been verified by electrical stimulation of afferents (Goldberg et al. 1984) and by a stochastic model of repetitive discharge (Smith and Goldberg 1986). Because the AHP of an afferent reflects the variety of ionic currents in its postsynaptic terminal, there is no reason to suppose that this property would have a morphological correlate. In their modeling study, Smith and Goldberg (1986) also had to assume that mEPSPs (synaptic quanta) arising from the hair cell also differed between regular and irregular afferents. In the theory, quantal size was much less influential in determining discharge regularity than the size and time course of AHPs. Nevertheless, quantal size was of interest because it might have a presynaptic morphological correlate. Synapses located relatively far away from the spike generator in fibers with extensive terminals might be expected to have smaller effective quanta. Some support for this idea comes from mammals because regular fibers have more extensive trees than irregular fibers (Baird et al. 1988; Goldberg et al. 1990b). But results in fish (Boyle et al. 1991) and frogs (Myers and Lewis 1990, 1991) show a contrary trend. In particular, regularly discharging fibers near the planum in the latter animals have very circumscribed terminals; units with the most extensive terminals are located near the longitudinal center of the crista and are irregularly discharging. The results in nonmammalian species emphasize the importance of AHPs, rather than quantal size, in determining discharge regularity.

5.3.2. Response Dynamics

Given the general validity of the torsion-pendulum model for semicircular canals (Rabbitt et al., Chapter 4), one might expect that the time constants of the model would reflect the geometry of the canal ducts and the cupula, as well as the physical properties of the cupula and endolymph. In a similar way, the macromechanics of the maculae might be expected to be reflected in such variables as the height and physical properties of the otolithic membrane. The stiffness of the hair bundles inserting into the accessory structure might also contribute to macromechanics. Presumably, these factors would influence the response dynamics of all hair cells and afferents in a similar manner. What about differences in response dynamics? Virtually every stage in the transduction chain could be involved. Baird's (1992) comparison of receptor potentials evoked by hair bundle displacement and by intracellular currents indicates that the high-frequency response dynamics of frog utricular cells may be determined by the adaptation of the mechanoelectric transduction (MET) channel. Evidence in mammals is contradictory. In one study in neonatal mice (Géléoc et al. 1997), MET currents of hair cells likely to be located in the extrastriola of the utricular macula did not show adaptation to step displacements of the hair bundle, whereas, in another study, adaptation was conspicuous (Holt et al. 1997). The first study would seem more congruent with studies in adult animals (Goldberg et al. 1990b), which show that extrastriolar afferents show little adaptation. Other evidence that differences in high-frequency response dynamics arise early in transduction comes from studies of the afferent responses to galvanic currents (Goldberg et al. 1984; Highstein et al. 1987) and of voltage-gated ionic currents recorded in turtle type II hair cells from different regions of the peripheral zone of the posterior crista (Brichta et al. 2002). A possible exception to the conclusion was found in type I and in some central type II hair cells from the turtle crista. In these cells, voltage-gated outward currents may contribute to the gain enhancements and phase leads of the corresponding afferents.

In contrast to differences in high-frequency dynamics, low-frequency adaptation may, at least in part, reflect processes taking place in the afferent terminal (Taglietti et al. 1977; Goldberg et al. 1982; Ezure et al. 1983).

From the perspective of the present chapter, none of these processes would have a morphological correlate in intraaxonally labeled fibers examined by light microscopy.

5.3.3. Response Gain

Any stage of the transduction chain up to and including spike encoding can contribute to the gain of an afferent. To keep the discussion manageable, we will concentrate on the influence of discharge regularity and response dynamics, as well as of hair cell type and number of afferent endings. Based on both experiment (Goldberg et al. 1984) and theory (Smith and Goldberg 1986), discharge regularity is related to the gain of the postsynaptic spike encoder, where gain may be expressed in units of spikes·s^{-1} mV, with the denominator expressing depolarization from any source (e.g., afferent or efferent synaptic input or galvanic currents).

Response dynamics also contribute to response gain in the following way. Vestibular afferents behave as minimal-phase systems (Bode 1945) in which a phase lead (in degrees) of $\phi = (90°/k)$ is associated with a relative gain enhancement equalling the kth power of frequency (f^k). Gain enhancement is easy to evaluate when it is confined to a frequency band with a definite lower limit, f_0, such as is the case for canal afferents in mammals (Fernández and Goldberg 1971; Baird et al. 1988). In that case, the absolute gain enhancement is $(f/f_0)^k$, but in those cases where phasic response dynamics is seen over a wide, ill-defined range of frequencies, the value of f_0 is difficult, if not impossible, to determine. Such a difficulty was encountered in attempting to analyze the gains of mammalian otolith (Goldberg et al. 1990b) and turtle posterior crista afferents (Brichta and Goldberg 2000a).

Hair cell types can help to determine gain. This is especially clear in comparing type I and type II cells in birds (Correia and Lang 1990), reptiles (Brichta et al. 2002; Goldberg and Brichta 2002), and mammals (Rennie and Correia 1994; Rüsch and Eatock 1996). Type I cells have a large current that lowers the input impedance of type I, as compared to type II, hair cells. Input impedance, the ratio of output voltage to input current, may be equated with hair cell gain because the input to the hair cell is the MET current and the output is the membrane voltage controlling the activation of voltage-gated calcium channels, which in turn control neurotransmitter release (Rossi et al. 1989, 1994; Martini et al. 2000). Type II cells have been characterized in the frog (Masetto et al. 1994), pigeon (Masetto and Correia 1997), and chick (Masetto et al. 2000). Cells differ in their shape, location, and ionic currents. The latter differences could also influence afferent gain.

It would be reasonable to suppose that afferent gain would be related to the number and types of afferent endings. Results have been equivocal. Myers and Lewis (1990, 1991) report that gains in the frog zones II and III may be higher in afferents with larger terminal fields and, presumably, more endings. In the case of the chinchilla crista, Baird et al. (1988) attempted by statistical means to eliminate the influence of discharge regularity and response dynamics. They further assumed that the residual gain was proportional to a linear combination of the number of calyx and bouton endings. Although it was possible to distinguish between calyx and dimorphic groups, there was little or no correlation between residual gain and number of endings when either group was considered separately. A similar conclusion was drawn in the turtle posterior crista, although here the two groups distinguished were calyx-bearing (calyx plus dimorphic) and bouton afferents. Finally, in the toadfish (*O. tau*), an attempt was made to predict conventional gains from several morphological variables, including (1) terminal location and (2) number and types of endings, both separately and in combination (Boyle et al. 1991). Predictions were no better when location and endings were considered together than when location was considered separately.

5.3.4. Afferent Responses to Electrical Stimulation of Efferents

In all animals studied, efferent responses are smallest in regularly discharging afferents with small afferent gains and small afferent phases. One might suppose that the efferent innervation might be relatively sparse in those regions where such afferents are located. Evidence is available only in mammals and refutes the suggestion (Lysakowski and Goldberg 1997; Purcell and Perachio 1997). One could speculate that regional differences in efferent responses are the result of differences in postsynaptic receptor mechanisms rather than the density of the efferent innervation. Another, not mutually exclusive, possibility is that efferents regulate metabotropic functions so that the size of a discharge rate variation may be a poor measure of efferent effectiveness.

5.4. Conclusions

Morphophysiological studies have now been done in four widely separated species. Based on these results, a narrative summarizing possible evolutionary changes in the functional organization of the cristae is presented. At the same time, the narrative has to remain speculative because so few species have been studied and certain structural features of the cristae of nonmammalian vertebrates need further evaluation. If anything, our understanding of the otolith organs is even less complete, being based on information from only two species, the frog and the chinchilla.

Because they involve a correlative approach, morphophysiological studies cannnot be used to show causal relations. Yet, the approach can be used to disprove hypotheses about the relation between structure and function. Despite the limitations of such studies, a common theme emerges. There is a diversity in the discharge properties of afferents innervating different regions of individual organs. In the final section of the chapter, we have summarized our understanding of the cellular mechanisms determining this afferent diversity. It seems clear that future studies along these lines would profit from morphophysiological studies by comparing transduction in hair cell and afferent populations located in various regions of an organ.

One subject not addressed in the study concerns the use made by central pathways of the diversity of afferents innervating individual organs. Although there have been several studies of this problem, a consensus has not emerged concerning the significance of this diversity in central processing. Various possibilities have been listed (Goldberg 2000). This remains a key problem in vestibular physiology. A second problem, which we have considered only tangentially, concerns potential functions of the efferent innervation (for reviews, see Highstein 1991; Goldberg et al. 2000).

Acknowledgments. Research in the authors' laboratories during the writing of this chapter was supported by the National Institute for Deafness and Communication Disorders and by NASA. The support of both institutes is gratefully acknowledged. We would also like to acknowledge the technical assistance of Mr. Steven Price and the contribution of trainees in our laboratories, Drs. J.C. Holt and Sapan Desai and Ms. Sohera Syeda, who commented on a previous version of the manuscript.

References

Ades HW, Engström H (1965) Form and innervation of the vestibular epithelia. In: Graybiel A (ed) First Symposium on the Role of Vestibular Organs in Space Exploration. NASA SP-77. Washington, DC: U.S. Government Printing Office, pp. 23–42.

Anastasio TJ, Correia MJ, Perachio AA (1985) Spontaneous and driven responses of semicircular canal primary afferents in the unanesthetized pigeon. J Neurophysiol 54:335–347.

Anderson JH, Blanks RHI, Precht W (1978) Response characteristics of semicircular canal and otolith systems in the cat. I. Dynamic responses of primary vestibular fibers. Exp Brain Res 32:491–507.

Art JJ, Fettiplace R, Fuchs PA (1984) Synaptic hyperpolarization and inhibition of turtle cochlear hair cells. J Physiol (Lond) 356:525–550.

Ashmore JF, Russell IJ (1982) Effect of efferent nerve stimulation on hair cells of the frog sacculus. J Physiol (Lond) 329:25P–26P.

Baird IL (1974) Anatomical features of the inner ear in submammalian vertebrates. In: Keidel WD, Neff WD (eds) Handbook of Sensory Physiology. Berlin: Springer-Verlag, pp. 159–212.

Baird RA (1992) Morphological and electrophysiological properties of hair cells in the bullfrog utriculus. Ann N Y Acad Sci 656:12–26.

Baird RA (1994) Comparative transduction mechanisms of hair cells in the bullfrog utriculus. II. Sensitivity and response dynamics to hair bundle displacement. J Neurophysiol 71:685–705.

Baird IL, Lowman GF (1978) A study of the structure of the papilla neglecta in the lizard, *Anolis carolinensis*. Anat Rec 191:69–90.

Baird RA, Lewis ER (1986) Correspondences between afferent innervation patterns and response dynamics in the bullfrog utricle and lagena. Brain Res 369:48–64.

Baird RA, Schuff NR (1994) Peripheral innervation patterns of vestibular nerve afferents in the bullfrog utriculus. J Comp Neurol 342:279–298.

Baird RA, Desmadryl G, Fernández C, Goldberg JM (1988) The vestibular nerve of the chinchilla. II. Relation between afferent response properties and peripheral innervation patterns in the semicircular canals. J Neurophysiol 60:182–203.

Baker CVH, Bronner-Fraser M (2001) Vertebrate cranial placodes. I. Embryonic induction. Dev Biol 232:1–61.

Bell CC (1981) Central distribution of octavolateral afferents and efferents in a teleost (*Mormyridae*). J Comp Neurol 195:391–414.

Bennett MVL (1972) A comparison of electrically and chemically mediated transmission. In: Pappas GD, Purpura DP (eds) Structure and Function of Synapses. New York: Raven Press, pp. 221–256.

Bernard C, Cochran SL, Precht W (1985) Presynaptic actions of cholinergic agents upon the hair cell-afferent fiber synapses in the vestibular labyrinth of the frog. Brain Res 338:225–236.

Blanks RH, Precht W (1976) Functional characterization of primary vestibular afferents in the frog. Exp Brain Res 25:369–390.

Blanks RH, Precht W (1978) Response properties of vestibular afferents in unanesthetized cats during optokinetic and vestibular stimulation. Neurosci Lett 10:225–229.

Blanks RH, Curthoys IS, Markham CH (1972) Planar relationships of semicircular canals in the cat. Am J Physiol 223:55–62.

Blanks RH, Curthoys IS, Bennett ML, Markham CH (1985) Planar relationships of the semicircular canals in rhesus and squirrel monkeys. Brain Res 340:315–324.

Bleckmann H, Niemann U, Fritzsch B (1991) Peripheral and central aspects of the acoustic and lateral line system of a bottom dwelling catfish, *Ancistrus* sp. J Comp Neurol 314:452–466.

Bode HW (1945) Network Analysis and Feedback Amplifier Design. New York: Van Nostrand Co., Inc.

Boyle R, Highstein SM (1990a) Resting discharge and response dynamics of horizontal semicircular canal afferents in the toadfish, *Opsanus tau*. J Neurosci 10:1557–1569.

Boyle R, Highstein SM (1990b) Efferent vestibular system in the toadfish: action upon horizontal semicircular canal afferents. J Neurosci 10:1570–1582.

Boyle R, Carey JP, Highstein SM (1991) Morphological correlates of response dynamics and efferent stimulation in horizontal semicircular canal afferents. J Neurophysiol 66:1504–1521.

Brichta AM, Goldberg JM (1998) The papilla neglecta of turtles: a detector of head rotations with unique sensory coding properties. J Neurosci 18:4314–4324.

Brichta AM, Peterson EH (1994) Functional architecture of vestibular primary afferents from the posterior semicircular canal of a turtle, *Pseudemys (Trachemys) scripta*. J Comp Neurol 344:481–507.

Brichta AM, Goldberg JM (2000a) Morphological identification of physiologically characterized afferents innervating the turtle posterior crista. J Neurophysiol 83:1202–1223.

Brichta AM, Goldberg JM (2000b) Responses to efferent activation and excitatory response intensity relations of turtle posterior crista afferents. J Neurophysiol 83:1224–1242.

Brichta AM, Aubert A, Eatock RA, Goldberg JM (2002) Regional analysis of whole-cell currents from hair cells of the turtle posterior crista. J Neurophysiol 88: 3259–3278.

Bronté-Stewart HM, Lisberger SG (1994) Physiological properties of vestibular primary afferents that mediate motor learning and normal performance of the vestibulo-ocular reflex in monkeys. J Neurosci 14:1290–1308.

Bruce LL, Kingsley J, Nichols DH, Fritzsch B (1997) The development of vestibulocochlear efferents and cochlear afferents in mice. Int J Neurosci 15:671–692.

Cajal SR (1911) Histology of the Nervous System. New York and Oxford: Oxford University Press.

Camis M (1930) The Physiology of the Vestibular Apparatus. Oxford: Clarendon Press.

Carney ME, Hoffman LF, Honrubia V (1990) Quantitative analysis of canalicular nerves in the chinchilla. Assoc Res Otolaryngol Abstr 13:364–365.

Carpenter MB, Chang L, Pereira AB, Hersh LB, Bruce G, Wu J-Y (1987) Vestibular and cochlear efferent neurons in the monkey identified by immunocytochemical methods. Brain Res 408:275–280.

Caston J, Precht W, Blanks RHI (1977) Response characteristics of frog's lagena afferents to natural stimulation. J Comp Physiol [A] 118:273–289.

Chang JSY, Popper AN, Saidel WM (1992) Heterogeneity of sensory hair cells in a fish ear. J Comp Neurol 324:621–640.

Christensen-Dalsgaard J, Narins PM (1993) Sound and vibration sensitivity of VIIIth nerve fibers in the frogs *Leptodactylus albilabris* and *Rana pipiens pipiens*. J Comp Physiol [A] 172:653–662.

Claas B, Fritzsch B, Munz H (1981) Common efferents to lateral line and labyrinthine hair cells in aquatic vertebrates. Neurosci Lett 27:231–235.

Coats AC, Smith SY (1967) Body position and the intensity of caloric nystagmus. Acta Otolaryngol (Stockh) 63:515–532.

Correia MJ, Lang DG (1990) An electrophysiological comparison of solitary type I and type II vestibular hair cells. Neurosci Lett 116:106–111.

Correia MJ, Lang DG, Eden AR (1985) A light and transmission electron microscope study of the neural processes within the pigeon anterior semicircular canal neuroepithelium. Prog Clin Biol Res 176:247–262.

Correia MJ, Perachio AA, Dickman JD, Kozlovskaya IB, Sirota MG, Yakushin SB, Beloozerova IN (1992) Changes in monkey horizontal semicircular canal afferent responses after spaceflight. J Appl Physiol 73:112S–120S.

Cortopassi KA, Lewis ER (1998) A comparison of the linear tuning properties of two classes of axons in the bullfrog lagena. Brain Behav Evol 51:331–348.

Corwin JT (1978) The relation of inner ear structure to the feeding behavior in sharks and rays. Scanning Electron Microsc 2:1105–1112.

Corwin JT (1981) Peripheral auditory physiology in the lemon shark: evidence of parallel otolithic and nonotolithic sound detection. J Comp Physiol [A] 142: 379–390.

Curthoys IS, Curthoys EJ, Blanks RHI, Markham CH (1975) The orientation of the semicircular canals in the guinea pig. Acta Otolaryngol (Stockh) 80:197–205.

de Vries H (1950) The mechanics of the labyrinth otoliths. Acta Otolaryngol (Stockh) 38:262–273.

Dechesne C, Raymond J, Sans A (1984) The efferent vestibular system in the cat: a horseradish peroxidase and fluorescent retrograde tracers study. Neuroscience 11:893–901.

Denison RH (1966) The origin of the lateral-line sensory system. Am Zool 6:369–370.

Denton EJ, Gray JAB (1979) The analysis of sound by the sprat ear. Nature 282:406–407.

Desai SS, Zeh C, Lysakowski A (2000) A comparative quantitative analysis of the striola in six rodent otolith organs. Abstr Soc Neurosci 26:6.

Desmadryl G, Dechesne CJ (1992) Calretinin immunoreactivity in chinchilla and guinea pig vestibular end organs characterizes the calyx unit subpopulation. Exp Brain Res 89:102–108.

Dickman JD (1996) Spatial orientation of semicircular canals and afferent sensitivity vectors in pigeons. Exp Brain Res 111:8–20.

Dowling JE, Boycott BB (1966) Organization of the primate retina: electron microscopy. Proc R Soc Lond B Biol Sci 166:80–111.

Eden AR, Correia MJ (1982) Identification of multiple groups of efferent vestibular neurons in the adult pigeon using horseradish peroxidase and DAPI. Brain Res 248:201–208.

Elgoyhen AB, Johnson DS, Boulter J, Vetter DE, Heinemann S (1994) Alpha 9: An acetylcholine receptor with novel pharmacologic properties expressed in rat cochlear hair cells. Cell 79:705–715.

Engström H (1958) On the double innervation of the sensory epithelia of the inner ear. Acta Otolaryngol (Stockh) 49:109–118.

Estes MS, Blanks RH, Markham CH (1975) Physiologic characteristics of vestibular first-order canal neurons in the cat. I. Response plane determination and resting discharge characteristics. J Neurophysiol 38:1232–1249.

Ewald JR (1892) Physiologische Untersuchungen über das Endorgan des Nervus Octavus. Wiesbaden: Bergmann.

Ezure K, Cohen MS, Wilson VJ (1983) Response of cat semicircular canal afferents to sinusoidal polarizing currents: implications for input–output properties of second-order neurons. J Neurophysiol 49:639–648.

Fay RR, Kendall JI, Popper AN, Tester AL (1975) Vibration detection by the macular neglecta of sharks. Comp Biochem Physiol 47A:1235–1240.

Fayyazuddin A, Brichta AM, Art JJ (1991) Organization of eighth nerve efferents in the turtle, *Pseudemys scripta*. Abstr Soc Neurosci 17:312.

Fernández C, Goldberg JM (1971) Physiology of peripheral neurons innervating semicircular canals of the squirrel monkey. II. Response to sinusoidal stimulation and dynamics of peripheral vestibular system. J Neurophysiol 34:661–675.

Fernández C, Goldberg JM (1976a) Physiology of peripheral neurons innervating otolith organs of the squirrel monkey. I. Response to static tilts and to long-duration centrifugal force. J Neurophysiol 39:970–984.

Fernández C, Goldberg JM (1976b) Physiology of peripheral neurons innervating otolith organs of the squirrel monkey. II. Directional selectivity and force-response relations. J Neurophysiol 39:985–995.

Fernández C, Goldberg JM (1976c) Physiology of peripheral neurons innervating otolith organs of the squirrel monkey. III. Response dynamics. J Neurophysiol 39:996–1008.

Fernández C, Goldberg JM, Abend WK (1972) Response to static tilts of peripheral neurons innervating otolith organs of the squirrel monkey. J Neurophysiol 35:978–987.

Fernández C, Baird RA, Goldberg JM (1988) The vestibular nerve of the chinchilla. I. Peripheral innervation patterns in the horizontal and superior semicircular canals. J Neurophysiol 60:167–181.

Fernández C, Baird RA, Goldberg JM (1990) The vestibular nerve of the chinchilla. III. Peripheral innervation patterns in the utricular macula. J Neurophysiol 63:767–780.

Fernández C, Lysakowski A, Goldberg JM (1995) Hair-cell counts and afferent innervation patterns in the cristae ampullares of the squirrel monkey with a comparison to the chinchilla. J Neurophysiol 73:1253–1281.

Fex J (1962) Auditory activity in centrifugal and centripetal cochlear fibres in cat. Study of a feedback system. Acta Physiol Scand Suppl 55:1–68.

Fex J (1967) Efferent inhibition in the cochlea related to hair cell d.c. activity: study of postsynaptic activity of the crossed olivo-cochlear fibres in the cat. J Acoust Soc Am 41:666–675.

Flock Å (1964) Structure of the macula utriculus with special reference to directional interplay of sensory responses as revealed by morphological polarization. J Cell Biol 22:413–431.

Flock Å (1971) Sensory transduction in hair cells. In: Loewenstein WR (ed) Handbook of Sensory Physiology. Berlin: Springer-Verlag, pp. 396–441.

Flock Å, Orman S (1983) Micromechanical properties of sensory hairs on receptor cells of the inner ear. Hear Res 11:249–260.

Flock Å, Russell IJ (1973) The post-synaptic action of efferent fibres in the lateral line organ of the burbot *Lota lota.* J Physiol (Lond) 235:591–605.

Flock Å, Russell IJ (1976) Inhibition by efferent nerve fibres: action on hair cells and afferent synaptic transmission in the lateral line canal organ of the burbot *Lota lota.* J Physiol (Lond) 257:45–62.

Fritzsch B (1981) Efferent neurons to the labyrinth of *Salamandra salamandra* as revealed by retrograde transport of horseradish peroxidase. Neurosci Lett 26:191–196.

Fritzsch B (1987) Inner ear of the coelocanth fish *Latimeria* has tetrapod affinities. Nature 327:153–154.

Fritzsch B (1996) Development of labyrinthine efferent system. Ann N Y Acad Sci 781:21–33.

Fritzsch B, Crapon de Caprona MD (1984) The origin of centrifugal inner ear fibres of gymnophions. Neurosci Lett 46:131–136.

Fritzsch B, Wake MH (1988) The inner ear of gymnophine amphibians and its nerve supply: a comparative study of regressive events in a complex sensory system (Amphibia, Gymnophiona). Zoomorph 108:201–217.

Fritzsch B, Dubuc R, Ohta Y, Grillner S (1989) Efferents to the labyrinth of the river lamprey (*Lampetra fluviatilis*) as revealed with retrograde tracing techniques. Neurosci Lett 96:241–246.

Fuchs PA, Murrow BW (1992a) Cholinergic inhibition of short outer hair cells in the chick's cochlea. J Neurosci 12:800–809.

Fuchs PA, Murrow BW (1992b) A novel cholinergic receptor mediates inhibition of chick cochlear hair cells. Proc R Soc Lond B Biol Sci 248:35–40.

Furukawa T (1981) Effects of efferent stimulation on the saccule of goldfish. J Physiol (Lond) 315:203–215.

Furukawa T, Ishii Y (1967) Neurophysiological studies on hearing in goldfish. J Neurophysiol 30:1377–1403.

Gacek RR, Lyon M (1974) The localization of vestibular efferent neurons in the kitten with horseradish peroxidase. Acta Otolaryngol (Stockh) 77:92–101.

Géléoc GSC, Lennan GWT, Richardson GP, Kros CJ (1997) A quantitative comparison of mechanoelectrical transduction in vestibular and auditory hair cells of neonatal mice. Proc R Soc Lond B Biol Sci 264:611–621.

Gioglio L, Congiu T, Quacci D, Prigioni I (1995) Morphological features of different regions in frog crista ampullaris (*Rana esculenta*). Arch Histol Cytol 58:1–16.

Goldberg JM (1979) Vestibular receptors in mammals: afferent discharge characteristics and efferent control. Prog Brain Res 50:353–367.

Goldberg JM (2000) Afferent diversity and the organization of central vestibular pathways. Exp Brain Res 130:277–297.

Goldberg JM, Fernández C (1971a) Physiology of peripheral neurons innervating semicircular canals of the squirrel monkey. I. Resting discharge and response to constant angular accelarations. J Neurophysiol 34:635–660.

Goldberg JM, Fernández C (1971b) Physiology of peripheral neurons innervating semicircular canals of the squirrel monkey. III. Variations among units in their discharge properties. J Neurophysiol 34:676–684.

Goldberg JM, Fernández C (1977) Conduction times and background discharge of vestibular afferents. Brain Res 122:545–550.

Goldberg JM, Fernández C (1980) Efferent vestibular system in the squirrel monkey: anatomical location and influence on afferent activity. J Neurophysiol 43:986–1025.

Goldberg JM, Fernández C (1984) The vestibular system. In: Smith ID (ed) Handbook of Physiology. Section 1. The Nervous System, Volume 3. Sensory Processes. Baltimore, MD: Williams and Wilkins, pp. 977–1022.

Goldberg JM, Fernández C, Smith CE (1982) Responses of vestibular-nerve afferents in the squirrel monkey to externally applied galvanic currents. Brain Res 252:156–160.

Goldberg JM, Highstein SM, Moschovakis AK, Fernández C (1987) Inputs from regularly and irregularly discharging vestibular nerve afferents to secondary neurous in the vestibular nuclei of the squirrel monkey. I. An electrophysiological analysis. J Neurophysiol 58:700–718.

Goldberg JM, Smith CE, Fernández C (1984) Relation between discharge regularity and responses to externally applied galvanic currents in vestibular nerve afferents of the squirrel monkey. J Neurophysiol 51:1236–1256.

Goldberg JM, Desmadryl G, Baird RA, Fernández C (1990a) The vestibular nerve of the chinchilla. IV. Discharge properties of utricular afferents. J Neurophysiol 63:781–790.

Goldberg JM, Desmadryl G, Baird RA, Fernández C (1990b) The vestibular nerve of the chinchilla. V. Relation between afferent discharge properties and peripheral innervation patterns in the utricular macula. J Neurophysiol 63:791–804.

Goldberg JM, Lysakowski A, Fernández C (1990c) Morphophysiological and ultrastructural studies in the mammalian cristae ampullares. Hear Res 49:89–102.

Goldberg JM, Lysakowski A, Fernández C (1992) Structure and function of vestibular nerve afferents in the chinchilla and squirrel monkey. Ann N Y Acad Sci 656:92–107.

Goldberg JM, Brichta AM, Wackym PA (2000) Efferent vestibular system: anatomy, physiology and neurochemistry. In: Anderson JH, Beitz AJ (eds) Neurochemistry of the Vestibular System. Boca Raton, FL: CRC Press, pp. 61–94.

Goldberg JM, Brichta AM (2002) Functional analysis of whole-cell currents from hair cells of the turtle posterior crista. J Neurophysiol 88:3279–3292.

Gonzalez MJ, Anadon R (1994) Central projections of the octaval nerve in larval lamprey: an HRP study. J Hirnforsch 35:181–189.

Grant JW, Huang CC, Cotton JR (1994) Theoretical mechanical frequency response of the otolithic organs. J Vestib Res 4:137–151.

Gulley RL, Bagger-Sjöbäck D (1979) Freeze-fracture studies on the synapse between the type I hair cell and the calyceal terminal in the guinea-pig vestibular system. J Neurocytol 8:591–603.

Guth PS, Fermin CD, Pantoja M, Edwards R, Norris CH (1994) Hair cells of different shapes and their placement along the frog crista ampullaris. Hear Res 73:109–115.

Guth PS, Perin P, Norris CH, Valli P (1998) The vestibular hair cells: post-transductional signal processing. Prog Neurobiol 54:193–247.

Haque A, Dickman JD (2001) Afferent innervation patterns of the horizontal crista ampullaris in pigeons. Assoc Res Otolaryngol Abstr 24:123.

Hartmann R, Klinke R (1980) Discharge properties of afferent fibers of the goldfish semicircular canal with high frequency stimulation. Pflügers Arch 388:111–121.

Highstein SM (1991) The central nervous system efferent control of the organs of balance and equilibrium. Neurosci Res 12:13–30.

Highstein SM, Politoff AL (1978) Relation of interspike baseline activity to the spontaneous discharges of primary afferents from the labyrinth of the toadfish, *Opsanus tau.* Brain Res 150:182–187.

Highstein SM, Baker R (1985) Action of the efferent vestibular system on primary afferents in the toadfish, *Opsanus tau.* J Neurophysiol 54:370–384.

Highstein SM, Baker R (1986) Organization of the efferent vestibular nuclei and nerves of the toadfish, *Opsanus tau.* J Comp Neurol 243:309–325.

Highstein SM, Goldberg JM, Moschovakis AK, Fernández C (1987) Inputs from regularly and irregularly discharging vestibular nerve afferents to secondary neurons in the vestibular nuclei of the squirrel monkey. II. Correlation with output pathways of secondary neurons. J Neurophysiol 58:719–738.

Highstein SM, Rabbitt RD, Boyle R (1996) Determinants of semicircular canal afferent response dynamics in the toadfish, *Opsanus tau.* J Neurophysiol 75: 575–596.

Hillman DE (1976) Morphology of peripheral and central vestibular systems. In: Llinás R, Precht W (eds), Frog Neurobiology. A Handbook. Berlin: Springer-Verlag, pp. 452–480.

Holt JC, Xue J-T, Goldberg JM (2001) A pharmacological analysis of the responses of turtle posterior crista sfferents to efferent activation. Abstr Soc Neurosci 27: #513.17.

Holt JR, Corey DP, Eatock RA (1997) Mechanoelectrical transduction and adaptation in hair cells of the mouse utricle, a low-frequency vestibular organ. J Neurosci 17:8739–8748.

Honrubia V, Kuruvilla A, Mamikunian D, Eichel JE (1987) Morphological aspects of the vestibular nerve of the squirrel monkey. Laryngoscope 97:228–238.

Honrubia V, Hoffman LF, Sitko S, Schwartz IR (1989) Anatomic and physiological correlates in bullfrog vestibular nerve. J Neurophysiol 61:688–701.

Hoshino T (1975) An electron microscopic study of the otolithic maculae of the lamprey (*Entosephenus japonicus*). Acta Otolaryngol (Stockh) 80:43–53.

Hudspeth AJ (1989) How the ear's works work. Nature 341:397–404.

Hunter-Duvar IM, Hinojosa R (1984) Vestibule: sensory epithelia. In: Friedemann I, Ballantyne J (eds) Ultrastructural Atlas of the Inner Ear. London: Butterworths, pp. 211–244.

Iurato S, Luciano L, Pannese E, Reale E (1972) Efferent vestibular fibers in mammals: morphological and histochemical aspects. Prog Brain Res 37:429–443.

Jørgensen JM (1974) The sensory epithelia of the inner ear of two turtles, *Testudo graeca* L. and *Pseudemys scripta* (Schoepff). Acta Zool (Stockh) 55:289–298.

Jørgensen JM (1975) The sensory epithelia in the inner ear of a lizard, *Calotes versicolor* Daudin. Vidensk Medd Dan Naturhist Foren Khobenhavn 138:7–19.

Jørgensen JM (1988) The number and distribution of calyceal hair cells in the inner ear utricular macula of some reptiles. Acta Zool (Stockh) 69:169–175.

Jørgensen JM, Anderson T (1973) On the structure of the avian maculae. Acta Zool (Stockh) 54:121–130.

Jørgensen JM, Locket NA (1995) The inner ear of the echidna *Tachyglossus aculeatus*: the vestibular sensory organs. Proc R Soc Lond B Biol Sci 260:183–189.

Kachar B, Parakkal M, Fex J (1990) Structural basis for mechanical transduction in the frog vestibular sensory apparatus: I. The otolithic membrane. Hear Res 45:179–190.

Kalmijn AJ (1988) Hydrodynamic and acoustic field detection. In: Atema J, Fay RR, Popper AN, Tavolga WN (eds) Sensory Biology of Aquatic Animals. New York: Springer-Verlag, pp. 84–130.

Keller EL (1976) Behavior of horizontal semicircular canal afferents in alert monkey during vestibular and optokinetic stimulation. Exp Brain Res 24:459–471.

Kevetter GA, Blumberg KR, Correia MJ (2000) Hair cell and supporting cell density and distribution in the normal and regenerating posterior crista ampullaris of the pigeon. Int J Dev Neurosci 18:855–867.

Koester DM (1983) Central projections of the octavolateralis nerve of the clearnose skate, *Raja eglanteria*. J Comp Neurol 221:199–215.

Koyama H, Lewis ER, Leverenz EL, Baird RA (1982) Acute seismic sensitivity in the bullfrog ear. Brain Res 250:168–172.

Koyama H, Kishida R, Goris RC, Kosunoki T (1989) Afferent and efferent projections of the VIIIth cranial nerve in the lamprey *Lampetra japonica*. J Comp Neurol 280:663–671.

Lanford PJ, Popper AN (1996) Novel afferent terminal structure in the crista ampullaris of the goldfish, *Carassius auratus*. J Comp Neurol 366:572–579.

Lanford PJ, Platt C, Popper AN (2000) Structure and function in the saccule of the goldfish (*Carassius auratus*): a model of diversity in the non-amniote ear. Hear Res 143:1–13.

Lapeyre P, Guilhaume A, Cazals Y (1992) Differences in hair bundles associated with type I and type II vestibular hair cells of the guinea pig saccule. Acta Otolaryngol 112:635–642.

Ledoux A (1949) Activite electrique des nerfs des canaux semicircularies, du saccule and de l'utricle chez la grenouille. Acta Otorhinolaryngol Belg 3:335–349.

Lenzi D, Runyeon JW, Crum J, Ellisman MH, Roberts WM (1999) Synaptic vesicle populations in saccular hair cells reconstructed by electron tomography. J Neurosci 19:119–132.

Lewis ER, Li CW (1975) Hair cell types and distribution in the otolithic and auditory organs of the bullfrog. Brain Res 83:35–50.

Lewis ER, Baird RA, Leverenz EL, Koyama H (1982) Inner ear: dye injection reveals peripheral origins of specific sensitivities. Science 215:1641–1643.

Lewis ER, Leverenz EL, Bialek WS (1985) The vertebrate inner ear. Boca Raton, FL: CRC Press.

Liberman MC, Dodds LW, Pierce S (1990) Afferent and efferent innervation of the cat cochlea: quantitative analysis with light and electron microscopy. J Comp Neurol 301:443–460.

Lim DJ (1973) Formation and fate of the otoconia. Scanning and transmission electron microscopy. Ann Otol Rhinol Laryngol 82:23–35.

Lim DJ (1984) The development and structure of the otoconia. In: Friedemann I, Ballantyne J (eds) Ultrastructural Atlas of the Inner Ear. London: Butterworths, pp. 245–269.

Lindeman HH (1969) Studies on the morphology of the sensory regions of the vestibular apparatus. Ergeb Anat Entwicklungsgesch 42:1–113.

Loe PR, Tomko DL, Werner G (1973) The neural signal of angular head position in primary afferent vestibular nerve axons. J Physiol (Lond) 230:29–50.

Lombard RE, Bolt JR (1979) Evolution of the tetrapod ear: an analysis and rein-terpretation. Biol J Linn Soc 11:19–76.

Lorente de Nó R (1926) Études sur l'anatomie et la physiologie du labyrinthe de l'oreille et du VIIIe nerf. II. Quelque données au sujet de l'anatomie des organes sensoriels du labyrinthe. Trav Lab Rech Biol Univ Madrid 24:53–153.

Louie AW, Kimm J (1976) The response of 8th nerve fibers to horizontal sinusoidal oscillation in the alert monkey. Exp Brain Res 24:447–457.

Lowenstein O (1936) The equilibrium function of the vertebrate labyrinth. Biol Rev Camb Philos Soc 11:113–145.

Lowenstein O (1950) The equilibrium function of the otolith organs of the thorn-back ray (*Raja clavata*). J Physiol (Lond) 110:392–415.

Lowenstein O (1956) Peripheral mechanisms of equilibrium. Br Med Bull 12:14–18.

Lowenstein O, Osborne MP (1964) Ultrastructure of the sensory hair cells in the labyrinth of the ammocete larva of the lamprey, *Lampetra fluviatilis*. Nature 204:97.

Lowenstein O, Roberts TDM (1951) The localization and analysis of the responses to vibration from the isolated elasmobranch labyrinth. J Physiol (Lond) 114:471–489.

Lowenstein O, Sand A (1936) The activity of the horizontal semicircular canal of the dogfish, *Scyllium canalicula*. J Exp Biol 13:416–428.

Lowenstein O, Sand A (1940a) The individual and integrated activity of the semi-circular canals of the elasmobranch labyrinth. J Physiol (Lond) 99:89–101.

Lowenstein O, Sand A (1940b) The mechanism of the semicircular canal. A study of the responses of single fiber preparations to angular accelerations and rota-tions at constant speed. Proc R Soc Lond B Biol Sci 129:256–273.

Lowenstein O, Thornhill RA (1970) The electrophysiological study of the responses of the isolated labyrinth of the lamprey (*Lampetra fluviatilis*) to angular acceler-ation, tilting and mechanical vibration. Proc R Soc Lond B Biol Sci 174:419–434.

Lowenstein O, Wersäll J (1959) A functional interpretation of the electron-microscopic structure of the sensory hairs in the cristae of the elasmobranch *Raja clavata* in terms of directional sensitivity. Nature 184:1807–1808.

Lowenstein O, Osborne MP, Wersäll J (1964) Structure and innervation of the sensory epithelia of the labyrinth in the thornback ray. Proc R Soc Lond B Biol Sci 160:1–12.

Lowenstein O, Osborne MP, Thornhill RA (1968) The anatomy and ultrastructure of the labyrinth of the lamprey, (*Lampetra fluviatilis* L.). Proc R Soc Lond B Biol Sci 170:113–134.

Lysakowski A (1996) Synaptic organization of the crista ampullaris in vertebrates. Ann N Y Acad Sci 781:164–182.

Lysakowski A (2001) Cytoarchitectural organization of the crista ampullaris in ver-tebrates. Abstr Soc Neurosci 27: #298.21.

Lysakowski A, Goldberg JM (1993) Regional variations in synaptic innervation of the squirrel monkey crista. Abstr Soc Neurosci 19:1578.

Lysakowski A, Goldberg JM (1997) A regional ultrastructural analysis of the cellu-lar and synaptic architecture in the chinchilla cristae ampullares. J Comp Neurol 389:419–443.

Lysakowski A, Singer M (2000) Nitric oxide synthase localized in a subpopulation of vestibular efferents with NADPH diaphorase histochemistry and nitric oxide synthase immunohistochemistry. J Comp Neurol 427:508–521.

Lysakowski A, Minor LB, Fernández C, Goldberg JM (1995) Physiological identification of morphologically distinct afferent classes innervating the cristae ampullares of the squirrel monkey. J Neurophysiol 73:1270–1281.

Lysakowski A, Alonto A, Jacobson L (1999) Peripherin immunoreactivity labels small-diameter vestibular "bouton" afferents in rodents. Hear Res 133:149–154.

Macadar O, Budelli R (1984) Mechanisms of sensory adaptation in the isolated utricle. Exp Neurol 86:147–159.

Macadar O, Wolfe GE, O'Leary DP, Segundo JP (1975) Response of the elasmobranch utricle to maintained spatial orientation, transitions and jitter. Exp Brain Res 22:1–12.

Marco J, Lee W, Suarez C, Hoffman LF, Honrubia V (1993) Morphologic and quantitative study of the efferent vestibular system in the chinchilla: 3-D reconstruction. Acta Otolaryngol (Stockh) 113:229–234.

Marlinsky V, Plotnik M, Goldberg JM (2000) Responses of vestibular-nerve afferents to electrical stimulation of brain stem efferent pathways in anesthetized chinchillas. Abstr Soc Neurosci 26:1491.

Martini M, Rossi, ML, Rubbini G, Rispoli G (2000) Calcium currents in hair cells isolated from semicircular canals of the frog. Biophys J 78:1240–1254.

Masetto S, Correia MJ (1997) Electrophysiological properties of vestibular sensory and supporting cells in the labyrinth slice before and during regeneration. J Neurophysiol 78:1913–1927.

Masetto S, Russo G, Prigioni I (1994) Differential expression of potassium currents by hair cells in thin slices of frog crista ampullaris. J Neurophysiol 72:443–455.

Masetto S, Perin P, Malusa A, Zucca G, Valli P (2000) Membrane properties of chick semicircular canal hair cells *in situ* during embryonic development. J Neurophysiol 83:2740–2756.

Mayr E (1997) This is Biology: The Science of the Living World. Cambridge, MA: The Belknap Press of Harvard University Press, p. 64.

Mazan S, Jaillard D, Baratte B, Janvier P (2000) Otx1 gene-controlled morphogenesis of the horizontal semicircular canal and the origin of the gnathostome characteristics. Evol Dev 2:186–193.

McCue MP, Guinan JJ Jr (1994) Influence of efferent stimulation on acoustically responsive vestibular afferents in the cat. J Neurosci 14:6071–6083.

Merchant SN, Velazquez-Villasenor L, Tsuji K, Glynn RJ, Wall C 3rd, Rauch SD (2000) Temporal bone studies of the human peripheral vestibular system. Normative vestibular hair cell data. Ann Otol Rhinol Laryngol Suppl 181:3–13.

Meredith GE (1988) Comparative view of the central organization of afferent and efferent circuitry for the inner ear. Acta Biol Hung 39:229–249.

Meredith GE, Roberts BL (1987a) Central organization of the efferent supply to the labyrinthine and lateral line receptors of the dogfish. Neuroscience 17:225–233.

Meredith GE, Roberts BL (1987b) Distribution and morphological characteristics of efferent neurons innervating end organs in the ear and lateral line of the European eel. J Comp Neurol 265:494–506.

Minor LB, Goldberg JM (1990) Influence of static head position on the horizontal nystagmus evoked by caloric, rotational and optokinetic stimulation in the squirrel monkey. Exp Brain Res 82:1–13.

Money KE, Myles WS (1974) Heavy water nystagmus and effects of alcohol. Nature 247:404–405.

Montandon P, Gacek RR, Kimura RS (1970) Crista neglecta in the cat and human. Ann Otol Rhinol Laryngol 79:105–112.

Mridhar Z, Huss D, Dickman JD (2001) Comparison of morphology and afferent innervation patterns between normal and regenerated saccular macula in pigeons. Assoc Res Otolaryngol Abstr 24:124.

Myers SF, Lewis ER (1990) Hair cell tufts and afferent innervation of the bullfrog crista ampullaris. Brain Res 534:15–24.

Myers SF, Lewis ER (1991) Vestibular afferent responses to microrotational stimuli. Brain Res 543:36–44.

Northcutt RG (1997) Evolution of gnathostome lateral line ontogenies. Brain Behav Evol 50:25–37.

Ohno K, Takeda N, Yamano M, Matsunaga T, Tohyama M (1991) Coexistence of acetylcholine and calcitonin gene related peptide in the vestibular efferent neurons in the rat. Brain Res 566:103–107.

Okano Y, Sando I, Myers EN (1978) Crista neglecta in man. Ann Otol Rhinol Laryngol 87:306–312.

O'Leary DP, Dunn RF (1976) Analysis of afferent responses from isolated semicircular canal of the guitarfish using rotational acceleration white-noise inputs. I. Correlation of response dynamics with receptor innervation. J Neurophysiol 39:631–644.

O'Leary DP, Dunn RF, Honrubia V (1974) Functional and anatomical correlation of afferent responses from the isolated semicircular canal. Nature 251:225–227.

Paige GD (1985) Caloric responses after horizontal canal inactivation. Acta Otolaryngol (Stockh) 100:321–327.

Parsons TD, Lenzi D, Almers W, Roberts WM (1994) Calcium-triggered exocytosis and endocytosis in an isolated presynaptic cell: capacitance measurements in saccular hair cells. Neuron 13:875–883.

Pellegrini M, Ceccotti F, Magherini P (1985) The efferent vestibular neurons in the toad (*Bufo bufo* L.): their location and morphology. A horseradish peroxidase study. Brain Res 334:1–8.

Perachio AA, Correia MJ (1983) Responses of semicircular canal and otolith afferents to small angle static head tilts in the gerbil. Brain Res 280:287–298.

Perachio AA, Kevetter GA (1989) Identification of vestibular efferent neurons in the gerbil: histochemical and retrograde labelling. Exp Brain Res 78:315–326.

Peterson EH, Cotton JR, Grant JW (1996) Structural variation in ciliary bundles of the posterior semicircular canal. Quantitative anatomy and computational analysis. Ann N Y Acad Sci 781:85–102.

Pickles JO, Corey DP (1992) Mechanoelectrical transduction by hair cells. Trends Neurosci 15:254–259.

Poljak S (1927) Über die nervenendigungen in den vestibulären sinnesorganen der säugetierre. Z Anat Entwicklungsgesch 84:131–144.

Popper AN, Platt C (1979) The herring ear has a unique receptor pattern. Nature 280:832–833.

Popper AN, Saidel WM, Chang JS (1993) Two types of sensory hair cell in the saccule of a teleost fish. Hear Res 64:211–216.

Popper KR (1968) The Logic of Scientific Discovery. New York and Evanston: Harper Torchbooks, Harper & Row, p. 61.

Popper KR (1993) Realism and the Aim of Science. London: Routledge.

Precht W, Llinás R, Clarke M (1971) Physiological responses of frog vestibular fibers to horizontal angular rotation. Exp Brain Res 13:378–407.

Prigioni I, Valli P, Casella C (1983) Peripheral organization of the vestibular efferent system in the frog: an electrophysiological study. Brain Res 269:83–90.

Prigioni I, Russo G, Marcotti W (1996) Potassium currents of pear-shaped hair cells in relation to their location in the frog crista ampullaris. Neuroreport 7:1841–1843.

Purcell IM, Perachio AA (1997) Three-dimensional analysis of vestibular efferent neurons innervating semicircular canals of the gerbil. J Neurophysiol 78:3234–3248.

Rabbitt RD (1999) Directional coding of three-dimensional movements by the vestibular semicircular canals. Biol Cybern 80:417–431. [Published erratum appears in Biol Cybern Biol Cybern 82:355 (2000)].

Ramprashad F, Landolt JP, Money KE, Laufer J (1986) Comparative morphometric study of the vestibular system of the vertebrata: reptilia, aves, amphibia, and pisces. Acta Otolaryngol Suppl 427:1–42.

Rao-Mirotznik R, Harkins AB, Buchsbaum G, Sterling P (1995) Mammalian rod terminal: architecture of a binary synapse. Neuron 14:561–596.

Reisine H, Simpson JI, Henn V (1988) A geometric analysis of semicircular canals and induced activity in their peripheral afferents in the rhesus monkey. Ann N Y Acad Sci 545:10–20.

Rennie KJ, Correia MJ (1994) Potassium currents in mammalian and avian isolated type I semicircular canal hair cells. J Neurophysiol 71:317–329.

Retzius G (1881) Das Gehörorgan der Wirbelthiere: Morphologisch-histologische Studien. Stockholm: in Commission bei Samson & Wallin.

Rosenhall U (1970) Some morphological principles of the vestibular maculae in birds. Arch Klin Exp Ohren Nasen Kehlkopfheilkd 197:154–182.

Ross DA (1936) Electrical studies on the frog's labyrinth. J Physiol (Lond) 86:117–146.

Ross MD (1985) Anatomic evidence for peripheral neural processing in mamalian graviceptors. Aviat Space Environ Med 56:338–343.

Ross MD, Rogers CM, Donovan KM (1986) Innervation patterns in rat saccular macula. A structural basis for complex sensory processing. Acta Otolaryngol (Stockh) 102:75–86.

Rossi ML, Martini M (1991) Efferent control of posterior canal afferent receptor discharge in the frog labyrinth. Brain Res 555:123–134.

Rossi ML, Prigioni I, Valli P, Casella C (1980) Activation of the efferent system in the isolated frog labyrinth: effects on the afferent EPSPs and spike discharge recorded from single fibres of the posterior nerve. Brain Res 185:125–137.

Rossi ML, Bonnifazzi C, Martini M, Fesce R (1989) Static and dynamic properties of synaptic transmission at the cyto-neural junction of frog labyrinth posterior canal. J Gen Physiol 94:303–327.

Rossi ML, Martini M, Pelucchi B, Fesce R (1994) Quantal nature of synaptic transmission at the cytoneural junction in the frog labyrinth. J Physiol (Lond) 478:14–35.

Rüsch A, Eatock RA (1996) A delayed rectifier conductance in type I hair cells of the mouse. J Neurophysiol 76:995–1004.

Rushton WAH (1951) A theory of the effects of fibre size in medullated nerve. J Physiol (Lond) 115:101–122.

Russell IJ (1968) Influence of efferent fibres on a receptor. Nature 219:177–178.

Russell IJ (1971) The pharmacology of efferent synapses in the lateral-line system of *Xenopus laevis.* J Exp Biol 54:643–658.

Saidel WM, Lanford PJ, Yan HY, Popper AN (1995) Hair cell heterogeneity in the goldfish saccule. Brain Behav Evol 46:362–370.

Sans A, Highstein SM (1984) New ultrastructural features in the vestibular labyrinth of the toadfish, *Opsanus tau.* Brain Res 308:362–370.

Schessel DA (1982) Chemical synaptic transmission between type I vestibular hair cells and the primary afferent nerve chalice: an intracellular study utilizing horseradish peroxidase. Ph.D. Thesis, Albert Einstein College of Medicine, Bronx, NY.

Schessel DA, Ginzberg R, Highstein SM (1991) Morphophysiology of synaptic transmission between type I hair cells and vestibular primary afferents. An intracellular study employing horseradish peroxidase in the lizard, *Calotes versicolor.* Brain Res 544:1–16.

Schmitz F, Konigstorfer A, Sudhof TC (2000) RIBEYE, a component of synaptic ribbons: a protein's journey through evolution provides insight into synaptic ribbon function. Neuron 28:857–872.

Schneider LW, Anderson DJ (1976) Transfer characteristics of first and second order lateral canal vestibular neurons in gerbil. Brain Res 112:61–76.

Schwarz DWF, Schwarz IE, Tomlinson RD (1978) Avian efferent vestibular neurons identified by axonal transport of 3H-adenosine and horseradish peroxidase. Brain Res 155:103–107.

Schwarz DWF, Satoh K, Schwarz IE, Hu K, Fibiger HC (1986) Cholinergic innervation of the rat's labyrinth. Exp Brain Res 64:19–26.

Schwarz IE, Schwarz DWF, Fredrickson JM, Landolt JP (1981) Efferent vestibular neurons: a study employing retrograde tracer methods in the pigeon (*Columba livia*). J Comp Neurol 196:1–12.

Segal BN, Outerbridge JS (1982) Vestibular (semicircular canal) primary neurons in bullfrog: nonlinearity of individual and population response to rotation. J Neurophysiol 47:545–562.

Sewell WF, Starr PA (1991) Effects of calcitonin gene-related peptide and efferent nerve stimulation on afferent transmission in the lateral line organ. J Neurophysiol 65:1158–1169.

Shotwell SL, Jacobs R, Hudspeth AJ (1981) Directional sensitivity of individual vertebrate hair cells to controlled deflection of their hair bundles. Ann N Y Acad Sci 374:1–10.

Si X, Zakir MM, Dickman JD (2003) Afferent innervation patterns of the utricular macula in pigeons. J Neurophysiol 89:1660–1677.

Silver RB, Reeves AP, Steinacker A, Highstein SM (1998) Examination of the cupula and stereocilia of the horizontal semicircular canal in the toadfish, *Opsanus tau.* J Comp Neurol 402:48–61.

Simon H, Lumsden A (1993) Rhombomere-specific origin of the contralateral vestibulo-acoustic efferent neurons and their migration across the embryonic midline. Neuron 11:209–220.

Sjöstrand FS (1958) Ultrastructure of retinal rod synapses as revealed by three-dimensional reconstructions from serial sections. J Ultrastruct Res 2:122–170.

Smith CA, Rasmussen GL (1968) Nerve ending in the maculae and cristae of the chinchilla vestibule with a special reference to the efferents. In: Graybiel A (ed),

Third Symposium on the Role of Vestibular Organs in Space Exploration, NASA SP-152. Washington, DC: U.S. Government Printing Office, pp. 183–201.

Smith CA, Sjöstrand FS (1961) A synaptic structure in the hair cells of the guinea pig cochlea. J Ultrastruct Res 5:185–192.

Smith CE, Goldberg JM (1986) A stochastic afterhyperpolarization model of repetitive activity in vestibular afferents. Biol Cybern 54:41–51.

Spoendlin H (1965) Ultrastructural studies of the labyrinth in squirrel monkeys. In: Graybiel A (ed), First Symposium on the Role of Vestibular Organs in Space Exploration, NASA SP-77. Washington, DC: U.S. Government Printing Office, pp. 7–22.

Sridhar TS, Liberman MC, Brown MC, Sewell WF (1995) A novel cholinergic "slow effect" of efferent stimulation on cochlear potentials in the guinea pig. J Neurosci 15:3667–3678.

Sridhar TS, Brown MC, Sewell WF (1997) Unique postsynaptic signaling at the hair cell efferent synapse permits calcium to evoke changes on two time scales. J Neurosci 17:428–437.

Stensiö EA (1927) The Downtonian and Devonian vertebrates of Spitsbergen. Part I. Family Cephalaspidae. A. Text. Skr Svalbard Nordishavet 12:1–391.

Strutz J (1981) The origin of centrifugal fibers to the inner ear in *Caiman crocodilus*. A horseradish peroxidase study. Neurosci Lett 27:95–100.

Strutz J (1982a) The origin of efferent fibers to the inner ear in a turtle (*Terrapene ornata*). A horseradish peroxidase study. Brain Res 244:165–168.

Strutz J (1982b) The origin of efferent vestibular fibres in the guinea pig. Acta Otolaryngol (Stockh) 94:299–305.

Strutz J, Schmidt CL (1982) Acoustic and vestibular efferent neurons in the chicken (*Gallus domesticus*). A horseradish peroxidase study. Acta Otolaryngol (Stockh) 94:45–51.

Strutz J, Schmidt CL, Sturmer C (1980) Origin of efferent fibers of the vestibular apparatus in goldfish. A horseradish peroxidase study. Neurosci Lett 18:5–9.

Strutz J, Spatz WB, Schmidt CL, Sturmer C (1981) Origin of centrifugal fibers to the labyrinth in the frog (*Rana esculenta*). A study with the fluorescent retrograde neuronal tracer "Fast blue". Brain Res 215:323–328.

Sugai T, Sugitani M, Ooyama H (1991) Effects of activation of the divergent efferent fibers on the spontaneous activity of vestibular afferent fibers in the toad. Jpn J Physiol 41:217–232.

Taglietti V, Rossi ML, Casella C (1977) Adaptive distortions in the generator potential of semicircular canal sensory afferents. Brain Res 123:41–57.

Tomko DL, Peterka RJ, Schor RH (1981a) Responses to head tilt in cat eighth nerve afferents. Exp Brain Res 41:216–221.

Tomko DL, Peterka RJ, Schor RH, O'Leary DP (1981b) Response dynamics of horizontal canal afferents in barbiturate-anesthetized cats. J Neurophysiol 45:376–396.

Usami S, Hozawa J, Shinkawa H, Tazawa M, Jin H, Matsubara A, Fujita S, Ylikoski J (1993) Immunocytochemical localization of substance P and neurofilament proteins in the guinea pig vestibular ganglion. Acta Otolaryngol Suppl 503:127–131.

Usami S, Matsubara A, Shinkawa H, Matsunaga T, Kanzaki J (1995) Neuroactive substances in the human vestibular end organs. Acta Otolaryngol Suppl 520:160–163.

Valli P, Botta L, Zucca G, Casella C (1986) Functional organization of the peripheral efferent vestibular system in the frog. Brain Res 362:92–97.

van Bergeijk WA (1966) Evolution of the sense of hearing in vertebrates. Am Zool 6:371–377.

Vetter DE, Mann JR, Wangemann P, Liu J, Marcus DC, Lazdunski M, Heinemann SF, Barhanin J (1996) Inner ear defects induced by null mutation of the IsK gene. Neuron 17:1251–1254.

von Gersdorff H (2001) Synaptic ribbons: versatile signal transducers. Neuron 29:7–10.

von Gersdorff H, Matthews G (1999) Electrophysiology of synaptic vesicle cycling. Annu Rev Physiol 61:725–752.

Wangemann P (1995) Comparison of ion transport mechanisms between vestibular dark cells and strial marginal cells. Hear Res 90:149–157.

Wangemann P, Shen Z, Liu J (1996) K^+-induced stimulation of K^+ secretion involves activation of the IsK channel in vestibular dark cells. Hear Res 100:201–210.

Warr WB (1975) Olivocochlear and vestibular efferent neurons of the feline brain stem: their location, morphology and number determined by retrograde axonal transport and acetylcholinesterase histochemistry. J Comp Neurol 161:159–181.

Werner CF (1933) Die differenzierung der maculae im labyrinth, insbesondere bei säugetieren. Z Anat Entwicklungsgesch 99:696–706.

Wersäll J (1956) Studies on the structure and innervation of the sensory epithelium of the cristae ampullaris in the guinea pig. A light and electron microscopic investigation. Acta Otolaryngol (Stockh) Suppl 126:1–85.

Wersäll J (1968) Efferent innervation of the inner ear. In: von Euler C, Skoglund C, Söderberg U (eds) Structure and Function of Inhibitory Neuronal Mechanisms. Oxford: Pergamon, pp. 123–139.

Wersäll J, Bagger-Sjöbäck D (1974) Morphology of the vestibular sense organ. In: Kornhuber HH (ed) Handbook of Sensory Physiology. Berlin: Springer-Verlag, pp. 123–170.

Whitehead MC, Morest DK (1981) Dual populations of efferent and afferent cochlear axons in the chicken. Neuroscience 6:2351–2365.

Wiederhold ML, Kiang NYS (1970) Effects of electric stimulation of the crossed olivocochlear bundle on single auditory-nerve fibers in the cat. J Acoust Soc Am 48:950–965.

Will U (1982) Efferent neurons of the lateral-line system and the VIII cranial nerve in the brain stem of anurans. Cell Tissue Res 225:673–685.

Xue J-T, Holt JC, Brichta AM, Bian J-T, Goldberg JM (2000) Synaptic transmssion to afferents in the turtle posterior crista. Abstr Soc Neurosci 26:1121.

Yagi T, Simpson NE, Markham CH (1977) The relationship of conduction velocity to other physiological proerties of the cat's horizontal canal neurons. Exp Brain Res 30:587–600.

Yamashita M, Ohmori H (1990) Synaptic responses to mechanical stimulation in calyceal and bouton type vestibular afferents studied in an isolated preparation of semicircular canal ampullae of chicken. Exp Brain Res 80:475–488.

Zakir M, Huss D, Dickman JD (2003) Afferent innervation patterns of the saccule in pigeons. J Neurophysiol 89:534–550.

4
Biomechanics of the Semicircular Canals and Otolith Organs

Richard D. Rabbitt, Edward R. Damiano, and J. Wallace Grant

1. Introduction

Modern auditory and vestibular hair cell end organs exhibit remarkable parallels on the molecular and cellular levels that underlie mechanotransduction, synaptic transmission, afferent response dynamics, and efferent control. These parallels are supplemented by species-specific specializations that serve to distinguish the physiological function of each individual organ. Striking morphological specializations are clearly evident at virtually all levels of observation. Without exception, biomechanical specializations play an important role in predetermining the sensitivity and selectivity of each end organ. On the gross anatomical level, the geometry of each organ is sculpted to enhance sensitivity to a particular mechanical stimulus such as sound, linear acceleration, or angular acceleration. The biomechanical response of the mammalian cochlea, for example, plays an important role in selectivity to acoustic stimuli and the spectral decomposition of sound into amplitude-coded and frequency-coded neural signals. Semicircular canal biomechanics is responsible for the selectivity to angular acceleration stimuli and to preferential response to a particular direction of motion. The biomechanical response of the otolith organs is critical to the ability to sense the direction of gravity as well as the time course and direction of three-dimensional linear acceleration. Biomechanics is therefore an important component of transduction by vestibular hair cell sensory organs. In this chapter, we examine the physical properties of end organ fluids and tissues as well as organ morphology in the context of their biomechanical roles in motion sensation by the vestibular semicircular canals and otolith organs.

2. Gross Morphology

The vestibular organs of the inner ear are phylogenetically ancient and, apart from cyclostomes, the gross morphology shows striking similarity throughout the vertebrates (Retzius 1881, 1884; Engström et al. 1966;

Igarashi 1966, 1967; Wersäll and Bagger-Sjöbäck 1974; Lowenstein and Saunders 1975; Igarashi et al. 1981; Blanks et al. 1985; Curthoys and Oman 1987; Ghanem et al. 1998). Membranous labyrinths generally consist, bilaterally, of three endolymph-filled semicircular canals, oriented in nearly orthogonal planes, and a series of endolymph-filled membranous sacs. Examples are shown in Figure 4.1. The primary sensory organs are the horizontal (lateral), anterior, and posterior semicircular canal ampullary organs and the utricular and saccular otolith organs. In nonmammals, these organs are typically supplemented with a lagena and/or papilla neglecta, which expand response dynamics and frequency dependence of motion-sensitive afferent inputs to the brain stem. The endolymph-filled membranous labyrinth is completely immersed in perilymphatic fluid and suspended by fine trabecular filaments within a cartilaginous or osseous labyrinth.

3. Labyrinthine Fluids

The inner ear contains two distinct extracellular fluids, endolymph and perilymph, separated by a membranous labyrinth. Compartmentalization of the fluids plays two roles. One is mechanical because it minimizes sensitivity of the inner ear to atmospheric pressure modulations (Yamauchi et al. 2002) and allows the semicircular canals to take advantage of endolymph fluid dynamics to sense angular motion. The second role is biophysical and provides for the electrochemical gradients between the endolymph, perilymph, and the intracellular compartments necessary for hair cell transduction and neural transmission. These two roles are partially coupled through regulatory mechanisms that control ionic gradients, transport, and fluid volume in the inner ear (Konishi 1982; Sziklai et al. 1992; Rask-Anderson et al. 1999; Salt and DeMott 2000; Salt 2001). The perilymph is rich in Na^+ ions and deficient in K^+ ions and has a chemical composition very similar to cerebrospinal fluid. The ionic composition of endolymph varies considerably from that of perilymph and generally has a K^+ concentration higher than that found in any other fluid in the body. Specific concentrations of endolymphatic Ca^{++}, K^+, and Na^+ vary considerably among species (Ghanem 2002).

All vestibular sensory epithelia are located on the inside surface of the membranous labyrinth, with hair cells oriented to project stereocilia into

--->

FIGURE 4.1. Schematics of the inner ear of a frog, a bony fish, a bird, and a mammal showing the three semicircular canals and otolith maculae (**A–D**) based on Retzius (1881, 1884) and sketches by Wersäll and Bagger-Sjöbäck (1974). Photomicrographs of fixed membranous labyrinths are shown for the oyster toadfish, *Opsanus tau* (**E**) (Provided 62 T.A. Ghanem) and a guinea pig (**F**) (from Engström et al. 1966).

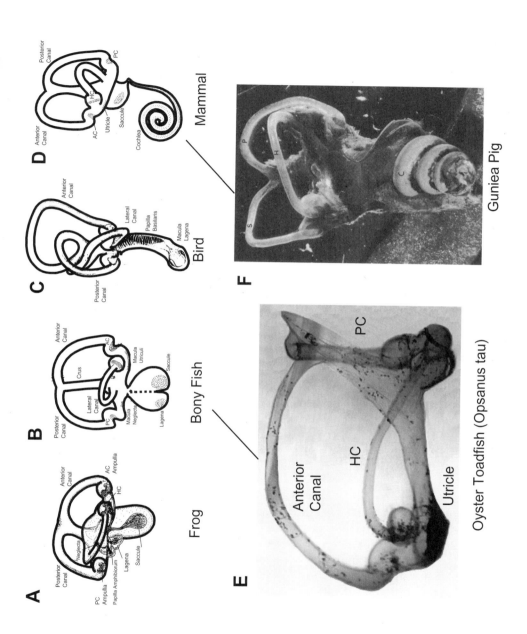

A   Frog

B   Bony Fish

C   Bird

D   Mammal

E   Oyster Toadfish (Opsanus tau)

F   Guniea Pig

the endolymph-filled lumen. Tight junctions at the apical face of hair cells and adjoining supporting cells form a barrier that separates the labyrinthine fluids such that the basolateral surfaces of hair cells are bathed in perilymph whereas the apical faces are bathed in endolymph. The electrochemical gradients present across the apical face are therefore distinct from those present across the basolateral surfaces. Apical (endolymph) electrochemical gradients are .critical to the function of stereociliary displacement-sensitive transduction channels, whereas basolateral (perilymph) electrochemical gradients are critical to hair cell tuning/resonance, synaptic transmission, and efferent control (Crawford et al. 1982; Art and Fettiplace 1984; Art et al. 1985, 1995; Fuchs and Evans 1988; Goodman and Art 1996a, 1996b; Dallos et al. 1997; He and Dallos 1999).

The mechanical properties of perilymph and endolymph are very similar to water (i.e., in terms of density, $\rho = 1.0\,g/cm^2$, and viscosity, $\mu = 0.0085$ dyne/cm^2, Steer et al. 1967). Using a microviscometer, Steer et al. (1967) observed that the viscosity of endolymph was essentially independent of shear rate over the range of physiologically relevant values. It is therefore reasonable to assume Newtonian behavior when considering the dynamics of endolymph. In otolith and canal organs, hair cell stereocilia project into anisotropic, hydrated, mucopolysaccharide structures having mechanical properties similar to the cochlear tectorial membrane. It is important to note that standard linear viscoelastic constitutive models (such as a Kelvin model) cannot capture the stress–strain behavior of these materials as the frequency is increased (Weiss and Freeman 1997; Freeman and Abnet 2000). This may be particularly important in the cochlea and in auditory responses of the saccule in nonmammalian species. The volumetric impedance of the cupula measured in the semicircular canals indicates that an elastic approximation may be adequate, at least for low frequencies of stimulation associated with volitional head movements (Yamauchi 2002).

4. The Semicircular Canals

The semicircular canals are the primary systems responsible for sensing angular acceleration of the head and transmitting the information to the brain stem. As originally described independently by Mach, Bruerer, and Crum-Brown in the latter part of the nineteenth century, angular motion sensation relies on inertial forces, caused by head accelerations, to generate endolymph fluid flow within the toroidal semicircular canals (Breuer 1874; Crum-Brown 1874; Mach 1875; Camis 1930). The site of mechanotransduction within the semicircular canals is localized to the crista ampullaris, which is a crestlike ridge in the ampullary wall that protrudes into the lumen of the ampulla. A sensory epithelium resides on the surface of the crista, which is encased by the cupula. The sensory epithelium consists of a monolayer of supporting cells and receptor cells. Hair cell bundles, consisting of several stereocilia and a single kinocilium, originate from the apical surface of the sensory hair cells and extend up to 80 μm into the cupula (Silver et al. 1998; Helling et al. 2000). The cupula spans the cross section of the ampulla and conforms approximately to the receptor region of the crista (Hillman 1974), as illustrated in Figure 4.2A. A cross-sectional view of the cupula–crista system from the horizontal ampulla of a squirrel monkey (Saimiri sciureus) is shown in Figure 4.3 (Igarashi 1966). McLaren and Hillman measured the displaced configuration of the cupula in response to a static transcupular pressure (Hillman and McLaren 1979; McLaren and Hillman 1979). They observed a displacement distribution that resembled a diaphragm attached to the ampullary wall along its entire perimeter, with a maximum displacement near the center of the ampulla that decreased monotonically to zero around the periphery (see Fig. 4.4). This diaphragmlike deformation field has since been confirmed to extend to dynamic stimuli *in vivo* (Yamauchi 2002; Yamauchi et al. 2002).

The membranous labyrinth enclosing the endolymph varies between 15 and 50 μm in thickness. It is comprised of fibrous ectodermal epithelium that consists of three layers: an outer layer of fibrous tissue containing blood vessels, a thick middle layer containing many papilliform projections, and an inner layer comprised of epithelial cells (Gray 1907, 1908). The stiffness of the membranous tissue is orders of magnitude greater than that of the cupula. Based on this, one would expect pressure gradients within the

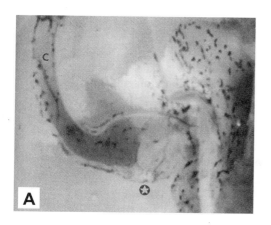
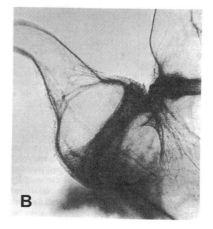

FIGURE 4.2. Photomicrograph (**A**) of the ampulla and horizontal canal in the bullfrog (*Rana catesbiana*) after injection with alcian blue dye shows that the cupula spans the entire cross section of the ampulla and illustrates the contour of the cupular leaflet contacting the endolymph (from Hillman and McLaren 1979). Photomicrograph (**B**) shows a plan view of the crista illustrating the orientation of the sensory epithelium and the thickness of the cupula (from Engström et al. 1966).

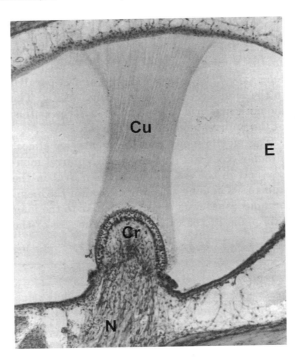

FIGURE 4.3. Photomicrograph showing a cross-sectional view of the cupula (Cu), crista (Cr), nerve bundle (N), and endolymphatic space (E) from the horizontal canal ampulla of a squirrel monkey (*Saimiri sciureus*) (from Igarashi 1966). Fibers in the cupula are identified in the image as running from the sensory epithelium to the apex of the ampulla.

ampulla to have a much larger impact on cupular deformations than on membranous duct deformations. This has been directly validated by Yamauchi et al. (2002) based on endolymphatic pressure measurements and afferent recordings. There are, however, some extreme conditions involving large accelerations, surgical canal manipulations, or canal dehiscence where membranous duct deformation appears to play a role (Rabbitt et al. 1999; Cremer et al. 2000; Hess et al. 2000; Minor et al. 2001; Rabbitt et al. 2001a), but such conditions are pathological. Interested readers should consult Rabbitt et al. (1999) for a model and discussion of the role of membranous duct deformability.

Another factor to consider is the extent to which motion of the membranous labyrinth follows the motion of the temporal bone. In birds and mammals, the perilymphatic space is enclosed by a nearly continuous osseous canal that is lined with periosteum (Curthoys et al. 1977). Fine connective tissue filaments, referred to as trabeculae, span the perilymphatic space and suspend the membranous endolymphatic duct from the

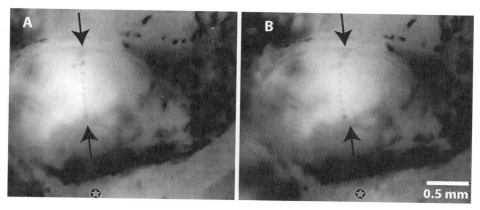

FIGURE 4.4. Photomicrograph showing resting (**A**) and displaced (**B**) configurations of the cupula in the horizontal ampulla of the bullfrog (*R. catesbiana*). The cupula was stained along a linear tract using oil droplets delivered with a micropipet (left). Following compression of the canal near the ampulla, the displacement of the cupula can be visualized (right). The displacement of the drops remains small near the crista and near the ampullary apex opposite the crista. The greatest displacement occurs near the center of the ampulla, suggesting that the cupula deforms like a diaphragm and is attached to the ampullary wall along its entire perimeter (from McLaren and Hillman 1979).

periosteum within the osseous canal. The filaments serve to anchor the membranous labyrinth to the temporal bone such that the gravitoinertial acceleration experienced by the sensory organs could be expected to be nearly identical to that experienced by the temporal bone. There are no experimental data to suggest significant relative motion between the temporal bone and the membranous labyrinth. In light of the structure, most mathematical models of semicircular canal mechanics treat the endolymphatic duct as a perfectly rigid series of interconnected tubes securely attached to, and moving with, the temporal bone (Van Buskirk et al. 1976; Oman et al. 1987; Rabbitt 1999). In the present review, we therefore restrict our attention to rigid membranous labyrinth models constrained to move in synchrony with the temporal bone—assumptions that appear to be reasonable for patent labyrinths subjected to volitional motion stimuli.

4.1. Macromechanical Model

The flow of endolymph within the semicircular canals is induced by angular acceleration of the head and is governed by Newton's second law of motion (i.e., conservation of momentum). For Newtonian incompressible fluids, such as endolymph, conservation of momentum takes the form of the Navier–Stokes equations, subject to appropriate viscous and kinematical boundary conditions arising from fluid–structure interactions within the endolymphatic canal. The first documented fluid-dynamical analysis of endolymph flow was carried out by Lorente de Nó (1927). He applied the creeping flow limit of the Navier–Stokes equations to the fluid in a single uniform toroidal duct and ignored the cupula. Under these assumptions, he was able to derive an approximate expression for transient endolymph flow in response to a step change in head velocity. His results indeed capture the integrating property of semicircular canal fluid mechanics. A more detailed version of this analysis that includes an elastic model of the cupula was developed by Van Buskirk and co-workers (Van Buskirk and Grant 1973; Van Buskirk et al. 1976) and is reviewed by Van Buskirk (1987). An alternative approach that includes the nonuniform geometry of the canal was presented by Oman et al. (1987). Damiano and Rabbitt (1996) and Damiano (1999) unified these approaches using asymptotic methods and expanded on the previous work to address the fluid–structure interaction within the ampulla. Differences between macromechanical predictions of these models are relatively small in the sense that all three approaches lead to an overdamped second-order differential equation describing canal macromechanics (with slight differences in morphological significance in the model parameters). More substantial differences are found when the analysis is extended to three-dimensional motions of the head by including fluid coupling and interactions between all three endolymphatic ducts in the canal triad (Rabbitt 1999). Single-canal and three-canal models are discussed below.

4.2. General Equations of Motion

In the present review, we use a simple control volume approach to derive equations governing the volume displacement of endolymph relative to the membranous duct. Readers interested in an alternative approach, starting with the three-dimensional Navier–Stokes equations, should consult Damiano and Rabbitt (1996). It is worth noting that both approaches arrive at the same equations governing endolymph volume displacement. Taking the control volume approach, we consider conservation of momentum for a short element of endolymph within the membranous duct. The fluid in contact with the membranous duct is assumed to move with the velocity of the duct (and therefore with head velocity) according to the no-slip condition, whereas fluid within the center of the duct, in general, has a velocity that lags or leads the velocity of the duct wall (see Fig. 4.5). Applying Newton's second law tangent to the duct centerline coordinate, s, provides the general expression for momentum conservation of an incompressible fluid given by

$$\rho\left(\frac{D(Au)}{Dt} + A\frac{\partial U}{\partial t}\right) = -A\frac{\partial P}{\partial s} - \tau C, \qquad (4.1)$$

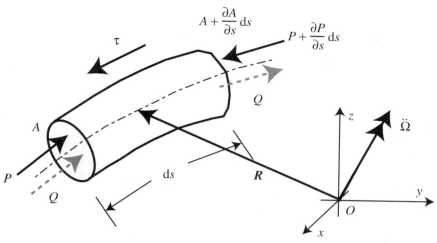

FIGURE 4.5. Free-body diagram of a short section of endolymph within a semicircular canal showing pressure acting within the fluid and shear stresses acting tangent to the curved centerline of the duct. These stresses act on their respective areas to generate forces that accelerate or decelerate the endolymph in inertial space.

where $P(s, t)$ is the pressure, $\tau(s, t)$ is the shear stress acting on the duct wall, $A(s)$ is the local cross-sectional area of the rigid duct, and $C(s)$ is the local inside circumference of the duct. Relative to the ground fixed coordinate system, the local velocity of the duct wall averaged over the duct cross section and projected in the direction tangent to s is denoted by $U(s, t)$. The spatially averaged tangential velocity component of the endolymphatic fluid *relative* to the duct wall is denoted by $u(s, t)$ and is in general referenced to a noninertial reference frame. Because Newton's second law applies to the absolute acceleration (i.e., relative to an inertial reference frame), the tangential acceleration of the rigid duct wall, $\partial U/\partial t$, must be accounted for in the total acceleration of the endolymphatic fluid on the left-hand side of Eq. (4.1). Convected acceleration of endolymph relative to the moving coordinate s must be accounted for when calculating the acceleration of the fluid. This is done in the usual way using the chain rule (i.e., the well-known material derivative from fluid mechanics, where $(Du/Dt) = (\partial u/\partial t) + (u(\partial u/\partial s))$. To simplify the equations, we replace the average fluid velocity, u, with $(1/A)\cdot(dQ/dt)$, where Q (cm^3) is the endolymph volume displacement. In terms of Q, Eq. (4.1) can be written as

$$\frac{\rho}{A}\frac{\partial^2 Q}{\partial t^2} + \frac{\tau C}{A} = -\frac{\partial P}{\partial s} - \rho\frac{\partial U}{\partial t}, \qquad (4.2)$$

where the convective nonlinearities, u^2 and $u(\partial u/\partial s)$, can be shown to be small and were neglected (Damiano and Rabbitt 1996).

To complete the formulation, we need to relate the wall shear stress, τ, to the kinematical variables pertaining to the flow. We divide the stress into two parts, one due to fluid viscosity, τ_μ, and the other due to cupular shear stiffness, τ_γ. Damiano and Rabbitt (1996) solved for the unsteady velocity profile in the canal to determine

$$\frac{\tau_\mu C}{A} = \frac{\mu \lambda_\mu}{A^2} \frac{dQ}{dt}, \tag{4.3}$$

where μ is the dynamic viscosity of endolymph (dyne s/cm^2) and λ_μ is a dimensionless frequency-dependent velocity profile factor that depends on the shape of the velocity distribution over the duct cross section. This factor reduces to 8π for low frequencies of head movement where the instantaneous velocity profile is approximately parabolic (Oman et al. 1987; Damiano and Rabbitt 1996). Following the same approach for an elastic material provides

$$\frac{\tau_\gamma C}{A} = \frac{\gamma \lambda_\gamma}{A^2} Q, \tag{4.4}$$

where γ is the cupular shear stiffness (dyne/cm^2) and λ_γ is a dimensionless displacement profile factor that depends on the shape of the cupular displacement distribution over the ampullary cross sections. If we assume a simple diaphragm displacement, then $\lambda_\gamma = 8\pi$ (McLaren and Hillman 1979). Inserting the endolymph drag and cupular restoring force into Eq. (4.2), we find

$$\frac{\rho}{A} \frac{d^2Q}{dt^2} + \frac{\mu \lambda_\mu}{A^2} \frac{dQ}{dt} + \frac{\gamma \lambda_\gamma}{A^2} Q = -\frac{\partial P}{\partial s} - \rho \frac{\partial U}{\partial t}. \tag{4.5}$$

This is the key equation governing endolymph and cupular volume displacements in the semicircular canals. Note that, by conservation of mass of an incompressible fluid in a rigid duct, $Q = Q(t)$ only, and thus the volume flow is instantaneously uniform along the entire length of the duct. In the following sections, we apply Eq. (4.5) first to model a single canal and then to model three interconnected canals.

4.3. Single-Canal Macromechanics

To model a single semicircular canal, we ignore fluid communication with the two sister canals and integrate Eq. (4.5) around the toroidal loop. The result is

$$m \frac{d^2Q}{dt^2} + c \frac{dQ}{dt} + kQ = f, \tag{4.6}$$

where the equivalent mass, damping, and stiffness parameters are

$$m = \oint \frac{\rho}{A} ds, \tag{4.7}$$

$$c = \oint \frac{\mu \lambda_\mu}{A^2} ds, \tag{4.8}$$

and

$$k = \oint \frac{\gamma \lambda_\gamma}{A^2} ds. \tag{4.9}$$

Equations (4.7)–(4.9) are appropriate for nearly circular endolymphatic duct cross sections but can easily be extended to other shapes by adjusting λ following Oman et al. (1987). The inertial forcing appearing on the right-hand side is

$$f = \oint \rho (\ddot{\boldsymbol{\Omega}} \times \boldsymbol{R}(s)) \cdot d\boldsymbol{s}. \tag{4.10}$$

In this we have written the magnitude of the local tangential acceleration, $\partial U / \partial t$, in terms of the angular acceleration, $\ddot{\boldsymbol{\Omega}}$, and the local position vector, $\boldsymbol{R}$. According to Kelvin's theorem, the linear acceleration can be shown to evaluate to zero on integration around the toroidal loop such that the rigid-duct model is only sensitive to angular acceleration (Rabbitt 1999). This is the reason why semicircular canals generally do not respond to linear accelerations. One exception is noted for cases when duct deformability is critical, such as in surgically plugged canals (Hess et al. 2000; Rabbitt 2000). The rigid-duct models presented here do not address this case. For rotation about a fixed axis, it is convenient to write the inertial forcing, Eq. (4.10), in the form

$$f = \ddot{\Omega} g, \tag{4.11}$$

where

$$g = \oint \rho \hat{\boldsymbol{m}} \times \boldsymbol{R}(s) \cdot d\boldsymbol{s}, \tag{4.12}$$

$\ddot{\Omega}$ is the magnitude of the angular acceleration, and $\hat{\boldsymbol{m}}$ is a unit vector in the direction of the angular acceleration vector.

Equation (4.6) provides a very simple second-order ordinary differential equation describing macromechanics of a single semicircular canal. This single-canal model is the same as that derived directly from the Navier–Stokes equations by Damiano and Rabbitt (1996) for the fluid dynamics in the long and slender regions of the canal and is closely related to the work of Oman et al. (1987). The model also has nearly the same mathematical form as Steinhausen's classical *torsion pendulum model* (Steinhausen 1933) and predicts essentially the same viscous drag as the analysis derived by Lorente de Nó (1927). The main difference between the classical model and the model presented here is that, in the case of the latter, the parameters are determined directly from the morphology of the canal and the physical properties of the endolymph, perilymph, and cupulae. The viscosity, density, and shear modulus appearing in the integrands of Eqs. (4.7)–(4.9) vary as functions of position, taking on values for the endolymph in the fluid and values for the mucopolysaccharide in the cupulae. These lumped coefficients take into account the distributed kinetic energy and dissipation of unsteady viscous endolymph flow and the elastic restoring force of the cupulae. Notice that the effective mass is proportional to the inverse of the integrated local cross-sectional area of the duct, whereas the effective damping is proportional to the squared inverse of the integrated local cross-sectional area. Because of this, slender regions of each duct dominate the determination of coefficients and therefore dominate the mechanical response.

4.3.1. Time Constants and Transfer Function

Equation (4.6) can easily be transformed to the Laplace domain to obtain the transfer function ($cm^3 s/rad$)

$$T_{ssc}(\tilde{s}) = \frac{\tilde{Q}}{\dot{\tilde{\Omega}}} = \frac{(g/m)\tilde{s}}{(\tilde{s}+1/\tau_1)(\tilde{s}+1/\tau_2)}, \tag{4.13}$$

which relates the transformed cupular volume displacement $\tilde{Q}$ (cm^3) to the transformed angular head velocity $\dot{\tilde{\Omega}}$ (rad/s). The two time constants appearing in the transfer function are

$$\frac{1}{\tau_1}, \frac{1}{\tau_2} = \frac{c}{2m}\left(1 \pm \sqrt{1 - \frac{4km}{c^2}}\right). \tag{4.14}$$

In all species studied to date, the semicircular canals are highly overdamped ($4km \ll c^2$), which results in real-valued time constants. Expanding Eq. (4.14) in a Taylor series for small $4km/c^2$ provides the approximate time constants (s) and corner frequencies (rad/s) given by

$$\frac{1}{\tau_1} = \omega_1 \approx \frac{k}{c} \tag{4.15}$$

and

$$\frac{1}{\tau_2} = \omega_2 \approx \frac{c}{m}. \tag{4.16}$$

4.3.2. Mechanical Response Dynamics

For sinusoidal stimulation with frequency ω, the response is found in the frequency domain by replacing the Laplace variable in Eq. (4.13) with $\tilde{s} = i\omega$, where $i = \sqrt{-1}$. The cupular volume displacement predicted by this simple model has a bandpass filter character when plotted relative to angular head velocity (see Fig. 4.6A,B). For the canals, the lower corner frequency is determined by the slow (long) time constant, τ_1, and the upper corner frequency is determined by the fast (short) time constant, τ_2 (Steinhausen 1933; Wilson and Jones 1979; Van Buskirk 1987). Between the two corner frequencies, the model predicts that the viscosity of the endolymph flowing within the slender duct mechanically integrates the angular acceleration of the head to produce a cupular volume displacement proportional to angular velocity of the head. Below the lower corner frequency, the response is predicted to be attenuated by the stiffness of the cupula. Above the upper corner frequency, the response is predicted to be attenuated by the inertia of the endolymph within the slender portion of the duct, where the kinetic energy is highest. Figure 4.7 illustrates these features and their dependence on model parameters. The lower corner frequency predicted by this simple model corresponds well with experimental data. The upper corner frequency, however, is likely to be in error due to the fact that this simple model ignores frequency dependence of the velocity profile and membranous duct deformability—both of which increase with increasing frequency (Rabbitt 1999).

Additional insight into semicircular canal mechanics can be gained by studying cupular responses in the time domain. For a step change in angular velocity of the head, the volume displacement of the cupula is predicted to be

$$Q(t) = \frac{\dot{\Omega} g \tau_1 \tau_2}{m(\tau_1 - \tau_2)} (e^{-t/\tau_1} - e^{-t/\tau_2}), \tag{4.17}$$

where $\dot{\Omega}$ (rad/s) is the magnitude of the step change in angular velocity of the head. For a step in angular acceleration of the head, the cupular volume displacement is predicted to be

$$Q(t) = \frac{\alpha g \tau_1 \tau_2}{m}\left(1 - \frac{\tau_1 e^{-t/\tau_1} - \tau_2 e^{-t/\tau_2}}{\tau_1 - \tau_2}\right), \tag{4.18}$$

where α (rad/s^2) is the magnitude of the step change in angular acceleration of the head. These responses to step stimuli are shown in Figure 4.6C,D. For a step change in angular velocity, the inertial force consists of a very brief impulse occurring at time $t = 0$. The impulse causes a rapid displacement of the cupula governed by the fast time constant followed by a period of slow recovery to zero governed by the slow time constant. For a step change in angular acceleration, the inertial force is maintained over time. This maintained force causes a maintained displacement of the cupula that recovers to zero only after the stimulus is discontinued.

Real stimuli are not ideal steps or steady sinusoids but rather consist of more complex wave forms. One simple example is the response to sinusoidal angular velocity of the head initiated at time $t = 0$. For this stimulus, the cupular volume displacement is

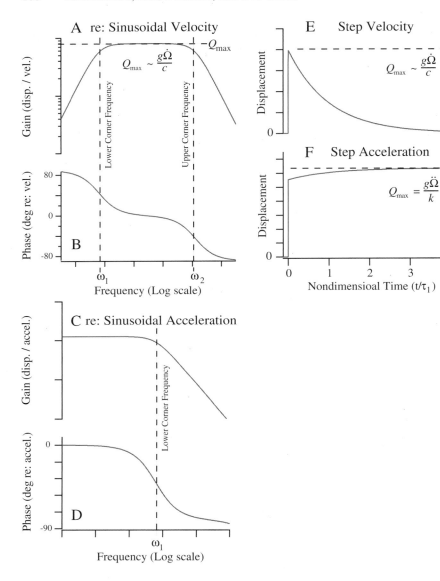

$$Q(t) = \dot{\Omega}\left(\frac{g\tau_1\tau_2\omega}{m(\tau_1 - \tau_2)}\left(\frac{\tau_2 e^{-t/\tau_2}}{1 + \tau_2^2\omega^2} - \frac{\tau_1 e^{-t/\tau_1}}{1 + \tau_1^2\omega^2}\right) + G_n \sin(\omega t - \phi)\right), \quad (4.19)$$

where G_n is the steady-state gain and ϕ_n is the steady-state phase as provided in the Bode plots (Figs. 4.6 and 4.7). For practical purposes, the transient involving the exponential is typically neglected, provided the stimulus frequency is sufficiently low (i.e., if $\omega \ll 1/\tau_2$).

4.3.3. Approximate Model Parameters Based on Morphology

Equations (4.7)–(4.10) can be integrated numerically on the basis of 3D morphological data or approximated in terms of key semicircular canal dimensions. Because the mass and stiffness terms are dominated by the long and slender portion of the semicircular canal duct, their values are given approximately by

$$m \approx \frac{\rho\ell}{A_d}, \quad (4.20)$$

and

$$c \approx \frac{8\pi\mu\ell}{A_d^2}, \quad (4.21)$$

where A_d is the cross-sectional area of the slender portion of the endolymphatic duct and ℓ is the length of the same portion of the duct measured along its curved centerline. In our approximation of the viscous coefficient, c, we have used the steady-flow approximation of 8π for the velocity profile

◄———————————————————————————————

FIGURE 4.6. Temporal response dynamics. Bode plots showing the volume displacement of the cupula relative to peak angular head velocity as a function of stimulus frequency in the form of magnitude (**A**) and phase (**B**). These frequency-dependent responses are replotted in (**C, D**) relative to angular acceleration rather than angular velocity. Volitional angular head movements fall primarily at frequencies between the upper and lower corner frequencies where the phase (**B**) is near zero. This illustrates the function of the semicircular canals as a band-pass angular velocity transducer. Cupular responses in the time domain are shown on the right for a step in angular velocity (**E**) and a step in angular acceleration (**F**). The initial rise time is governed by the fast time constant. The slow time constant governs the recovery to zero for a step in angular velocity and the asymptote to steady state for a step in angular acceleration. These figures also apply to the linear acceleration stimulation of the otolith organs with the following change in variables: $Q \rightarrow u, m \rightarrow m_o, c \rightarrow c_o, k \rightarrow k_o$, and $g\dot{\Omega} \rightarrow g_o a$ (see Section 5). For the otolith organs, the primary physiological frequencies fall below τ_1, corresponding to linear acceleration sensitivity where the response is proportional to, and in phase with, gravitoinertial acceleration.

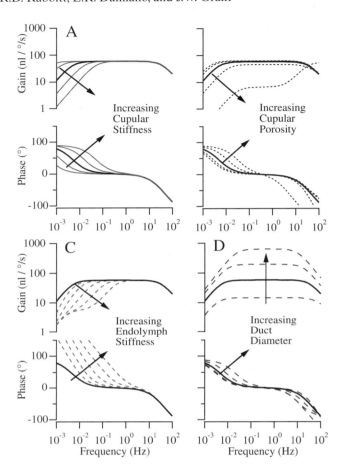

FIGURE 4.7. Parameter study. The family of Bode curves illustrates the influence of changing the equivalent mass, stiffness, and damping parameters on the gain and phase of the cupular volume displacement with regard to head angular velocity. Results are based on a parameter set for the oyster toadfish (*O. tau*) and include the influence of cupular porosity (from Rabbitt et al. 1999).

factor, λ_μ. The cupular stiffness term can be approximated in a similar way using

$$k \approx \frac{8\pi\gamma h}{A_c^2},\tag{4.22}$$

where A_c is the area of the cupula and h is its thickness. If we approximate the duct as lying in a single plane, then the inertial forcing coefficient simplifies to

$$g \approx 2\pi\rho R^2 \cos(\theta), \tag{4.23}$$

where R is the average radius of the semicircular canal measured relative to its centroid and θ is the angle between the normal to the canal plane and the angular rotation vector. The slow and fast time constants obtained using these approximations are

$$\tau_1 \approx \frac{c}{k} \approx \frac{\mu \ell A_c^2}{\gamma h A_d^2} \tag{4.24}$$

and

$$\tau_2 \approx \frac{m}{c} \approx \frac{\rho A_d}{8\pi\mu}. \tag{4.25}$$

It is notable that the slow time constant, τ_1, is relatively insensitive to uniform changes in the size of the labyrinth because the geometric scale factors enter equally in the numerator and denominator of Eq. (4.24). This scale invariance does not hold for the fast time constant. Table 4.1 provides estimates based on the equations above for the three species illustrated in Figure 4.8 for the angle $\theta = 0$. To obtain these estimates, we used morphological data from a variety of sources throughout the literature (Igarashi 1966, 1967; Igarashi et al. 1981; Curthoys and Oman 1987; Ghanem et al. 1998). Physical parameters were approximated using endolymph density, $\rho = 1\,\mathrm{g/cm^3}$, viscosity, $\mu = 0.0085\,\mathrm{dyne\,s/cm^2}$, and cupular shear stiffness, $\gamma = 3.7\,\mathrm{dyne/cm^2}$. The area of the cupula, A_c, was based on the ampullary luminal area above the sensory epithelium, and the thickness of the cupula was taken to be $h = 0.7\sqrt{A_c/\pi}$. The long time constant, τ_1, listed in Table 4.1 for the toadfish (*Opsanus tau*), is consistent with the afferent data reported for that species (Boyle and Highstein 1990; Highstein et al. 1996), whereas the value listed for the squirrel monkey (*S. sciureus*) is slightly higher than the afferent data reported for that species (Fernández and Goldberg 1971). The difference, in this case of the latter, between the mechanical model and the afferent data may be due to geometrical estimates (see Fig. 4.8), the cupular shear stiffness estimate, or nonmechanical factors associated with afferent/hair-cell signal processing (i.e., adaptation). Direct measurements of the mechanical lower corner frequency have not yet been reported, and therefore the numerical values should be viewed cautiously.

TABLE 4.1. Horizontal canal model parameters.

Species	m $(\mathrm{g/cm^4})$	c $(\mathrm{dyne\,s/cm^5})$	k $(\mathrm{dyne/cm^5})$	g $(\mathrm{g/cm})$	τ_1 (s)	τ_2 (s)
Human	1070	179,000	13,320	0.76	13.2	0.0060
Squirrel monkey (*S. sciureus*)	1960	950,600	105,450	0.24	8.9	0.0021
Oyster toadfish (*O. tau*)	1500	330,000	29,230	1.20	11.4	0.0042

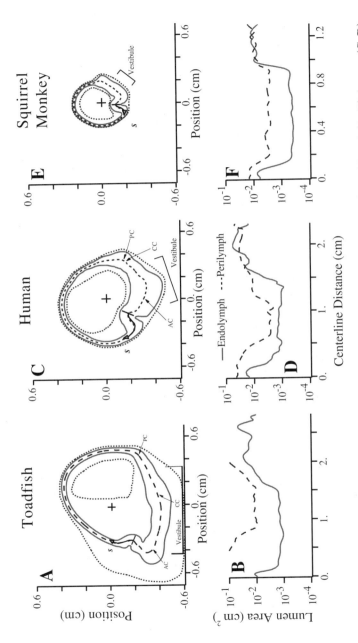

FIGURE 4.8. Model geometries for the horizontal semicircular canal of the oyster toadfish (*O. tau*) (**A**, **B**), human (**C**, **D**), and squirrel monkey (*S. sciureus*) (**E**, **F**) used to estimate model parameters in Table 4.1. The toadfish (*O. tau*) geometry is based on Ghanem et al. (1998), the human geometry is based on Curthoys and Oman (1987), and the squirrel monkey (*S. Sciureus*) geometry is based on Igarashi (1966, 1967) and Igarashi et al. (1981).

4.4. Three-Canal Macromechanics

The single-canal model provides a reasonably accurate prediction of the macromechanical frequency response associated with temporal coding of angular head movements. A shortcoming of the model, however, is its inability to address three-canal interactions. It is well-established that the direction of angular acceleration that elicits the maximal response of a canal nerve is not the same as the anatomical canal plane (Estes et al. 1975; Ezure and Graf 1984; Reisine et al. 1988; Dickman 1996). Part of the reason for this difference is fluid coupling between the canals—flow in one canal can induce flow in the sister canals by entrainment and induced pressure gradients. These features cannot be addressed by single-canal models. Therefore, to understand the role of canal mechanics in directional coding, it is necessary to consider all three canals as one interacting system. To do this, we model the labyrinth by dividing it into a set of N curved tubular segments, each of which is modeled using Eq. (4.5). The endolymph volume displacement, Q_n (cm^3), within each segment is governed by

$$m_n \frac{d^2 Q_n}{dt^2} + c_n \frac{dQ_n}{dt} + k_n Q_n = P_n(l_n) - P_n(0) + f_n, \qquad (4.26)$$

where the parameters m_n, c_n, and k_n are analogous to Eqs. (4.7)–(4.9) but with the integration limited here to the duct segment of interest. In particular,

$$m_n = \int_0^{l_n} \frac{\rho}{A} ds, \qquad (4.27)$$

$$c_n = \int_0^{l_n} \frac{\mu \lambda_\mu}{A^2} ds, \qquad (4.28)$$

and

$$k_n = \int_0^{l_n} \frac{\gamma \lambda_\gamma}{A^2} ds. \qquad (4.29)$$

The coordinates, s_n, run along the curved centerlines of each duct segment, and all other parameters are defined as in the single-canal model. Again, this approach could be generalized for noncircular duct cross sections using the approach of Oman et al. (1987). The inertial forcing appearing on the right-hand side is

$$f_n = \int_0^{l_n} \rho(\ddot{\mathbf{\Omega}} \times \mathbf{R}(s)) \cdot ds. \qquad (4.30)$$

where $\ddot{\mathbf{\Omega}}$ is the angular acceleration relative to the inertial frame resolved into vectorial components in the head-fixed system, and the vector $\mathbf{R}(s)$ runs from the head-fixed origin to the local duct centerline. As before, closed contour integrations of the terms associated with linear head velocity and acceleration vanish and have therefore been omitted from the model equations (Rabbitt 1999).

For three-dimensional motions, we must also take into account the difference between angular acceleration resolved in the head-fixed system and the inertial ground-fixed system. The components of the angular acceleration vector in the head-fixed system are determined from the inertial acceleration vector by application of the orthonormal rotation matrix $\mathbf{N}$ following

$$\ddot{\mathbf{\Omega}} = \mathbf{N}\ddot{\mathbf{\Omega}}_{\text{inertial}}, \tag{4.31}$$

where $\ddot{\mathbf{\Omega}}_{\text{inertial}}$ is the angular acceleration written in the ground-fixed inertial reference frame, and $\ddot{\mathbf{\Omega}}$ is the same vector resolved into the head-fixed system. The matrix $\mathbf{N}$ is not needed for rotations about fixed axes because the coordinate systems can be selected to render $\mathbf{N}$ the identity matrix.

As an example of how to apply these equations, consider a labyrinth composed of six individual segments—the horizontal canal (HC) and ampulla, the anterior canal (AC) and ampulla, the posterior canal (PC) and ampulla, the common crus (CC), the anterior section of the utricular vestibule (UA), and the posterior section of the utricular vestibule (UP). An illustration of the toadfish (*O. tau*) labyrinth divided into these segments is provided in Figure 4.8 (Ghanem et al. 1998; Rabbitt 1999). After some algebra, Eqs. (4.26)–(4.30), along with conservation of mass at the bifurcation points, provide

$$\mathbf{M}\frac{d^2\vec{Q}}{dt^2} + \mathbf{C}\frac{d\vec{Q}}{dt} + \mathbf{K}\vec{Q} = \vec{F}, \tag{4.32}$$

where the vector

$$\vec{Q} = \begin{pmatrix} Q_{\text{HC}} \\ Q_{\text{AC}} \\ Q_{\text{PC}} \end{pmatrix} \tag{4.33}$$

contains the cupular volume displacements. The mass, damping, and stiffness matrices are

$$\mathbf{M} = \begin{pmatrix} m_{\text{HC}} & -m_{\text{UA}} & -m_{\text{UP}} \\ -m_{\text{UA}} & m_{\text{AC}} & -m_{\text{CC}} \\ -m_{\text{UP}} & -m_{\text{CC}} & m_{\text{PC}} \end{pmatrix}, \tag{4.34}$$

$$\mathbf{C} = \begin{pmatrix} c_{\text{HC}} & -c_{\text{UA}} & -c_{\text{UP}} \\ -c_{\text{UA}} & c_{\text{AC}} & -c_{\text{CC}} \\ -c_{\text{UP}} & -c_{\text{CC}} & c_{\text{PC}} \end{pmatrix}, \tag{4.35}$$

and

$$\mathbf{K} = \begin{pmatrix} k_{\text{HC}} & -k_{\text{UA}} & -k_{\text{UP}} \\ -k_{\text{UA}} & k_{\text{AC}} & -k_{\text{CC}} \\ -k_{\text{UP}} & -k_{\text{CC}} & k_{\text{PC}} \end{pmatrix}. \tag{4.36}$$

Elements of these matrices are computed using Eqs. (4.27)–(4.29), where the subscripts (HC, AC, PC) denote closed contour integrals around each canal and the subscripts (UA, UP, CC) denote integrations over segments indicated by the subscripts. The inertial forcing vector is computed using Eq. (4.30) as follows:

$$\bar{F} = \begin{pmatrix} -f_{HC} & -f_{UA} & -f_{UP} \\ -f_{AC} & -f_{CC} & -f_{UA} \\ -f_{PC} & -f_{CC} & +f_{UP} \end{pmatrix}. \tag{4.37}$$

These equations are a simplified version of the model presented by Rabbitt (1999) that use a simple Kelvin viscoelastic model of the cupulae rather than a poroelastic model. Based on theoretical studies, cupular porosity is not expected to contribute significantly to the macromechanical behavior within the most physiologically relevant range of stimulus frequencies but may contribute to mechanical adaptation at very low frequencies. Interested readers should refer to Damiano (1999) and Rabbitt (1999) for additional analysis and discussion of cupular porosity.

4.4.1. Time Constants and Transfer Matrix

In order to determine the time constants for the three-canal model, it is useful to recast the equations as a symmetric system of six first-order differential equations in Hamilton's canonical form

$$\mathbf{C}^* \frac{d\vec{Q}^*}{dt} + \mathbf{K}^* \vec{Q}^* = \vec{F}^*, \tag{4.38}$$

where $\vec{Q}^*$ is a six-element vector containing the cupulae volume displacements in the first three elements and the corresponding velocities in the last three elements. The symmetric $\mathbf{C}^*$ and $\mathbf{K}^*$ matrices are

$$\mathbf{C}^* = \begin{pmatrix} \mathbf{0} & \mathbf{M} \\ \mathbf{M} & \mathbf{C} \end{pmatrix} \tag{4.39}$$

and

$$\mathbf{K}^* = \begin{pmatrix} -\mathbf{M} & \mathbf{0} \\ \mathbf{0} & \mathbf{K} \end{pmatrix}. \tag{4.40}$$

The forcing vector $\vec{F}^*$ is

$$\vec{F}^* = \begin{pmatrix} \vec{0} \\ \vec{F} \end{pmatrix}. \tag{4.41}$$

Solution of the eigenvalue problem, $[(\mathbf{C}^{*-1}\mathbf{K}^*) + \alpha\mathbf{I}]\,\vec{E} = \vec{0}$, provides six real-valued time constants, τ_n, and the corresponding eigenvectors, $\vec{E}_n$. The time constants are analogous to the two time constants arising in the single-canal model and provide the transient exponential time course of the response.

Once again, if the rotation is about a fixed axis, we can use Eqs. (4.11) and (4.12) instead of Eq. (4.30) and arrive at the vectorial analog to Eq. (4.13) given by

$$\tilde{\bar{Q}}* = \mathbf{T}\tilde{\bar{G}}*, \tag{4.42}$$

where the transfer function matrix, $\mathbf{T}$, is

$$\mathbf{T} = \tilde{s}(\mathbf{C}*\,\tilde{s} + \mathbf{K}*)^{-1}. \tag{4.43}$$

Although the response of any given canal depends on all six time constants, Rabbitt (1999) demonstrated that the macromechanical response is dominated by the three lowest time constants. Furthermore, the lower corner frequency of any given canal is dominated primarily by only one of the three time constants. The dominant time constant is similar, but not identical, to that predicted by a single-canal model.

The three-canal model predicts a *maximal response direction* (denoted by the unit vector $\hat{\boldsymbol{n}}_{max}$) for each canal. Rotation about an axis collinear with one of these directions maximizes the volume displacement of the respective cupula. In accordance with this, rotation about any axis orthogonal to $\hat{\boldsymbol{n}}_{max}$ nulls the volume displacement of the respective cupula. The set of vectors perpendicular to $\hat{\boldsymbol{n}}_{max}$ defines the null plane of the associated canal. This prediction is consistent with experimental data (Estes et al. 1975; Reisine et al. 1988; Dickman 1996). The gain of the HC cupula is illustrated in Figure 4.9 as a function of the direction of the canal rotational axis. A vector drawn from the center of rotation (origin in A–C) to the circle (B; sphere in A–C) provides the gain for a head rotation about the particular axis. The maximal response direction is predicted for rotations about the long axis of the double bubble. The corresponding minimal response plane is perpendicular to the maximal response direction (gray square). Orthographic projections are shown in Figure 4.9 along with the centerlines of the HC (solid), AC (short dashed), and PC (long dashed). The maximal direction and a normal vector perpendicular to the geometrical canal plane of the HC, AC, and PC are predicted by the model to differ by an average of ~10°. Hence, anatomical canal planes, regardless of computational technique, are predicted to be only approximate indicators of the maximal response directions of individual canals.

The three-canal model also predicts that rotation about a maximal response direction, $\hat{\boldsymbol{n}}_{max}$, of one canal does not minimize the response of the two sister canals. Null response planes for the AC and PC, for example, lie perpendicular to the maximal response directions sandwiched between the response bubbles. These two null planes intersect along a line denoted $\hat{\boldsymbol{n}}'_{HC}$, the *prime direction* of the HC. Rotation about this direction simultaneously nulls the response of the AC and the PC while maintaining a large (but not maximum) HC gain. For the particular geometry used in Figure 4.9, the prime directions predicted by the model differ from the maximal response directions and from the anatomical canal planes by an average of 28° and

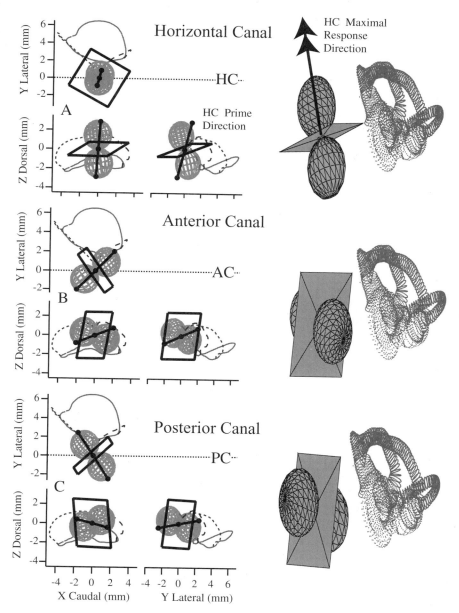

FIGURE 4.9. The response of each semicircular canal end organ to rotations in three-dimensional space is well-approximated by a cosine rule. The figure illustrates the magnitude of the cupular volume displacement as a function of the direction of rotation for the HC (**A**), AC (**B**), and PC (**C**). Left-hand panels show isometric views of the null plane (gray square) for each canal and double bubble indicating the directional sensitivity of each canal relative to the labyrinth surface (cloud of points). The gain for a given direction of rotation (using the right-hand rule) is the distance from the null plane to the surface of the double bubble. The maximal response direction corresponds to a line perpendicular to the null plane and passing through the long axis of the double bubble. Also shown in the form of orthographic projections (left) are canal prime directions (dumbbells).

$29°$, respectively. It is significant to note that rotation about a prime direction is predicted to excite only one canal nerve. For general three-dimensional angular movements, the motion is decomposed into base vectorial components relative to the basis formed by the three prime directions. Each component, which can be regarded as a projection onto this basis, is predicted to be carried by a single canal nerve.

4.4.2. Three-Canal Directional Coding

A very simple model can be used to approximate how the semicircular canals decompose angular head movements at low frequencies into separate vectorial components carried by each canal nerve. The three-dimensional angular acceleration of the head, $\ddot{\Omega}$, written in terms of unit vectors in the prime directions, is

$$\ddot{\Omega} = \ddot{\Omega}_{HC}\hat{n}'_{HC} + \ddot{\Omega}_{AC}\hat{n}'_{AC} + \ddot{\Omega}_{PC}\hat{n}'_{PC}, \tag{4.44}$$

where $\hat{n}'_{HC}$, $\hat{n}'_{AC}$, and $\hat{n}'_{PC}$ are unit vectors in the prime directions and $\ddot{\Omega}_{HC}$, $\ddot{\Omega}_{AC}$, and $\ddot{\Omega}_{PC}$ are the scalar components of acceleration, which excite each canal separately. This equation reproduces all of the directional behavior of the three-canal model, including maximal response directions and null planes. To use the equation, it is necessary to experimentally determine the prime directions. The equation is valid only if the base vectors are the prime directions. It is important to note that the $\hat{n}'$ unit vectors are nonorthogonal. For this nonorthogonal set of base vectors, individual components are *not* maximized for a stimulus in the base direction; they are instead maximized for a stimulus skewed at an angle relative to the base vector.

Equation (4.44) determines how a three-dimensional angular head acceleration is parsed to excite each of the three canals. Forming three inner products of Eq. (4.44) with $\hat{n}'_{HC}$, $\hat{n}'_{AC}$, and $\hat{n}'_{PC}$ provides a system of three scalar equations. Expressing this system of equations in matrix form and inverting, we obtain

$$\begin{pmatrix} \ddot{\Omega}_{HC} \\ \ddot{\Omega}_{AC} \\ \ddot{\Omega}_{PC} \end{pmatrix} = \mathbf{P}^{-1} \begin{pmatrix} \ddot{\Omega} \cdot \hat{n}'_{HC} \\ \ddot{\Omega} \cdot \hat{n}'_{AC} \\ \ddot{\Omega} \cdot \hat{n}'_{PC} \end{pmatrix}, \tag{4.45}$$

where $\mathbf{P}$ is the matrix of direction cosines defined by inner products of the prime unit vectors given by

$$\mathbf{P} = \begin{pmatrix} 1 & \hat{n}'_{AC} \cdot \hat{n}'_{HC} & \hat{n}'_{PC} \cdot \hat{n}'_{HC} \\ \hat{n}'_{HC} \cdot \hat{n}'_{AC} & 1 & \hat{n}'_{PC} \cdot \hat{n}'_{AC} \\ \hat{n}'_{HC} \cdot \hat{n}'_{PC} & \hat{n}'_{AC} \cdot \hat{n}'_{PC} & 1 \end{pmatrix}. \tag{4.46}$$

Equation (4.45) embodies the directional decomposition carried out by canal macromechanics (at least at relatively low accelerations, where the membranous labyrinth moves without deformation and in synchrony with the head). It can be used to convert the angular acceleration vector, $\ddot{\Omega}$, into three components, where each component excites one canal ampullary organ separately (Rabbitt 1999).

4.5. Cupular Micromechanics

Although semicircular canal macromechanics is reasonably well-understood, we do not yet know the relationship between individual micro-mechanical hair bundle movements and macromechanical cupular volume displacement. It is generally assumed that hair bundles are entrained by the cupula and move with it, but this has not yet been shown experimentally. Histological examination of the cupula reveals a highly hydrated anisotropic structure consisting, in part, of mucopolysaccharides that coat collagen-containing connective tissue fibers (Dohlman 1971; Lim 1971; Silver et al. 1998). These connective fibers are embedded in a gel matrix that extends from the sensory epithelium on the surface of the crista to the apex of the ampulla. In the oyster toadfish (O. tau), the connective tissue fibers are restricted to the so-called cupular shell region that surrounds the central region referred to as the cupular antrum (Silver et al. 1998). The cupular antrum itself has been described as devoid of connective tissue fibers and the cupular shell as consisting of a utricular-side and a canal-side central pillar linked through regions of loose fibrous connective tissue to the lateral wings (Silver et al. 1998). Using confocal light microscopy, Silver et al. (1998) observed that the cupular antrum appeared to be uniformly dense and concluded that the so-called "subcupular space" identified by Dohlman (1971) was not present in fresh tissue. Furthermore, cupular tubes, canals, and channels described in other species (Dohlman 1971; Lim 1971, 1973; Lim and Anniko 1985) were not observed by Silver et al. (1998) in any portion of the fish cupular antrum. Silver et al. (1998) described the cupular antrum as a homogeneous aqueous environment in which the stereocilia reside that is surrounded and supported by the anisotropic fibrous cupular shell. These somewhat mixed histological observations do not provide sufficient evidence at present to determine the mechanism(s) through which hair bundle movement is linked to cupular volume displacement. It is clear, however, that the cupula is highly inhomogeneous and divided into anisotropic regions. These morphological features may contribute to regional differences in the activation of hair bundles and may partially contribute to response dynamic maps known to exist in the sensory epithelium (Goldberg et al. 1992). Damiano (1999) has shown that cupular porosity may contribute to extremely low-frequency responses and to mechanical adaptation. In addition to clues obtained from cupular morphology, theoretical considerations indicate that the geometry of the sensory organ and dynamic effects may also contribute to spatial maps and to changes in spatial activation as a function of stimulus frequency. These factors are briefly discussed below.

4.5.1. Static Cupular Deflection and Strain

The morphology of the ampullary organ is well-suited to convert cupular volume displacements into displacements of hair bundles. Figure 4.10 shows results of a simple finite-element model in response to a static pressure gradient applied across the cupula (Yamauchi et al. 2002). Figure 4.10A shows the predicted displacement field as a grayscale contour plot ranging from largest magnitude (black) to smallest (white). The largest magnitude displacement was predicted to be near the center of the cupula, approximately halfway between the sensory epithelium and the apex of the ampulla. Displacement, however, is not the key variable leading to transduction by hair cells. Although hair cell data are often reported in terms of displacement of the hair bundle tip, it is actually the shear strain between adjacent hair bundles that is the key mechanical quantity leading to opening of transduction channels.

If individual cilia move with the cupula, then the shear strain between cilia would follow the shear strain in the cupula. Figure 4.10B shows model predictions for the shear strain in the toadfish (*O. tau*) cupula (Rabbitt et al. 2001b; Yamauchi et al. 2002). It is notable that the maximum shear strain occurs near the surface of the sensory epithelium precisely where the hair bundles are located. This indicates that the organ may be structured to maximize the shear strain acting on the hair bundles. The model also predicts a nonuniform distribution of shear strain across the surface of the crista. Similar results are expected in the human (Njeugna et al. 1992).

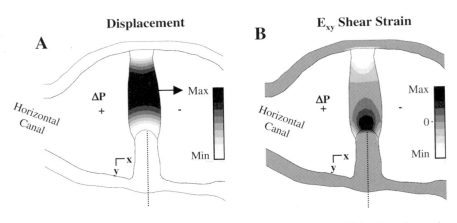

FIGURE 4.10. Results of a simple finite-element model of the toadfish (*O. tau*) cupula showing the displacement (**A**) and shear strain (**B**) in response to a static pressure gradient. The cupular shear strain is predicted to be maximal near the surface of the sensory epithelium, indicating effectiveness of the geometry in converting macro-mechanical stimuli into a form appropriate for activation of hair cell transduction channels (based on Yamauchi 2002; Yamauchi et al. 2002).

4.5.2. Dynamic Cupular Shear Strain

Results of the finite-element model shown in Figure 4.10 illustrate the importance of cupular shear strain in activating sensory hair cells. Macromechanical models of the semicircular canals, such as Eq. (4.6) or Eq. (4.32), only predict the volumetric displacement of the cupula and not the cupular shear strain. At low frequencies, the shear strain would be expected to be linearly related to the cupular volume displacement, as indicated in the finite-element model. A simple relationship of this type is not expected to extend to high frequencies due to the influence of unsteady inertia (Damiano and Rabbitt 1996). To gain some insight into the relationship between cupular shear strain and volumetric displacement, we consider a simple viscoelastic model of the cupula of the form

$$\rho_c \frac{\partial^2 u}{\partial t^2} - \nabla^2 \left(\mu_c \frac{\partial u}{\partial t} + \gamma u \right) = \frac{\Delta P}{h}, \tag{4.47}$$

where $u(r, \theta, t)$ is the cupular displacement field, (r, θ) is the polar coordinate of a point lying in the plane of the cupula, ΔP is the transcupular differential pressure, and h, ρ_c, and μ_c denote the cupular thickness, density, and dynamic viscosity, respectively. Although this simple model ignores cupular anisotropy, porosity, and inhomogeneity, it nevertheless does provide some insight into the dynamical behavior. Considering the special case of a circular ampullary cross section, having area A_c, and a sinusoidal stimulus, then, after start-up transients decay, the axisymmetric solution to Eq. (4.47) is written as

$$u = B e^{i\omega t} \sum_{n=1}^{\infty} \zeta_n J_{0_n}, \tag{4.48}$$

where

$$\zeta_n = \frac{\langle 1, J_{0_n} \rangle}{(\beta_n^2 + i(S_t - \beta_n^2 K)) \langle J_{0_n}, J_{0_n} \rangle}. \tag{4.49}$$

Here, $S_t = A_c \rho_c \omega / \pi \mu_c$ is the Stokes number characterizing the ratio of unsteady inertial to viscous forces in the cupula and $K = \gamma / \omega \mu_c$ is a dimensionless quantity characterizing the ratio of elastic to viscous forces in the cupula. For this simple axisymmetric model, the eigenfunctions are given by $J_{0_n} = J_0(\beta_n r / \sigma)$, where J_0 is the zeroth-order Bessel function of the first kind, σ is the cupular radius, and β_n are eigenvalues required to meet the no-slip boundary condition at $r = \sigma$. The inner product $\langle f, g \rangle$ denotes integration of the product fg over the cross-sectional area of the ampulla, and B is a constant determined by matching the macromechanical cupula volume displacement with endolymph volume displacement. The shear strain evaluated at the surface of the sensory epithelium is

$$\varepsilon = \left.\frac{\partial u}{\partial r}\right|_{r=\sigma} = Be^{i\omega t}\sum_{n=1}^{\infty}\frac{-\beta_n\zeta_n J_1(\beta_n)}{\sigma}. \tag{4.50}$$

Integration Eq. (4.48) over the ampullary cross section and using Eq. (4.50), we obtain the transfer function from macromechanical cupular volume displacement to cupular shear strain given by

$$T_2 = \frac{\varepsilon}{Q} = -\frac{\displaystyle\sum_{n=1}^{\infty}\beta_n\zeta_n J_1(\beta_n)}{2\pi\sigma^3\displaystyle\sum_{n=1}^{\infty}\frac{\zeta_n}{\beta_n}J_1(\beta_n)}. \tag{4.51}$$

The parameters ζ_n appearing in this result are complex-valued when written in the frequency domain and single-pole transfer functions when written in the Laplace domain. Equation (4.51) is therefore a ratio of two infinite series of terms consisting of simple transfer functions. These infinite series can be approximated as *fractional derivatives* or fractional Laplace operators. Although Eq. (4.51) is not considered a good model for the cupula, because it ignores most of the structure, the mathematical form does indicate that, at high frequencies, cupular micromechanics may give rise to a frequency-dependent phase shift between hair bundle movement and cupular volume displacement. Consistent with this view, a small phase shift present in hair cell receptor potentials and currents has been reported *in vivo* in the semicircular canals (Highstein et al. 1996; Rabbitt 2000). The effect, however, appears to be relatively small and may not be a major factor in most treatments of canal mechanics. The larger micromechanical contribution is likely to be associated with regional differences in mechanical hair bundle activation and is a question that remains open to investigation.

5. Otolith Mechanics

It was shown in the previous sections that the interconnected toroidal structure of the labyrinth renders each of the semicircular canals insensitive to linear acceleration and/or the direction of gravity. Vestibular sensitivity to rectilinear gravitoinertial acceleration is therefore restricted to the otolith organs. As their names imply, the utricular and saccular otoliths reside in the membranous utricle and saccule, respectively. Connective tissue firmly anchors the substrate of these organs to the membranous labyrinth and ultimately to the temporal bone of the skull. The otolithic organs are flat layered structures that lie between the endolymphatic fluid and the membranous labyrinth substrate. Figure 4.11 illustrates the orientation of the utricular and saccular otolith organs in the temporal bone (B,D) (Spoendlin 1966; Wilson and Jones 1979). As discussed later, both the orientation of hair cells within the organ and the gross orientation relative to the tempo-

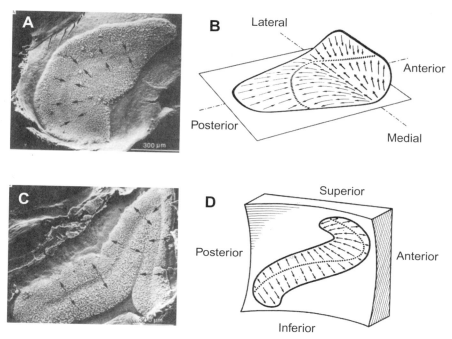

FIGURE 4.11. Spatial polarization of hair bundle orientations on utricular (**A, B**) and saccular (**C, D**) maculae. Figures B and D are from Spoendlin (1966), and figures A and C are from Lim (1979), both as reported by Wilson and Jones (1979).

ral bone play important roles in directional coding by the otolith organs. The layered structure plays an important role in temporal response dynamics. In mammals, the top layer of the otolithic structure consists of calcium carbonate crystals, called otoconia, which are bound together in a conglomerate by a saccharide gel. The otoconia have a density of $2.71\,g/cm^3$ (Carlström 1963), whereas the conglomerate otoconial layer as a whole has a density of 1.32–$1.39\,g/cm^3$ (Steinhausen 1934). The otoconial layer has a surface area of approximately $2\,mm^2$ and is approximately 20–30 μm thick (Igarashi 1966, 1967; Igarashi et al. 1981). The middle layer, called the gel layer, is approximately 20–30 μm thick and consists of a highly deformable, viscoelastic, gelatinous saccharide gel having a density of $1.0\,g/cm^3$. Most of the viscous damping in the otolithic system is attributed to the dissipation that arises in this layer. The bottom layer, referred to as the sensory base, is bound together with connective tissue and contains innervated receptor hair cells that have stereocilia that extend into the gel layer. During linear accelerations of the head, the inertia of the dense otoconial layer gives rise to relative displacements between it and the sensory base. These displacements result in shearing deformation (see Fig. 4.12) and hair bundle displacements that ultimately lead to spike train modulation in afferents. In accordance with Einstein's equivalency, the system responds equally to gravitational and inertial accelerations.

5.1. Otolith Macromechanics

Jaeger et al. (2002) presented a comprehensive analysis of human otolith biomechanics that includes the three-dimensional orientation and geometry of the utricle and saccule. Their results clearly illustrate the importance of the geometry and a spatially distributed deformation field in the generation of the detailed activation pattern of hair cells and subsequent neural activity transmitted to the brain stem. Although the fine-scale biomechanics is key to understanding the spatially distributed activation of individual hair bundles, the gross-scale behavior of the otolith organs can be described in terms of a relatively simple lumped-parameter system. Motion of the otolithic mass relative to the substrate is governed by the conservation of momentum (Steinhausen 1934; De Vries 1950; Grant and Cotton 1991; Cotton and Grant 2000; Kondrachuk 2001a, 2001b). The simple one-dimensional analysis presented here treats the otoliths as an overdamped, second-order spring-mass damper, where the otoconial layer is modeled as a rigid solid mass, the gel layer as an isotropic viscoelastic material, and the endolymph as a Newtonian fluid with uniform viscosity. The free-body diagram shown in Figure 4.12 depicts the externally applied forces acting on the otoconial layer in the $\hat{n}$ direction, where u corresponds to the relative displacement of the otoconial layer measured tangent and with respect to the sensory base, which is assumed to be rigid and firmly attached to the temporal bone. The x components of the forces that act on the otoconial layer are:

1. $F_x^W = \rho_o V_o \, \hat{n} \cdot \mathbf{g}$, the $\hat{n}$ component of the weight of the otoconial layer, where ρ_o is the density of the layer, V_o is the volume of the layer, and $\mathbf{g}$ is the gravitational acceleration vector;

FIGURE 4.12. Free-body diagram (**A**) of the otoconial layer showing applied forces acting on the layer. The interfacial forces include the shear elastic force, F_x^E, exerted by the gel layer and the viscous drag force, F_x^D, which represents the sum of the viscous shear forces, F_x^e and F_x^g, acting on the otoconial layer due to the endolymph and gel layer, respectively. The body force includes the influence of buoyancy and gravitoinertial acceleration acting on the otoconial mass in the direction tangent to the layer. Panels B–D show four configurations illustrating the system in equilibrium (**B**), under a purely tilt stimulus (**C**), under a purely inertial stimulus (**D**), and under a gravitoinertial acceleration (**E**).

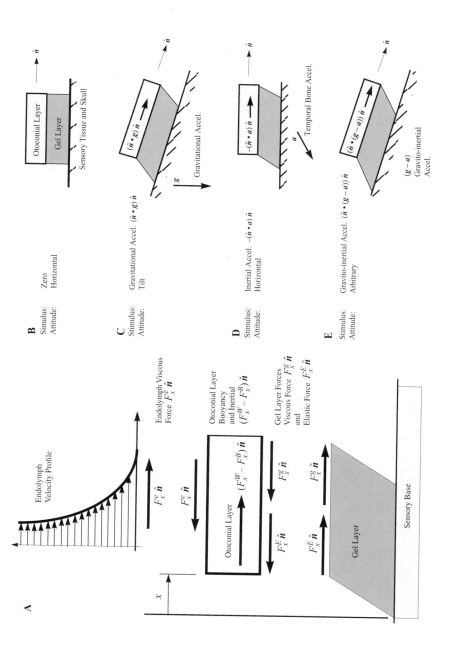

2. $F_x^B = -\rho_e V_o\,\hat{\boldsymbol{n}} \cdot (\boldsymbol{g} - \ddot{\boldsymbol{X}})$, the $\hat{\boldsymbol{n}}$ component of the buoyancy force acting on the otoconial layer, where ρ_e is the density of the endolymph and $\ddot{\boldsymbol{X}}$ is the acceleration of the substrate (i.e., the temporal bone);

3. $F_x^D = c_o\dot{u}$, where c_o is the effective drag coefficient associated with the endolymph and the gel layer and $\dot{u}$ is the velocity of the otoconial layer measured with respect to the sensory base; and

4. $F_x^E = k_o u$, where k_o is the effective stiffness coefficient associated with the gel layer.

The coefficient c_o accounts for the combined effects of the shear stresses that arise between the endolymph fluid and the otoconial layer and between the gel and otoconial layers. The viscous shear stress exerted by the gel layer is reported as being 5–10 times greater than the fluid shear stress exerted by the endolymph (Cotton and Grant 2000).

Application of Newton's second law of motion within the plane of the otoconial layer and in the $\hat{\boldsymbol{n}}$ direction provides

$$F_x^W - F_x^B - F_x^D - F_x^E = \rho_o V_o(\ddot{u} + \hat{\boldsymbol{n}} \cdot \ddot{\boldsymbol{X}}), \qquad (4.52)$$

where $\hat{\boldsymbol{n}} \cdot \ddot{\boldsymbol{X}}$ is the component of the inertial acceleration of the temporal bone within the otoconial plane and u is the displacement of the otoconial layer in the $\hat{\boldsymbol{n}}$ direction relative to the moving substrate. In terms of the relative displacement u, velocity $\dot{u}$, and acceleration $\ddot{u}$, Eq. (4.52) becomes

$$m_o\ddot{u} + c_o\dot{u} + k_o u = g_o a, \qquad (4.53)$$

where $g_o = m_o(1 - \rho_e/\rho_0)$ is the effective inertial mass and $a = \hat{\boldsymbol{n}} \cdot (\boldsymbol{g} - \ddot{\boldsymbol{X}})$ is the component of the gravitoinertial acceleration in the plane of the organ and in the $\hat{\boldsymbol{n}}$ direction. Notice that, with a slight change in variables, the form of this differential equation is identical to the equation describing cupular volume displacement in the semicircular canals given by Eq. (4.6). It is also noteworthy that if the otoconial layer were neutrally buoyant, such that $\rho_o = \rho_e$, the forcing term would vanish. The fact that $\rho_o > \rho_e$ leads to the inertial force responsible for movement of the otolith mass relative to its substrate. An otolith model similar to this was first derived by De Vries (1950).

5.2. Otolith Transfer Function and Time Constants

The macromechanical temporal response dynamics of the otoconial layer, governed by Eq. (4.53), is characterized by two time constants. These time constants are analogous to those found in the semicircular canals but are shifted relative to the physiological time scales of motion present in linear versus angular head movements. The semicircular canals experience a wide range of volitional angular motion stimuli at frequencies between the two characteristic times, whereas the otolith organs experience additional low-frequency stimuli arising primarily from slow tilts of the head relative to gravity. For these low-frequency stimuli, it is natural to report the otoconial layer displacement relative to the component of the gravitoinertial acceleration in the plane of the otolith. Equation (4.53) transformed to the Laplace domain provides the transfer function (s^2) given by

$$T_o(\tilde{s}) = \frac{\tilde{u}}{\tilde{a}} = \frac{(g_o/m_o)}{(\tilde{s} + 1/\tau_1)(\tilde{s} + 1/\tau_2)},\tag{4.54}$$

which relates the transformed otoconial layer displacement, $\tilde{u}(\tilde{s})$ (cm), to the transformed gravitoinertial linear acceleration of the head, $\tilde{a}(\tilde{s})$ (cm/s^2). In the case of the canals, the analogous transfer function T_{ssc} was written using angular velocity as the input. The form of T_{ssc} given by Eq. (4.13) illustrates the angular-velocity-sensing character of the canals, whereas the form of T_o given by Eq. (4.54) illustrates the linear-acceleration-sensing character of the otolith organs.

The two time constants appearing in the otolith transfer function, τ_1 and τ_2, are related to the parameters m_o, c_o, and k_o according to Eq. (4.14) (where we note the change in model parameters needed to convert results from the semicircular canal form to the otolith form: $Q \to u$, $m \to m_o$, $c \to c_o$, $k \to k_o$, and $g \to g_o$). As with the semicircular canals, the otolith organs have been found to be highly overdamped (De Vries 1950), resulting in the long and short time constants predicted by Eqs. (4.15) and (4.16). In humans, $0.1 < \tau_2 < 4\,\mu s$ and $5 < \tau_1 < 40\,\mu s$ (Grant and Cotton 1991; Grant et al. 1994). From these values, it is clear that transient decay in the system response is dominated by the long time constant, τ_1.

5.3. Macromechanical Response Dynamics

Temporal responses shown in Figure 4.16 for angular motion stimulation of the semicircular canals can be applied directly to linear motion stimulation of the otolith organs with the appropriate change in model parameters. A step change in linear velocity of the head produces an impulse of acceleration and is predicted to generate a displacement of the otoconial layer relative to the substrate of the form

$$u(t) = \frac{vg_o\tau_1\tau_2}{m_o(\tau_1 - \tau_2)}\left(e^{-t/\tau_1} - e^{-t/\tau_2}\right),\tag{4.55}$$

where v (cm/s) is the magnitude of the step change in linear velocity of the head. For a step change in linear gravitoinertial acceleration, the displacement of the otoconial layer is predicted to be

$$u(t) = \frac{ag_o\tau_1\tau_2}{m_o}\left(1 - \frac{\tau_1 e^{-t/\tau_1} - \tau_2 e^{-t/\tau_2}}{\tau_1 - \tau_2}\right),\tag{4.56}$$

where a (cm/s^2) is the magnitude of the step change in linear acceleration of the head. We see that the system response begins with a fast rise to a displacement magnitude that is proportional to the magnitude of the step change in velocity and then decays at a rate dictated by the long time constant, τ_1. Thus, the otolith transduces an impulse in acceleration for a step change in linear head velocity. These responses to step stimuli are shown in Figure 4.6C,D. For a step change in angular velocity, the inertial force consists of a very brief impulse occurring at time $t = 0$.

The displacement of the otoconial layer is proportional to the magnitude of the step change in gravitoinertial acceleration as $e^{-t/\tau_l} \to 0$. It is clear that, once the transient exponentials decay, the displacement of the otoconial layer is proportional to the magnitude of the linear or gravitoinertial acceleration. If the long time constant is short, we see that the otolith displacement is proportional to acceleration.

From the step responses, it is clear that static (zero frequency, DC) otolith deflection, u, is proportional to, and in phase with, the component of gravitoinertial acceleration projected into the plane of the otoconial layer and in the $\hat{n}$ direction. The Bode plots shown in Figure 4.6 provide the magnitude (Fig. 4.6C) and phase (Fig. 4.6D) of the otolith displacement as functions of frequency. The flat response from DC up to the first corner frequency illustrates the biomechanics of this system as a linear acceleration sensor over this frequency range. This corresponds to the range typical of physiological motion environments in adult humans where the otolith is expected to function.

5.4. Otolith Micromechanics and Directional Coding by the Otoliths

The displacement of the otolithic mass relative to the sensory base is the major macromechanical variable leading to hair bundle displacements and transduction. The transformation from macromechanical displacement to hair bundle deflection falls into the realm of otolith micromechanics. Although micromechanics of the otolith organs has not yet been investigated experimentally, some clues can be drawn from the morphology. Figure 4.13 shows the micromechanical structure of the saccular macula from the American bullfrog (*Rana catesbianas*) (Kachar et al. 1990; Kurc et al. 1999; Lins et al. 2000). The peripheral otolithic mass responsible for imparting the gravitoinertial force on the organ was removed prior to fixation and is not shown in this figure (Hillman and Lewis 1971; Lim 1984). The image shows the two layers separating the otolithic mass from the apical surfaces of hair cells. The gelatin membrane (GL) forms cavities above stereociliary bundles and attaches directly to the kinocilium. A layer of fine columnar filaments forms a second layer, which spans the space between the apical surfaces of hair cells and the basal surface of the gelatin layer. Based on this structure, Kachar et al. (1990) suggested that the shear strain in the columnar filament layer may directly relate to hair bundle shear strain and

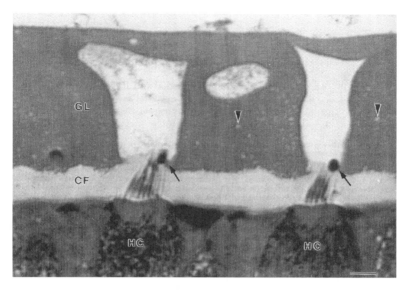

FIGURE 4.13. Otolith membrane. Light micrograph of fixed tissue showing hair cells (HC) surrounded by supporting cells and covered by two layers of extracellular gelatinous material: CF, columnal filament layer (light); and GL, otolithic gelatin membrane (dark). The kinocilium of each hair bundle is connected to the GL at the edge of a void in the GL (white). Scale bar: 5 μm (from Kachar et al. 1990).

thereby be a primary determinant of micromechanical transduction. It is not yet known to what extent local deformation of the gelatin layer near connections to the kinocilium might modify the direction and magnitude of hair bundle displacements globally or locally.

The otolith organs sense the direction of linear gravitoinertial acceleration using a combination of macromechanics and micromechanics. The otolithic mass primarily moves within the plane of the macula in the direction of the applied linear gravitoinertial acceleration (projected onto the plane). The component of acceleration perpendicular to the macula is not sensed. Directional sensitivities of afferent neurons innervating the utricle, for example, are sensitive to stimuli within the plane of the organ and can be nulled for stimuli approximately perpendicular to the organ (Lowenstein and Saunders 1975; Fernández and Goldberg 1976a, 1976b; Goldberg et al. 1990). All hair cells in the organ receive nearly the same hair bundle displacement oriented in the direction of otolithic mass displacement. Directional sensitivity within the plane comes from the polarization and orientation of hair cells relative to the otolithic mass. Hair cells oriented with their kinocilium aligned with the direction of displacement respond maximally, whereas hair cells oriented with their kinocilium perpendicular to this direction do not respond. The orientations of saccular and utricular hair cells are illustrated in Figure 4.11. This arrangement allows the utricular otolith organ to sense two components of gravitoinertial acceleration within the plane of the organ. Sensing the third component out of the utricular plane requires a second organ, the saccular macula. Responses of the two organs taken together provide both the direction and time course of gravitoinertial acceleration. This "encoding plane" description captures the major features of directional sensitivity of the otolith organs (Goldberg et al. 1990).

In mammals, the otolithic mass consists of numerous otoconia. It is generally assumed that the otoconia adhere tightly to each other and move as a single rigid unit, a claim that is based on the structure but has not yet been verified experimentally. Even if the otoconia do move as a single rigid unit, theoretical considerations suggest that the motion may not be a simple linear trajectory. Euler's equations describing the three-dimensional dynamics of rigid bodies uncouple only if the mass is completely symmetric. In the case of the otolith organs, Euler's equations tell us that linear acceleration in one direction will induce a compensatory moment and slight rotation due to the fact the the mass is not uniformly distributed in space. For sinusoidal stimuli, this would be manifested as an orbital wobble of the otolithic mass superimposed on a linear translation. Although motions of this type have not yet been directly recorded in the otolith organs, there is some evidence from afferent recordings in fish that such motions may be present even for a rigid otolithic mass (Fay and Edds-Walton 1997). In addition, theoretical analysis indicates that some deformation of the otoconial layer is likely present in humans (Jaeger et al. 2002). These considerations indicate that not all regions of the macula may experience exactly the same direction and magnitude of hair bundle deflection. As a result, directional sensitivities may be somewhat more complex than would be implied on the basis of hair bundle orientation and may have some weak frequency dependence. Based on correlations between afferent directional sensitivity and morphological structure, these effects are likely to be small relative to linear rigid-body translation.

6. Hair Cell Mechanics

6.1. Hair Bundle Micromechanics

It has been shown in otolith hair cells that the ion channels responsible for mechanotransduction reside at the tips of the cilia and are gated by the relative motion between adjacent cilia (Hudspeth 1989; Hudspeth and Logothetis 2000). Hair bundle micromechanics plays the crucial role of converting gross forces and displacements into a form appropriate for

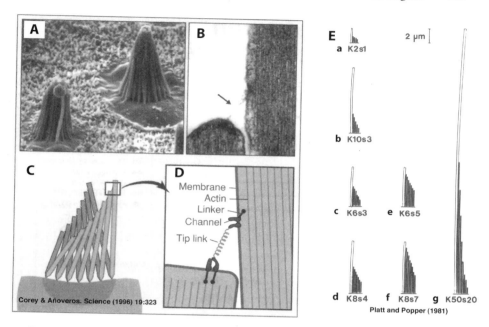

FIGURE 4.14. Hair bundle micromechanics. Scanning electron micrograph of hair bundles (**A**), electron micrograph of tip links (**B**), schematic illustration of hair bundle deflection (**C**), and tip link activation of mechanosensitive channel (**D**) (A–D from Corey and Garcia-Añoveros 1996). The right-hand panel (**E**) illustrates the relative length of hair bundles from various hair cell end organs of fish (Platt and Popper 1981).

gating the transduction channels. The process is illustrated in the left and center panels of Figure 4.14 (Corey and Garcia-Añoveros 1996). Hair bundle mechanical properties (Duncan and Grant 1987; Pickles 1992; Szymko et al. 1992; Cotton 1998; Cotton and Grant 2000), electrostatic interactions (Dolgobrodov et al. 2000a, 2000b), and tip link mechanics (Assad et al. 1991; Denk et al. 1995; Corey and Garcia-Añoveros 1996; Kachar et al. 2000; Tsuprun and Santi 2000), all contribute to the micromechanics. The net result is that the short cilia of otolithic hair cells (relative to the longer cilia of semicircular canal hair cells) appear to deflect like rigid rods pivoting at their insertions (Fig. 4.14C). This mode of deformation implies that each of the cilia is relatively soft in shear but stiff in extension, which is consistent with mechanical properties derived from an actin core ultrastructure (Cotton 1998; Cotton and Grant 2000). The bundle deformation pattern causes adjacent cilia to shear relative to each other, thus inducing a stretch of the tip links as the bundle is deflected. Tip link stretch leads to opening of transduction channels (Hudspeth 1989; Assad et al. 1991; Gillespie et al. 1993; Denk et al. 1995; Corey and Garcia-Añoveros 1996; Kachar et al. 2000). The tip links are subjected to a resting tension. This resting tension influences bundle mechanical properties and is believed to arise from the actin-myosin adaptation motor present in the stereocilia (Hocohen et al. 1989; Hudspeth 1989; Gillespie et al. 1993; Jaramillo and Hudspeth 1993; Duncan et al. 1999; Holt and Corey 2000; Hudspeth and Logothetis 2000). There are currently no experimental data available to determine whether the same pattern of deformation observed for short utricular hair bundles (Fig. 4.14E, a) extends to long semicircular canal hair bundles (Fig. 4.14E, g). Theoretical considerations indicate that the long bundles may exhibit significant bending along their length. Because the specific locations of the transduction channels in semicircular canal hair bundles are not yet known, the functional role of hair bundle mechanics in short versus long bundles remains speculative. One established observation is that the length of the hair bundles generally correlates with the best frequency of transduction—hearing and vibration-sensing organs express short stereocilia, whereas semicircular canals express long stereocilia (Platt and Popper 1981, 1984; Brownell 1982; Saunders and Dear 1983). A common hypothesis is that hair bundle lengths are tuned to optimize interaction/coupling with the adjacent fluids and tissues and, in some cases, to maximize the dynamic response within a specific frequency range (Wright 1984; Patuzzi and Yates 1987; Freeman and Weiss 1990a, 1990b).

6.2. Motility and Transduction Nanomechanics

The preceding sections have focused on the passive biomechanics of the vestibular semicircular canals and otolith organs. Since the 1980s, it has become evident that active forces generated by the sensory hair cells also contribute to micromechanical and nanomechanical responses of hair cell organs. Active force generation and amplification is particularly important to sensitivity and selectivity of the mammalian cochlea and nonmammalian hearing organs (Brownell et al. 1985; Gillespie et al. 1993; Hudspeth and Gillespie 1994; Murugasu and Russell 1996; Manley and Gallo 1997; Garcia et al. 1998; Hudspeth and Logothetis 2000; Manley 2000; Zheng et al. 2000). The energy provided by cochlear hair cells is quite large and is responsible for otoacoustic emissions present in the ear canal (Kemp 1978; Brownell 1990; Lonsbury-Martin et al. 1990; Burns et al. 1998). In mammals, at least part of the energy is attributed to outer hair cell somatic electromotility associated with the unique properties of the multilayered cell wall and the presence of the protein prestin (Brownell et al. 1985; Spector 1999; Zheng et al. 2000; Ludwig et al. 2001). Prestin is not extensively expressed in vestibular hair cells—an observation consistent with the apparent lack of somatic electromotility in these cells. There are, however, two forms of active hair bundle movements that have been observed in vestibular hair cells that are not derived from somatic electromotility. The first is associated with myosin-based adaptation of the hair bundle tip links and their associated transduction channels (Eatock et al. 1987; Hudspeth 1989; Gillespie and Corey 1997; Holt et al. 1997; Garcia et al. 1998; Eatock 2000; Holt and Corey 2000), and the second is associated with a rapid twitch of hair bundles triggered by the onset of the transduction current (Benser et al. 1996; Holt and Corey 2000; Ricci et al. 2000). The myosin-based mechanism is typically termed "adaptation" and the calcium-sensitive twitch is typically termed "fast adaptation." For low-frequency gravitoinertial stimuli, such as is present during locomotion or during changes in head orientation relative to gravity, the myosin-based adaptation would be expected to be the dominant active motile mechanism in vestibular hair bundles. One exception might be vestibular organs involved in sensing high frequencies. such as hearing in fish, where fast adaptation may be involved in enhancing frequency-specific tuning (Benser et al. 1996; Ricci et al. 2000). Both of these adaptation mechanisms generate active forces that feed back to influence the micromechanics. The cascade is illustrated in Figure 4.15. Measurements of the response of the lateral line cupula have provided at least

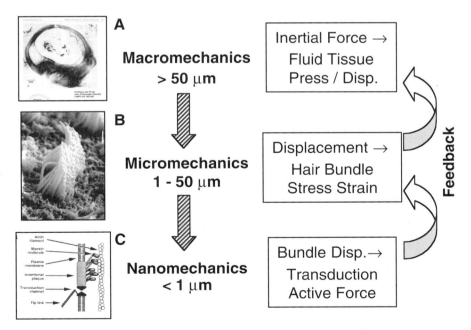

FIGURE 4.15. Schematic illustrating feedback of electromechanical forces generated by hair bundles to influence canal macromechanics. Three levels are indicated: macromechanics (**A**, Curthoys et al. 1987), micromechanics (**B**, Peterson et al. 1996), and nanomechanics (**C**, Hudspeth and Gillespie 1994).

some evidence that the nanomechanics associated with opening and closing of the hair cell transduction channel influences the micromechanics of cupular motion (van Netten and Khanna 1994). Although evidence of active hair bundle motility continues to mount, the physiological/biomechanical role of this motility in vestibular organs *in vivo*, including the extent to which active movements feed back to alter the micromechanics and macromechanics, remains open to investigation. When considering active feedback by hair cells in the labyrinth, it may be important to note the role of the end organs as inputs to essentially linear neural control systems such as the VOR. Low-gain vestibular afferents, in particular, have a very large linear operating range. These vestibular responses are in contrast with the mammalian cochlea, where a strong nonlinear behavior is present in both the afferent responses and the mechanics, a behavior that is attributed to hair cell active feedback in the cochlea. In addition to this difference, some vestibular afferents are essentially unaffected by activation of the efferent vestibular system. These facts taken together indicate that forces owing to active feedback by hair cells in the vestibular system may be quite small relative to forces owing to passive mechanics, at least for regions of the sensory epithelium mapping to low-gain afferents.

7. Relationship to Afferent Coding

As described above, nature has sculpted the vestibular labyrinth in three-dimensional space to mechanically decompose complex acceleration stimuli into individual components, where each component excites a unique subset of sensory hair cells and afferent fibers. This biomechanical specialization is responsible for selectivity of vestibular afferents to linear or angular motion and is responsible for coding the direction of three-dimensional linear and angular accelerations (Goldberg et al. 1990; Rabbitt 1999). Biomechanics is also responsible for processing (i.e., filtering) the temporal wave form of angular and linear motion stimuli. This temporal processing occurs prior to mechanotransduction by hair cells. Temporal responses of all afferents therefore are influenced by mechanical filtering. The extent to which the signal is further modified by hair cell/afferent dynamics varies significantly between individual afferent fibers (see Lysakowski and Goldberg, Chapter 3). A distinct subset of semicircular canal afferents, for example, respond with temporal dynamics that closely reflect canal macromechanics (Fig. 4.6). These are the velocity-sensitive semicircular canal afferents, which faithfully follow the macromechanical response of canal cupulae such that the signal transmitted to the brain reflects this response. Low-gain afferents that reflect the macromechanical response are also present in the otolith organs. These otolith afferents respond with dynamics closely aligned with otolith macromechanics (Fig. 4.6). Some of the acceleration-sensitive low-gain otolith afferents also exhibit a low-frequency phase advance and gain decrease not present in the macromechanics. A more notable difference between the mechanical responses of vestibular organs and afferent responses is observed at higher frequencies, where an increased gain and advanced phase are commonly observed in afferent responses. These differences are primarily the result of hair cell/afferent signal processing. The gain and phase enhancements have been described as resulting from a process of adaptation where maintained hair bundle displacements result in a rapid rise followed by a period of decay in the rate of neurotransmitter release (Fernández and Goldberg 1976a, 1976b; Hudspeth and Gillespie 1994; von Gersdorff and Matthews 1999). The fact that hair cell receptor potentials and transduction currents follow the mechanical response much more closely than they follow the afferent response indicates that a majority of the nonmechanical signal processing may arise from hair cell/afferent biophysics and synaptic complexes (Highstein et al. 1996; Rabbitt 2000). From a signal processing point of view, these nonmechanical events can be viewed as adding a fractional derivative, or fractional Laplace operator, to the response dynamics of individual afferents. This fractional derivative is associated with a phase lead and increased gain not present in the macromechanical response.

8. Summary

The semicircular canals and otolith organs are inertial sensors that respond to the magnitude and direction of head movement in three-dimensional space. The canals are sensitive to angular acceleration, whereas the otolith organs are sensitive to linear gravitoinertial acceleration. Macromechanics of the canals is responsible for parsing the angular acceleration vector into three components, one carried by each canal nerve. This biomechanical decomposition underlies directional coding of angular motion stimuli. Temporal response dynamics is governed primarily by fluid viscosity within the slender duct that serves to integrate angular acceleration stimuli to produce a mechanical response proportional to angular velocity (in the mid-frequency band). Stiffness of the cupula and mass of the endolymph cause the angular velocity sensitivity to be attenuated at low and high frequencies, respectively, such that the overall mechanical behavior of each canal can be described as a bandpass angular velocity sensor. Macromechanics of the otolith organs can be described as a low-pass linear acceleration sensor that converts the linear component of gravitoinertial acceleration into deflection of the otolithic mass (at low frequencies). Viscosity and inertia serve to attenuate the response and shift the phase at higher stimulus frequencies. Directional coding by the otolith organs relies on the macroscale orientation of the macula in the temporal bone as well as the microscale orientation of hair bundles within each epithelium. Biomechanics leading to the displacement of hair bundles and nanomechanical gating of the transduction channels plays an important role in predetermining the directional sensitivity of the end organs and in contributing to the temporal response dynamics of individual afferent nerves. Mechanics alone, however, cannot account for the entire range of temporal response dynamics represented in the population of vestibular afferents. For this it is necessary to also include processing by hair cell/afferent biophysics and synaptic complexes. Chapter 3 by Lysakowski and Goldberg provides a review of vestibular afferent response dynamics.

References

Art JJ, Fettiplace R (1984) Efferent desensitization of auditory nerve fibre responses in the cochlea of the turtle *Pseudemys scripta elegans*. J Physiol 356: 507–523.

Art JJ, Crawford AC, Fettiplace R, Fuchs PA (1985) Efferent modulation of hair cell tuning in the cochlea of the turtle. J Physiol 360:397–421.

Art JJ, Wu YC, Fettiplace R (1995) The calcium-activated potassium channels of turtle hair cells. J Gen Physiol 105:49–72.

Assad JA, Shepherd GM, Corey DP (1991) Tip-link integrity and mechanical transduction in vertebrate hair cells. Neuron 7:985–994.

Benser ME, Marquis RE, Hudspeth AJ (1996) Rapid, active hair bundle movements in hair cells from the bullfrog's sacculus. J Neurosci 16:5629–5643.

Blanks RHI, Curthoys IS, Bennett ML, Markham CH (1985) Planar relationships of the semicircular canals in rhesus and squirrel monkeys. Brain Res 340: 315–324.

Boyle R, Highstein SM (1990) Resting discharge and response dynamics of horizontal semicircular canal afferents of the toadfish, *Opsanus tau*. J Neurosci 10:1557–1569.

Breuer J (1874) Uber die funktion der bogengange des ohrlabyrinthes. Med Jahrb (Wien) 4:72–124.

Brownell WE (1982) Cochlear transduction: an integrative model and review. Hear Res 6:335–360.

Brownell WE (1990) Outer hair cell electromotility and otoacoustic emissions. Ear Hear 11:82–92.

Brownell WE, Bader CR, Bertrand D, de Ribaupierre Y (1985) Evoked mechanical responses of isolated cochlear outer hair cell. Science 227:194–196.

Burns EM, Keefe DH, Ling R (1998) Energy reflectance in the ear canal can exceed unity near spontaneous otoacoustic emission frequencies. J Acoust Soc Am 103:462–474.

Camis M (1930) The Physiology of the Vestibular Apparatus. Oxford: Clarendon Press.

Carlström D (1963) Crystallographic study of the vertebrate otoliths. Biol Bull 125:441–463.

Corey DP, Garcia-Añoveros J (1996) Mechanosensation and the deg/enac ion channels. Science 273:323–324.

Cotton JR (1998) Mechanical models of vestibular hair cell bundles. Ph.D. Thesis, Virginia Polytechnic Institute and State University, Blacksburg, VA.

Cotton JR, Grant JW (2000) A finite element method for mechanical response of hair cell ciliary bundles. J Biomech Eng 144:44–50.

Crawford JJ, Art AC, Fettiplace R, Fuchs PA (1982) Efferent regulation of hair cells in the turtle cochlea. Proc R Soc Lond B Biol Sci 216:377–384.

Cremer PD, Minor LB, Carey JP, Della Santina CC (2000) Eye movements in patients with superior canal dehiscence syndrome align with the abnormal canal. Neurology 55:1833–1841.

Crum-Brown A (1874) On the sense of rotation and the anatomy and physiology of the semicircular canals of the inner ear. J Anat Physiol 8:327–331.

Curthoys IS, Oman CM (1987) Dimensions of the horizontal semicircular duct, ampulla and utricle in the human. Acta Otolaryngol (Stockh) 103:254–261.

Curthoys IS, Markham CH, Curthoys EJ (1977) Semicircular canal duct and ampulla dimensions in cat, guinea pig and man. J Morphol 151:17–34.

Dallos P, He DZ, Lin X, Sziklai I, Mehta S, Evans BN (1997) Acetylcholine, outer hair cell electromotility, and the cochlear amplifier. J Neurosci 17:2212–2226.

Damiano ER (1999) A poroelastic continuum model of the cupula partition and the response dynamics of the vestibular semicircular canal. J Biomech Eng 121: 449–461.

Damiano ER, Rabbitt RD (1996) A singular perturbation model for fluid dynamics in the vestibular semicircular canal and ampulla. J Fluid Mech 307: 333–372.

Denk W, Holt JR, Shepherd GM, Corey DP (1995) Calcium imaging of single stereocilia in hair cells: localization of transduction channels at both ends of tip links. Neuron 15:1311–1321.

De Vries HL (1950) Mechanics of the labyrinth organs. Acta Otolaryngol 38:262–273.

Dickman JD (1996) Spatial orientation of semicircular canals and afferent sensitivity vectors in pigeons. Exp Brain Res 111:8–20.

Dohlman GF (1971) The attachment of the cupula, otolith and tectorial membranes to the sensory cell areas. Acta Otolaryngol 71:89–105.

Dolgobrodov SG, Lukashkin AN, Russell IJ (2000a) Electrostatic interaction between stereocilia: I. Its role in supporting the structure of the hair bundle. Hear Res 150:94–103.

Dolgobrodov SG, Lukashkin AN, Russell IJ (2000b) Electrostatic interaction between stereocilia: II. Influence on the mechanical properties of the hair bundle. Hear Res 150:273–285.

Duncan RK, Grant JW (1987) A finite element model of inner ear hair bundle micromechanics. Hear Res 104:15–26.

Duncan RK, Eisen MD, Saunders JC (1999) Distal separation of chick cochlear hair cell stereocilia: analysis of contact-constraint models. Hear Res 127:22–30.

Eatock RA (2000) Adaptation in hair cells. Annu Rev Neurosci 23:285–314.

Eatock RA, Corey DP, Hudspeth AJ (1987) Adaptation of mechanoelectrical transduction in hair cells of the bullfrog's sacculus. J Neurosci 7:2821–2836.

Engström H, Lindeman HH, Ades HW (1966) Anatomical features of the auricular sensory organs. In: Second Symposium on the Role of the Vestibular Organs in Space Exploration, SP-115. Moffett Field, CA: NASA Ames Research Center, pp. 33–46.

Estes MS, Blanks RHI, Markham CH (1975) Physiologic characteristics of vestibular first-order neurons in the cat. I. Response plane determination and resting discharge characteristics. J Neurophysiol 38:1239–1249.

Ezure K, Graf W (1984) A quantitative analysis of the spatial organization of the vestibuloocular reflexes in lateral- and frontal-eyed animals. I. Orientation of semicircular canals and extraocular muscles. Neuroscience 12:85–93.

Fay RR, Edds-Walton PL (1997) Directional response properties of saccular afferents of the toadfish, *Opsanus tau*. Hear Res 111:1–21.

Fernández C, Goldberg JM (1971) Physiology of peripheral neurons innervating semicircular canals of the squirrel monkey. II. Response to sinusoidal stimulation and dynamics of peripheral vestibular system. J Neurophysiol 34:661–675.

Fernández C, Goldberg JM (1976a) Physiology of peripheral neurons innervating otolith organs of the squirrel monkey. I. Response to static tilts and to long-duration centrifugal force. J Neurophysiol 39:970–984.

Fernández C, Goldberg JM (1976b) Physiology of peripheral neurons innervating otolith organs of the squirrel monkey. II. Directional selectivity and force-response relations. J Neurophysiol 39:985–995.

Freeman DM, Abnet CC (2000) Deformation of the isolated mouse tectorial membrane produced by oscillatory forces. Hear Res 144:29–46.

Freeman DM, Weiss TF (1990a) Hydrodynamic forces on hair bundles at low frequencies. Hear Res 48:17–30.

Freeman DM, Weiss TF (1990b) Hydrodynamic forces on hair bundles at high frequencies. Hear Res 48:31–36.

Fuchs PA, Evans MG (1988) Voltage oscillations and ionic conductances in hair cells isolated from the alligator cochlea. J Comp Physiol 164:151–163.

Garcia JA, Yee AG, Gillespie PG, Corey DP (1998) Localization of myosin-i beta near both ends of tip links in frog saccular hair cells. J Neurosci 18:8637–8647.

Ghanem TA (2002) Semicircular canal fluid compartment morphology, ionic composition and regulation in the oyster toadfish, *Opsanus tau*. Ph.D. Thesis, University of Utah, Salt Lake City, UT.

Ghanem TA, Rabbitt RD, Tresco PA (1998) Three-dimensional reconstruction of the membranous vestibular labyrinth in the toadfish, *Opsanus tau*. Hear Res 124:27–43.

Gillespie PG, Corey DP (1997) Myosin and adaptation by hair cells. Neuron 19:8637–8647.

Gillespie PG, Wagner MC, Hudspeth AJ (1993) Identification of a 120 kd hair-bundle myosin located near stereociliary tips. Neuron 11:581–594.

Goldberg JM, Desmadryl G, Baird RA, Fernández C (1990) The vestibular nerve of the chinchilla. V. Peripheral relation between afferent discharge properties and peripheral innervation patterns in the utricular macula. J Neurophysiol 63: 791–804.

Goldberg JM, Lysakowski A, Fernández C (1992) Structure and function of the vestibular nerve fibers in the chinchilla and squirrel monkey. Ann N Y Acad Sci 656:93–107.

Goodman MB, Art JJ (1996a) Positive feedback by a potassium-selective inward rectifier enhances tuning in vertebrate hair cells. Biophys J 72:430–442.

Goodman MB, Art JJ (1996b) Variations in the ensemble of potassium currents underlying resonance in turtle hair cells. J Physiol 497:395–412.

Grant JW, Cotton JR (1991) A model for otolith dynamic response with viscoelastic gel layer. J Vestib Res 1:139–151.

Grant JW, Huang CC, Cotton JR (1994) Theoretical mechanical frequency response of the otolithic organs. J Vestib Res 4:137–151.

Gray AA (1907) The Labyrinth of Animals. London: J. and A. Churchill.

Gray AA (1908) The Labyrinth of Animals. London: J. and A. Churchill.

He DZ, Dallos P (1999) Development of acetylcholine-induced responses in neonatal gerbil outer hair cells. J Neurophysiol 81:1162–1170.

Helling K, Clarke AH, Watanabe N, Scherer H (2000) Morphological studies of the form of the cupula in the semicircular canal ampulla. HNO 48:822–827.

Hess BJ, Lysakowski A, Minor LB, Angelaki DE (2000) Central versus peripheral origin of vestibulo-ocular reflex recovery flowing semicircular canal plugging in rhesus monkeys. J Neurophysiol 84:3078–3082.

Highstein SM, Rabbitt RD, Boyle R (1996) Determinants of semicircular canal affe ent response dynamics in the toadfish, Opsanus tau. J Neurophysiol 75:575–596.

Hillman DE, Lewis ER (1971) Morphological basis for a mechanical linkage in otolithic receptor transduction in the frog. Science 174:416–418.

Hillman DE, McLaren JW (1979) Displacement configuration of semicircular canal cupulae. Neuroscience 4:1989–2000.

Hocohen N, Assad JA, Smith WJ, Corey DP (1989) Regulation of tension on hair-cell transduction channels: displacement and calcium dependence. J Neurosci 9:3988–3997.

Holt JR, Corey DP (2000) Two mechanisms for transducer adaptation in vertebrate hair cells. Proc Natl Acad Sci U S A 97:11730–11735.

Holt JR, Corey DP, Eatock RA (1997) Mechanoelectrical transduction and adaptation in hair cells of the mouse utricle, a low-frequency vestibular organ. J Neurosci 17:8739–8748.

Hudspeth AJ (1989) How the ear's works work. Nature 341:397–404.

Hudspeth AJ, Gillespie PG (1994) Pulling springs to tune transduction: adaptation by hair cells. Neuron 12:1–9.

Hudspeth AJ, Logothetis NK (2000) Sensory systems. Curr Opin Neurobiol 10:631–641.

Igarashi M (1966) Dimensional study of the vestibular end organ apparatus. In: Second Symposium on the Role of the Vestibular Organs in Space Exploration, SP-115. Moffett Field, CA: NASA Ames Research Center, pp. 47–54.

Igarashi M (1967) Dimensional study of the vestibular apparatus. Laryngoscope 77:1806–1817.

Igarashi M, O-Uchi T, Alford BR (1981) Volumetric and dimensional measurements of vestibular structures in the squirrel monkey. Acta Otolaryngol 91:437–444.

Jaeger R, Takagi A, Haslwanter T (2002) Modeling the relation between head orientations and otolith responses in humans. Hear Res 173:29–42.

Jaramillo F, Hudspeth AJ (1993) Displacement-clamp measurements of the forces exerted by gating springs in the hair bundle. Proc Natl Acad Sci U S A 90:1330–1334.

Kachar B, Parakkal M, Fex J (1990) Structural basis for mechanical transduction in the frog vestibular sensory apparatus. I. The otolithic membrane. Hear Res 45:179–190.

Kachar B, Parakkal M, Kurc M, Zhao Y, Gillespie PG (2000) High-resolution structure of hair-cell tip links. Proc Natl Acad Sci U S A 97:13336–13341.

Kemp DT (1978) Stimulated acoustic emissions from within the human auditory system. J Acoust Soc Am 64:1386–1391.

Kondrachuk AV (2001a) Finite element modeling of the 3d otolith structure. J Vestib Res 11:13–32.

Kondrachuk AV (2001b) Models of the dynamics of otolithic membrane and hair cell bundle mechanics. J Vestib Res 11:33–42.

Konishi T (1982) Ion and water control in cochlear endolymph. Am J Otolaryngol 3:434–443.

Kurc M, Farina M, Linus U, Kachar B (1999) Structural basis for mechanical transduction in the frog vestibular sensory apparatus. I. The organization of the otoconial mass. Hear Res 131:11–21.

Lim DJ (1971) Vestibular sensory organs. A scanning electron microscopic investigation. Arch Otolaryngol 94:69–76.

Lim DJ (1973) Ultrastructure of the otolithic membrane and the cupula. A scanning electron microscopic observation. Adv Otorhinolaryngol 19:35–49.

Lim DJ (1979) Fine morphology of the otoconial membrane and its relationship to the sensory epithelium. Scanning Electron Microsc 3:929–938.

Lim DJ (1984) The development and structure of the otoconia. In: Friedmann I, Ballantine J (eds) Ultrastructural Atlas of the Inner Ear. London: Butterworths.

Lim DJ, Anniko M (1985) Developmental morphology of the mouse inner ear. A scanning electron microscopic observation. Acta Otolaryngol Suppl 422:1–69.

Lins U, Farina M, Kurc M, Riordan G, Thalmann R, Thalmann I, Kachar B (2000) The otoconia of the guinea pig utricle: internal structure, surface exposure, and interactions with the filament matrix. J Struct Biol 131:67–78.

Lonsbury-Martin BL, Harris FP, Stagner BB, Hawkin MD, Martin GK (1990) Distortion-product emissions in humans. I. Basic properties in normally hearing subjects. Ann Otol Rhinol Laryngol 99:3–42.

Lorente de Nó R (1927) Contribucion al estudio matematico del organo del equilibrio. Trab Publ En La 7:202–206.

Lowenstein O, Saunders RD (1975) Otolith-controlled response from the first order neurons of the labyrinth of the bullfrog to changes in linear acceleration. Proc R Soc Lond B Biol Sci 191:475–505.

Ludwig J, Oliver D, Frank G, Llocker N, Gummer WW, Fakler B (2001) Reciprocal electromechanical properties of rat prestin: the motor molecule from rat outer hair cells. Proc Natl Acad Sci U S A 98:4178–4183.

Mach E (1875) Grundlinien Der Lehre Von Den Bewegungsempfindugen. Leipzig: Engelmann.

Manley GA (2000) Cochlear mechanisms from a phylogenetic viewpoint. Proc Natl Acad Sci U S A 97:11736–11743.

Manley GA, Gallo L (1997) Otoacoustic emissions, hair cells and myosin motors. J Acoust Soc Am 102:1049–1055.

McLaren JW, Hillman DE (1979) Displacement of the semicircular canal cupula during sinusoidal rotation. Neuroscience 4:2001–2008.

Minor LB, Cremer PD, Carey JP, Della Santina CC, Streubel SO, Weg N (2001) Symptoms and signs in superior canal dehiscence syndrome. Ann N Y Acad Sci 942:259–273.

Murugasu E, Russell IJ (1996) The effect of efferent stimulation on basilar membrane displacement in the basal turn of the guinea pig cochlea. J Neurosci 16:325–332.

Njeugna E, Eichhorn JL, Kopp C, Harlicot P (1992) Mechanics of the cupula: effects of its thickness. J Vestib Res 2:227–234.

Oman CM, Marcus EN, Curthoys IS (1987) The influence of semicircular canal morphology on endolymph flow dynamics. Acta Otolaryngol 103:1–13.

Patuzzi RB, Yates GK (1987) The low-frequency response of inner hair cells in the guinea pig cochlea: implications for fluid coupling and resonance of the stereocilia. Hear Res 30:83–98.

Peterson EH, Cotton JR, Grant JW (1996) Structural variation in ciliary bundles of the posterior semicircular canal. Quantitative anatomy and computational analysis. Ann N Y Acad Sci 781:85–102.

Pickles JO (1992) A model for the mechanics of the stereociliary bundle on acousticolateral hair cells. Hear Res 68:159–172.

Platt C, Popper AN (1981) Fine structure and function of the ear. In: Tavolga WN, Popper AN, Fay RR (eds) Hearing and Sound Communication in Fishes. New York: Springer-Verlag, p. 3.

Platt C, Popper AN (1984) Variations in lengths of ciliary bundles on hair cells along the macula of the sacculus in two species of teleost fishes. Scanning Electron Microsc 4:1915.

Rabbitt RD (1999) Directional coding of three-dimensional movements by the vestibular semicircular canals. Biol Cybern 80:417–431.

Rabbitt RD, Highstein SM, Boyle R (eds) (2000) Adaptation to maintained cupular displacements in semicircular canal hair cells US. Afferent nerves of the toadfish, opsanus tau. Midwinter Meeting of the Association for Research in Otolaryngology. Mt. Royal, NJ: Association for Research in Otolaryngology, p. 5664.

Rabbitt RD, Highstein SM, Boyle R (1999) Influence of surgical plugging on horizontal semicircular canal mechanics and afferent response dynamics. J Neurophysiol 82:1033–1053.

Rabbitt RD, Highstein SM, Boyle R (2001a) Physiology of the semicircular canals after surgical plugging. Ann N Y Acad Sci 942:274–286.

Rabbitt RD, Yamauchi AM, Highstein SM, Boyle R (2001b) How endolymph pressure modulates semicircular canal primary afferent discharge. Ann N Y Acad Sci 942:313–321.

Rask-Anderson H, DeMott JE, Bagger-Sjöbäck D, Salt AN (1999) Morphological changes of the endolymphatic sac induced by microinjection of artificial endolymph into the cochlea. Hear Res 138:81–90.

Reisine H, Simpson JI, Henn V (1988) A geometric analysis of semicircular canals and induced activity in their peripheral afferents in the rhesus monkey. Ann N Y Acad Sci 445:163–172.

Retzius G (1881) Das Gehörorgan der Wirbeltiere. I. Das Gehörorgan der Fische und Amphibien. Stockholm: Centraldruckerei.

Retzius G (1884) Das Gehörorgan der Wirbeltiere. II. Das Gehörorgan der Reptilien, der Vögel und der Säugetiere. Stockholm: Centraldruckerei.

Ricci AJ, Crawford AC, Fettiplace R (2000) Active hair bundle motion linked to fast transducer adaptation in auditory hair cells. J Neurosci 20:7131–7142.

Salt AN (2001) Regulation of endolymphatic fluid volume. Ann N Y Acad Sci 942:306–312.

Salt AN, DeMott JE (2000) Ionic and potential changes of the endolymphatic sac induced by endolymph volume changes. Hear Res 149:46–54.

Saunders JC, Dear SP (1983) Comparative morphology of stereocilia. In: Fay RR, Gourevitch G (eds) Hearing and Other Senses: Presentations in Honor of EG Wever. Groton: Amphora Press, p. 175.

Silver RB, Reeves AP, Steinacker A, Highstein SM (1998) Examination of the cupula and stereocilia of the horizontal semicircular canal in the toadfish, *Opsanus tau*. Comp Neurol 402:48–61.

Spector AA (1999) A nonlinear electroelastic model of the cochlear outer hair cell. Appl Mech Am 6:19–22.

Spoendlin HH (1966) The ultrastructure of the vestibular sense organ. In: Wolfson RJ (ed) The Vestibular System and Its Diseases. Philadelphia: University of Pennsylvania Press, pp. 39–68.

Steer RW, Li YT, Young LR (1967) Physical properties of the labyrinthine fluids and quantification of the phenomenon of caloric stimulation. In: Third Symposium on the Role of the Vestibular Organs in Space Exploration, SP-152. Moffett Field, CA: NASA Ames Research Center, pp. 409–420.

Steinhausen W (1933) Über die beobachtungen der cupula in der bognegangsampullen des labyrinthes des libenden hechts. Pflügers Arch 232:500–512.

Steinhausen W (1934) Über die durch die otolithen ausgelösten kräfte. Pflügers Arch 235:538–544.

Sziklai I, Ferrary E, Horner KC, Sterkers O, Amiel C (1992) Timerelated alteration of endolymph composition in an experimental model of endolymphatic hydrops. Laryngoscope 102:431–438.

Szymko Y, Dimitri P, Saunders J (1992) Stiffness of hair bundles in the chick cochlea. Hear Res 59:241–249.

Tsuprun V, Santi P (2000) Helical structure of hair cell stereocilia tip link in the chinchilla cochlea. J Assoc Res Otolaryngol 21:224–231.

Van Buskirk W (1987) Vestibular mechanics. In: Skalak R, Chien S (eds) Handbook of Bioengineering. New York: New York Academy of Sciences, pp. 31.1–31.17.

Van Buskirk WC, Grant JW (1973) Biomechanics of the semicircular canals. Biomechanics Symposium. New York: American Society of Mechanical Engineers, pp. 53–54.

Van Buskirk WC, Watts RG, Liu YK (1976) The fluid mechanics of the semicircular canals. J Fluid Mech 78:87–98.

van Netten SM, Khanna SM (1994) Stiffness changes of the cupula associated with the mechanics of hair cells in the fish lateral line. Proc Natl Acad Sci U S A 91:1549–1553.

von Gersdorff H, Matthews G (1999) Electrophysiology of synaptic vesicle cycling. Annu Rev Physiol 61:725–752.

Weiss TF, Freeman DM (1997) Equilibrium behavior of an isotropic polyelectrolyte gel model of the tectorial membrane: effect of pH. Hear Res 111:55–64.

Wersäll J, Bagger-Sjöbäck D (1974) Morphology of the vestibular sense organ. In: Kornhuber HH (ed) Handbook of Sensory Physiology: Vestibular System. New York: Springer-Verlag, pp. 123–170.

Wilson VJ, Jones GM (1979) Mammalian Vestibular Physiology. New York: Plenum Press.

Wright A (1984) Dimensions of the cochlear stereocilia in man and guinea pig. Hear Res 13:89–98.

Yamauchi AM (2002) Cupular Micromechanics and Motion Sensation in the Toadfish Vestibular Semicircular Canals. Salt Lake City: University of Utah.

Yamauchi AM, Rabbitt RD, Boyle R, Highstein SM (2002) Relationship between inner-ear fluid pressure and semicircular canal afferent nerve discharge. J Assoc Res Otolaryngol 3:26–44.

Zheng J, Shen W, He DZ, Long KB, Madison LD, Dallos P (2000) Prestin is the motor protein of cochlear outer hair cells. Nature 405:149–155.

5
Sensory Processing and Ionic Currents in Vestibular Hair Cells

A<small>NTOINETTE</small> S<small>TEINACKER</small>

1. Introduction

Sensory hair cells of acousticolateralis systems are polarized, with transduction taking place at the apical pole and transmitter release occurring at the basal pole. Interposed between the transduction current and transmitter secretion are the ionic currents of the basolateral membrane of the hair cell. It is the delicate interplay between the properties of these currents that shapes the transduction current to pass to the afferent fiber specific characteristics of the sensory stimulus. In the present review, the properties of these currents in vertebrate vestibular hair cells will be examined with the aim of delineating those features that may be clues to the role that these different mixtures of ionic currents and their kinetics play in shaping transduction current. Although it is now clear that there are many variations in stereociliary properties, including active movements of the hair cell bundle itself, that must be designed to optimize the transduction process for specific sensory signals (Peterson et al. 1996; Silver et al. 1998; Ricci et al. 2000), it is also clear that the variation in basolateral ionic currents serves to process the sensory signal further.

This review will deal with the properties of basolateral ionic currents in vestibular hair cells, with emphasis on properties unique to the vestibular system that may serve to further understanding of the coding of the characteristics of movement in space. The literature on auditory hair cells will be referred to only when it will serve to illuminate features of the vestibular hair cell or when the end organ is thought to serve auditory and vestibular functions. Efferent mediated ionic currents have been excluded because this problem is complicated by the need to infer efferent function from ionic currents activated by synapses on both the afferent fiber and/or hair cell directly. For a treatment of vestibular efferent function at the afferent fiber or systems level, refer to Highstein (1991, 1992) and, more recently, Bricta and Goldberg (2000).

2. Potassium Currents

There is a diverse set of K^+ channels in hair cells, and the problem in analyzing them and understanding their function is made more difficult by the variety of these ionic channels and their kinetics within and between end organs. These differences are sufficiently marked, not only in the current kinetics but also in the voltage dependence of activation, that it is a foregone conclusion that they have been configured to transmit specific aspects of head movements. Channel inactivation properties also should provide clues in the analysis of channel function in hair cell sensory coding. In recent years, it has been shown that most voltage-gated K^+ channels inactivate with sufficiently prolonged depolarization (Lopez-Barneo et al. 1993). This inactivation is classed as N-type or C-type, depending on the time scale and the inherent structure of the K^+ channel. N-type inactivation has a time constant of less than 10 ms and is accomplished by a cytoplasmic particle found on the amino terminal of the channel protein (Zagotta et al. 1990). It is this class of inactivation that is so obvious in the hair cell transient K^+ current (**IA**). C-type inactivation occurs on a slower time scale and involves a conformational change in the channel protein near the outer end of the channel (Liu et al. 1996). C-type inactivation requires a longer time for removal and can hence accumulate in the channels with repeated openings, with consequences for signal coding at high frequencies or amplitudes. These two types of inactivation can coexist in the same channel protein, as is seen in the **IA** of hair cells or as a single C-type inactivation of the delayed rectifier. In the paragraphs below, properties of the diverse K^+ currents in hair cells will be reviewed with an attempt to understand possible functional roles in the different vestibular end organs.

2.1. The Transient Potassium Current (**IA**)

The original description of this current (Hagiwara et al. 1961; Conner and Stevens 1971) stressed its transient nature as a rapidly activating and inactivating current at subthreshold membrane potentials. Following a depolarization, this current can only be activated after a period of hyperpolarization; thus, with a maintained highly depolarized level of membrane potential, this current is not functional. The fast phase of **IA** in canal hair cells is quite easy to identify by its fast activation and inactivation. Around the level of membrane potential that is subthreshold for **IKCa** or **IKv** activation, **IA** is capable of entraining the cell to high-stimulus frequencies. **IA** activation is steep and, in crista hair cells, begins at potentials positive to $-65–70$ mV, while the voltage dependence of inactivation is also steep and results in the removal of this current at maintained potentials between -40 mV and -50 mV. These two parameters shape a window of **IA** action between -70 mV and -40 mV. Stimulus frequency coding is accomplished by the **IA** rapid activation and inactivation following a stimulus-induced depolarization. **IA** activation produces a hyperpolarization and removal of **IA** inactivation. The depolarization of the second stimulus input again activates **IA**, repeating a cycle that allows the membrane to encode the frequency of the stimulus, subject only to the amplitude of the stimulus and filtering by **IA** kinetics. This current can be separated from other potassium currents in the hair cell membrane by the use of negative and sufficiently positive prepulses, followed by a test pulse to positive values sufficient to activate all available channels (Fig. 5.1). Pharmacological separation is possible by using its sensitivity to low levels of 4-AP and relative insensitivity to TEA (Housley et al. 1989; Steinacker and Romero 1991; Goodman and Art 1996). The concentration of 4-AP used is critical because the other current frequently coexisting with **IA** is **IKv**, which is also blocked in hair cells by 4-AP but usually with less efficacy. To properly separate **IA** and **IKv**, it is necessary to use, in addition to the pharmacology, the marked difference in time and voltage dependence of inactivation processes of the two currents.

IA has been reported in a certain percentage of type II hair cells from all vestibular end organs but never from type I hair cells. The **IA** in type II hair cells from the semicircular canals shows features not found in either of the otolithic end organs in which it has been studied, the utricle and saccule (Fig. 5.1). The most obvious of these is the dominance of the **IA**, in the percentage of hair cells possessing this current, its amplitude as a percentage of the total outward current, and in the voltage dependence of inactivation (Housley et al. 1989; Masetto and Correia 1997; Steinacker et al. 1997). In the frog crista, 62% of maximal **IA** could still be activated at −60 mV (Housley et al. 1989; Norris et al. 1992). In the toadfish (*Opsanus tau*) crista, the half-inactivation voltage was −53 mV (Steinacker et al. 1997). In the guinea pig crista, this depolarized $V_{1/2}$ for **IA** inactivation also exists in hair cells where **ICKa** appears to be the only other outward current (Grigeur et al. 1993a). In the amniote pigeon crista (Lang and Correia 1989), half-inactivation for **IA** takes place in type II hair cells around −70 mV. In the frog saccule, $V_{1/2}$ of **IA** is −83 mV (Hudspeth and Lewis 1988), similar to that found in the toadfish (*O. tau*) saccule (A. Steinacker, unpublished data). Thus, the **IA** with the very positive range for inactivation may be unique to the crista type II hair cell because it has not been described in otolithic end organs, type I hair cells, or auditory hair cells. This current is capable of producing regular interpulse intervals of depolarization and glutamate release. Any repetitive timing information on a fast time scale in sufficiently negative membrane potential ranges will involve **IA** as the potassium current with the fastest activation and inactivation. If **IA** is the first basolateral current to activate, it will produce a pause in transmitter release in the cell's response to a depolarizing transduction current. The length of this delay will be determined by the amplitude and temporal properties of the transduction current. In a glutaminergic sensory synapse in the

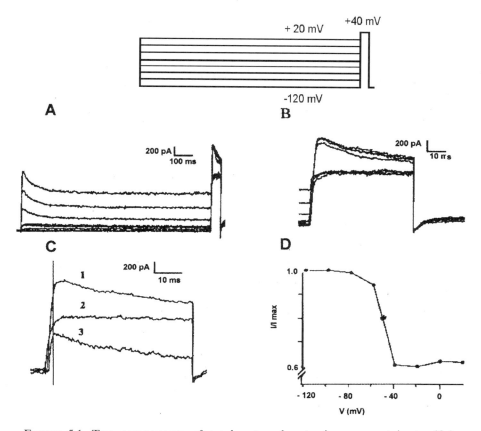

FIGURE 5.1. Two components of total outward potassium current in toadfish (*Opsanus tau*) crista hair cells. (**A**) In response to voltage clamp command protocol seen above, given from a holding potential of −60 mV, inactivating and noninactivating components are seen. Note that in response to the prepulse voltage commands from a holding potential of −60 mV, a large fraction of the inactivating current is still available. (**B**) Expanded time-scale view of responses to +40 mV test pulses. Test pulses to +40 mV following 1 s prepulses from −120 mV to +20 mV show that it is possible to evoke the inactivating current in response to prepulses negative to −40 mV. (**C**) Selected traces from B plotted on a faster time scale. Trace 1 is the maximum current response to the test pulse, which includes inactivating and noninactivating currents. Trace 2 is the noninactivating current response to the test pulse after removal of the inactivating component by more positive prepulses. The lack of a current component with a slower inactivation rate rules out **IKv** as the second current component. By elimination, this leaves **IKCa** as the slower current component because activation and deactivation kinetics are too rapid for I_h. Trace 3 is the inactivating current obtained by subtraction of trace 2 from trace 1. The vertical line passes through the peak of the current showing the fastest activation rate (trace 3); trace 2 peak activation is considerably slower, and the activation of trace 1 (maximum current evoked in response to test pulse) is, of course, intermediate. This illustrates the primacy of the inactivating current **IA** in response to voltage commands. (**D**) Data in B plotted against prepulse voltage seen in A. (Taken from Steinacker et al. 1997.)

olfactory system, **IA** was shown to be responsible for the rapid synaptic timing through the synapse (Schoppa and Westbrook 1999).

2.2. The Calcium Activated Potassium Current (IKCa)

The **IKCa** in semicircular canal hair cells has not received the attention given to this same current in pure auditory or saccular hair cells, perhaps due to the fact that there has not been the emphasis on sharp frequency coding by **IKCa** and **ICa** coupling in processing vestibular signals. Often, this current is neglected in recordings from the type II semicircular canal hair cells because of the predominance of **IA**, apparent by its large amplitude, rapid activation, and high voltage threshold for inactivation. However, **IA** is usually followed by a more slowly activating, noninactivating, or sustained current (Housley et al. 1989; Norris et al. 1992; Griguer et al. 1993a; Steinacker and Zuazaga 1994; Steinacker et al. 1997). Whether this is **IKv** or **IKCa** has not been systematically studied. Griguer et al. (1993a) showed that a large part of both the rapidly inactivating and the more sustained component of outward current in guinea pig crista hair cells was diminished by 5 mmol 4-AP. They classified the inactivating current as a delayed rectifier because of its positive $V_{1/2}$ of inactivation ($-50 mV$). The rapid time course of inactivation of this current makes it much more likely to be **IA**, which is also blocked by even lower concentrations of 4-AP. In that case, this would be an example in a mammalian (guinea pig) crista type II hair cell of an **IA** with a positive inactivation $V_{1/2}$. Ten mmol TEA, which should be relatively specific for hair cell **IKCa** BK channels at this concentration (Housley et al. 1989; Steinacker and Romero 1991; Goodman and Art 1996), blocked 60% of the sustained current in some of these cells, as did zero Ca^{2+} solution, suggesting the coexistence of **IKCa** with **IA** in guinea pig crista type II hair cells.

In many of these hair cells, **IA** was followed by a more slowly activating, noninactivating, or sustained current (Housley et al. 1989; Norris et al. 1992; Griguer et al. 1993a; Steinacker and Zuazaga 1994) that has been treated as C-type inactivation of **IA** (Norris et al. 1992). Evidence that this current is a separate and distinct, noninactivating current underlying the inactivating **IA** in crista hair cells was given by Steinacker and Zuazaga (1994), where the importance of **IKCa** was unmasked at voltages where it coexists with **IA** by increasing the external Ca^{2+} (Fig. 5.2A). The same procedure demonstrates the dominant role of **IKCa** following inactivation of **IA**. In normal Ringer (4 mmol Ca^{2+}), the initial outward current is a rapidly activating and inactivating current that is clearly **IA**. That is followed by a low-level slowly inactivating outward current that could be **IKCa**, **IKv**, and/or C-type **IA**. Replacement of the normal external Ringer of 4 mmol Ca^{2+} with one containing 10 mmol Ca^{2+} did not affect the fast activation or the inactivation of **IA** in response to the test pulses but greatly increased the steady-state outward current in response to the conditioning and test pulses, identifying the current as **IKCa**.

This current was only a small fraction of the outward current in 4 mmol Ca^{2+} and may be only minimally activated in the cell *in vivo* in the absence of sustained higher-amplitude transducer currents because of the rapid hyperpolarization of the membrane by **IA** activation. In 10 mmol Ca^{2+}, not only did the steady-state outward current increase in amplitude, but its rate of activation was greatly increased at more positive potentials, leading to the conclusion that this current could be modifiable under normal conditions that increase internal Ca^{2+}, such as high-intensity stimuli. All of the properties of this current (high conductance, Ca^{2+} and voltage dependence of activation and lack of inactivation) suggest that the current underlying **IA** in these crista hair cells is **IKCa**. This current will displace **IA** and activate first in response to a stimulus under the proper conditions. The functional consequence is a shift of the hair cell from one coding process that is Ca^{2+}-independent and rapidly inactivating to another Ca^{2+} and voltage-dependent noninactivating coding process. This use of elevated external calcium mimics natural states of the hair cell during a high and/or maintained transduction current and when under the influence of the efferent transmitter acetylcholine, which has been shown to increase intracellular Ca^{2+} (Ohtani et al. 1994; Yoshida et al. 1994). Increasing external Ca^{2+}, as opposed to the use of Ca^{2+} free media, avoids problems associated with the increase in conductance coupled with loss of specificity for K^+ produced by the omission of extracellular Ca^{2+} (Armstrong and Lopez-Barneo 1987). The lowering of external Ca^{2+} has also been shown, in auditory hair cells, to produce loss of specificity of Ca^{2+} channels, which then conduct monovalent cations nonspecifically (Art et al. 1993). Increasing external Ca^{2+} has no effect on, or increases, C-type inactivation in a variety of voltage-gated K^+ channels (Grissmer and Cahalan 1989; Lopez-Barneo et al. 1993). In fact, removal of external Ca^{2+} decreases the rapid inactivating phase of the IA (Salkoff 1983).

Not all type II hair cells of the crista or otolithic end organs express multiple K^+ currents. In the toadfish (*O. tau*) crista and saccule, one small population of hair cells exists with **IKCa** as the only potassium current (Fig. 5.2B). These cells show no inward current negative to –60 mV and no inactivating outward current (Steinacker and Romero 1991; Steinacker and Zuazaga 1994; Steinacker 1996; Steinacker et al. 1997). The **IKCa**-only hair cells in the toadfish (*O. tau*) crista are also noteworthy for their high-frequency, high-Q-factor resonance. These cells are identical to a population of hair cells of the toadfish (*O. tau*) saccule, where they make up a much larger percentage of the hair cell population but have lower resonant frequencies. They are also similar to the "classic" auditory hair cell in which primarily **IKCa** and **ICa** are found and produce the high Q resonance believed to code the characteristic or bet frequency of the hair cell (Art and Fettiplace 1987; Hudspeth and Lewis 1988).

The **IKCa** in anamniote vestibular hair cells has best been studied in the saccular hair cells, where the cell's function may be either auditory or vestibular but cannot be defined from ionic currents alone (Hudspeth and

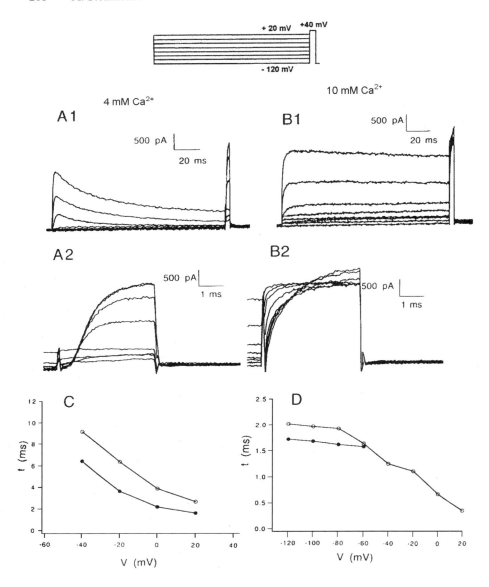

Lewis 1988; Sugihara and Furukawa 1989, 1995; Steinacker and Romero 1991, 1992; Steinacker and Perez 1992). This current appears to be largely from BK channels, and its kinetics of activation and inactivation are related to resonant frequency. It is sensitive to external Ca^{2+} levels and to low concentrations of TEA (2 mmol). The properties of **IKCa** in these cells are very similar to that recorded from a purely auditory organ, the turtle basilar papilla (Art and Fettiplace 1987). Separation of **IKCa** into BK and SK currents with associated coding functions is not as clearly defined in vestibular hair cells as in auditory hair cells, in which an SK of low conductance and high Ca^{2+} sensitivity was described in the turtle cochlea (Art et al. 1995). Here it is thought to contribute to low-frequency transduction and the action of the efferent transmitter (Art et al. 1995). In a study of single-channel recordings of **IKCa** in goldfish (*Carassius auratus*) saccular hair

---

FIGURE 5.2A. Demonstration of an underlying virtually invisible **IKCa** in current records from a cell that might have been described as having **IA** with N- and C-type inactivation as the sole outward current. Voltage command protocol shown above. (**A1**) Outward current response in 4 mmol Ca^{2+}. (**B1**) Outward current response of same cell with 6 mmol additional Ca^{2+} added to the Ringer. Note the slow rise of current at the peak of the prepulses. (**A2**) The response to the test pulse in A1 on an expanded time scale. Note the almost complete inactivation of outward current following the -40 mV prepulse but large **IA** component following the -60 mV prepulse. Small, almost steady-state test pulse current following prepulses of -40 mV to $+2$ mV test pulses is instantaneous current component from prepulses plus small additional noninactivating current evoked by test pulse. (**B2**) Six mmol additional Ca^{2+} adds a noninactivating current that grows with positive prepulses. The two highest-current test pulse responses are the **IA** apparent in A2 following -120 and -100 mV prepulses plus an increasing Ca^{2+}-activated component, which grows as **IA** inactivates. The next two test pulse responses from -80 and -60 mV prepulses show the same decrease in **IA** due to inactivation but with an increasing noninactivating component. Test pulse response to $+40$ mV arising from prepulses positive to -60 mV shows the addition of a noninactivating current and an increase in the activation rate of the test pulse current. Both of these features can be explained by an increased Ca^{2+} current positive to -60 mV that increases both the amplitude and the activation rate of an **IKCa**. (**C**) Rise time of the prepulse current from the -60 mV holding potential in 4 mmol Ca^{2+} as seen in A1 (filled circles) and rise time of prepulse current with the additional 6 mmol Ca^{2+} in B1 (open circles). When outward prepulse current is evoked from a minus -60 mV holding potential, additional non-inactivating current is added, but the activation rate of this current is still slower (note rounded curve of peak prepulse current in B1). (**D**) As the activation of the test pulse current arising from prepulse potentials positive to -60 mV is measured, the effect of the incoming Ca^{2+} on **IKCa** activation becomes apparent as a large increase in **IKCa** activation rate. Test pulse activation rate in 4 mmol Ca^{2+} (filled circles) shows the removal of **IA** by inactivation in response to prepulses positive to -40 mV. (Taken from Steinacker and Zuazaga 1994.)

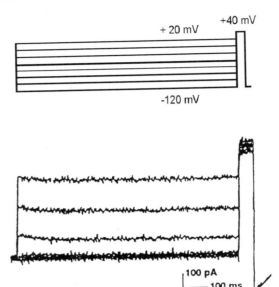

FIGURE 5.2B. Total outward current profile from a horizontal semicircular canal hair cell. This cell (comprising 10% of the total crista hair cell population) shows no inactivating current and no inward current negative to the potassium equilibrium potential in response to the voltage command protocol above. This class of cells are those of the crista that show high-Q, high-frequency resonance. (Taken from Steinacker et al. 1997.)

cells, only a single form of the **IKCa** channel, the BK channel, was found, although four separate conductance levels were seen. The Ca^{2+} sensitivity of these BK channels was low (Sugihara and Furukawa 1994).

2.3. The Delayed Rectifier (*IKv*)

The delayed rectifier is the name originally given to the K^+ conductance in the squid axon because of the delayed activation in the conductance of the potassium ion after the initial rapid sodium conductance had begun. Today it is apparent that there are a large variety of "delayed rectifiers" that differ in their kinetics and pharmacology. The voltage and time dependence of inactivation of a delayed rectifier (Kv2.1) has been characterized with study of the cumulative inactivation process (Klemic et al. 1998). What has become accepted in type II vestibular hair cell literature as the "delayed rectifier" is a voltage-sensitive, Ca^{2+}-insensitive, slowly inactivating C-type outward K^+ current found in a certain fraction of vestibular hair cells, often in combination with **IA** and/or **IKCa** (Lang and Correia 1989; Steinacker and Romero 1991; Goodman and Art 1996). Its inactivation is slow, being apparent only after several hundred ms (Lang and Correia 1989; Steinacker and Romero 1991). It can be separated from **IKCa** by the greater sensitivity of **IKCa** to TEA and from **IA** by the rapid inactivation and the greater sensitivity of the latter current to 4-AP (Lang and Correia 1989; Steinacker and Romero 1991). It was not found in toadfish (*O. tau*) horizontal semicircular canal hair cells (Steinacker et al. 1997) or hair cells of the frog crista (Housley et al. 1989; Norris et al. 1992). It was, however, a large component of the outward current in toadfish (*O. tau*) saccular hair cells (Fig. 5.3) (Steinacker and Romero 1991; Steinacker and Perez 1992).

In a slice preparation of the pigeon crista, Russo et al. (1995) and Masetto and Correia (1997) describe an **IKv** with the "normal" voltage activation range from the center of the crista in visually identified type II hair cells. The quotation marks around normal are used to call attention to the dif-

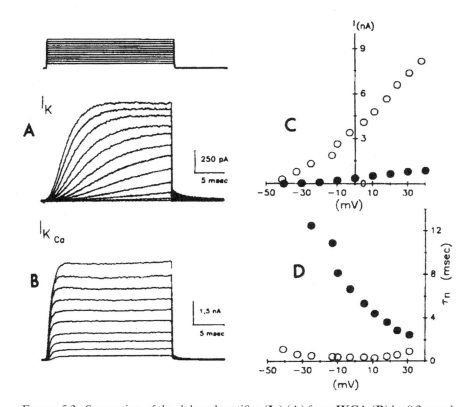

FIGURE 5.3. Separation of the delayed rectifier (I_K) (**A**) from **IKCA** (**B**) by 0.2 mmol cadmium in a hair cell from the toadfish (*O. tau*) saccule. At this concentration, cadmium blocks **IKCa** and **ICa** quite selectively. Cadmium at concentrations >1.0 mmol will also slowly block **IK**. (**C**) illustrates the steady-state current voltage relationship for the two currents (**IK**, filled circles; **IKCA**, open circles). (**D**) The voltage dependence of the activation rates of the same two currents. (Taken from Steinacker and Perez 1992.)

ferences in this current later reported for type I hair cells (see below). This **IKv** was also described for mouse utricular hair cells (Rusch et al. 1998). In turtle auditory hair cells, the **IKv** is believed to be a primary determinant of sensory coding of low-frequency hair cells, with fairly clear-cut evidence in the experimental records (Art et al. 1995). In toadfish (*O. tau*) saccular hair cells, the cells displaying the "normal" delayed rectifier did not show the symmetrical higher frequency and high Q resonance of cells expressing only **IKCa** (Steinacker and Perez 1992). This suggests that they also may code the lower frequency in the toadfish (*O. tau*) saccule, believed to be a dual function auditory and vestibular end organ (Popper and Fay 1973). At this point, it is of interest to note that the state of phosphorylation has been shown to affect the voltage gating of the **IKv** channel (Perozo and Bezanilla 1990), and the variation among laboratories in the field in the presence or concentration of ATP in recording pipettes could account for some of the variations found in the voltage dependence of the delayed rectifier in hair cells.

2.4. Inward Rectifiers

2.4.1. The "Classic" Inward Rectifier ($IK_{ir} = IRK = I_{KI}$)

This current has properties that distinguish it from other K^+ channels in that it conducts, with a steep voltage dependence, only negative to the K^+ equilibrium potential and its voltage dependence for gating shifts with the concentration of extracellular K^+ following the Nernst equation for this ion (Hagiwara and Takahashi 1974; Hagiwara et al. 1976). This equilibrium potential may be moved to more positive voltages by an increase in extracellular K^+ during physiological activity. It is thus a stabilizing current that depends on the extracellular K^+ level. The inward rectification is due to the highly voltage-dependent block of outward current through this channel by intracellular cations, primarily Mg^{2+}, that compete with K^+ for access to the pore to reduce outward current at membrane potentials positive to the K^+ equilibrium potential. On removal of Mg^{2+} from the intracellular media, this current becomes ohmic. This flickering open channel block by Mg^{2+} increases the steady-state current (positive to the K^+ equilibrium potential) by preventing the closure of the channels. Polyamines, such as spermine, also produce a flickering block of these channels, perhaps due to their resemblance to organic Mg^{2+} analogs (Lopatin et al. 1994). External Ba^{2+} and Cs^+ at micromolar concentrations are effective blockers of this inward current, with Cs^+ being the most effective. In addition, the neuronal inward rectifiers have for some time been associated with intracellular mediators, G-proteins (Lesage et al. 1995), and protein kinases (Kubo et al. 1993) and modulated by intracellular ATP, fluoride, and vanadate (Iyengar and Birnbaumer 1990; Huang et al. 1998). This makes them a particularly vulnerable current to use for comparison of hair cell data from different laboratories, given the variability in the use of these components in the recording pipette intracellular media in various laboratories.

The inward rectifier was first described in hair cells by Ohmori (1984) from both whole cell and single-channel recordings, with both recording methods providing data indicating that this current followed standard criteria for the inward rectifier I_{KI}. This current is always found in some population of hair cells within an end organ, often with a regional distribution (Griguer et al. 1993c; Masetto et al. 1994; Steinacker et al. 1994; Sugihara and Furukawa 1996; Masetto and Correia 1997; Steinacker et al. 1997). In the goldfish (*C. auratus*) saccule, only the spiking caudal hair cells showed the I_{KI} (Sugihara and Furukawa 1996). In the crista of the toadfish (*O. tau*) (Steinacker et al. 1994, 1997), the frog (Masetto et al. 1994), and the pigeon (Masetto and Correia 1997), the I_{KI} was found only in type II hair cells located in the center of the crista, where it was colocalized with **IA**. The properties of I_{KI} have been expertly illustrated in the goldfish (*C. auratus*) spike-type hair cells (Fig. 5.4A), which cover the spectrum of the properties of this current.

There have been two reports on the effect of I_{KI} on resonance in hair cells. In auditory hair cells of the turtle, the I_{KI} produced an enhancement of resonance (Art and Goodman 1996). The explanation given for this is that, in voltage regions in which I_{KI} is active and outward, the I/V curve has a negative slope due to inhibition of this current with depolarization. The authors term this effect "positive feedback" because removal of I_{KI} and activation of **ICa** by depolarization provides a feedback that depolarizes the cell further. This was demonstrated by resonance recording with and without external Cs^+ (5 mmol) to block I_{KI}. Data from auditory hair cells should apply equally well to vestibular hair cells. However, these data are in contrast to those recorded from frog saccular hair cells, where it was shown, by data and by modeling of the membrane response to this current, that cells with the I_{KI} had more negative resting potentials and lower input resistances, which required more current to evoke resonance (Holt and Eatock 1995). The latter data agree with what was found in current clamp recordings from goldfish (*C. auratus*) saccular hair cells. Two classes of hair cells are found these; one class shows a spike-type response to injected current, and the other class responds to current with an oscillatory response of the membrane potential. The former class of hair cells possessed the I_{KI} thought by the authors to be responsible for the low negative plateau potential following the spike produced by current commands (Sugihara and Furukawa 1996).

2.4.2. The Hyperpolarization-Activated Mixed Sodium and Potassium Current ($I_h = I_H = I_f$)

In heart cells and in several classes of neurons, a hyperpolarization-activated current with a mixed Na^+/K^+ conductance was described some

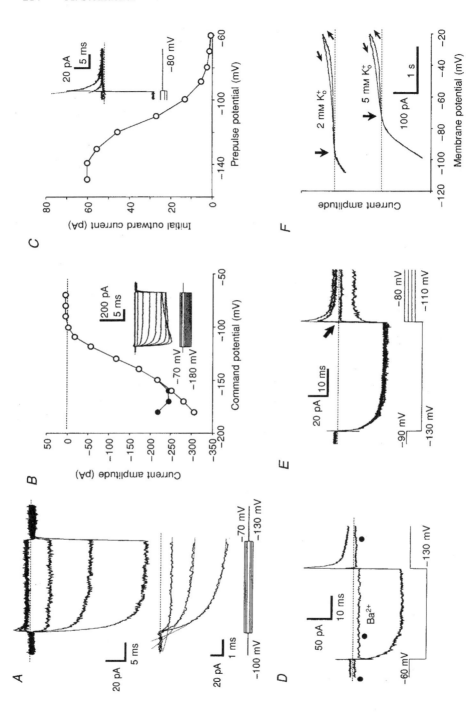

time ago. In heart cells, this "pacemaker current" was titled I_f when it was first described by Noble and Tsien (1968) in a multicellular preparation. It was later defined in patch clamp studies (DiFrancesco et al. 1986) to show many of the properties later found in nodose neurons that were labeled I_h (Doan and Kunze 1999). The latter authors found that this current, contrary to earlier reports, was present in all neonatal rat nodose sensory neurons but differed in its amplitude and activation time constant in A-and C-type neurons. It activated with a slow time constant at potentials negative to $-50\,mV$ and was inhibited by 1–5 mmol external Cs^+. During current clamp experiments, this current produced a time-dependent rectification (depolarization) in response to hyperpolarizing current.

In hair cells, an inwardly rectifying current having properties distinctly different from the more commonly described I_{KI} was first described by Sugihara and Furukawa (1989) and its similarity to I_h in the rod inner segment and the sinoatrial node cells noted. This current was later comprehensively investigated in the saccule of the leopard frog by Holt and Eatock (1995). All hair cells in the leopard frog saccule expressed this current, carried by Na^+ and K^+, with a slow sigmoidal activation negative to -50 and full activation by $-130\,mV$. The current was not sensitive to external K^+ concentrations but could, like I_{KI}, be blocked by external Cs^+ in concentrations of 5 mmol or less and by 2 mmol Cd^{+2}. Cells with I_h responded in current clamp to positive current commands from their zero current potentials with a higher resonant frequency and Q factor than cells with both I_{KI} and I_h at their zero current potentials. This would be expected if I_h produced higher Vz values due to the Na^+ conductance that would place the membrane potential at voltage levels where **IKCa** and **ICa** are more

---

FIGURE 5.4A. Inwardly rectifying K^+ current (I_{KI}) in spike-type hair cells of goldfish (*Carassius auratus*) saccule. (**A**) I_{KI} recorded in a slender spike-type hair cell from the caudal saccule. In the bottom inset, the initial portions of the inward current activated in response to potential steps are shown. Single-exponential curves (time constant 1.5, 1.8, and 1.6 for -10, -20, and $-30\,mV$ steps, respectively) fitted to the region from 0.5 to 3.5 ms after the onset of the step are shown. (**B**) I–V relationship of I_{KI} (peak value, O; value at the end of 25 ms pulse, obtained from the same cell as shown in A). (**C**) Initial amplitude of outward current measured by applying pulses of different amplitudes followed by a command pulse of $-80\,mV$. (**D**) Current traces recorded before and after adding Ba^{2+} (1.6 mmol) to the bath of normal Ringer solution. (**E**) Reversal of the decaying current of I_{KI}. (**F**) I–V relationship in different $[K^+]_o$ measured by applying slow ramps from a holding potential of $-60\,mV$ before and after increasing $[K^+]_o$ from 2 to 5 mmol. In D, E, and F, tetrotoxin (0.3 nmol) was added to the bath to prevent activation of the Na^+ current. The zero current levels are indicated by dashed lines. (Taken from Sugihara and Furukawa 1996.)

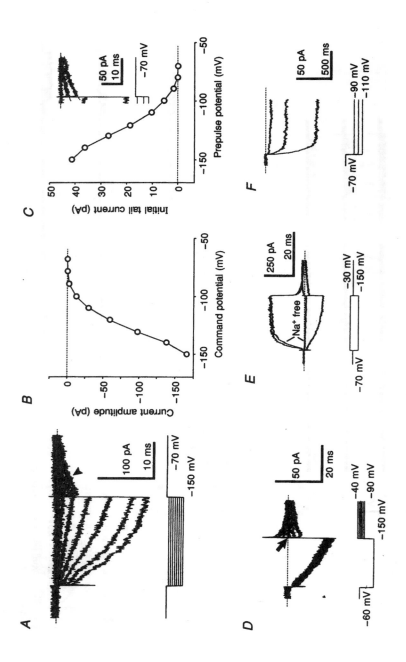

active. A current resembling I_h can be seen in Figure 4 of Hudspeth and Lewis (1988). That current is in response to conditioning prepulses and post test-pulse voltage commands in the range from $-50\,mV$ to $-120\,mV$ as a steady-state current in prepulses and a slowly activating current after the test pulse is returned to prepulse values.

In the goldfish (*C. auratus*) saccule, differences in I_h distribution are marked (Fig. 5.4B) (Sugihara and Furukawa 1996). The rostral and caudal hair cells that oscillate in current clamp are those with a short soma and large-diameter S1 nerve fiber innervation, whereas a second group of caudal hair cells are longer and innervated with smaller-diameter S2 fibers. They also have longer hair bundles. Only the oscillatory type hair cell showed the I_h, and 86% of those cells displayed this current. The other inward rectifier, I_{KI}, was found only in spiking hair cells, which had more negative resting potentials than the rostral hair cells. The authors note the similarity of this distribution with the presence of the latter current in only cylindrical small width–length ratio cells in the leopard frog saccule (Holt and Eatock 1995). Resonance of the goldfish (*C. auratus*) oscillatory hair cells was not of the symmetrical high Q values found for leopard frog saccular hair cells. However, this could be a function of the holding potential because resonant frequency and Q factor are both clearly a function of membrane potential.

This current is not a product of the hair cell isolation procedure because it is seen in type II hair cells in slices of pigeon crista (Masetto and Correia 1997). It is also not induced by the pipette ATP because Sugihara and Furukawa (1996) and Masetto and Correia (1997) did not use ATP in the recording pipettes. A current resembling I_h can be seen in records from the frog saccular hair cells (Hudspeth and Lewis 1988, Fig. 4) and in guinea pig crista hair cells (Rennie and Ashmore 1991, Fig. 8). However, both of these recordings used fluoride, in the form of KF for the former and CsF for the latter, isotonically substituted for Cl⁻. Modulation of K^+ channels by

---

FIGURE 5.4B. Hyperpolarization-activated potassium–sodium current (**Ih**) in oscillatory-type hair cells of goldfish (*C. auratus*) saccule. (**A**) **Ih** evoked by hyperpolarizing steps (amplitude -10 to $-80\,mV$) applied from a holding potential of -70 mV. The arrow indicates tail current. (**B**) I–V curve measured at the end of potential steps from the records in A. (**C**) Amplitude of the initial tail current plotted against the prepulse potential measured using the data depicted in A. (**D**) Reversal of the tail current of **Ih**. The trace with the flattest tail current is marked by an arrow. (**E**) Before and after replacing Na^+ in the bath with choline. (**F**) Activation of **Ih** with a slow time scale. Potential steps of 1s duration were applied. Records were filtered at 100 Hz and sampled at 500 Hz. All of the records in this figure were obtained in short oscillatory-type hair cells in the caudal sacculus. KCl internal solution was used in A and C, and potassium citrate internal solution was used in D, E, and F. (Taken from Sugihara and Furukawa 1995.)

fluoride is well-documented, particularly for inward rectifiers and channels associated with G-proteins (Iyengar and Birnbaumer 1990; Huang et al. 1998).

3. The Calcium Current (*ICa*)

Unlike the literature on **ICa** in auditory hair cells, there has been no systematic study of Ca^{2+} channels in relation to outward K^+ currents in a purely vestibular cell population. It will be assumed here, on the basis of the prevailing literature, that Ca^{2+} channels in vestibular hair cells resemble the L type found in auditory end organs because of similarities of basic kinetics of activation and deactivation, the lack of inactivation, dihydropyridine sensitivity, and in the nature of graded transmitter release properties in both classes of hair cells. Among the ionic currents of all of the acousticolateralis end organs, the data on the Ca^{2+} current show the most uniformity. With exceptions (Rennie and Ashmore 1991; Su et al. 1995; Yamoah 1997; Martini et al. 2000), these rapidly activating currents appear to be classic L-type channels. However, Rennie and Ashmore (1991) show a very convincing record of a rapidly inactivating inward current with the voltage characteristics of **ICa** that could be blocked by 250 µmol Cd^{2+}. This current resembles the **ICa** found in nudibranch (*Hermissenda*) hair cells (Yamoah 1997). In a comparison of hair cell properties from enzymatically and mechanically isolated hair cells (see Armstrong and Roberts 1998), the **ICa** was of the L type but showed a more negative activation range in the mechanically isolated hair cells. Additionally, this current in mechanically isolated hair cells had a faster activation rate than in enzymatically isolated cells.

Data on the relationship of **ICa** and hair cell function come primarily from work on auditory hair cells in the turtle and chick, where higher-frequency hair cells had a larger number of voltage-gated Ca^{2+} channels and larger numbers of synaptic bodies (Sneary 1988; Martinez-Dunst et al. 1997). Given that Ca^{2+} channels are clustered at release sites (Roberts et al. 1990), that larger numbers of channels are associated with higher-frequency hair cells (Art et al. 1995), and that these data from auditory cells should also hold true for vestibular hair cells, the clustering of synaptic bodies in hair cells in the center of the toadfish (*O. tau*) crista (Perez et al. in manuscript) would predict a larger number of Ca^{2+} channels there, making this a possible candidate for the production of the high-frequency gain enhancement seen in toadfish (*O. tau*) crista afferents (Boyle and Highstein 1990).

4. Sodium Currents

Sodium currents are not a common feature of hair cells. They have not been reported in any population of type I hair cells. A subpopulation of type II hair cells of the developing rat utricle showed a TTX-sensitive (100 mmol), rapidly inactivating inward current at postnatal day 3. This current was not seen at postnatal day 1 or postnatal day 12, the two other developmental stages at which hair cell current recordings were made (Lennan et al. 1999). This current showed half-activation values of −50.5 mV and half-inactivation at −25.4 mV, as measured by the response to a protocol of conditioning pulses followed by a test pulse to −10 mV. A sodium conductance is believed to underlie the spiking hair cells of the goldfish (*C. auratus*) saccule (Sugihara and Furukawa 1989). A similar current was found in alligator A. mississipprensis (*Alligator*) cochlear hair cells (Fuchs and Evans 1988). A specific role for this current has not been shown, but it may clearly enable rapid phasic transmitter release properties driven by action potentials.

5. Resonance in Vestibular Hair Cells

Resonance with high Q values has not generally been found in hair cells of the vestibular system (Housley et al. 1989; Correia and Lang 1990; Rennie and Ashmore 1991; Baird 1992, 1994; Norris et al. 1992). Nonetheless, a small number of toadfish (*O. tau*) horizontal semicircular canal hair cells that possessed only **IKCa** and **ICa** showed, in current clamp responses to small current command pulses, high Q-factor resonance from 44 to 360 Hz at 12°C (Fig. 5.5) (Steinacker et al. 1992). If a Q_{10} of 2 is used (Rusch and Eatock 1996) to compare resonant frequencies from the crista recorded at 12°C with those from the toadfish (*O. tau*) saccule that were taken at 22°C, the equivalent resonant frequency range of the crista hair cells would be from 88 to 720 Hz.

These high resonant frequencies found in the toadfish (*O. tau*) crista are in agreement with the rapid activation of ionic currents in these cells reported by all laboratories working with the crista hair cells (Housley et al. 1989; Masetto et al. 1994; Steinacker et al. 1997). The function of such high resonant frequencies in vestibular hair cells, a system whose dynamics are slow (compared to those of the auditory system), can only be speculated on at this moment. Certainly, detection of transients by the vestibular canals will necessitate high-frequency transduction. It may be that the use of pulse or trapezoidal stimuli in place of the usual sine waves for vestibular stimuli would bring out this aspect of crista function in afferent fibers. Even sine wave commands in a current clamp of isolated gerbil crista hair cells elicit faithful frequency following up to 1 kHz (Rennie et al. 1996).

A thorough study of resonance and ionic currents was not carried out in toadfish (*O. tau*) crista recordings, but those cells (N = 7) that showed high Q resonance expressed only one outward current, which was a Ca^{2+}-dependent, noninactivating TEA-sensitive (2 mmol), 4-AP current-insensitive (1 mmol) current, characteristic of **IKCa**. A clear *inward* current can be seen in these cells before the outward **IKCa** activated in voltage ranges between −30 and +10 mV. This is definitely not the case with **IA-**

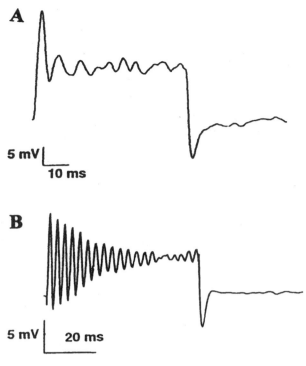

FIGURE 5.5. Hair cells of the semicircular canal are capable of high-frequency, high-Q resonance in current clamp mode. Current command pulse was 100 pA, and membrane potential is (**A**) –55 mV; (**B**) –47 mV. Frequency and quality factor Q are (A) 147 Hz and 4.05; (B) 270 Hz and 14.5. In voltage clamp mode, these cells respond as in Figure 5.3 with only one K$^+$ current, an outward **IKCa**. (Take from Steinacker et al. 1997.)

dominated cells, where the outward current activation *always* obscures any inward current that may be present, implying that the **IA** activation rate is either equivalent to or faster than that of the inward current. In cells with **IA**, in voltage and temporal ranges where **IA** can be activated, resonance would not be expected because the rapid hyperpolarization produced by **IA** before **ICa** activation will prevent the positive swing of the membrane potential necessary for initiation of resonance. However, as the membrane potential passes into more positive voltage ranges for sufficient time to inactivate **IA**, there is little argument that the **IKCa** and **ICa** should produce membrane potential resonance, controlled only by the rapid kinetics and amplitudes of these two currents. This implies that resonance does not have to be a property of the membrane at the zero current potential but may only come into play when the transduction current carries the membrane potential into certain voltage levels for sufficient time to remove **IA** by inactivation. This would clearly move the cell from one signal-processing mode to another. This may explain why, when using Fourier analysis of the current recorded from toadfish (*O. tau*) crista hair cells with intracellular electrodes and a switch clamp amplifier, Highstein et al. (unpublished data) recorded low resonant frequencies from 32 to 58 Hz at resting potential. Those cells may have been cells with **IA** as the predominant current at resting membrane potentials where inactivation is near its $V_{1/2}$. From patch clamp recordings from crista hair cells in all laboratories in this field, **IA** is the dominant current at resting potential in all but a small percentage of **IKCA**-only hair cells, as discussed above. It is also similarly agreed that high Q resonance requires **IKCa** and/or **IK$_V$**.

Thus, the failure of many labs using patch clamp technology to find resonance in crista hair cells, the high Q and high frequency resonance found by this author, and the low frequency resonance found by Highstein et al. (unpublished data) using switch clamp sharp electrode recording from the same cells is currently an unresolved point. This issue was not resolved by the publication by Masetto et al. (1999), which compares sharp microelectrode recordings of resonance with those using patch clamp amplifiers and attributes the difference to poorly tuned patch clamp amplifiers. In the experience of this author, most researchers in this field have long since learned to tune up their amplifier and head stages. Using the same methods and amplifier, high Q but lower frequency resonance from the toadfish (*O. tau*) saccule was found over a frequency range of 107–175 Hz (Steinacker and Romero 1991; Steinacker and Perez 1992). The cells showing this resonance had, in addition to an inward **ICa**, either **IKCa** alone or **IKCa** and **IK$_V$**. The toadfish (*O. tau*) saccule has been proposed to carry out both static vestibular and auditory functions. Weiner kernel analysis, noise stimulation, and reverse correlation analysis from toadfish (*O. tau*) saccular auditory afferents yielded two best frequencies at 74 Hz and 140 Hz (Fay and Edds-Walton 1997), values close to the resonant frequencies found with patch clamp analysis of the same cells. Thus, the high frequency resonance from toadfish (*O. tau*) hair cells should be real.

Spiking, as opposed to resonating, in response to current commands has been reported for a number of hair cell types. However, only in the goldfish (*C. auratus*) saccule has a regionally consistent population of hair cells shown spiking. There, the caudal hair cells show spiking whereas the rostral cells resonate under the same conditions. The spiking cells had an **I$_{KI}$** with properties of the "classic" inward rectifier, whereas the oscillatory cells of the caudal saccule expressed an inward potassium current with properties similar to the **I$_h$**. In the frog utricle, an otolithic end organ with a clear anatomical separation between hair cell types, with striolar and extrastriolar regions, hair cells were divided on the basis of hair bundle morphology. Only one class of these cells showed resonance, and they were the type E, located in the innermost part of the striola. Other cells showed either passive responses or spiking (Baird 1992, 1994).

An interesting publication by Armstrong and Roberts (1998) using hair cells from the frog saccule demonstrated an effect of papain used to isolate hair cells on the currents involved in membrane potential resonance. The authors used a crude papain 5125 from Calbiochem (La Jolla, CA) in combination with 2.5 mmol cysteine (thought to activate papain by reducing disulfide bonds). The use of the disulfide reducing agent cysteine is not a common feature in many laboratories using papain. However, the authors performed the obvious control of using cysteine without papain and did not find the current property modifications seen with papain. Papain treatment was followed by an hour in low-Ca^{2+} solution containing bovine serum albumin to remove enzyme activity. The mechanical dissection of hair cells without the use of enzymes was done in the semiintact saccule, with basal hair cell surfaces exposed by tearing a hole in the epithelium with a dog hair. The use of enzymatic isolation has worried many workers in this field and has led to the use of the semiintact end organ (Rusch and Eatock 1996; Lennan et al. 1999) or slice preparations (Masetto et al. 1994; Masetto and Correia 1997). However, the kinetics in these cells did not differ greatly from those of other publications using these same cell types and using enzymatic isolation. There is a long history in the general electrophysiological literature, quoted by Armstrong and Roberts (1998), of the changes in membrane proteins following enzyme treatment, giving good reason to be wary of this issue.

Armstrong and Roberts (1998) found, in saccular hair cells isolated with enzyme treatment, resonant frequencies for any given current command amplitude were approximately three times the value found for the mechanically isolated hair cells. For enzymatically isolated hair cells, frequencies ranged from 58 to 271 Hz, whereas those from mechanically isolated hair cells ranged from 19 to 170 Hz. A slow, partial inactivation of **IKCa** in mechanically isolated hair cells was found that was not present in enzymatically isolated hair cells. However, a similar slowly inactivating current (evidenced as a "sag" in the current at high-positive-voltage command pulses) was often found in recordings from toadfish (*O. tau*) saccular hair cells isolated by papain treatment (Fluka, Switzerland) before application of any of the usual channel blockers used for hair cell potassium currents (Steinacker and Romero 1991). It may be that the papain obtained from Fluka, which is a more crude preparation, does not contain the enzyme activity responsible for the removal of this transient **IKCa**. However, this same putative **IKCa** with a slight "sag" was also reported in goldfish (*C. auratus*) saccular hair cells after isolation by the Sigma papain P 3125 by Furukawa and Sugihara (1990). An inactivating **IKCa** has been reported that is sensitive to the protease trypsin (Solero and Lingle 1992). Because a crude papain has a number of mostly unidentified proteases, it is possible that the irregular appearance of this inactivation in hair cell **IKCa** is due to differences in time of incubation or a variation in proteases in the papain obtained from different sources.

Several laboratories have used mechanical removal of hair cells from the end organ and have not shown any difference in current activation or deactivation rates. Hair cells from toadfish (*O. tau*) saccule and crista hair cells have been taken by mechanical as well as enzyme treatment (0.5 mg/mL crude papain from Fluka, Switzerland or, alternatively, with 0.25 mg/mL of collagenase (Sigma type A) and/or a 0.03 mg/mL protease (Sigma type XIV) (Sigma, St. Louis, MO). No consistent differences in ionic current properties were seen with any of these methods. The disappearance of **IKv** found by Armstrong and Roberts (1998) in enzymatically isolated hair cells was not seen in enzymatically isolated toadfish (*O. tau*) saccular hair cells, where this current is prominent (Steinacker and Romero 1991). However, it is still possible that differences would be found if a study was done using larger numbers of cells and looking specifically at this issue, as was done by Armstrong and Roberts (1998). Certainly, the issue of enzyme use bears watching because the source and level of purity of papain differ among laboratories and the components of crude papain used by many labs include a complex of proteases that are not necessarily the same even for different batches of papain from the same source.

6. The Type I Hair Cell

The type I hair cell, unique to amniote vestibular end organs, coexists with the type II hair cell throughout the end organ but is concentrated in the center of the crista and in striolar regions of the otolithic maculae. In addition to the easily recognized constriction of the area under the cuticular plate in the type I hair cell, there are also differences in the hair cell bundle (Peterson et al. 1996; Rusch et al. 1998) and mitochondria of the two hair cell types (Wersäll 1960; Wersäll and Bagger-Sjöbäck 1974; Rusch et al. 1998). As in type II hair cells, there is no unique distribution of ionic currents found in all type I hair cells. The one current that is unequivocally not found in type I hair cells is the **IA**, a current found in a large percentage of type II hair cells. (In this fact must lay a clue to the function of **IA** and the type I hair cell.)

In addition to the delayed rectifier described above in type II hair cells, a second low threshold variant (**I$_{KI}$**) has been described for type I hair cells (Griguer et al. 1993c; Rennie and Correia 1994; Rusch and Eatock 1996). In gerbil and pigeon semicircular canal hair cells, **I$_{KI}$** was blocked by 20 mmol internal TEA and by 5 mmol external 4-AP or 4 mmol Ba^{2+}. Five-mmol external Cs$^+$ blocked the inward component of the current (Rennie and Correia 1994). This current was named the major conductance in type I hair cells of pigeon semicircular canal, lagenar, utricular, and saccular hair cells (Ricci et al. 1996), although see Rennie and Correia (1994, Fig. 10). The deactivation of this current was labile, and it shifted to hyperpolarized values over time. Both magnitude and rate of deactivation varied over time for a constant voltage step. This shift of the voltage dependence occurred with perforated patch as well as whole cell recordings, indicating that it was not due to dialysis of an intracellular component by pipette media. No difference in kinetics between the different vestibular end organs was found for this type I hair cell current. In the mouse utricle, this current was found in type I hair cells, where it is characterized by an unusually negative activation range, with $V_{1/2}$ varying from −62 to −88 mV, and a large conductance. It is sensitive to external Ba^{2+}, Cs^+, and low concentrations of 4-AP, highly selective for K^+ over Na^+ but permeable to Cs^+. This current coexisted in some type I hair cells with the "normal" delayed rectifier (Rusch and Eatock 1996).

Curiously, this current with a labile activation range was sought but not found in type I utricular hair cells of mice of a similar age by another laboratory (Lennan et al. 1999). Because both laboratories used recordings from hair cells in an excised utricle, rather than isolated hair cells, some factor in the isolation process cannot be the causative agent. Some of the utricles used by Rusch and Eatock (1996) were organ cultured for up to eight days, but acutely excised utricles were also used, leaving out that explanation. The external media were identical in the two laboratories. The pipette media differed only by the use of 10 mmol Na^+ in Lennan et al. (1999) versus 2.5 mmol used by Rusch and Eatock (1996). ATP, as Na_2ATP (2.5 mmol), was routinely used by Rusch and Eatock (1996) and only in a fraction of the cells of Lennan et al. (1999). Mg^{2+} is a cofactor for the activity of ATP (Iyengar and Birnbaumer 1990) but can presumably be supplied by the $MgCl_2$ used in the Ringer. The fact that the half-activation voltage varied from −62 to −88 mV (Rusch and Eatock 1996) to −70 mV (Rennie and Correia 1994) and −30 to −40 mV (Lennan et al. 1999), with considerable lability in all laboratories, suggests that there may be an intracellular factor susceptible to modulation. Nitric oxide has been shown to decrease this conductance (Chen and Eatock 1994). A decrease in conductance by cGMP of a delayed rectifier termed I_{KI} has been reported in mammalian type I hair cells. cGMP was tested because prior work had shown modification of this current by nitric oxide, and cGMP is known to mediate many of the effects of nitric oxide (Schwark et al. 1997).

The existence of IK_V in more negative voltage ranges is not unique to hair cells because it has been well-described in blowfly (Calliphoridae) photoreceptors (Weckstrom et al. 1991). In that system, this current was designated a delayed rectifier because of its reversal at the potassium equilibrium potential, its sensitivity to internal TEA, and its slow and partial inactivation. There are two components to the current, one that activates rapidly with a midpoint of activation at −65 mV and a second more slowly activating component with a midpoint voltage of activation at −50 mV. The former current is more sensitive to quinidine and the second to 4-AP. In the photoreceptor, this current is said to increase coding efficiency by modulating the gain and frequency response of the cell during light and dark adaptation.

Type I hair cells with the "delayed rectifier" labeled I_{KI} in the gerbil crista, despite a poor current clamp response to depolarizing current pulse injection, followed sine wave current injection quite well, with gain nearly stable between 1 Hz and 100 Hz (Rennie et al. 1996). The fidelity of the wave form was preserved at 1 kHz, with only a 25% gain decrease in amplitude, a not insignificant accomplishment for a hair cell of the semicircular canal. These cells expressed the I_{KI} described previously for type I canal hair cells (Rennie and Correia 1994; Rusch and Eatock 1996) with concomitant low input resistances. The authors conclude that type I hair cells have conferred upon them, by the possession of this large conductance, a reduced input resistance, which produces an extended frequency response at the expense of lower gain. The extended frequency response is due to the reduced time constant of the hair cell membrane produced by I_{KI} (or **IKL**).

Although **IKCa** is not emphasized in type I hair cell data, it can make up a considerable part of the outward current in these cells. Type I hair cells of gerbil semicircular canal hair cells show a large **IKCa** when 20 mmol TEA is used in the pipette to block I_{KI} or by a current subtraction done after use of low-Ca^{2+} external media. This current could not be **IKv** because it shows a decreased conductance positive to +40 mV and no inactivation over time periods where **IKv** shows a C-type inactivation (Rennie and Correia 1994, Fig. 8). Another cell that displayed a current that meets criteria for **IKCa** (insensitivity to 20 mmol internal Cs^+ and reduction by lowered external Ca^{2+}) had a peak amplitude of over 3 nA in response to voltage steps from −39 mV to an unidentified positive voltage (Rennie and Correia 1994, Fig. 10). For this reason, it seems unusual that a higher Q resonance has not been reported in these cells because it should be evoked at membrane potentials above −50 mV, the voltage at which I_{KI} is less dominant. This voltage range may be a switchover point for dominance of coding by I_{KI} to **IKCa** in these cells, just as cells with **IA** and **IKCa** could switch over when the transduction current met requirements for **IA** inactivation. After that point, a high Q resonance in response to a current clamp command or the transduction current should be possible.

7. Summary

In attempting to assign functional roles to the ionic currents in hair cells, one is faced with the problem that there are no data on these cells done with these cells *in situ* with synaptic and metabolic influences intact. Even a brief review of the literature in other neural systems will show that an ionic current in isolation often functions quite differently when even a single synaptic input modulates it. Thus, removing the transduction current, the efferent currents, and possible calyx and autoreceptor input from a hair cell and using injected sine waves or rectangular pulses as quite unnatural stimuli may mean that the response of the ionic currents bears no resemblance to *in vivo* reality. It may therefore be useful to ask what functions the vestibular periphery must perform and relate these to ionic current properties. Foremost among these is the fact that transmitter release in sensory hair cells takes place as a graded process. Although graded potentials cannot transmit over the distance possible with action potentials, they allow a higher bandwidth and increase the spectrum of information transmission possible. This subject has been concisely analyzed and reviewed for the visual system by Juusola et al. (1996) and De Ruyter van Steveninck and Laughlin (1996). Graded potential transmission requires an electrophysiological and a biochemical synaptic machinery dedicated to maintaining a low tonic level of transmitter release while preserving the potential for response to all levels of evoked release. High-frequency gain enhancement has been cited as one of the properties of synapses with graded transmission and the presence of multiple synaptic bodies (Juusola et al. 1996). **IKCa, ICa,** and also $\mathbf{I_h}$ are ideally suited to this function because they show little or no inactivation and are active over membrane potentials where the hair cell often resides during sustained transduction. These currents are operated at a huge metabolic cost to the cell, and their function must be necessary to the sustained transduction by the cell. They are therefore most suited to confer gain and frequency over sustained periods of time.

There is a large and fairly unanimous body of literature on the interaction of **IKCa** and **ICa** in hair cells. This class of cell, either with **IKCA** as the only K^+ current or in combination with another K^+ current, is found in a certain percentage of the type II hair cells in every end organ, suggesting that it confers some property possessed by all end organs. Auditory end organs appear to show the highest development and largest percentage of this cell type, specialized for high-fidelity transduction of signal frequency and intensity. In the auditory system, the interaction of the voltage swings of resonance with the voltage sensitivity of transducer adaptation and hair bundle stiffness may have these cells operating at optimum (Ricci et al. 2000). In this regard, it is interesting that a certain small proportion of type II hair cells of the toadfish (*O. tau*) crista express only **IKCa** and **ICa** (Steinacker 1996) and exhibit resonant frequencies much higher than the auditory end organ in the same animal (Steinacker and Perez 1992) as measured by the same technology. Does this resonance in crista hair cells serve to tune up a certain percentage of those crista long hair cell bundles to optimally pick up a certain signal, as an example of periodic frequency forcing well-known to physicists? What aspect of the vestibular signal requires this resonance? Perhaps high-frequency gain enhancement. This link of frequency coding and **IKCa** may have an exception in type I hair cells, with their low threshold K^+ current and concurrent low input impedance, which have been deemed specialized for high-frequency transduction. Here too, however, at more positive membrane potentials, data from some type I hair cells show voltage regions with **IKCa** activation. Even the type I hair cells of Figure 8 of Rennie et al. (1996), showing high-fidelity frequency following, may be in an **IKCa**-dominated region because the membrane potential is not specified and 20 mmol TEA internally will not block **IKCa**. This would be another example in which the cell can switch from dominance by the low threshold potassium current to that of the unexplored **IKCa** of these cells, ideally suited for transmitting high frequency, another example of the switching of coding modes between **IA** and **IKCa** so clearly seen in toadfish (*O. tau*) crista type II hair cells (see Section 2 above).

Because vestibular hair cells are seldom, if ever, inactive, their channels must also be configured for metabolic economy. In fly (Diptera) photoreceptors, the ionic currents are tailored to metabolic costs of coding different light levels and retinal image velocities. Flies differ in their flight speed, and the currents in their photoreceptors are configured to these flight speeds, producing differences in both gain and frequency responses (Weckstrom et al. 1991). Fast-flying flies have strong outward rectification of a voltage-gated K^+ current (**IKv**), whereas slower flies have an **IKv** that shows rapid inactivation because the latter do not require the fast membrane time constants produced by the large conductances in fast-flying fly photoreceptors (Laughlin and Weckstrom 1993; Weckstrom and Laughlin 1995). Obviously, the dominant **IA** of crista hair cells is not suited to tonic release or response gain. **IA** is, however, well-suited to producing high-frequency interpulse intervals of depolarization and glutamate release until the transduction current is of sufficient amplitude and duration to move the membrane potential into the voltage range sufficient to remove **IA** and replace it by **IKCa** and **ICa**. This may offer a clue to the function of this class of type II hair cell because any repetitive timing information on a fast time scale in sufficiently negative membrane potential ranges will involve **IA** as the potassium current with the fastest activation and inactivation. If **IA** is the first basolateral current to activate in response to a depolarizing transduction current, it will permit an initial spike, which will be followed by a pause in transmitter release. (This may also be the function of the enigmatic **IA** in auditory hair cells because they must produce such a spike to transduce signal onset for sound localization.) The length of this **IA**-induced silence will be determined by **IA** kinetics and the amplitude and temporal properties of the transduction current. The **IA** of the crista hair cells is able to enter and exit the coding process with rapid kinetics and high conductances in physiological membrane potential ranges. This tailors **IA** in crista type II hair cells aptly for rapid changes, such as coding signal onset. Unlike the tight coupling of **IKCA** and **ICa**, there is at present no quantitative experimental evidence addressing the relationship between activation rates of **IA** and **ICa**. If **IA** has, faster activation than **ICa** in that small but critical voltage range between −70 and −40 mV, as available data appear to indicate the inevitable consequence of a more rapid **IA** activation will be the introduction of timed pauses in transmitter release in the response of the cell to high-frequency sensory stimuli because **IA** is able to hyperpolarize the cell before **ICa** has begun to activate. Only with a maintained stimulus will a depolarization be seen at the synaptic body to allow transmitter release. Additionally, when a head movement in the on direction is preceded by one in the off direction for that canal, the hair cell will be hyperpolarized, **IA** inactivation will be removed, and **IA** will again be primed the first current to be activated by the on direction stimulus. The same removal of **IA** inactivation will occur with efferent activated hyperpolarization. **IA** activates at voltages negative to those for **ICa** and **IKCa**, and its activation rate is faster (Steinacker and Zuazaga 1994). In that voltage range below −40 mV, **IA** is king, or perhaps queen, in type II hair cells lacking **Ih** and/or the delayed rectifier with a low voltage threshold. The response of this cell must therefore provide rapid timing of signal onset and then produce a delay in the subsequent signal by at least the inactivation time of **IA**.

To understand the ionic currents of the type I hair cell, it may help to consider that these cells and their calyx appear with the appearance of the flexible neck of terrestrial animals, which has placed on the vestibular system an additional task of the transduction of high-frequency, high-fidelity coding for analysis of voluntary and involuntary rapid head movements and for the central separation and coordination of movements of the head and body. The development of the neck and the type I hair cell–calyx complex had to develop hand in hand and slowly over many years of adjustment to life on land, where the buoyancy of water to support the head is lacking. The type I hair cell, and particularly its calyx, must serve the more rapid head and neck movements of amniotes. The output of the type I hair cells travels to the first vestibular relay nucleus via the most rapidly conducting axons innervating the hair cells. There it is integrated with the sensory input coding neck and body movements (Highstein and McCrea 1988) and reafference information of intended neck movement (Roy and Cullen 2001). The single initial output signal sent by an **IA**-dominated hair cell is ideally suited for integration with the reafference signal at the level of the first central vestibular synapse. In amniotes, the low threshold K^+ current in type I hair cells is well-suited to transmit the onset of head movement. Where a similar current exists in central nuclei of the auditory and vestibular neurons, the data show clearly that a single spike at stimulus onset is produced (Gamkrelidze et al. 1998; Rathouz and Trussell 1998) because the current prevents the neuron from firing high-frequency spikes in the face of prolonged depolarizing current. This signal in the auditor system provides the speed and precision necessary for sound localization. In the chick, the signal from the calyx is carried to the principal cells of the tangential nucleus via the rapidly conducting colossal fibers (Peusner and

Morest 1977), where, early in development, a low-voltage K^+ current permits only a single spike at stimulus onset. In the vestibular type I hair cell, this is, in essence, the same solution used by the auditory system for sound localization; in vestibular analysis, it transmits the spatial axis of the stimulus, discriminated in the high-frequency domain.

There exists an insightful set of experiments on a presynaptic calyx enclosing a large area of postsynaptic membrane area that may be useful for understanding the type I hair cell, calyx function, and presynaptic glutaminergic feedback in a system configured to transmit high-frequency signals with high fidelity (Forsythe and Barnes-Davies 1993; Forsythe 1994; Barnes-Davies and Forsythe 1995). The calyx of Held in the medical nucleus of the trapezoid body includes an autoreceptor modulation of glutamate release, not through an ionic channel or changes in intracellular Ca^{2+} but through a direct action on the exocytotic mechanism (Barnes-Davies and Forsythe 1995). These data are also consistent with the clustering of synaptic bodies facing the calyx (Sneary 1988; Martinez-Dunst et al. 1997) and at the enlarged terminals of hair cells in the center of the toadfish (*O. tau*) crista (Perez et al., in manuscript), where they may be responsible for the high-frequency gain enhancement (Boyle and Highstein 1990).

Few of these concepts of ionic current function or data on graded release and calyces in other cell systems have been applied to the peripheral vestibular system but could, without doubt, be profitably integrated into hair cell research. This requires a more "holistic" integrated systems approach to the experimental work on the peripheral vestibular system, a task that is not easy given the limitations of patch clamp recording and the constraints of access to hair cells in an *in vivo* semiintact animal preparation. Until that can be done, the function of many basic properties of the hair cell will remain uncertain.

Acknowledgments. The author wishes to thank Dr. Izumi Sugihara for his kindness in his early reading of the manuscript and Dr. Juan Ribas Serna for his support and help with the graphics. Supported by NSF Grant VIBNS 0077880.

References

Armstrong CM, Lopez-Barneo J (1987) External calcium ions are required for potassium channel gating in squid neurons. Science 230:712–714.

Armstrong CE, Roberts WM (1998) Electrical properties of frog saccular hair cells: distortion by enzymatic dissociation. J Neurosci 18:2962–2973.

Art JJ, Fettiplace R (1987) The calcium activated potassium channels of turtle hair cells. J Physiol 356:525–550.

Art JJ, Goodman MB (1996) Ionic conductances and hair cell tuning in the turtle cochlea. Ann N Y Acad Sci 781:103–122.

Art JJ, Fettiplace R, Wu Y-C (1993) The effects of low calcium on the voltage-dependent conductance involved in tuning of turtle hair cells. J Physiol 470: 109–126.

Art JJ, Wu Y-C, Fettiplace R (1995) The calcium-activated potassium channels of turtle hair cells. J Gen Physiol 105:49–72.

Baird RA (1992) Morphological and electrophysiological properties of hair cells in the bullfrog utriculus. Ann N Y Acad Sci 656:12–26.

Baird A (1994) Comparative transduction mechanisms of hair cells in the bullfrog utriculus. I. Responses to intracellular current. J Neurophysiol 71:666–684.

Barnes-Davies M, Forsythe ID (1995) Pre- and postsynaptic glutamate receptors at a giant excitatory synapse in rat auditory brainstem slices. J Physiol 488: 387–406.

Boyle, R, Highstein SM (1990) Resting discharge and response dynamics of horizontal semicircular canal afferents in the toadfish, *Opsanus tau*. J Neurosci 10:1557–1569.

Brichta AM, Goldberg JM (2000) Responses to efferent activation and excitatory response-intensity relations of turtle posterior-crista afferents. J Neurophysiol 83:1224–1242.

Chen W, Eatock RA (1994) Nitric oxide inhibits a low voltage activated potassium conductance in mammalian type I hair cells. Biophys J 66:A430.

Conner JA, Stevens CF (1971) Voltage clamp studies of a transient outward membrane current in gastropod neuronal somata. J Physiol 213:21–30.

Correia MJ, Lang DG (1990) An electrophysiological comparison of solitary type I and type II vestibular hair cells. Neurosci Lett 116:106–111.

De Ruyter van Steveninck RR, Laughlin SB (1996) The rate of information transfer at graded-potential synapses. Nature 379:642–645.

DiFrancesco D, Ferroni A, Mazzanti M, Tromba C (1986) Properties of the hyperpolarizing activated current (i_f) in cells isolated from the rabbit sino-atrial node. J Physiol 377:61–88.

Doan TN, Kunze DL (1999) Contribution of the hyperpolarization-activated current to the resting potential of rat nodose sensory neurons. J Physiol 514: 125–138.

Fay R, Edds-Walton P (1997) Diversity of frequency response properties of saccular afferents of the toadfish (*Opsanus tau*). Hear Res 113:235–246.

Forsythe I (1994) Direct patch recording from identified presynaptic terminals mediating glutamatergic EPSCs in the rat CNS in vitro. J Physiol 479:381–387.

Forsythe I, Barnes-Davies M (1993) The binaural auditory pathway: membrane currents limiting multiple action potential generation in the rat medial nucleus of the trapezoid body. Proc R Soc Lond B Biol Sci 251:143–150.

Fuchs PA, Evans MG (1988) Voltage oscillations and ionic conductances in hair cells isolated from the alligator cochlea. J Comp Physiol A 164:151–163.

Furukawa T, Sugihara I (1990) Multiplicity of ionic currents underlying the oscillatory-type activity of isolated goldfish hair cells. Neurosci Res Suppl 12:S27–38.

Gamkrelidze G, Guiame C, Peusner KD (1998) The differential expression of low threshold sustained potassium current contributes to the distinct firing patterns in embryonic central vestibular neurons. J Neurosci 18:1449–1464.

Goodman MB, Art JJ (1996) Variations in the ensemble of potassium currents underlying resonance in turtle hair cells. J Physiol 497:395–412.

Griguer C, Kros CJ, Sans A, Lehouelleur J (1993a) Potassium currents in type II hair cells isolated from the guinea pig's crista ampullaris. Pflügers Arch 425: 344–352.

Griguer C, Kros CJ, Sans A, Lehouelleur J (1993c) Non-typical K^+-current in cesium-loaded guinea pig type I vestibular hair cell. Pflügers Arch 422:407–409.

Grissmer S, Cahalan M (1989) Divalent trapping inside potassium channels of human T lymphocytes. J Gen Physiol 93:609–630.

Hagiwara S, Takahashi K (1974) The anomalous rectification and cation selectivity of a starfish egg cell. J Membr Biol 18:61–80.

Hagiwara S, Kusano K, Saito N (1961) Membrane changes in *Onchidium* nerve cell in potassium rich media. J Physiol 155:470–489.

Hagiwara S, Miyazaki S, Rosenthal NP (1976) Potassium current and the effect of cesium on this current during the anomalous rectification of the egg cell membrane of a starfish. J Gen Physiol 67:621–638.

Highstein SM (1991) The central nervous system efferent control of the organs of balance and equilibrium. Neurosci Res 12:13–30.

Highstein SM (1992) The efferent control of the organs of balance and equilibrium in the toadfish, *Opsanus tau*. Ann N Y Acad Sci 656:108–123.

Highstein SM, McCrea RA (1988) The anatomy of the vestibular nuclei. Rev Oculomot Res 2:177–202.

Holt JR, Eatock RA (1995) Inwardly rectifying currents of saccular hair cells from the leopard frog. J Neurophysiol 73:1484–1502.

Housley G, Norris CH, Guth PS (1989) Electrophysiological properties and morphology of hair cells isolated from the semicircular canal of the frog. Hear Res 38:259–276.

Huang C-L, Feng S, Hilgemann DW (1998) Direct activation of inward rectifier potassium channels by PIP_2. Nature 391:803–806.

Hudspeth AJ, Lewis RS (1988) Kinetic analysis of voltage and ion dependent conductances in saccular hair cells of the bull-frog, *Rana catesbeiana*. J Physiol 400:237–274.

Iyengar R, Birnbaumer L (eds) (1990) G Proteins. London: Academic Press, p. 454.

Juusola M, French AS, Uusitalo RO, Weckstrom M (1996) Information processing by graded-potential transmission through tonically active synapses. Trends Neurosc 19:292–297.

Klemic KG, Shieh C-C, Kirsch GE, Jones SW (1998) Inactivation of Kv2.1 potassium channels. Biophys J 74:1779–1789.

Kubo Y, Reuveny E, Slesinger PA, Jan YN, Yan LY (1993) Primary structure and functional expression of a rat G-protein-coupled muscarinic potassium channel. Nature 364:802–806.

Lang DG, Correia MJ (1989) Studies of solitary semicircular canal hair cells in the adult pigeon. II. Voltage dependent ionic conductances. J Neurophysiol 62:935–945.

Laughlin SB, Weckström M (1993) Fast and slow photoreceptors—a comparative study of the functional diversity of coding and conductances in the Diptera. J Comp Physiol A 172:593–609.

Lennan GWT, Steinacker A, Lehouelleur J, Sans A (1999) Ionic currents and current clamp depolarizations of type I and type II hair cells from the developing rat utricle. Pflügers Arch 438:40–46.

Lesage F, Guillemare E, Fink M, Duprat F, et al. (1995) Molecular properties of neuronal G-protein-activated inwardly rectifying K⁺ channels. J Biol Chem 270(48): 28660–28667.

Lopatin AN, Makhina EN, Nichols CG (1994) Potassium channels block by cytoplasmic polyamines as the mechanism of intrinsic rectification. Nature 372:366–369.

Lopez-Barneo J, Hoshi T, Heinemann SF, Aldrich RW (1993) Effects of external cations and mutations in the pore region on C-type inactivation of Shaker potassium channels. Receptors Channels 1:61–71.

Liu Y, Jurman ME, Yellen G (1996) Dynamic rearrangement of the outer mouth of a K⁺ channel during gating. Neuron 16:859–867.

Martini M, Rossi ML, Rubbini G, Rispoli G (2000) Calcium currents in hair cells isolated from semicircular canals of the frog. Biophys J 78:1240–1254.

Martinez-Dunst C, Michaels RL, Fuchs PA (1997) Release sites and calcium channels in hair cells of the chick's cochlea. J Neurosci 17:1–16.

Masetto S, Russo G, Prigioni I (1994) Differential expression of potassium currents by hair cells in thin slices of frog crista ampullaris. J Neurophysiol 72:443–455.

Masetto S, Correia MJ (1997) Electrophysiological properties of vestibular sensory and supporting cells in the labyrinth slice before and during regeneration. J Neurophysiol 78:1913–1927.

Masetto S, Weng T, Valli P, Correia MJ (1999) Artifactual voltage response recorded from hair cells with patch-clamp amplifiers. Neuroreport 10:1837–1841.

Noble D, Tsien RW (1968) The kinetics and rectifier properties of the slow potassium current in calf Purkinje fibres. J Physiol 195:185–214.

Norris CH, Ricci AJ, Housley GD, Guth PS (1992) The inactivating potassium currents of hair cells isolated from the crista ampullaris of the frog. J Neurophysiol 68:1642–1653.

Ohmori H (1984) Studies of ionic currents of isolated vestibular hair cells of the chick. J Physiol 350:561–581.

Ohtani M, Vevau G, Lehouelleur J, Sans A (1994) Cholinergic agonists increase intracellular calcium in frog vestibular hair cells. Hear Res 80:167–173.

Perozo E, Bezanilla F (1990) Phosphorylation affects voltage gating of the delayed rectifier K⁺ channel. Neuron 5:685–690.

Peterson EH, Cotton JR, Grant JW (1996) Structural variation in ciliary bundles of the posterior semicircular canal. Ann N Y Acad Sci 781:150–163.

Peusner KD, Morest DK (1977) The neuronal architecture and topography of the nucleus vestibularis tangentialis. Neuroscience 2:189–207.

Popper AN, Fay RR (1973) Sound detection and processing by teleost fishes: a critical review. J Acoust Soc Am 53:1515–1529.

Rathouz M, Trussell L (1998) Characterization of outward currents in neurons of the avian nucleus magnocellularis. J Neurophysiol 80:2824–2835.

Rennie KJ, Ashmore JF (1991) Ionic currents in isolated vestibular hair cells from the guinea pig crista ampullaris. Hear Res 51:279–292.

Rennie KJ, Correia MJ (1994) Potassium currents in mammalian and avian isolated type 1 semicircular canal hair cells. J Neurophysiol 71:1–13.

Rennie KJ, Ricci AJ, Correia MJ (1996) Electrical filtering in gerbil isolated type I semicircular canal hair cells. J Neurophysiol 75:2117–2123.

Ricci AJ, Rennie KJ, Correia MJ (1996) The delayed rectifier **IK1** is the major conductance in type 1 vestibular hair cells across vestibular end organs. Pflügers Arch 432:34–42.

Ricci AJ, Crawford AC, Fettiplace R (2000) Active hair bundle motion linked to fast transducer adaptation in auditory hair cells. J Neurosci 20:7131–7142.

Roberts WM, Jacobs RA, Hudspeth AJ (1990) Co-localization of ion channels involved in frequency selectivity and synaptic transmission at presynaptic active zones of hair cells. J Neurosci 10:3664–3684.

Roy JE, Cullen KE (2001) Selective processing of vestibular reafference during self-generated head motion. J Neurosci 21:2131–2142.

Rusch A, Eatock RA (1996) A delayed rectifier conductance in type I hair cells of the mouse utricle. J Neurophysiol 76:995–1004.

Rusch A, Lysakowski A, Eatock RA (1998) Postnatal development of type I and type II hair cells in the mouse utricle: acquisition of voltage-gated conductances and differentiated morphology. J Neurosci 18:7487–7501.

Russo G, Masetto S, Prigioni I (1995) Isolation of A type K^+ current in hair cells of the frog crista ampullaris. Neuroreport 6:425–428.

Salkoff L (1983) *Drosophila* mutants reveal two components of fast outward current. Nature 302:249–251.

Schoppa NE, Westbrook GL (1999) Regulation of synaptic timing in the olfactory bulb by an A type potassium current. Nat Neurosci 2:1106–1113.

Schwark O, Kunihiro T, Strupp M (1997) Cyclic GMP inhibits the activation curve of the delayed-rectifier (I[KI]) of type I mammalian vestibular hair cells. Neuroreport 8:2687–2689.

Silver RB, Reeves AP, Steinacker A, Highstein SM (1998) Examination of the cupula and stereocilia of the horizontal semicircular canal in the toadfish, *Opsanus tau*. J Comp Neurol 402:48–61.

Sneary M (1988) Auditory receptor of the red-eared turtle: II. Afferent and efferent innervation patterns. J Comp Neurol 276:588–608.

Solaro CR, Lingle CJ (1992) Trypsin-sensitive, rapid inactivation of a calcium-activated potassium channel. Science 257:694–698.

Steinacker A (1996) Ionic current contributions to signal processing by vestibular hair cells. Ann N Y Acad Sci 781:150–163.

Steinacker A, Perez L (1992) Sensory coding in the saccule: patch clamp study of ionic currents in isolated cells. Ann N Y Acad 656:27–49.

Steinacker A, Romero A (1991) Characterization of voltage-gated and calcium-activated potassium currents in toadfish saccular hair cells. Brain Res 556:2–32.

Steinacker A, Romero A (1992) Voltage gated potassium current and resonance in the toadfish saccular hair cell. Brain Res 574:229–236.

Steinacker A, Zuazaga DC (1994) Calcium and voltage sensitivity of potassium current activation in toadfish semicircular canal hair cells. Biol Bull 187:267–268.

Steinacker A, Monterrubio J, Perez R, Highstein SM (1992) Potassium current composition and kinetics in toadfish semicircular canal hair cells. Biol Bull 183:346–347.

Steinacker A, Highstein SM, Rabbitt R (1994) Contribution of potassium currents of the toadfish semicircular canal hair cells to afferent fiber properties. Biophys J 66:A169.

Steinacker A, Monterrubio J, Perez R, Mensinger A, Marin A (1997) Characterization of outward potassium currents in the horizontal semicircular canal of the toadfish. Hear Res 109:11–20.

Su Z-L, Jiang S, Gu R, Yang W (1995) Two types of calcium channels in the bullfrog saccular hair cells. Hear Res 87:62–68.

Sugihara I, Furukawa T (1989) Morphological and functional aspects of two different types of hair cells in the goldfish sacculus. J Neurophysiol 62:1330–1343.

Sugihara I, Furukawa T (1994) Calcium-activated potassium channels in goldfish hair cells. J Physiol 476:373–390.

Sugihara I, Furukawa T (1995) Potassium currents underlying the oscillatory response in hair cells of the goldfish sacculus. J Physiol 489:445–453.

Sugihara I, Furukawa T (1996) Inwardly rectifying currents in hair cells and supporting cells in the goldfish sacculus. J Physiol 495:665–679.

Weckström M, Laughlin SB (1995) Visual ecology and voltage-gated ion channels in insect photoreceptors. Trends Neurosci 18:17–21.

Weckström M, Hardie RC, Laughlin SB (1991) Voltage-activated potassium channels in blowfly photoreceptors and their role in light adaptation. J Physiol 440:635–657.

Wersäll J (1960) Vestibular receptor cells in fish and mammals. Acta Otolaryngol (Stockh) Suppl 163:25–29.

Wersäll J, Bagger-Sjöbäck D (1974) Morphology of the vestibular sense organ. In: Kornhuber HH (ed) Handbook of Sensory Physiology. Vestibular System. Basic Mechanisms. New York: Springer, pp. 123–170.

Yamoah EN (1997) Potassium currents in presynaptic hair cells of *Hermissenda*. Biophys J 72:193–203.

Yoshita N, Shigimoto T, Sugai T, Ohmori H (1994) The role of inositol triphosphate on ACh-induced outward currents in bullfrog saccular hair cells. Brain Res 644:90–100.

Zagotta WN, Hoshi T, Aldrich RW (1990) Restoration of inactivation in mutants of *Shaker* K^+ channels by a peptide derived from ShB. Science 250:568–571.

6
The Physiology of the Vestibuloocular Reflex (VOR)

BERNARD COHEN and THEODORE RAPHAN

1. Introduction

Angular and linear accelerations are experienced with every head and body movement. The vestibular system senses these accelerations and generates eye, head, and body movements that stabilize gaze and maintain posture in three-dimensional space. Vestibular control is produced by a number of well-defined subsystems. The vestibuloocular reflex (VOR) is an inertial stabilization mechanism that subserves vision, functioning to move the eyes to oppose or "compensate" for head and body movement. Such compensation minimizes the slip of images across the retina, enabling clear sight. The VOR also "orients" the eyes toward the direction of net linear acceleration to maintain the position of the retina with regard to the spatial or gravitational vertical and to align the eyes with the linear acceleration generated during movement. The vestibular, visual, and somatosensory systems also control head movement by activating vestibulocollic (neck) and vestibulospinal reflexes. Together, these reflexes provide postural control during standing and help stabilize gaze and the body while walking and running.

The visual and somatosensory systems also interact with the vestibular system to produce an internal representation of body position and movement. From the combined inputs of these sensory systems and from the motor commands that control gaze and body position, the brain fashions a sense of spatial orientation.[1] There are also important vestibular autonomic functions that respond to rapid changes in posture with regard to gravity. Heart and respiratory rates are rapidly altered by the vestibular system every time we stand erect to maintain orthostatic tolerance by increasing blood pressure while the slower baroreceptor reflex system is being recruited.

---

[1] There are many definitions of spatial orientation. (See Howard and Templeton 1966; Howard 1982; Schöne 1984). In this chapter, we will define spatial orientation as aligment of the body or a part of the body to gravity or to the sum of all linear accelerations acting on the body.

This chapter provides an overview of one component of this complex control system: the vestibuloocular reflex (VOR). We describe the three-dimensional organization of the VOR and its physiological underpinnings and briefly show how the visual system interacts with the vestibular system to support ocular compensation. Modeling has been of great value in quantitatively predicting VOR characteristics and adding insight as to how these functions are produced by central neural mechanisms. In this chapter we present a model that is capable of implementing the compensatory and orienting functions of the VOR.

1.1. Historical Perspective

The vestibular system is composed of the semicircular canals and the otolith organs, the saccule and utricle, embedded in the labyrinth in the petrous portion of the temporal bone. The receptors and their central connections in the brain stem and cerebellum work at an unconscious level to sense and respond to angular and linear motion automatically with great speed and efficiency. It is not surprising, therefore, that disease of the end organs or lesions in central pathways that transmit and process this motion information can cause frightening, disabling vertigo and imbalance. Such disorders can persist for long periods of time, and physicians have been intrigued with the causes and treatment of vertigo since antiquity (Darwin 1796; Camis and Creed 1930; Cohen 1984; Grüsser 1984; Balaban and Jacob 2001). The sensation of motion, the sixth sense, was the last to be discovered, thousands of years after the five primary sensations were well-known (Kornhuber 1974, Introduction; Cohen 1984; Berthoz 1993; Balaban and Jacob 2001). Only in the late eighteenth and early nineteenth centuries did it become clear that separate sensory organs existed to perceive the basic attributes of movement (i.e., angular and linear acceleration, angular and linear velocity, and position with regard to gravity). Erasmus Darwin, Charles Darwin's grandfather and a leading physician of the late 1800s, considered vertigo at some length in Zoonomia, a two-volume set in which he classified all diseases and treatments known at the time (Darwin 1796; Cohen 1984). In the second edition (1796), he describes an experiment, later repeated by Purkinjě (1820), that strongly implies the presence of separate motion receptors (See Grüsser 1984 for a description). Darwin rotated himself with his head back while looking at a point on the ceiling directly above him. He then stopped and, while experiencing rotatory vertigo, he tipped his head forward. The rotatory vertigo was maintained as an aftersensation well beyond the time of head pitch. Darwin was primarily interested in comparing the movement aftersensations with the afterimages caused by bleaching of retinal pigments with bright light. Purkinjě (1820), however, recognized the importance of this experiment in demonstrating that the motion had excited discrete receptors for movement around different axes and repeated it about 20 years later. Despite his great sophistication, however, Purkině thought that movement of the cerebellum in the cranial cavity was the means for sensing the movements.

Darwin accepted the dogma of the times that there was no torsion of the eyes, although, shortly before, Hunter had postulated from the anatomy that the oblique muscles tort the eyes (Hunter 1786). The question of whether the eyes had independent torsion generated a polemic between Darwin and Charles E. Wells that led Wells to conceive an experiment that proved the existence of torsional eye movements beyond doubt and also gave a conclusive demonstration of torsional nystagmus (Wade 2000a, 2000b). Wells imposed a linear afterimage on the retina, using a shiny metal ruler and a candle. He then repeated Darwin's experiment and induced rotatory vertigo. Wells describes how the afterimages oscillated in roll about a central point, which he correctly assumed could only have come from torsional movements of the eyes. He also described the rotatory nystagmus by noting the slow deviation of the eyes in one direction, with rapid resets in the opposite direction.

By the mid-nineteenth century, Flourens had established that angular head movements could arise from the semicircular canals (Flourens 1824). He extirpated the semicircular canals in pigeons and produced continuous postlesional, rotatory head movements. This suggested that there were separate receptors for angular motion of the head in the inner ear. At that time, "swinging" was used widely in Europe as a treatment for psychiatric disease (Cox 1804; Cohen 1984; Grüsser 1984). Patients were passively rotated about a vertical axis, either centered (Fig. 6.1) or positioned at some distance from the axis of rotation (Fig. 6.2). The first stimulus (Fig. 6.1) provided pure angular acceleration at the beginning and end of rotation. Because the subjects were rotated in light, the "treatment" also provided continuous movement of the visual surround. The second stimulus added another factor, centripetal (inward) linear acceleration (Ac, Fig. 6.2), which combined with the upward linear acceleration of gravity (Ag) to tip the net gravitoinertial acceleration (GIA) vector[2] in toward the axis of rotation. Both stimuli would be accompanied by motion aftereffects. Although used for therapeutic purposes, these conditions undoubtedly provided additional demonstrations that there were independent receptors for sensing rotation.

It remained for Breuer (1874), Crum-Brown (1874), and Mach (1875) to discover, almost simultaneously, that the semicircular canals perceived angular motion. Mach further inferred that *the canals were transducers of angular acceleration, not angular velocity* from experiments in which

---

[2] All linear accelerations are equivalent (Einstein's Equivalence Principle). Therefore, the linear acceleration of gravity, translation, and centripetal acceleration are equivalent. Gravitoinertial acceleration (GIA) is the sum of all linear accelerations sensed by the head and body. In the absence of translation or centripetal accelerations, the GIA is the acceleration due to gravity.

FIGURE 6.1. During "swinging," which was widely used to treat mental disease in the early 1800s, the patient was rotated about a vertical axis, as shown by the superimposed circular arrow. If turned at a constant velocity, the patient would experience angular acceleration and deceleration along the long body or "Z"-axis at the onset and end of stimulation, respectively, which would excite receptors in the lateral and vertical semicircular canals. During constant-velocity rotation in light, however, the input from the canals would cease and nystagmus and the sense of rotation would be provided by optokinetic stimulation through the visual system. (From Hayner 1818; cited in Grüsser 1984.)

FIGURE 6.2. Swinging that involved centrifugation while supine. The patient lay with the head out on a bed that rotated around a heavy beam. During rotation, the patient experienced angular acceleration at the beginning and end of rotation around the nasooccipital or "X"-axis. In addition, there would be centripetal linear acceleration from the head toward the feet (Ac) as shown in the diagram above the head of the subject. The centripetal acceleration would combine with the upward linear acceleration of gravity (Ag) to form the gravitoinertial acceleration vector (GIA). If rotated in darkness, the patient would take the GIA to be the upright and have a somatogravic illusion that he or she was being tilted backward, almost upside-down. (From Horn 1818; cited in Grüsser 1984.)

subjects were rotated in darkness and in light. The subjects sensed angular movement only at the beginning and end of rotation in darkness and felt that they were stationary while rotating at a constant velocity. In contrast, if subjects were rotating in light, they never lost the sensation of motion. This demonstrated that vestibular receptors in the semicircular canals were activated only at the beginning and end of rotation and that the visual system supplied the activity during the constant velocity portion of the rotation to maintain the sense of motion.

Mach (1875) also determined that the otolith organs sensed linear acceleration. While riding in a train around a long curve that was not appropriately banked, he noted that the telegraph poles appeared to be tilted. From this, he inferred that the brain was taking the vector sum of the upward gravitational and inward centripetal accelerations as the spatial upright, hence the perception that the telegraph poles were tilted. During centrifugation in darkness, subjects invariably sense this "somatogravic illusion" and feel tilted relative to the GIA, which they take as the spatial vertical (Guedry 1974); the magnitude of the centripetal linear acceleration determines the angle of tilt. For example, when experiencing 0.5g of centripetal acceleration, subjects feel tilted about 22–25°, whereas 1.0g of centripetal acceleration causes them to feel tilted about 35–40° (Clément et al. 2001). Mach also determined that the visual system played a critical role in sensing motions, and he was the first to systematically induce a sense of self-motion (circularvection) by rotating the visual surround around stationary subjects. Young, Henn, and Scherberger (2001) have recently published an excellent translation of Mach's original study that can be consulted for further details.

Shortly after the basic discovery of where angular motion was sensed, Ewald (1892) determined how the semicircular canals worked. Using microsurgery and an air-filled syringe implanted into the semicircular canals of the pigeon (a "pneumatic hammer"), Ewald initiated flow of endolymph in the canals. From this, he discovered that the induced head movements were in the planes of the canals that were activated. He also inferred that there were excitatory and inhibitory directions of fluid flow (Ewald's Second Law) and devised "canal-plugging" to study canal function, instead of labyrinthectomy, which destroyed the entire labyrinth. Bárány (1907) then made a major contribution to clinical neurootology. He inferred that the nystagmus induced by irrigation of the external ear canals with cool and warm water was largely due to convection currents that were set up by local cooling or warming of the canals. He postulated that this thermal stimulus caused the endolymph to flow toward or away from gravity, thereby deflecting the cupula and hair cells. This discovery resulted in the caloric test, a Nobel Prize for Bárány in 1916, and his release from a Russian prison, where he was languishing as a prisoner of war. A century later, evaluation of the nystagmus induced by caloric stimulation still remains one of the most important ways to assess the sensitivity of the individual lateral canals (see Arai et al. 2002 for a review). At about that time, Dodge (1903) developed a photographic technique to measure eye movements. This led to the classification of the five different types of oculomotor activity. Vestibular movements were used to stabilize vision, saccades to "foveate" or fixate stationary targets, pursuit movements and optokinetic nystagmus (OKN) to track moving targets, and vergence to fix and follow targets moving toward or away from the subject.

By the turn of the twentieth century, eye movements from the semicircular canals that compensated for head movement were well-known. Orienting eye movements that position the retina with regard to the visual world and to the spatial vertical, however, were studied in detail only after World War I. An early elegant experiment demonstrating ocular orientation is shown in Fig. 6.3 (Benjamins 1918). The eye position of a pike was recorded with a camera that moved with the head while the animal was respired by a tube of water flowing into its mouth (Fig. 6.3A). This provided a picture of the eye in head coordinates with the animal in various positions in pitch with regard to gravity (Fig. 6.3B). The angular deviation of the eyes in pitch (about ±30°) was related to the tilt of the head and body by a sinusoidal function (Fig. 6.3C). Because gravitational acceleration is a sinusoidal function of the angle of tilt, this experiment showed that the angular deviation of the eyes was a linear function of the projection of gravitational acceleration onto the plane of the utricles, which are approximately horizontal in head positions B1 and B5 of Fig. 6.3. Similar counterpitch and counterroll that orient the retina to the visual world and to the spatial vertical were then found across a wide range of species (Benjamins 1918; de Kleijn and Magnus 1921; Magnus 1924; Lorente de Nó 1932), including humans (see Cohen et al. 2001 for a review).

The three-neuron vestibuloocular reflex that links the semicircular canals to the eye muscles was first described by Lorente de Nó (1933). Later, Szentágothai (1950), confirming earlier work of Högyes (1880), determined that each canal primarily activates one muscle in each eye. Ter Braak (1936) then demonstrated that optokinetic nystagmus (OKN) induced by the visual system could activate the vestibular system and that visual and vestibular aftereffects could interact to cancel each other. The ability to record the activity of single afferent axons in the vestibular nerve led to the classic studies of Lowenstein on the semicircular canals (Lowenstein and Sand 1940) and otolith organs (Lowenstein and Roberts 1950). Camis and Creed (1930) have provided a fascinating review of the early period of vestibular research up to 1931, and Lowenstein (1936) reviewed the then current concepts of the compensatory and orienting functions of the semicircular canals and otolith organs.

World War II effectively stopped vestibular research for about 10 years, but with the remarkable growth of electronic technology and computers and with increasing government support, the field has flourished over the last 50 years. Some of the relevant references are listed in the bibliography. This list is not inclusive, and interested readers can find additional information and references in a wide range of textbooks, symposia, and reviews that include but are not limited to Brodal et al. (1962), Bender (1964), Bach-Y-Rita et al. (1971), Kornhuber (1974, Parts 1 and 2); Baker and Berthoz (1977), Wilson and Melvill Jones (1979), Henn et al. (1980), Brodal (1981), Cohen (1981), Ito (1984), Freund et al. (1986), Cohen and Henn (1988),

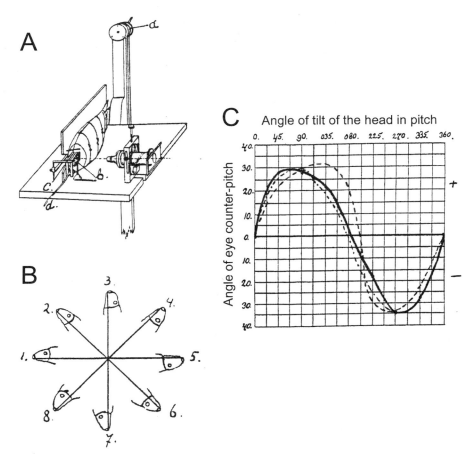

FIGURE 6.3. Orienting counterpitch eye movements in the pike and carp. Eye movements are expressed here in head coordinates. Therefore, in a lateral-eyed animal, counterpitch is torsion in ocular coordinates. (**A**) The pike was placed against a board on a table that could be tilted or rolled. A camera fixed relative to the eyes recorded eye positions in pitch in different head positions with regard to gravity. (**B**) Position of the head in each of the eight pitch positions, 45° apart, in which the photographs were taken. (**C**) Graph of angles of eye deviation in pitch in the pike (solid line) and carp (dashed and dotted lines). The peak deviations were about 30°, and there was a sinusoidal relationship between the linear acceleration of gravity and the counterpitch of the eyes. (Adapted from Benjamins 1918.)

Cohen et al. (1992a), Baloh and Halmagyi (1996), Highstein et al. (1996a), Fetter et al. (1997), Brandt (1999), Cohen and Hess (1999), Leigh and Zee (1999), Goldberg (2000), Goebel and Highstein (2001), Curthoys (2002), Kaminsky and Leigh (2002), and Raphan and Cohen (2002).

2. Organization of the Vestibular System

Essentially, two types of reflex responses, *compensatory* and *orienting*, are produced by the VOR (Camis and Creed 1930; Lowenstein 1936; Lowenstein and Roberts 1950; Raphan and Cohen 1996; Cohen et al. 2001; Raphan and Cohen 2002). Compensatory responses can be further divided into two subclasses. Angular rotation of the head is sensed by the semicircular canals, which drive the eyes to maintain gaze along a given direction in space by activation of the compensatory angular vestibuloocular reflex (aVOR; Fig. 6.4A). Similarly, the linear acceleration associated with side-to-side, front–back, and up–down translation of the head activates the otolith organs, which initiate the compensatory linear vestibuloocular reflex (lVOR), to maintain gaze fixed on an invariant point in space (Fig. 6.4B). The lVOR also orients the eyes to the sum of linear acceleration (i.e., to the GIA) acting on the head and body. Thus, when subjects are stationary and upright (Fig. 6.4C) or when the head is tilted (Fig. 6.4D), the vertical axes of the eyes tend to align with gravity. When linearly accelerating sideways, the orienting component of the lVOR drives the eyes to align with the tilted GIA (Fig. 6.4E) as it would to gravity when stationary (Fig. 6.4C,D). The mechanisms that produce these dual modes of response of the lVOR from activation of the otolith organs are generalizations of those that have been termed "tilt and translation" responses (Mayne 1974; Angelaki and Hess 1996a; Merfeld et al. 1996; Paige and Seidman 1999). The central vestibular system codes the vestibular information in three-dimensional head and body coordinates and in spatial coordinates, which are realized by convergence of otolith- and canal-based signals. These mechanisms are paralleled by vestibulocollic reflexes that perform analogous head rotations to stabilize head pointing in space and orient the head to correspond to alterations in the GIA in space (Magnus 1924; Wilson and Melvill Jones 1979; Peterson et al. 1985; Imai et al. 2001) as well as by vestibulospinal reflexes that readjust the position of the limbs to counter the accelerations and maintain postural stability.

2.1. Coordinate Frames in the Vestibular System

Three semicircular canals in each labyrinth—the anterior (AC), posterior (PC), and lateral (LC) canals—sense angular head rotations in three dimensions (Fig. 6.5B). The three canals are approximately orthogonal, and the planes of the lateral canals on the two sides of the head are close to parallel, as are the planes of one anterior canal and the contralateral posterior canal (Fig. 6.5C). Thus, the canal system can be considered to be composed of three planar structures, the left and right lateral canals (LLC/RLC), the left anterior/right posterior (LARP) canals, and the right anterior/left posterior (RALP) canals. The canals in each plane work in reciprocal fashion;

COMPENSATORY GAZE MECHANISMS

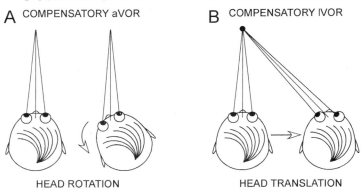

A   COMPENSATORY aVOR

B   COMPENSATORY lVOR

HEAD ROTATION   HEAD TRANSLATION

ORIENTING GAZE MECHANISMS

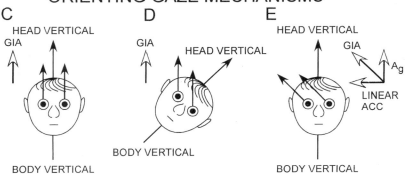

C   D   E

HEAD VERTICAL   HEAD VERTICAL

GIA   GIA   GIA

HEAD VERTICAL A_g

LINEAR ACC

BODY VERTICAL

BODY VERTICAL   BODY VERTICAL

FIGURE 6.4. Compensatory (A,B) and orienting (C–E) responses of the aVOR and lVOR that govern the maintenance of gaze stability. (**A**) The compensatory angular vestibuloocular reflex (aVOR) tends to maintain gaze in a given direction incrementally in response to head rotation. For brief angular accelerations, the aVOR maintains the angle of gaze in space. For larger or continuous rotations, nystagmus is generated. Then, the angle of gaze is held constant during the slow phases of nystagmus and moves forward in saccadic jumps during the quick phases. (**B**) The compensatory linear VOR (lVOR) tends to maintain gaze at a fixed point in space in response to incremental head translations. (**C**) When the head is upright, the head vertical and eye yaw axes, denoted by the arrows on top of the eyes, are aligned with the GIA. In this position, with the subject stationary, the GIA is along the equivalent acceleration of gravity. (**D**) When the head is tilted, the eyes roll so that their yaw axes move toward alignment with the GIA. (**E**) Similar roll of the eyes occurs during linear acceleration, which tilts the GIA relative to the head vertical. Thus, in all instances, the eyes roll toward the GIA.

when one canal is excited by rotation around the normal to the canal plane, the canal in the same plane on the opposite side is inhibited. The beauty of this organization is that the central nervous system can code all angular head movements with only three vectors, greatly simplifying central processing. This planar organization is highly conserved and is present over a wide range of phylogeny. Thus, the same canal structure is present from cartilaginous fish such as sharks and rays (Lee 1893, 1894–1895; Lowenstein 1936; Lowenstein and Sand 1940) to mammals (Fig. 6.5B) (Curthoys et al. 1977; Blanks et al. 1985; Curthoys et al. 1999). Even dinosaurs had a similar three-canal structure, as shown by a three-dimensional reconstruction of the left labyrinth of a brachiosaurus (*Brachiosaurus*) by Andrew Clarke (Fig. 6.6) (personal communication).

The semicircular canal system in three dimensions is best viewed in conjunction with the coordinate frames for the head and eyes. The head coordinate system (Fig. 6.5B) is defined with the X_h axis along the nasooccipital axis, the Y_h axis along the interaural axis, and the Z_h axis pointing out of the top of the head and orthogonal to the other two. The Z_h axis is perpendicular to the horizontal stereotaxic plane (the plane made by the line between the inferior orbital ridge of the eye and the external auditory canal) or from the plane defined by Reid's line in humans.[3] The head frame is coincident with the spatial frame when the head is upright (Fig. 6.5A; X_s, Y_s, Z_s). The eye coordinate frame (Fig. 6.5D) is similarly defined by the unit vectors along X_e, Y_e, and Z_e, which are aligned with the head coordinate frame when the eye is looking forward in the head. Eye rotations can then be given as a rotation matrix relative to these head coordinates (Helmholz, 1867).

The semicircular canals can be described in terms of a coordinate frame whose axes are the normals to the canal planes. The normals to the reciprocal canal pairs can also be viewed as a single coordinate axis sensing head angular movement about axes that are approximately orthogonal. The positive axes of head rotation that would excite the canals on the left-hand side of the head of the brachiosaurus (*Brachiosaurus* sp.) (insert) are shown by the three arrows in Figure 6.6. These positive axes were derived using a right-hand rule[4]: Xc for the left anterior canal, Yc for the left posterior canal, and Zc for the left lateral canal. To excite the left anterior canal, the head would pitch down and roll to the left; to excite the left posterior canal, the head would pitch up and roll to the left; and to activate the left lateral canal, the head would rotate to the left in yaw with a small roll to the left. Head movements in the opposite directions would cause inhibition in these

---

[3] Reid's line extends from the lateral canthus of the eye through the external auditory meatus. (World Federation of Neurology, 1962, Brit. J. Radiol. 35:501–503).
[4] The right-hand rule is a convention established for defining vectors associated with rotations. If the fingers of the right hand are curled around the direction of rotation, the extended thumb points in the direction of the vector.

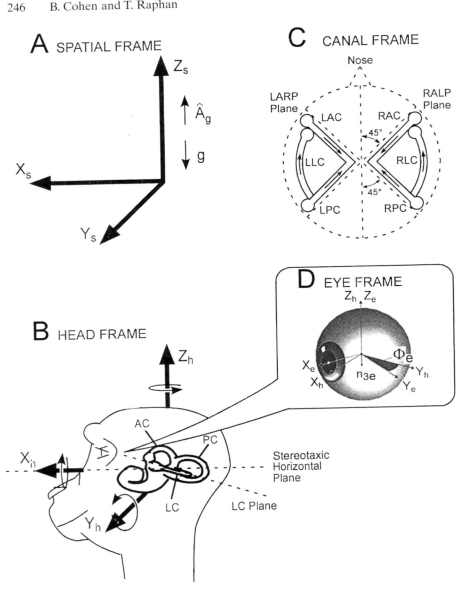

A SPATIAL FRAME

Z_s

$\hat{A}_g$

g

X_s

Y_s

C CANAL FRAME

Nose

LARP Plane

RALP Plane

LAC   RAC

45°

LLC   RLC

45°

LPC   RPC

D EYE FRAME

Z_h, Z_e

X_e

X_h

Φ_e

Y_h

n_{3e}

Y_e

B HEAD FRAME

Z_h

AC

PC

X_h

Stereotaxic Horizontal Plane

LC

LC Plane

Y_h

canals as well as excitation in the canals on the opposite side of the head. The flow of endolymph produced by head rotation around the positive axis for the canals on the left side would be just opposite to these head rotations. Thus, flow in the canals that caused an increase in neural activity would be away from the ampullae of the anterior and posterior canals (ampullofugal flow) and toward the ampulla of the lateral canal (ampullopetal flow). These excitatory directions of endolymph flow are shown by the small arrows in the canals in Fig. 6.5C.

The coordinate axes of the semicircular canals are not aligned with either of the coordinate frames of the head or eyes. The lateral canal axis (LC, Fig. 6.5A) is tilted back about 30° from the Z_h axis, whereas the vertical canal axes tilted back about 40° (AC and PC, Fig. 6.5) relative to the negative X_h axis. Therefore, they are about 10° from being perpendicular to the lateral canals. As a result, all of the semicircular canals receive a component of angular acceleration during all horizontal head rotations and actively participate in producing horizontal compensatory movements (Cohen et al. 1964; Suzuki and Cohen 1964; Robinson 1982; Baker et al. 1984; Yakushin et al. 1995, 1998). There is very little participation of the lateral canals during pitch head movements. Nevertheless, the weak activation or inhibition of

---

FIGURE 6.5. Coordinate frames of the head, semicircular canals, and eyes with relation to the spatial coordinate frame. The arrows in each diagram point in the positive direction, according to a right-hand rule. Therefore, the positive directions in the spatial frame (A) are rotation to the left around the Z-axis, clockwise for the Y-axis from the subject's point of view, and down for the Z-axis. (A) Spatial frame. The vertical axis Z_s is coincident with the spatial vertical, and X_s and Y_s are orthogonal and coincident with the spatial horizontal. (B) The position of the semicircular canals in the head coordinate frame. Shown are the stereotaxic horizontal plane and the plane of the lateral canals. The three canals, LC, AC, and PC, are indicated. The arrows show the positive directions of rotation about each of the three head axes, yaw Z_h, pitch Y_h, and roll X_h. The arrowheads point to the positive direction. The small curved arrows show the positive direction of rotation for each of these axes. The canals do not align with the head frame, although the canals on the two sides are approximately parallel and perpendicular. (C) Diagram looking from above showing the relative positions of the semicircular canals on both sides of the head. Note that LAC and RPC form the LARP plane, RAC and LPC the RALP plane, and LLC and RLC the lateral canal plane. The arrows inside each canal show the direction of endolymph flow in the canals that produces excitation in the vestibular nerve. (D) Coordinate frame of the eyes (X_e, Y_e, and Z_e). X_e is torsion, Y_e is pitch, and Z_e is yaw. If the eye is looking straight ahead, the eyes coordinates can be given relative to the head coordinates (X_h, Y_h, and Z_h). Also shown is the axis of rotation of the eye (n_{3e}). The difference between the visual angle (X_e) and the nasooccipital axis (X_h) is given by the solid angle Φ_e. In humans and monkeys, this angle is about 5°.

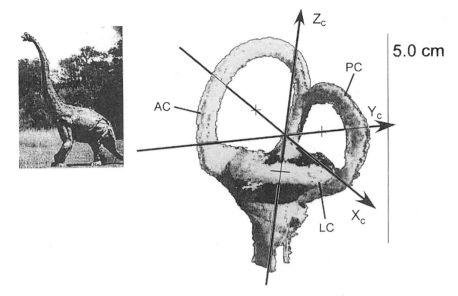

Brachiosaurus labyrinth

FIGURE 6.6. Semicircular canals of a dinosaur, (*Brachiocephalus*) (inset), now extinct but carbon-dated to have existed about 150 million years ago in the Cretaceous Period. The central diagram shows a reconstruction of the left labyrinth by Andrew Clarke of the Frei Universität of Berlin. With the exception of the cochlea, the configuration of the semicircular canals and the portion of labyrinth just below the canals (i.e., the vestibule) is similar to that of the monkey in Figure 6.5B. The anterior (AC), lateral (LC), and posterior canals (PC) are marked. The arrows point in the direction around which head rotation would cause excitation of that canal. The arrows are X_c for the left anterior canal, Y_c for the left posterior canal, and Z_c for the left lateral canal. (Courtesy of Andrew Clarke.)

muscles with close to orthogonal pulling directions are essential for maintaining the compensatory eye movements along canal planes. The nonorthogonality of the canals can be treated analytically using matrix transformations (Robinson, 1982; Yakushin et al. 1995, 1998).

The positive directions for the coordinate axes are in agreement with the excitatory directions for head and eye movements induced by electrical stimulation of the individual canal nerves, shown here for the monkey (Fig. 6.7A, small arrows) (Cohen et al. 1964, 1966; Suzuki and Cohen 1964; Suzuki et al. 1964). When the head is restrained, electric stimulation of the anterior canal nerve (LAC) causes the eyes to move up and roll clockwise from the animal's perspective. Left lateral canal (LLC) stimulation causes the

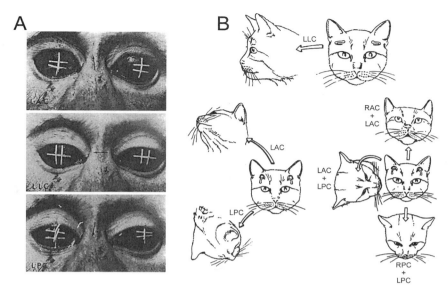

FIGURE 6.7. (**A**) Eye movements of the monkey induced by stimulation of the left semicircular canals. Flashes were made at the beginning and end of short trains of electrical impulses to the semicircular canal nerves to capture eye position at the beginning and end of stimulation. The fine arrows over the nose show the approximate plane of eye movement. Top: The eyes moved up and rolled clockwise (from the animal's point of view) for left anterior canal nerve stimulation (LAC), to the left for left lateral canal (LLC) nerve stimulation, and down and rolled clockwise for left posterior canal (LPC) nerve stimulation. (**B**) Head movements induced in the cat by stimulation of the semicircular canals on the left side. LLC stimulation caused a head deviation to the right. LAC caused the head to move up and roll clockwise to the right, and LPC caused the head to move down and roll clockwise to the right. Figures on the right: By combining the stimuli to the various canals, head movements could be generated about any head axis. Shown are combined stimulations for RAC + LAC, which caused the head to move up; RPC + LPC, which caused the head to move down; and LAC + LPC, which caused the head to roll clockwise to the right. (From Suzuki and Cohen 1964; Suzuki et al. 1964.)

eyes to move to the right, and left posterior canal (LPC) stimulation causes eye movements down and counterclockwise. When the head is free and allowed to move, stimulation of individual canal nerves moves the head in the same direction as the eyes had moved when the head was fixed (Fig. 6.7B, arrows). During large saccades made during lateral gaze shifts, however, the VOR is inhibited and the eyes and head move together (Laurutis and Robinson 1986). By combining the canal nerve stimulations, eye and head movements can be produced about any head axis. Examples are shown on the right side of Figure 6.7B for RAC + LAC, RPC + LPC, and LAC + LPC stimulation in the cat. The fact that any combination of eye and/or head movements can be induced by stimulation of the individual canal nerves is verification that all head and eye movements are coded as a superposition of the vectors along the canal axes and that the canal coordinates form the basis vectors that drive head and eye movements.

Each canal strongly activates one muscle and inhibits one muscle in each eye (Högyes 1880; Szentágothai 1950; Cohen et al. 1964), termed primary activation or inhibition. Because the eyes move so precisely in canal coordinates when the head rotates, there has been considerable interest in understanding how the torque axes of the eye muscles are related to the canal axes. The pulling directions of the primary activated and inhibited muscles are closely aligned with the planes of the canals (See Graf and Simpson 1981; Ezure and Graf 1984; Simpson et al. 1986 for reviews), and there is a close correspondence between the canal planes and the torque axes of the eye muscles that are primarily activated or inhibited by those canals. Examples from the rabbit and cat are shown in Figure 6.8. This establishes a common coordinate frame for vestibular input and oculomotor output that allows the torque axes for reciprocal muscles to work directly in synergy with the canals without having to compute a different set of muscle activations for every eye position (Cohen et al. 1966; Graf and Simpson 1981; Ezure and Graft 1984). It should be emphasized, however, that the "primary" activation or inhibition is accompanied by activation and inhibition of each of the other four muscles in each eye (secondary and tertiary activation) (Cohen et al. 1964) and that every movement is made by all six eye muscles.

Visual information that reaches the vestibular system directly through the subcortical visual system (i.e., through the accessory optic system) is also coded in canal coordinates (Simpson 1984; Simpson et al. 1988). Activity related to movement of the visual fields in the plane of the lateral canals is coded in the nucleus of the optic tract (NOT) and the dorsal terminal nuclei (DTN), whereas activity related to vertical and oblique movement of the visual field is coded in or close to canal planes in the medial and lateral terminal nuclei (Hoffmann and Schoppmann 1975; Simpson 1984; Schiff et al. 1988, 1990; Simpson et al. 1988; Mustari and Fuchs 1990; Yakushin et al. 2000a, 2000b). The synchrony between visual and vestibular inputs and oculomotor output in canal and/or eye muscle coordinates is also extensively coded in the vestibular nuclei and vestibulocerebellum (for a review, see Cohen and Henn 1988). The fact that this canal-based organization is so widely represented throughout the brain stem and cerebellum and in the anatomic organization of the eye muscles is an indication of the importance of vestibular stabilization of gaze for the functioning of the organism. Information about eye and head movement and position in other spatial frames must also be embedded in central structures, however. The central vestibular system codes vestibular information in three-dimensional

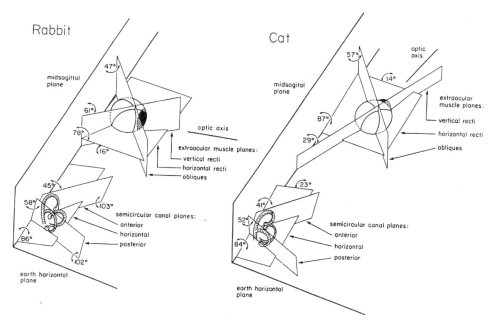

FIGURE 6.8. Relation between the planes of the semicircular canals and the eye muscles activated by the canals on the same side for the rabbit and cat. In both animals, the right anterior canal lies close to the summed pulling directions of the right superior and inferior (vertical) recti and to the left superior and inferior obliques (not shown). The right posterior canal lies close to the summed pulling direction of the right superior and inferior obliques and the left superior and inferior recti (not shown), and the right lateral canal is close to the summed pulling direction of the lateral and medial recti (horizontal recti) in both eyes. (From Simpson and Graf 1981; Ezure and Graf 1984.)

head and body coordinates (Cohen and Henn 1988) and in spatial coordinates, which are realized by convergence of otolith- and canal-based signals (Markham and Curthoys 1972). This information can also come from the visual system and from graviceptor sensors such as the otolith organs as well as from somatosensory receptors (Fredrickson et al. 1965; Solomon and Cohen 1992a; Weber et al. 2000; Yates et al. 2000; Jian et al. 2002).

There is a basic difference in frontal-eyed primates between the semicircular canal coordinate frame of the vestibular system and a major coordinate frame of the visual system. Because the fovea is small and the retina around it, where acuity is greatest, is approximately flat, the three-dimensional visual world must be reconstructed from two-dimensional images, with depth being introduced, to a large extent, by the disparity in the position of targets on the retina in the two eyes. Thus, movement processing in the superior colliculus for saccades that bring the fovea onto the target is essentially two-dimensional, and roll is not represented, although roll is an integral part of every movement produced by the semicircular canals. The mapping functions that relate the two-dimensional processing of visual information onto the three-dimensional oculomotor and vestibular systems have received considerable attention since the late 1980s (Tweed and Vilis 1987, 1990; Raphan 1997; Tweed 1997; Raphan 1998; Thurtell et al. 2000) but are beyond the scope of this chapter. For a more complete description of this mapping, see Raphan and Cohen (2002) for a review and references.

2.2. Dynamic Properties of Canal–Ocular Responses

When the head turns, angular acceleration induces movement of the endolymph relative to the walls of the membranous canals. The fluid movement causes deflection of the cupula, causing it to bend the stereocilia and kinocilia of the hair cells that are embedded in it. When the stereocilia are bent toward or away from the kinocilia, ion channels at the tips of the stereocilia open or close, generating K^+, Ca^{++}, and Cl^- currents in the hair cells (Hudspeth and Corey 1977; Hudspeth and Jacobs 1979; Hudspeth 1982, 1985). These currents raise or lower the resting potential of the hair cells, activating synapses between the hair cells and afferent nerve fibers in the vestibular portion of the eighth nerve (Goldberg et al. 1987). The subsequent increases or decreases in discharge rate from the resting level of the afferent fibers signal to the brain that the head is moving. The transduction process at the hair cell receptors is rapid, with activation of the afferents occurring with a minimum delay of about 0.7 ms after the onset of pulses of current to the endolymph (Highstein 1996). This is close to the estimate of the synaptic delay between the hair cells and the afferent terminals obtained using natural stimulation in the pigeon (Dickman and Correia 1989). The transfer function describing the afferent frequencies relative to head velocity in the squirrel monkey (*Saimiri sciureus*) is comprised of approximately three dynamic modes: 5.7 s, which corresponds to the dominant time constant of the cupula; 0.003 s, the short time constant of the cupula, and an 80 s time constant, which corresponds to an adaptation process (Goldberg and Fernández 1971).

Because only angular acceleration is sensed by the receptors in the semicircular canals and the motoneurons code position of the eyes, two integrations are necessary to convert head acceleration to eye position (Robinson 1977). One of these integrations is done mechanically at the hair cells and cupula of the semicircular canals so that the afferent activity induced in the canal afferents in the vestibular nerve closely follows the angular velocity of the head (Goldberg et al. 1987). Thus, for frequencies of head movement between 0.05 and 8 Hz, which is the normal physiological range of head movements, activity in the vestibular nerve from the semicircular canals is predominantly related to angular head velocity, not to head position or to head acceleration. A signal related to head velocity is also the major information used by the canal-recipient portions of the central vestibular system to produce compensatory eye movements over the aVOR and head movements over the angular vestibulocollic reflex (aVCR). At lower frequencies of head rotation (<0.05 Hz), the phase of the afferent firing rate is shifted closer to angular acceleration and eye velocities induced over the aVOR are greatly diminished.

The second integration required for the aVOR is in the brain stem, close to the motor nuclei. This integration converts the neural signal related to angular velocity of the head, observed in the vestibular nuclei (Waespe and Henn 1977, 1978), to the signal observed in the motoneurons, which is closely related to eye position (Skavenski and Robinson 1973). Because of the overdamped nature of the oculomotor plant (i.e., the eye and surrounding tissue), direct activation of the eye muscles is necessary to reduce delays and maintain the dynamics of the aVOR, saccades, and smooth eye movements (Robinson, 1981). The time constant of the velocity–position integrator is about 30 s, but its value is under cerebellar control (Robinson 1974, 1976). In addition to being an important component of the VOR, velocity–position integration is required by a number of subsystems related to oculomotor control, such as the saccadic and smooth pursuit systems, and is common to each of these oculomotor subsystems (Robinson 1964, 1965, 1968, 1971, 1975). There has recently been considerable work and controversy about how the velocity—position integrator is organized in three dimensions (Tweed and Vilis 1987; Raphan 1998). These three-dimensional approaches are beyond the scope of this review. For a more complete treatment of this issue, see Raphan and Cohen (2002).

2.3. Behavioral Dynamics of the aVOR; Vestibular and Optokinetic Nystagmus

The rapid dynamics of the canals are reflected in ocular dynamics through the generation of short-latency eye movements (Cohen et al. 1964; Baker et al. 1969; Highstein 1973; Collewijn and Smeets 2000). Eye movements that compensate for head movements of short duration ($\approx \leq 1$ s) are generally smooth and not are often interrupted by saccades. For head movements of longer duration, however, compensatory eye movements are frequently interspersed with repetitive, oppositely directed rapid eye movements. Together they form a pattern known as nystagmus. The slow phases represent the sensory portion of the movement, and vestibular or visual input related to head motion is processed during the slow phases. Consequently, the velocity of the slow phases is a function of the velocity of movement of the head or visual surround. The quick phases are largely restorative, resetting the eyes for the next slow phase. Their characteristics are the same as those of saccades. That is, the metrics of the movements are set by their size and are not directly related to the magnitude of the vestibular or visual inputs that produced them. For a more complete description of the characteristics and generation of saccades and quick phases of nystagmus, see Hepp et al. (1989) and Curthoys (2002).

During rotation in darkness for short periods (3–10s), eye velocity compensates well for head velocity, and when the rotation is stopped, there is little or no afterresponse (Fig. 6.9A,B). If the head is rotated for a longer period, however, the eye velocity of the induced nystagmus declines steadily to zero, and at the cessation of rotation, there is a substantial afterresponse in the opposite direction (Fig. 6.9C). During constant-velocity rotation, the hair cells are initially deflected but soon return to their resting position. The return to the resting position and the decline in activity in the vestibular nerve have been modeled by a second-order system with a time constant of 3.5–5s (Steinhausen 1933; Goldberg 1971; see Dai et al. 1999 for a review), which is represented in the second trace of Figure 6.9C by the heavy curved line. The fact that there is good ocular compensation for up to 10s after the onset of rotation (Fig. 6.9A,B) indicates that a central mechanism has prolonged the canal response. This central mechanism has been termed "velocity storage" because it holds or stores activity from the receptor and discharges it over its own time course (Raphan et al. 1979). One functional consequence of this central mechanism is to enhance compensation for lower frequencies of head movement than would occur from cupula deflection alone.

For longer durations of rotation in light, the visual input related to the continuous surround motion generates optokinetic nystagmus (OKN), which is also compensatory. Typical slow phase velocity of horizontal OKN of the rhesus monkey (*Macaca mulatta*) in response to a horizontally moving, visual stimulus is shown in Figure 6.9D. Eye velocity initially jumps quickly to about 0.6 of the velocity of the visual surround and then climbs toward the actual velocity of the visual surround to generate compensatory eye velocity throughout rotation in light (Cohen et al. 1977). In humans, the initial jump in velocity is larger, bringing eye velocity close to that of the visual surround immediately (Cohen et al. 1981).

An important aspect of the OKN response is that the visual input activates the velocity storage network in the vestibular system. This helps maintain the response to rotation in light. Velocity storage also produces optokinetic after-nystagmus (OKAN) when the lights are extinguished (Fig. 6.9D; Cohen et al. 1977; Raphan et al. 1979), which sums with activity from the semicircular canals to cancel vestibular after-responses (Fig. 6.9E; Ter Braak 1936; Cohen 1974; Raphan et al. 1979). In recent years, it has been shown that velocity storage is also tied to spatial orientation in three dimensions and to otolith function (Dai et al. 1991; Raphan and Sturm 1991). The nature of this orientation will be considered in more detail following a parametric description of the aVOR and a review of otolith-ocular transformations.

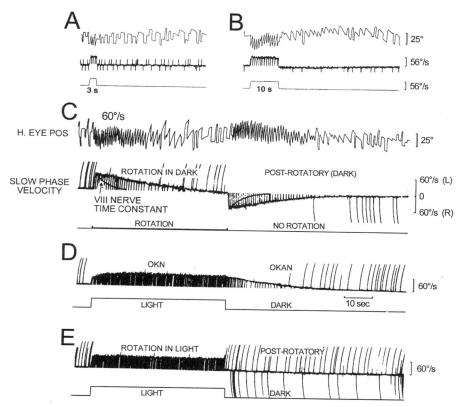

FIGURE 6.9. Eye positions and velocities in response to constant-velocity rotation at 60°/s in the monkey. (**A,B**) Horizontal eye position (top trace) and eye velocity (second trace) in response to constant-velocity rotations (bottom trace) of 3 (A) and 10 (B) s, respectively in darkness. Note that there was good compensation for 3 s of rotation and a slight decline in velocity, followed by a weak afterresponse after 10 s of rotation. (**C**) If rotation at 60°/s was continued for a longer period (50 s, third trace), the nystagmus (top trace) and eye velocity (second trace) gradually declined to zero, and at the end of rotation, there was a vigorous afternystagmus in the opposite direction. The solid lines show the cupula time constant, which is approximately 5 s. Note that the nystagmus considerably outlasted the decline in activity reaching the CNS from the peripheral labyrinth. This is evidence for activation of a central mechanism, termed "velocity storage," which stores activity from the labyrinth and discharges it over its own time course (see the text for details). (**D**) Eye velocity during optokinetic nystagmus (OKN) induced by a step of surround velocity in the monkey. Note the initial rapid rise to a percentage of the stimulus velocity, followed by a slower rise to the velocity of the stimulus. At the end of stimulation, there was vigorous optokinetic afternystagmus (OKAN) in darkness, during which eye velocity gradually declined to zero over the same time course as the afternystagmus after rotation in darkness in C. (**E**) When the two were combined, during rotation in light, eye velocity rose rapidly to the velocity of the visual surround at the onset of stimulation, was maintained at this compensatory velocity for the duration of stimulation, and there was no afterresponse when the animal was again in darkness at the end of rotation (Raphan et al. 1979; Raphan and Cohen 1980).

Functionally, the direction of gaze is approximately fixed in space when head rotations are small and of short duration (Halmagyi and Curthoys 1988; Halmagyi et al. 1990; Tabak and Collewijn 1995; Halmagyi et al. 2001). For larger rotations of the head and body, however, ocular nystagmus compensates incrementally to maintain a stable gaze direction. As the head or visual surround rotates, gaze stability is maintained during the slow phases, but the fixations are interrupted by quick phases of nystagmus produced by the saccadic system. These rapid movements successively refixate the eyes and move gaze forward into the direction of rotation. From this, it can be inferred that the aVOR attempts to match the velocity of the eyes to the velocity of the head or to the velocity of movement of the visual surround. As a result, stabilization of eye position is a consequence of the "velocity" commands in the canal-related portions of the central vestibular system.

2.4. Parametric Descriptors of the aVOR

Because of its input–output and dynamic properties, control system methodologies have been particularly useful for characterizing aVOR behavior. One important and useful parameter for assessing the compensatory aVOR response is its gain (G_{VOR}), generally expressed as a ratio of induced slow-phase eye velocity (V_e) divided by the stimulus (head) velocity (V_h):

$$G_{VOR} = V_e/V_h.$$

Gain can also be expressed in terms of head position or head acceleration as long as the terms of the numerator and denominator are consistent. For the angular VOR, the vestibular system is linear within broad limits. Thus, the gains calculated from eye position, velocity, and acceleration are all the same, and the responses induced by the various frequencies of rotation superpose. One consequence of this is that the gain of the aVOR, calculated at one frequency or velocity, can be assumed to apply to all other frequencies or velocities as well. The gain of the aVOR of normal humans during rotation in the dark generally lies between 0.5 and 0.7. This is low considering that the velocity of the eyes must be close to that of targets moving with relation to the head if objects are to be fixated on the retina (Steinman and Collewijn 1980). There are various mechanisms to enhance ocular compensation during head turns that help maintain ocular stability. Ocular compensation is enhanced during active head rotation or rotation in light, and the visual input contributes additional eye velocity to compensate for head velocity more accurately. Ocular compensation is also enhanced if one attempts to fixate an imaginary, space-fixed target while in darkness, demonstrating the powerful effect of cognitive "set" in enhancing vestibular reflexes (Barr et al. 1976; Barnes 1993). In addition, ocular compensation can be enhanced or suppressed while moving linearly while rotating (Paige et al. 1998). Thus, the definition of gain can be extended to compensatory gain, G_{comp}:

$$G_{comp} = V_e/V_h,$$

which is a measure of compensation taking into account the effects of vision and other sensory modalities, whereas G_{VOR} is related only to the gain of the pure aVOR.

Because stabilization of visual objects is the ultimate aim of the coordinated function of these systems, a unity compensatory gain (G_{comp}) may not always be desirable. If subjects are moving toward or away from objects, for example, the aVOR gains will be lower or higher to take the combined linear and angular accelerations into account (Solomon and Cohen 1992b; Virre and Demer 1996). In addition, viewing distance is of great importance in determining the gain of both the aVOR and the lVOR. If the head is rotated with a target at optical infinity, a value of $G_{comp} = 1$ is ideal. However, as targets come closer and the eyes converge, higher gains are necessary to account for the offset of the axis of eye rotation from the axis of head rotation, which causes translation as well as angular rotation during head movement (Virre et al. 1986; Snyder and King 1992; Telford et al. 1996; Virre and Demer 1996). Consequently, the gain rises to meet the demand for larger eye movements to fixate near targets.

Another important parameter associated with the aVOR is its dominant time constant (Raphan et al. 1979). For horizontal eye movements, the time constant can be defined as the time over which the eye velocity response to a step of head velocity decays to 63% of its peak value (Fig. 6.9C). The time constant can also be determined from Bode plots of the response to different frequencies of sinusoidal oscillation. Because the aVOR has a multiplicity of dynamic modes leading to a plateau in the response, a single time constant often is not adequate to define aVOR decay behavior (Raphan et al. 1979; Cohen et al. 1992b; Dai et al. 1999). In these cases, a better description is obtained from the use of two time constants, one associated with decay in eighth nerve activity (Fig. 6.9C, arrows) and another with the time constant of the velocity storage integrator, which is considerably longer. The time constant of the eighth nerve activity depends solely on cupula dynamics and hair cell transduction, but the time course of central processing of the aVOR is critically dependent on previous motion experience (i.e., habituation) (Cohen et al. 1992b) and on head tilt with regard to the GIA. The latter, in turn, is dependent on otolith processing and the linear vestibuloocular reflex (Raphan and Cohen 1986, 1988; Dai et al. 1991; Raphan and Sturm 1991). How the otoliths are activated impacts both G_{comp} and the time constant of velocity storage (Dai et al. 1991; Raphan and Sturm 1991).

Therefore, an understanding of otolith—ocular reflexes is essential for understanding aVOR behavior over a wide range of behavioral tasks.

FIGURE 6.10. (**A**) Orientation of the utricles and saccules in the head. The utricles lie close to the plane of the lateral canals, and the saccules are approximately orthogonal. The insert in (**B**) shows the tetrahedronal calcite otoconia that are embedded in a mucopolysaccaride matrix atop each of the utricles and saccules. Note that interaural translation or head tilts to the side would cause an inertial response in the utricles, whereas when assuming an upright position or moving vertically along the median sagittal plane the saccules are displaced. (Adapted from Miller 1962.) (**C**) Compensatory IVOR demonstrating importance of near fixation on the amplitude of horizontal eye movement in response to interaural acceleration. The top graph shows the sinusoidal oscillation of a monkey along its interaural axis over an amplitude of about 5 mm at 4 Hz. The solid line in the top graph shows the appearance of a near target (trace movement up). The horizontal position of the eyes is shown in the bottom trace by the thin solid lines and the vergence of the eyes by the heavy solid line. There were small (0.5 mm) horizontal oscillations in response to the oscillating linear acceleration when the target was not visible and the eyes diverged (down position of the solid trace). When the *near target* appeared (top trace), the eyes converged (upward movement of the solid trace) and the amplitude of response became about 2.5–3.0 times as large as at baseline on the left. As convergence relaxed after the target was extinguished (*far target*), the amplitude of the vergence gradually decreased, only to become large again when the near target appeared again. Thus the amplitude of horizontal eye movement produced by the IVOR was closely related to the position of the target relative to the subject (From Paige et al. 1996).

━━━▶

2.5. Origanization of the Otoliths and Otolith–Ocular Reflexes

Otolith–ocular reflexes originate in the maculae of the otolith organs, the utricle and saccule, which behave as inertial accelerometers. Each consists of hair cells, nerve fibers that innervate them, and an overlying gelatinous substance in which the otoliths are embedded (Fig. 6.10A). The structure of the otolith organs differs significantly from the cristae of the semicircular canals. The otoconia are small tetrahedrons, which are composed of calcium carbonate $Ca(CO_3)_2$ crystals in a protein matrix (Fig. 6.10B). Because the specific gravity of these tetrahedrons is greater than that of the endolymph, they are displaced in the direction of the gravitational field when the head is tilted, bending the hair cells. Hair cells on the maculae have a wide range of polarization vectors[5] (Fernández and Goldberg 1976a, 1976b, 1976c). The cells in the utricular macula are oriented roughly in the horizontal plane, and the saccular maculae are roughly vertical, parallel to the midsagittal plane (Fig. 6.10A). Utricular cells are also dominantly

---

[5] A polarization vector is the direction in three-dimensional space relative to the head in which bending of the stereocilia causes maximal excitation of the afferents from that hair cell.

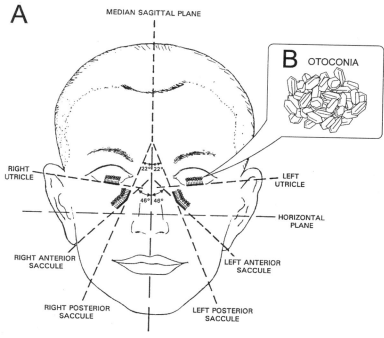

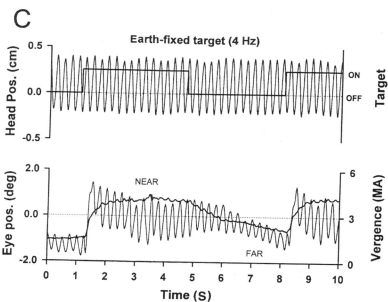

polarized toward the ipsilateral interaural direction, whereas saccular units are polarized approximately at right angles, tending to be clustered in the up (+Z) and down (−Z) directions (Fernández and Goldberg 1976a, 1976b, 1976c). The directions of the polarization vectors are also reversed across a central zone in each macula, called the striola (Wersäll and Bagger-Sjöbäck 1974). Thus, the otoliths have a more complex structure than the semicircular canals both in spatial organization (Curthoys et al. 1999) and in distribution of polarization vectors, which has complicated understanding of how the lVOR is implemented. However, Uchino and colleagues have demonstrated a double inhibitory loop, which causes the hair cells on both sides of the striola to cause coordinated excitation or inhibition in central vestibular neurons for changes in head position or for translational movements in the same direction (Uchino et al. 1997, 1999) which would simplify some aspects of otolith function.

The otoliths also have dynamic attributes. Their inertia and the fluid in which they are embedded cause the underlying hair cells to bend away from the direction of linear acceleration when the head is translated (moved linearly) forward and back, side-to-side, or up and down. Similiar to the semicircular canals, the transformation of head translation to a neural signal representing linear acceleration is done by the mechanicoelectrical properties of the hair cells, responding to movement of otoconia (Hudspeth and Corey 1977; Hudspeth 1982, 1985; Fettiplace et al. 1992). As for canal afferents, the firing rates of otolith afferents range from regular (low coefficient of variation) to irregular (high coefficient of variation) (Ferández and Goldberg 1976a, 1976b, 1976c) and are diverse in their innervation patterns (Goldberg 2000). The characteristics of the regular otolith afferents indicate that they act essentially as linear accelerometers, but the contribution of the irregular afferents to the dynamics of the lVOR is still not well-understood (Goldberg et al. 1987, 1990; Chen-Huang et al. 1997). An appropriate combination of the activity of irregular and regular units, however, may help realize the filtering characteristics of the lVOR and implement its compensatory and orienting functions (Raphan and Cohen 1996, 2002; Raphan et al. 1996).

2.6. Ocular Compensation from the Otoliths

The compensatory and orienting functions of the lVOR operate over distinctly different frequency bands (Mayne 1974). At high frequencies of head translation (1–5 Hz), the lVOR produces compensatory conjugate eye movements for side-to-side and up–down translation of the head. The key to understanding the translational or compensatory lVOR has been the understanding that the lVOR function to subserve vision (Schwarz et al. 1989; Miles and Busettini 1992; Miles et al. 1992). Consequently, compensatory lVOR responses are close to zero when subjects view target at a distance because translation of the head causes little or no movement of targets on the retina. As target are brought nearer, however, much larger eye movements are needed to stabilize vision during translational head and body movements, and the demands on the lVOR gain become much higher. In response, the gain of the compensatory lVOR rises to meet these demands. Thus, the amplitude of the ocular compensatory movements is dependent on the distance of the target relative to the observer (Schwarz et al. 1989; Bronstein and Gresty 1991; Busettini et al. 1991; Paige and Tomko 1991b; Telford et al. 1997; Paige et al. 1998). An example of side-to-side oscillation that produced interaural linear acceleration is shown in Figure 6.10C (from Paige et al. 1996). When the eyes focused on objects nearer to the subject, the eyes converged and the amplitude of the eye position oscillation produced by the lVOR was dramatically increased.

To model the compensatory lVOR to linear translation, it is critical to determine the type of central signal processing that is used to produce this response. Early (Niven et al. 1966) and more recent (Hain 1986; Hess and Dieringer 1990; Angelaki 1992; Hess and Angelaki 1993) workers have proposed that, "jerk" the derivative of linear acceleration, might be the critical signal that produces the lVOR. This formulation, however, does not consider the requirement that although the end organs respond to head acceleration, it is changes in eye position that compensate for changes in head position. By analogy to the aVOR, this implies that two mathematical integrations of the acceleration signal transduced by the hair cells are necessary to generate a position-related signal to drive the motoneurons. If jerk were the critical central signal, three central integrations would be necessary to generate the appropriate eye position signal in oculomotor motoneurons, and this seems unlikely.

How the activity from the otolith receptors is transformed from head linear acceleration into head linear velocity and then recoded into eye position is not clear. The lVOR probably shares one integration with the canal system, the velocity–position integrator that converts a velocity command to the position signals that are necessary to drive the motoneurons (Skavenski and Robinson 1973; Robinson 1975). However, an earlier integration would be necessary to convert the otolith signal related to linear acceleration to a signal related to the eye velocity command. It would not be surprising if processing in the compensatory lVOR is also in terms of velocity, as in the aVOR, because activity from the otoliths combines on central vestibular neurons with activity from the semicircular canals related to head velocity. It is a general principle that neurons must receive information coding similar parameters if they are to make sense of the various inputs. In addition, syndromes occurring after vestibular system lesions are usually associated with a sense of motion and nystagmus, not with feelings of position change or tilt. Objective manifestation of a disorder involving velocity-driven signals would be an ongoing nystagmus with slow phase velocity. One hypothesis, which has been modeled, is that the first integration is done by centrally superposing the signals from the large array of regular and irregular afferents that have the integrative process as a common mode of their dynamic response (Raphan and Cohen 1996). Other models have suggested that there is central filtering of otolith signals (Paige and Seidman 1999). The velocity–position integrator performs the second integration in these models, as for activity that arises in semicircular canals (Raphan and Cohen 2002).

2.7. Orienting Ocular Movements

Orienting eye movements also subserve vision in that they tend to align the yaw axis of the eye with the GIA and maintain the retina in a stable position in three dimensions with regard to gravity or to the GIA (Fig. 6.4C–E). They operate over a much lower frequency band (DC–0.1 Hz) than the compensatory movements (Telford et al. 1996; Paige and Seidman 1999; Cohen et al. 2001; Raphan and Cohen 2002). The best-studied of the orienting movements is ocular counterroll (OCR). As demonstrated in humans, OCR is similar, regardless of how the acceleration is produced, whether by static tilt, translation on a linear sled, rotation about axes tilted from the vertical, centrifugation, or when turning corners (see Cohen 2001; Moore 2001 for reviews). Approximately 4–6° of OCR is elicited by 1g of interaural linear acceleration during static tilt and 6–10° of counterroll during centrifugation (Moore et al. 2001). Examples of OCR from a study using centrifugation (Fig. 6.11B) 15 days before (L – 15), during (FD5), and two days after space flight (R + 1) are shown in Figure. 6.11C. Changes in roll eye position were detected by comparing iral contrasts in video sequences (Fig. 6.11A) (Zhu et al. 1999; Moore et al. 2001). During constant linear acceleration (Fig. 6.11C), the eyes torted toward the GIA and maintained the position in roll for as long as the acceleration was applied. A dynamic component of OCR is also present in humans at the onset of constant-velocity centrifugation, but it disappears in the steady state. Because the direction of the initial roll component reverses with the direction of rotation, Curthoys and colleagues have attributed the dynamic OCR to activation of the semicircular canals (MacDougall et al. 1999).

2.8. Off-Vertical Axis Rotation (OVAR)

OVAR is a particularly useful tool for studying orienting eye movements because it provides a relatively pure stimulus to the otoliths and body tilt receptors that can be averaged to obtain reliable measures of eye position and because it elicits clearly demonstrable orienting reflexes in yaw, pitch and roll (Cohen et al. 2001; Kushiro et al. 2002). Guedry (1965, 1974) and Benson (Benson and Bodin 1966; Benson 1974) almost simultaneously discovered that continuous horizontal nystagmus with superimposed oscillations was induced by OVAR. Young and Henn (1975) showed that there were also modulations in eye position and eye velocity around all three axes

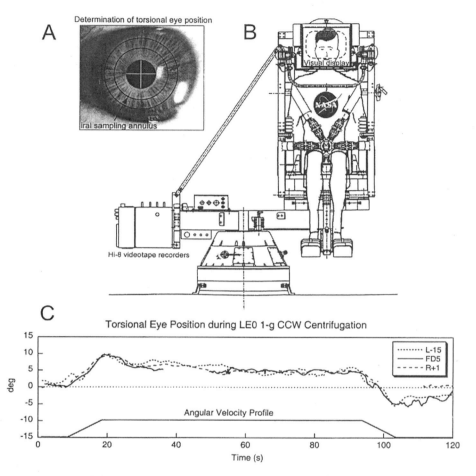

FIGURE 6.11. (**A**) Measurement of ocular counterroll (OCR). Torsional eye position was measured from video images of the eye using pattern matching of iral gray-level information sampled from an annular region centered on the pupil. The annulus used for measurement is defined by the two concentric circles over the iris. The cross shows the center of the pupil, which was used to measure horizontal and vertical positions of the eye. (**B**) Schematic of the flight centrifuge from the Neurolab Mission. The subject was held firmly in place by a five-point safety harness, foot rests, and Velcro handgrips. A visual display and miniature video cameras were mounted directly in front of the subject's face. Video images of the subjects's eyes were recorded onto two Hi-8 videotape recorders mounted on the opposite end of the rotator beam. (**C**) Torsional eye position during 1 g centrifugation on Earth. At the onset of rotation, there was a dynamic ocular torsional component whose direction was dependent on the direction of rotation. When the subject reached a constant angular velocity, this dynamic component decayed, and static OCR was generated in response to the tilted GIA. (From Moore et al. 2001.)

during such stimulation and that similar kinds of continuous and oscillatory components were produced by rotation about pitch axes. More recently, Angelaki and Hess (1996b) have shown that such nystagmus and three-dimensional oscillations are also produced by rotation about a roll axis.

During OVAR, the semicircular canals are initially activated at the onset of rotation, but as rotation continues at a constantly velocity, semicircular canal activity disappears. A projection of the gravity vector continues to circle the head and body, however, moving in a direction opposite to the direction of rotation. During rotation around a yaw axis tilted 30° from the vertical, for example, $0.5g$ is sequentially projected along all axes that lie in the plane orthogonal to the axis of rotation, whereas rotation about an axis tilted 90° projects $1g$ along these same axes. Such rotation causes sequential activation of otolith hair cells and body tilt receptors, and this activity is converted into a velocity signal by the central vestibular system through velocity storage (Raphan et al. 1981; Cohen et al. 1983; Dai et al. 1994). *As a result, the brain is able to determine how fast the head and body are rotating solely from information coming from the otoliths and body tilt receptors.*

Across a wide range of species, steady-state compensatory nystagmus and modulations in eye position in roll, pitch, and yaw as a function of head position with regard to gravity are produced by OVAR (Fig. 6.12A). The

FIGURE 6.12. Off-vertical axis rotation (OVAR). (**A**) Roll, pitch, and yaw eye positions (second to fourth traces) induced by rotation of a monkey at 30°/s about an axis tilted 90° from the Earth vertical. Eye movements to the left, down, and clockwise from the animal's point of view are positive. The position of the animal with regard to the axis of rotation, shown in the top trace, was recorded by a potentiometer, which reset each 360°. The animal's positions with regard to gravity at various times during the rotation cycle are shown by the figures in the inserts below, and the vertical lines show the animal's position with regard to gravity for each of the recorded parameters. The rotation induced horizontal nystagmus (fourth trace) and regular modulations in each component of eye movement as the animal was in different head positions. OCR and horizontal deviations were maximal in side-down positions, and vertical deviations were maximal when the animal was nose-down and nose-up. Note also that saccades were superimposed on the deviations in pitch. (Adapted from Kushiro et al. 2002.) (**B**) Vergence modulation in a monkey during counterclockwise OVAR at 60°/s around an Earth-horizontal axis. The top trace (EYE POS) shows the position of both eyes (left eye, solid line; right eye, dotted line). The straight horizontal line is the baseline position for each eye at rest in darkness. At points a, b, and c, there were maximal eye excursions to the left (a,c) and right (b). At these points, the positions of the left and right eyes were synchronized (i.e., the vergence was close to baseline). The middle trace (LT-RT) is vergence modulation. The straight line across is baseline vergence. Upward deflection of the yaw axis position of turntable rotation (bottom trace) occurred when the head was in the left-side-down position. (Adapted from Dai et al. 1996.)

Off-Vertical Axis Rotation (OVAR)

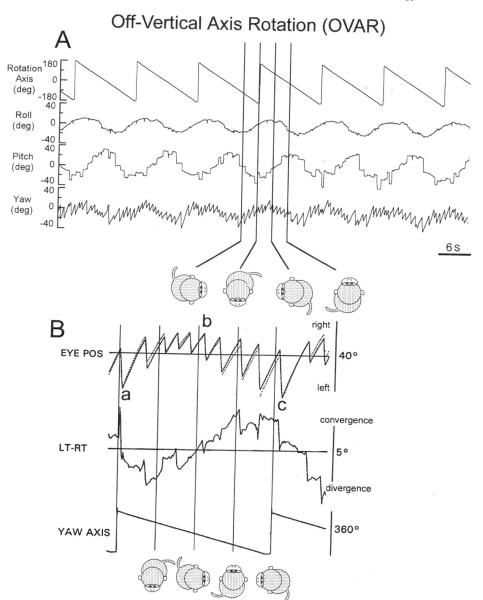

sensitivity of the system is great and a tilt of as little as 5° of the axis of rotation is sufficient to induce otolith-driven compensatory eye velocities (Kushiro et al. 2002). About 8° of vertical change is produced by OVAR about an axis tilted 30° from the vertical in humans (Haslwanter et al. 2000). This is comparable to the changes in vertical position produced by OVAR in the rabbit (Maruta et al. 2001) and monkey (Kushiro et al. 2002). From this, it is clear that otolith activation produces orientation of the eyes to the GIA when the head is pitched with regard to gravity across a broad range of species (Cohen et al. 2001). Forward saccades are frequently interposed on the slow position changes in pitch to achieve eye positions in the orbit during OVAR (Fig. 6.12A, third trace). Thus, the saccadic system is also involved in orienting the eyes in the orbit in response to dynamic otolith stimuli. Eye velocities related to the pitch and roll eye position changes during yaw axis OVAR lead the position phases by approximately 90° in both the rabbit and monkey, indicating that if saccades are excluded, the pitch and roll velocities are largely a differentiation of the position modulations (Kushiro et al. 2002).

Horizontal orienting movements are also clearly demonstrated during OVAR. Continuous horizontal nystagmus appears with slow phases in the direction of the rotating gravity vector, and the beating field ("schlagfeld") is cyclically modulated as a function of head position with regard to gravity (Fig. 6.12A, fourth trace). Across animals, the amplitude of the modulations is similar to that induced by equivalent static tilts (Cohen et al. 2001), and the maximum lateral deviations of the beating field occur in the same head positions as the maximal yaw deviations during static tilt. Paige and Tomko (1991a) compared the interaural component of the lVOR induced by translation on a linear sled to the oscillations of horizontal eye velocity during OVAR. They concluded that both of these responses had consistent amplitude and phase characteristics, suggesting that they were produced by the same mechanism.

2.9. Vergence and Visual Orientation

Vergence in response to head tilt (i.e., to nasooccipital linear acceleration) was originally noted in the rabbit by Magnus (1924) and Lorente de Nó (1932). In the rabbit, each eye deviates by 11–14°/g for 90° of upward pitch of the head. Because convergence and divergence are the combined movements of both eyes, the vergence angle of the eyes changes by 22–28° in the rabbit when 1g of linear acceleration is imposed along the nasooccipital axis. The eyes also converge and diverge during forward or backward nasooccipital linear acceleration across a wide range of species (Lorente de nó 1932; Paige and Tomko 1991b; Dai et al. 1996; Telford et al. 1997; Cohen et al. 2001; Maruta et al. 2001; Kushiro et al. 2002).

One or both eyes can verge, depending on the position of the target relative to the head (Paige 1991; Paige and Tomko 1991b). For example, if a target is moved parallel to the nasooccipital axis along the visual axis of one eye, only the contralateral eye verges. If the target is moved along the nasooccipital axis, both eyes verge. In the monkey and human, such vestibularly induced vergence supports the "near response"—to help select targets in near space when moving forward. In the rabbit, such vergence would substantially increase the binocular field in front of the animal, also supporting vision (Maruta et al. 2001).

From its characteristics in primates, vergence induced by the lVOR has often been considered a high-frequency compensatory response because its gain falls sharply at lower frequencies (Paige and Tomko 1991a). Nevertheless, large angles of vergence are induced in response to both static tilts and low frequencies of rotation during OVAR in the rabbit (Maruta et al. 2001) and monkey (Fig. 6.11B) (Dai et al. 1996). Thus, lVOR-induced vergence subserves both compensatory and orienting functions. An understanding of convergence behavior in three dimensions or how the compensatory and orienting convergence information is processed centrally is still largely unknown.

2.10. Central Organization of the VOR; Velocity Storage and Spatial Orientation

Both the aVOR and optokinetic reflexes can be modeled as being comprised of two components. One component is the short latency response that stabilizes gaze for high frequency head movements or for movements of the visual surround over rapid or 'direct' pathways. A second, slower component has been modeled as a leaky integrator that stores activity from the eighth nerve related to head velocity and outputs a velocity command to produce slow phase eye velocity (Raphan et al. 1977, 1979; Robinson 1977). This is the 'velocity storage' integrator', which is separate from the integrative neural network that produces velocity-to-position integration (Robinson 1968). Velocity storage is also activated by the somatosensory system during circular locomotion (Solomon and Cohen 1992a) and by otolithic input during OVAR (Raphan et al. 1981; Cohen et al. 1983; Kushiro et al. 2002) to provide a signal related to head and body velocity in space. Such activity is used to generate compensatory gaze velocity during locomotion when there is an angular component (Solomon and Cohen 1992a).

The VOR has been modeled in three dimensions by combining the components of the aVOR and lVOR described above. Underlying the model are basic postulates that define the functional components of the VOR in terms of compensatory and orienting functions (Fig. 6.13). In the model, the aVOR is driven by the angular head acceleration vector, which is

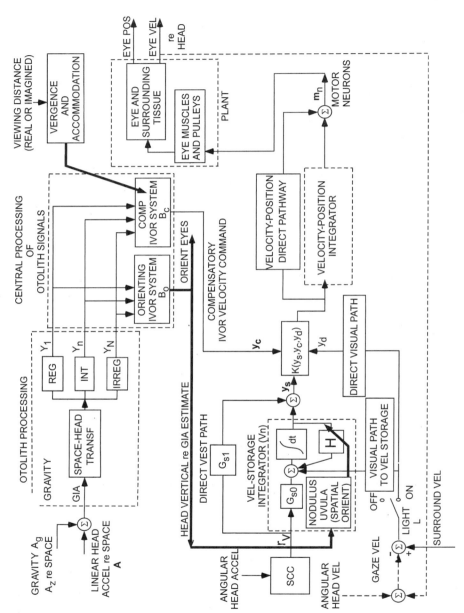

FIGURE 6.13. Model of VOR with visual vestibular interaction and orienting responses. See the text for details.

transduced by the spatial and dynamic semicircular canal transformation. This producers the neural command vector, $\mathbf{r}_v$, which drives the velocity storage integrator (VEL-STORAGE INTEGRATOR (Vn)) and a direct vestibular pathway (DIRECT VEST PATH). These pathways combine to generate the angular eye velocity command, $\mathbf{y}_s$. The aVOR command, $\mathbf{y}_s$, combines with the linear command, $\mathbf{y}_c$, generated form the lVOR and $\mathbf{y}_d$, a signal form the visual system via a direct visual pathway, bypassing velocity storage. This is done through a system denoted as $K(\mathbf{y}_s, \mathbf{y}_c, \mathbf{y}_d)$, whose output signal drives the velocity–position integrator and its direct pathway. This combination drives the motoneurons, which in turn drive the ocular plant, including the eye muscles and surrounding pulleys, to orient the eyes (Raphan and Cohen 1996, 2002; Raphan 1997, 1998; Thurtell et al. 2000).

The lVOR portion of the model (dashed-line boxes marked OTOLITH PROCESSING and CENTRAL PROCESSING OF OTOTLITH SIGNALS) has both compensatory and orientating subsystems. Both subsystems are driven by the otoliths, which are inertial sensors responding to the composite linear acceleration vector, GIA (the vector sum of linear head acceleration with regard to space and gravity). Because the otolith organs are fixed in the head, there is an implicit space-to-head coordinate transformation (SPACE–HEAD TRANF), which represents the coding of the linear accelerations acting on the head (the GIA) by the otolith afferents. The afferents, $y_l, \ldots, y_N$, have widely varying frequency characteristics. In the model, this afferent activity is combined by central filters, which implement the dynamics of the compensatory and orienting functions, respectively. The novel aspect of this model is that the compensatory and orienting filters can be implemented by a weighted sum of the otolith afferents $(y_l, \ldots, y_N)$ based on the actual characteristics of otolith units (Raphan and Cohen 1996, 2002). The compensatory lVOR is constrained by the functional requirement that changes in head position be countered by oppositely directed changes in eye position. This constraint requires that the acceleration signal sensed by the otolith organs be integrated twice to generate the position-related neural signals, which drive the motoneurons. The first integration is achieved as part of the filtering characteristics of the COMP lVOR SYSTEM. The second is produced by the velocity–position integrator (Raphan and Cohen 1996, 2002). The orienting mechanism functions to align the yaw axis of the eyes with the direction of the GIA. Thus, in the model, frequency segregation is implemented by differentially weighting the common otolith afferent information in the orientation and compensatory subsystems. The three-dimensional properties of the system are achieved by implementing these systems as matrices, which distribute the lVOR drive along the various axes of the head. If the orientation drive on the eyes is modeled as a cross product between the body (head) vertical with the GIA, then the sine of the angle between the GIA and body vertical becomes the important parameter for determining the orienting component of the lVOR.

An interesting aspect of velocity storage is that it is spatially oriented (Raphan and Cohen 1986, 1996; Dai et al. 1991; Raphan and Sturm 1991; Angelaki and Hess 1994b). This spatial orientation is present for all modes of vestibular or optokinetic activation of velocity storage and is present in all mammalian species that have been tested, including humans, monkeys, and rabbits. Thus, despite weak OKAN (Lafortune et al. 1986), spatial orientation of OKN is robust in humans (Gizzi et al. 1994). When subjects are upright, so that their yaw or body axis is aligned with gravity, visual (Fig. 6.14A) or vestibular stimuli around the long axis of the body produce pure horizontal eye movements (Fig. 6.14B). Because eye movements are rotations of a globe in the orbit, these are rotations of the eyes around axes that are close to the spatial vertical (Fig. 6.14C). If subjects are tilted and drum rotation is maintained around the yaw axis of the head (Fig. 6.14D), however, a vertical component appears (Fig. 6.14E). This shifts the axis of rotation of the eyes so that the eye velocity vector tends to align with the spatial vertical (Fig. 6.14F, dashed vertical line) rather than with the yaw axis of the head and body (Fig. 6.14F, solid axes). This

---

FIGURE 6.14. (A–F) Effect of head tilt with regard to gravity on horizontal optokinetic nystagmus (OKN) in a human subject. In A–C, movement of the visual surround at 35°/s about the head yaw axis when this axis was aligned with gravity (A) produced the horizontal OKN shown in B. There was no vertical component (B, bottom trace), and the eyes rotated about a spatially vertical axis. Thus, the slow-phase velocity vectors (dots) aligned with the spatial vertical (C). In D–F, when the same horizontal stimulus was given with regard to the head but the head was tilted 45° with regard to gravity (D), a vertical component appeared in the nystagmus (E, bottom trace). This caused the axis of eye rotation and the vector of eye velocity to shift toward the gravitational vertical. In F, the tilt of the head is shown by the directions of yaw and pitch eye velocity axes in a head coordinate frame (solid lines) with regard to the spatial vertical (dashed vertical line). The dots represent the slow-phase velocity vectors of the nystagmus in D with regard to both the head tilt and spatial vertical. The solid oblique line shows the mean vector. Note that the slow-phase eye velocity vector tended to align with the GIA (spatial vertical). (G–I) Model simulations of the spatial orientation of OKN. G shows the yaw eye velocity (EYE VEL) and activation of the velocity storage integrator (INT) during horizontal OKN at 35°/s. Eye velocity rose to about 26°/s, and the integrator was charged to about 16°/s, which is close to the saturation velocity of human velocity storage. In H, vertical optokinetic stimulation produced an eye velocity of about 26°/s, largely due to activation of the direct visual pathway. The vertical integrator time constant was short in the upright position and made little contribution to the overall response. In I, with the head tilted, the yaw axis time constant was shorter than when the head was upright, and the integrator was charged to only about 11°/s (top graph). The pitch time constant was longer in the tilted than in the upright condition and produced a vertical eye velocity that tilted the eye velocity vector 18° from the axis of rotation (insert). (From Gizzi et al. 1994.)

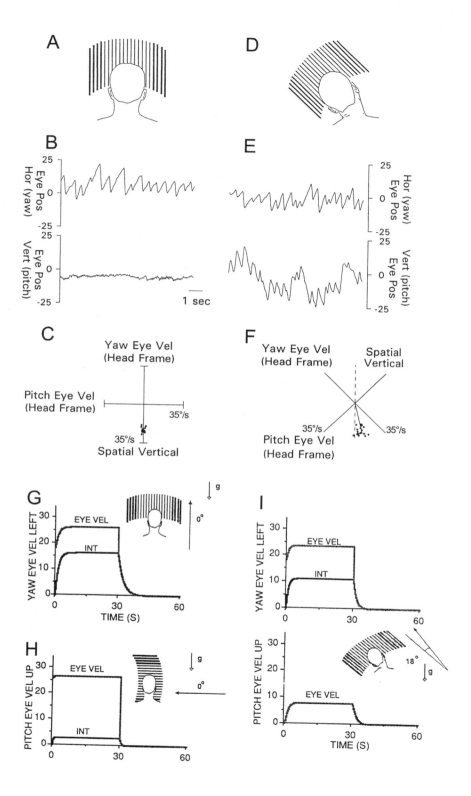

can only however, in the absence of orthogonal contrasts to the vertical stripes so that the eyes can move up and down along the stripes without constraint.

The model of OKN–OKAN, which is included in the model shown in Fig. 6.13, predicts these spatial-orientation properties (Gizzi et al. 1994). For horizontal or vertical OKN in the upright condition, the model generated eye velocity that was along the respective horizontal (Fig. 6.14G) and vertical (Fig. 6.14H) directions. The time constant of vertical velocity storage is short in the upright position, so that the integrator (INT) contributed little to the overall response to the vertically moving stripes, and eye velocity fell precipitously at the end of stimulation. In contrast, the time constant of the storage integrator was longer for the horizontal component, and the integrator contributed a larger portion of the total eye velocity during yaw axis OKN (Fig. 6.14G). When viewing the same horizontal stimulus as in Figure 6.14G but with the head tilted, the yaw axis time constant and the integrator contribution to the total response were reduced (Fig. 6.14I). Because the head was tilted with regard to gravity, however, cross coupling from yaw to pitch was introduced. This generated a pitch component of eye velocity that rotated the eye velocity vector toward the spatial vertical by 18° (insert), fitting the experimental data (Gizzi et al. 1994). Thus, the model simulation supports the hypothesis that the system is capable of generating significant spatial components in response to pure horizontal stimulation in humans, despite the short time constants and weaker velocity storage in humans than in monkeys. The critical factor for the appearance of the spatial components is that orientation is determined by the ratio of the cross coupling between horizontal and vertical states and the difference between the inverse of the time constants (eigenvalues) of vertical and horizontal modes of velocity storage (Raphan and Sturm 1991). As long as the time constants of the horizontal and vertical eye velocities produced by velocity storage are close to each other, the orientation vector relative to the head vertical can be large and eye velocity can be shifted toward the spatial vertical.

In monkeys, the time constants of OKAN and saturation velocities of velocity storage are relatively large, and the orientation vectors can be demonstrated in a more direct fashion. When optokinetic stimulation is given along an orientation vector[5], for example, both horizontal and vertical eye velocities are generated (Fig. 6.15A). When the eye velocity trajectory of the OKAN for this lateral tilt condition was plotted in the pitch/yaw plane (Fig. 6.15B), eye velocity declined from the value at the

---

[5] An eigenvector is a vector in three-dimensional space such that the decline in eye velocity is along the same direction as the stimulus. We have also termed this an "orientation vector" (Dai et al., 1991; Raphan and Sturm 1991).

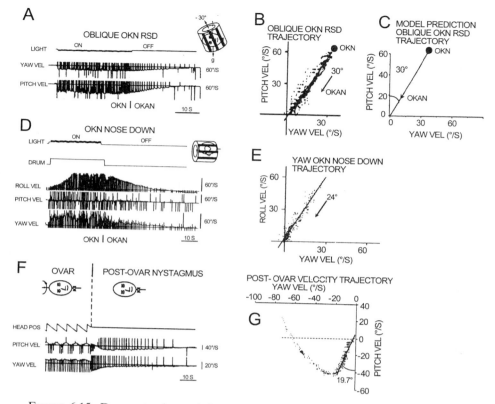

FIGURE 6.15. Demonstrations of the orientation of velocity storage to the GIA, which in this case was gravity (**g**), for different monkeys and different experimental paradigms that activate velocity storage. With the head right side down (RSD), yaw and pitch eye velocity during oblique OKN was followed by decline in pitch and yaw eye velocity with the same time constant (**A**). The eye velocity trajectory in the pitch-yaw plane during OKAN was along a straight line, which was tilted approximately 30° from the vertical (g) and was the yaw orientation vector of velocity storage (**B**). When the appropriate cross coupling and time constants were incorporated into the 3-D model of OKN-OKAN, the model predicted the straight line trajectory (**C**), and could be used to predict the eye velocity trajectory during other stimulus modalities such as centrifugation (Wearne et al. 1999). If the stimulus was along the yaw axis while the animal was prone, the OKAN developed a cross-coupled roll component, which then declined to zero (**D**). The trajectory of the roll-yaw OKAN velocity approached a straight line close to the yaw orientation vector of velocity storage (**E**). Similar cross coupling of yaw to pitch eye velocity and orientation of the yaw-pitch velocity vector to the GIA occurred during vestibular stimulation when animals were stopped in a side down position after off-vertical axis rotation (**F**). The terminal portions of the trajectory of the cross-coupled, pitch-yaw response declined toward zero in a straight line close to the yaw orientation vector predicted from the OKAN. (**G**). The data show that the yaw orientation vector of velocity storage is a combination of the spatial and body vertical directions and that the eye velocity trajectory approaches this orientation vector when the stimulus has a component along the body yaw direction. While the contributions of each component vary from monkey to monkey, the spatial component plays a dominant role in orienting eye velocity for any orientation of the animal. From (Raphan and Cohen 1988; Dai et al. 1991).

end of OKN (filled circle, Fig. 6.15B) toward the origin in a linear fashion at an angle of 30°. A similar decline was predicted for this condition by the model (Fig. 6.15C). This demonstrates that this direction is an orientation vector for the system. Similar orientation vectors were found by plotting the trajectory of yaw axis OKAN when the head was in the prone position (Fig. 6.15D). In this instance, the decline in OKAN velocity was along the direction of the orientation vector for that head position, which was in the yaw/roll plane (Fig. 6.15E). Note that the angle at which eye velocity approached the origin (24°) was similar to that in Fig. 6.15B, although the latter, from another monkey, was in the pitch/yaw plane. Model predictions, plotted on the same graph in the Fig. 6.15E, overlay the experimental data.

Alignment toward the vector sum of the linear accelerations also occurs during vestibularly-induced nystagmus (Raphan et al. 1992; Raphan and Cohen 1996), since velocity storage is common to optokinetic and vestibular responses (Raphan et al. 1979). Thus, when monkeys were stopped in side down positions after constant velocity rotation about a tilted axis (Fig. 6.15F), the post-rotatory eye velocity trajectory (Fig. 6.15G) approached the orientation vectors established during OKAN (Fig. 6.15B, E). This shows that velocity storage is the primary orienting component of velocity in the aVOR. This conclusion is supported by studies utilizing centrifugation (Raphan and Cohen 1996; Wearne et al. 1999), and by the orientation of eye velocity to gravity during caloric stimulation in canal-plugged animals (Arai et al. 2002). The spatial orientation of velocity storage is under control of the nodulus and uvula of the vestibulo-cerebellum, and the eyes no longer reorient to the GIA vector when these structures are damaged (Angelaki and Hess 1994a; Wearne et al. 1996, 1998).

3. Summary

In this chapter, we have introduced the physiology of the vestibuloocular reflex and its compensatory and orienting functions. Advances in technology have frequently resulted in advances in basic understanding of the system. Means are now available to record the movements of the eyes, body, and limbs in freely moving subjects and to study vestibular function more precisely at a cellular level. From this, new understanding of the physiology of the vestibular system will emerge, particularly at the cellular level, which will probably dwarf all previous efforts. The clinical significance of this system is great because the reflexes that provide postural control during standing and help stabilize gaze and the body while walking and running are vital to normal function and can easily be disordered, both by disease and by aging. Thus, the physiology warrants deeper investigation solely for its clinical importance. Beyond that, however, the vestibular system sits at the heart of the motion-sensing apparatus of the brain, and it can serve as a model system for studying how the brain integrates information from many sensory systems to generate a motor output whose function is both well-known and vital to normal behavior and survival.

Acknowledgments. Supported by NIH Grants EY11812, EY04148, DC03284, DC04222, by NASA Cooperative Agreement NC9-58 with the National Space Biomedical Research Institute (NSBRI), and by Core Center Grants DC05204 and EY01867.

References

Angelaki DE (1992) Two-dimensional coding of linear acceleration and the angular velocity sensitivity of the otolith system. Biol Cybern 67:511–521.

Angelaki DE, Hess BJM (1994a) The cerebellar nodulus and ventral uvula control the torsional vestibulo-ocular reflex. J Neurophysiol 72:1443–1447.

Angelaki DE, Hess BJM (1994b) Inertial representation of angular motion in the vestibular system of rhesus monkeys. I. Vestibuloocular reflex. J Neurophysiol 71:1222–1249.

Angelaki DE, Hess BJM (1996a) Organizational principles of otolith and semicircular canal–ocular reflexes in rhesus monkeys. Ann N Y Acad Sci 781:332–347.

Angelaki DE, Hess BJM (1996b) Three-dimensional organization of otolith-ocular reflexes in rhesus monkeys. I. Linear acceleration responses during off-vertical axis rotation. J Neurophysiol 75:2405–2424.

Bach-Y-Rita P, Collins CC, Hyde JE (eds) (1971) The Control of Eye Movements. New York: Academic Press.

Baker J, Goldberg J, Hermann G, Peterson B (1984) Optimal response planes and canal convergence in secondary neurons in vestibular nuclei of alert cats. Brain Res 294:133–137.

Baker R, Berthoz A (eds) (1977) Control of Gaze by Brain Stem Neurons. Amsterdam: Elsevier.

Baker R, Mano N, Shimazu H (1969) Postsynaptic potentials in abducens motoneurons induced by vestibular stimulation. Brain Res 15:577–580.

Balaban CD, Jacob RG (2001) Background and history of the interface between anxiety and vertigo. J Anxiety Disord 15:27–51.

Baloh RW, Halmagyi M (eds) (1996) Disorders of the Vestibular System. New York: Oxford University Press.

Bárány R (1907) Physiologie und Pathologie des Bogengangsapparates beim Menchen. Wien: Franz Deutick.

Barnes GR (1993) Visual–vestibular interaction in the control of head and eye movement: the role of visual feedback and predictive mechanisms. Prog Neurobiol 41:435–472.

Barr CC, Schultheis LW, Robinson DA (1976) Voluntary non-visual control of the human vestibulo-ocular reflex. Acta Otolaryngol (Stockh) 81:365–375.

Bender MB (1964) The Oculomotor System. New York: Hoeber Medical Division, Harper and Row.

Benjamins CE (1918) Contribution à la connaissance des réflexes tonique des muscles de l'oeil. Neerl Phys l'Homme Animaux 2:536–544.

Benson AJ (1974) Modification of the response to angular accelerations by linear accelerations. In: Kornhuber H (ed) Handbook of Sensory Physiology, Volume VI. Berlin: Springer, pp. 281–320.

Benson AJ, Bodin MA (1966) Interaction of linear and angular accelerations on vestibular receptors in man. Aerosp Med 37:144–154.

Berthoz A (1993) The Brain's Sense of Movement. New York: Arno.

Blanks RHI, Curthoys IS, Bennet M, Markham CH (1985) Planar relationships of the semicircular canals in rhesus and squirrel monkeys. Brain Res 340:315–324.

Brandt T (1999) Vertigo. London: Springer-Verlag.

Breuer J (1874) Über die funktion de bogengange des ohrlabyrinths. Med Jahrb 4:72–124.

Brodal A (1981) Neurological Anatomy: In Relation to Clinical Medicine. New York: Oxford University Press.

Brodal A, Pompeiano O, Walberg F (1962) The vestibular nuclei and their connections. Springfield, IL: Charles C. Thomas.

Bronstein AM, Gresty MA (1991) Compensatory eye movements in the presence of conflicting canal and otolith signals. Exp Brain Res 86:697–700.

Busettini C, Miles FA, Schwarz U (1991) Ocular responses to translation and their dependence on viewing distance. II. Motion of scene. J Neurophysiol 66:865–878.

Camis M, Creed RS (1930) The Physiology of the Vestibular Apparatus. Oxford, UK: Clarendon Press.

Chen-Huang C, McCrea RA, Goldberg JM (1997) Contributions of regularly and irregularly discharging vestibular-verve inputs to the discharge of central vestibular neurons in the alert squirrel monkey. Exp Brain Res 114:405–422.

Clément G, Moore ST, Raphan T, Cohen B (2001) Perception of tilt (somatogravic illusion) in response to sustained linear acceleration during space flight. Exp Brain Res 138:410–418.

Cohen B (1974) The vestibulo-ocular reflex arc. In: Kornhuber HH (ed) Handbook of Sensory Physiology. Vestibular System. Basic Mechanisms, Volume VI, Part 1. Berlin: Springer-Verlag, pp. 477–540.

Cohen B (ed) (1981) Vestibular and Oculomotor Physiology: International Meeting of the Bárány Society. New York: New York Academy of Sciences.

Cohen B (1984) Erasmus Darwin's observations on rotation and vertigo. Hum Neurobiol 3:121–128.

Cohen B, Maruta J, Raphan T (2001) Orientation of the eyes to gravito-inertial acceleration. N Y Acad Sci 942:241–258.

Cohen B, Henn V (1988) Representation of Three-Dimensional Space in the Vestibular, Oculomotor, and Visual Systems. New York: New York Academy of Sciences.

Cohen B, Hess B (eds) (1999) Otolith Function in Spatial Orientation and Movement. New York: New York Academy of Sciences.

Cohen B, Suzuki J, Bender MB (1964) Eye movements from semicircular canal nerve stimulation in the cat. Ann Otol Rhinol Laryngol 73:153–169.

Cohen B, Tokumasu K, Goto K (1966) Semicircular canal verve eye and head movements. Arch Ophthalmol 76:523–531.

Cohen B, Matsuo V, Raphan T (1977) Quantitative analysis of the velocity characteristics of optokinetic nystagmus and optokinetic after-nystagmus. J Physiol (Lond) 270:321–344.

Cohen B, Henn V, Raphan T, Dennett D (1981) Velocity storage, nystagmus, and visual vestibular interactions in humans. Ann N Y Acad Sci 374:421–433.

Cohen B, Suzuki JI, Raphan T (1983) Role of the otolith organs in generation of horizontal nystagmus: effects of selective labyrinthine lesions. Brain Res 276:159–164.

Cohen B, Tomko D, Guedry F (eds) (1992a) Sensing and Controlling Motion—Vestibular and Sensorimotor Function. New York: New York Academy of Sciences.

Cohen H, Cohen B, Raphan T, Waespe W (1992b) Habituation and adaptation of the vestibulo-ocular reflex: a model of differential control by the vestibulo-cerebellum. Exp Brain Res 90:526–538.

Cohen B, Maruta J, Raphan T (2001) Orientation of the eyes to gravito-inertial acceleration. Ann N Y Acad Sci 942:241–258.

Collewijn H, Smeets BJ (2000) Early components of the human vestibulo-ocular response to head rotation: latency and gain. J Neurophysiol 84:376–389.

Cox JM (1804) Practical Observations on Insanity. London: L & R Baldwin.

Crum-Brown (1874) On the sense of rotation and the anatomy and physiology of the semicircular canals of the inner ear. J Anat Physiol (Lond) 8:327–331.

Curthoys IS (2002) Generation of the quick phase of horizontal vestibular nystagmus. Exp Brain Res 143:397–405.

Curthoys IS, Blanks RHI, Markham CH (1977) Semicircular canal functional anatomy in cat, guinea pig, and man. Acta Otolaryngol 83:258–265.

Curthoys IS, Betts GA, Burgess AM, MacDougall HG, Cartwright AD, Halmagyi GM (1999) The planes of the utricular and saccular maculae of the guinea pig. Ann N Y Acad Sci 871:27–34.

Dai M, Raphan T, Cohen B (1991) Spatial orientation of the vestibular system: dependence of optokinetic afternystagmus on gravity. J Neurophysiol 66:1422–1438.

Dai M, McGarvie L, Kozlovskaya IB, Raphan T, Cohen B (1994) Effects of space-flight on ocular counterrolling and spatial orientation of the vestibular system. Exp Brain Res 102:45–56.

Dai M, Raphan T, Kozlovskaya I, Cohen B (1996) Modulation of vergence by off-vertical yaw axis rotation in the monkey: normal characteristics and effects of space flight. Exp Brain Res 111:21–29.

Dai M, Klein A, Cohen B, Raphan T (1999) Model-based study of the human cupular time constant. J Vestib Res 9:293–301.

Darwin E (1796) Zoonomia. London: J. Johnson.

de Kleijn A, Magnus R (1921) Ueber die funktion der otolithen. Pflügers Arch Ges Physiol 186:6–81.

Dickman JD, Correia MJ (1989) Responses of pigeon horizontal semicircular canal afferent fibers. II, High-frequency mechanical stimulation. J Neurophysiol 62:1102–1112.

Dodge R (1903) Five types of eye movements in the horizontal meridian plane of the field of regard. Am J Physiol 8:307–329.

Ewald JR (1892) Physiologische Untersuchungen über das Endorgan des Nervus Octavus. Wiesbaden: Bergmann.

Ezure K, Graf W (1984) A quantitative analysis of the spatial organization of the vestibulo-ocular reflexes in lateral- and frontal-eyed animals. I. Orientation of semicircular canals and extra-ocular muscles. Neuroscience 12:85–93.

Fernández C, Goldberg JM (1976a) Physiology of peripheral neurons innervating otolith organs of the squirrel monkey. I. Response to static tilts and to long duration centrifugal force. J Neurophysiol 39:970–984.

Fernández C, Goldberg JM (1976b) Physiology of peripheral neurons innervating otolith organs of the squirrel monkey. II. Directional selectivity and force response relations. J Neurophysiol 39:985–995.

Fernández C, Goldberg JM (1976c) Physiology of peripheral neurons innervating otolith organs of the squirrel monkey. III. Response dynamics. J Neurophysiol 39:996–1008.

Fetter M, Haslwanter T, Misslisch H, Tweed D (eds) (1997) Three-Dimensional Kinematics of Eye, Head, and Limb Movements. Amsterdam: Harwood Academic Publishers.

Fettiplace R, Crawford AC, Evans MG (1992) The hair cell's mechanoelectrical transducer channel. Ann N Y Acad Sci 656:1–11.

Flourens MJP (1824) Recherches expérimentales sour les propriétés et les fonctions du systeme nerveux dans les animaux vertébrés. Paris: Crevot.

Fredrickson JM, Schwartz D, Kornhuber HH (1965) Convergence and interaction of vestibular and deep somatic afferents upon neurons in the vestibular nuclei of the cat. Acta Otolaryngol (Stockh) 61:168–188.

Freund H-J, Buettner U, Cohen B, Noth J (eds) (1986) The Oculomotor and Skeletalmotor System: Differences and Similarities. Amsterdam: Elsevier Science Publishers.

Gizzi M, Raphan T, Rudolph S, Cohen B (1994) Orientation of human optokinetic nystagmus to gravity: a model based approach. Exp Brain Res 99:347–360.

Goebel J, Highstein SM (eds) (2001) The Vestibular Labyrinth in Health and Disease. New York: New York Academy of Sciences.

Goldberg JM, Fernandez C (1971) Physiology of peripheral neurons innervating semicircular canals of the squirrel monkey. I. Resting discharge and response to angular accelerations. J Neurophysiol 34:635–660.

Goldberg JM (2000) Afferent diversity and the organization of central vestibular pathways. Exp Brain Res 130:277–297.

Goldberg JM, Fernández C (1971) Physiology of peripheral neurons innervating semicircular canals of the squirrel monkey. I. Resting discharge and response to angular accelerations. J Neurophysiol 34:635–660.

Goldberg JM, Desmadryl G, Baird RA, Fernández C (1990) The vestibular nerve of the chinchilla. IV. Discharge properties of utricular afferents. J Neurophysiol 63:781–790.

Goldberg JM, Highstein SM, Moschovakis AK, Fernández C (1987) Inputs from regularly and irregularly discharging vestibular nerve afferents to secondary neurons in the vestibular nuclei of the squirrel monkey. I. An electrophysiological analysis. J Neurophysiol 58:700–718.

Graf W, Simpson JI (1981) Relation between the semicircular canals, the optic axis, and the extraocular muscles in lateral eyed and frontal eyed animals. In: Fuchs AF, Becker W (eds) Progress in Oculomotor Research. Amsterdam: Elsevier, pp. 409–417.

Grüsser O-J (1984) J.E. Purkyne's contributions to the physiology of the visual, the vestibular and the oculomotor system. Hum Neurobiol 3:129–144.

Guedry FE (1965) Orientation of the rotation axis relative to gravity: its influence on nystagmus and the sense of rotation. Acta Otolaryngol 60:30–48.

Guedry FE (1974) Psychophysics of vestibular sensation. In: Kornhuber HH (ed) Handbook of Sensory Physiology, Volume 6. Berlin, Heidelberg, New York: Springer-Verlag, pp. 3–154.

Hain TC (1986) A model of the nystagmus induced by off vertical axis rotation. Biol Cybern 54:337–350.

Halmagyi GM, Curthoys IS (1988) A clinical sign of canal paresis. Arch Neurol 45:737–739.

Halmagyi GM, Curthoys IS, Cremer PD, Henderson CJ, Todd MJ, Staples MJ, D'Cruz DM (1990) The human horizontal vestibulo-ocular reflex in response to high-acceleration stimulation before and after unilateral vestibular neurectomy. Exp Brain Res 81:479–490.

Halmagyi GM, Aw ST, Cremer PD, Curthoys IS, Todd MJ (2001) Impulsive testing of individual semicircular canal function. Ann N Y Acad Sci 942:192–200.

Haslwanter T, Jaeger R, Mayr S, Fetter M (2000) Three-dimensional eye-movement responses to off-vertical axis rotations in humans. Exp Brain Res 134:96–106.

Helmholz HV (1867) *Handbuch der Physiologichen Optik*, Leipzig: Leopold Voss.

Henn V, Cohen B, Young LR (1980) Visual-Vestibular Interaction in Motion Perception and the Generation of Nystagmus. Cambridge: MIT Press.

Hepp K, Henn V, Vilis T, Cohen B (1989) Brainstem regions related to saccade generation. In: Wurtz R, Goldberg M (eds) The Neurobiology of Saccadic Eye Movements. Amsterdam: Elsevier, pp. 105–211.

Hess BJM, Angelaki DE (1993) Angular velocity detection by head movements orthogonal to the rotation plane. Exp Brain Res 95:77–83.

Hess BJM, Dieringer N (1990) Spatial organization of the maculo-ocular reflex of the rat: responses during off-vertical axis rotation. Eur J Neurosci 2:909–919.

Highstein SM (1973) The organization of the vestibulo-ocular and trochlear reflex pathways in the rabbit. Exp Brain Res 17:285–300.

Highstein SM, Cohen B, Buettner-Ennever JA(eds) (1996a) New Directions in Vestibular Research. New York: New York Academy of Sciences.

Highstein SM, Rabbitt, RD, Boyle R (1996b) Determinants of semicircular canal afferent response dynamics in the toadfish, *Opsanus tau*. J Neurophysiol 75:575–596.

Hoffmann KP, Schoppmann A (1975) Retinal input to direction selective cells in the nucleus tractus opticus of the cat. Brain Res 99:359–366.

Högyes A (1880) On the nervous mechanism of the involuntary associated movements of the eyes. Orv Hetil 23:17–29.

Howard IP (1982) Human Visual Orientation. New York: John Wiley & Sons.

Howard IP, Templeton WB (1966) Human Spatial Orientation. New York: Wiley.

Hudspeth AJ (1982) Extracellular current flow and the site of transduction by vertebrate hair cells. J Neurosci 2:1–10.

Hudspeth AJ (1985) The cellular basis of hearing: the biophysics of hair cells. Science 230:745–752.

Hudspeth AJ, Corey DP (1977) Sensitivity, polarity and conduction change in the response of vertebrate hair cells to controlled mechanical stimuli. Proc Natl Acad Sci U S A 74:2407–2411.

Hudspeth AJ, Jacobs R (1979) Stereocillia mediate transduction in vertebrate hair cells. Proc Natl Acad Sci U S A 76:1506–1509.

Hunter J (1786) The use of the oblique muscles. In: Observations on Certain Parts of the Animal Oeconomy, London: G. Nicol, 209–212.

Imai T, Moore ST, Raphan T, Cohen B (2001) Interaction of the body, head and eyes during walking and turning. Exp Brain Res 136:1–18.

Ito M (1984) The Cerebellum and Neural Control. New York: Raven Press.

Jian BJ, Shintani T, Emanual BA, Yates BJ (2002) Convergence of limb, visceral, and vertical semicircular canal or otolith inputs onto vestibular nucleus neurons. Exp Brain Res 144:247–257.

Kaminsky HJ, Leigh RJ (2002) The Neurobiology of Eye Movements: From Molecules to Behavior. Cleveland, OH: Case Western Reserve University, vol. 956.

Kornhuber HH (1974a) Handbook of Sensory Physiology, Vestibular System Part 1, Basic Mechanisms. Springer-Verlag, Berlin.

Kornhuber HH (ed) (1974b) Handbook of Sensory Physiology, Volume 6: Vestibular System. Psychophysics, Applied Aspects, and General Interpretations. Berlin: Springer-Verlag.

Kornhuber HH (1974b) Handbook of Sensory Physiology, Vestibular System Part 2, Psychophysics, Applied Aspects and General Interpretations. Springer-Verlag, Berlin.

Kushiro K, Dai MJ, Kunin M, Yakushin SB, Cohen B, Raphan T (2002) Compensatory and orienting eye movements induced by off-vertical axis rotation (OVAR) in monkeys. J Neurophysiol 88:2445–2462.

Lafortune S, Ireland DJ, Jell RM, Du Val L (1986) Human optokinetic afternystagmus: stimulus velocity dependence of two-component decay model and involvement of pursuit. Acta Otolaryngol (Stockh) 101:183–192.

Laurutis VP, Robinson DA (1986) The vestibulo-ocular reflex during human saccadic eye movements. J Physiol 373:209–233.

Lee FS (1893) A study of the sense of equilibrium in fishes. I. J Physiol (Lond) 15:311–348.

Lee FS (1894–1895) A study of the sense of equilibrium in fishes. II. J. Physiol (Lond) 17:192–210.

Leigh RJ, Zee DS (1999) The Neurology of Eye Movements. Philadelphia: F.A. Davis.

Lorente de Nó R (1932) The regulation of eye positions and movements induced by the labyrinth. Laryngoscope 42:233–332.

Lorente de Nó R (1933) Vestibular ocular reflex arc. Arch Neur Psychiatry 30:245–291.

Lowenstein O (1936) The equilibrium function of the vertebrate labyrinth. Biol Rev 11:113–145.

Lowenstein O, Roberts TDM (1950) The equilibrium function of the otolith organs of the thornback ray (*Raja clavata*). J Physiol (Lond) 110:392–415.

Lowenstein O, Sand A (1940) The mechanism of the semicircular canal. Proc R Soc Lond 129:256–275.

MacDougall HG, Curthoys IS, Betts GA, Burgess AM, Halmagyi GM (1999) Human ocular counterrolling during roll-tilt and centrifugation. Ann N Y Acad Sci 871:173–180.

Mach E (1875) Grundlinean der Lehre von den Bewegungsemphindungen. Leipzig: Engelmann.

Mach E (2001) Fundamentals of the Theory of Movement Perception. New York: Kluwer Academic/Plenum Publishers.

Magnus R (1924) Körperstellung [Body Posture]. Berlin: Verlag von Julius Springer.

Markham CH, Curthoys IS (1972) Convergence of labyrinthine influences on units in the vestibular nuclei of the cat. II. Electrical stimulation. Brain Res 43:383–396.

Maruta J, Simpson JI, Raphan T, Cohen B (2001) Orienting otolith-ocular reflexes in the rabbit during static and dynamic tilts and off-vertical axis rotation. Vision Res 41:3255–3270.

Mayne R (1974) A systems concept of the vestibular organs. In: Kornhuber HH (ed) Handbook of Vestibular Physiology. Vestibular System, Volume VI, Part 2. New York: Springer-Verlag, pp. 493–580.

Merfeld DM, Teiwes W, Clarke AH, Scherer H, Young LR (1996) The dynamic contribution of the otolith organs to human ocular torsion. Exp Brain Res 110:315–321.

Miles FA, Busettini C (1992) Ocular compensation for self-motion: visual mechanisms. Ann N Y Acad Sic 656:220–232.

Miles FA, Busettini C, Schwarz U (1992) Ocular responses to linear motion. In: Shimazu H, Shinoda Y (eds) Vestibular and Brain Stem Control of Eye, Head, and Body Movements. Tokyo: Japan Scientific Societies Press, pp. 379–395.

Miller EF (1962) Counterrolling of the human eyes produced by head tilt with respect to gravity. Acta Otolaryngol (Stockh) 54:479–501.

Moore ST, Clement G, Raphan T, Cohen B (2001) Ocular counter-rolling (OCR) induced by centrifugation during orbital spaceflight. Exp Brain Res 137:323–335.

Moore ST, Clement G, Raphan T, Cohen B (2001) Ocular counterrolling (OCR) induced by centrifugation during orbital spaceflight. Exp Brain Res 137:323–335.

Mustari M, Fuchs AF (1990) Discharge patterns of neurons in the pretectal nucleus of the optic tract. J Neurophysiol 64:77–90.

Niven JI, Hixson WC, Correia MJ (1966) Elicitation of horizontal nystagmus by periodic linear acceleration. Acta Otolaryngol 26:429–441.

Paige GD (1991) Linear vestibulo-ocular reflex (LVOR) and modulation by vergence. Acta Otolaryngol (Stockh) Suppl 481:282–286.

Paige GD, Tomko DL (1991a) Eye movement responses to linear head motion in the squirrel monkey. I. Basic characteristics. J Neurophysiol 65:1170–1182.

Paige GD, Tomko DL (1991b) Eye movement responses to linear head motion in the squirrel monkey. II. Visual–vestibular interactions and kinematic considerations. J Neurophysiol 65:1183–1196.

Paige GD, Seidman SH (1999) Characteristics of the VOR response to linear acceleration. Ann N Y Acad Sci 871:123–135.

Paige GD, Barnes GR, Telford L, Seidman SH (1996) Influence of sensori-motor context on the linear vestibulo-ocular reflex. Ann N Y Acad Sci 781:322–331.

Paige GD, Telford L, Seidman SH, Barnes GR (1998) Human vestibuloocular reflex and its interactions with vision and fixation distance during linear and angular head movement. J Neurophysiol 80:2391–2404.

Peterson BW, Goldberg J, Bilotto G, Fuller J (1985) Cervicocollic reflex: its dynamic properties and interaction with vestibular reflexes. J Neurophysiol 54:90–109.

Purkině JE (1820) Beiträge zur näheren Kenntnis des Schwindels aus heutognostischen Daten. Med J B (Wien) 6:79–125.

Raphan T (1997) Modeling control of eye orientation in three dimensions. In: Fetter M, Haslwanter T, Misslisch H, Tweed D (eds) Three-Dimensional Kinematics of Eye, Head, and Limb Movements. Amsterdam: Harwood Academic Publishers, pp. 359–374.

Raphan T (1998) Modeling control of eye orientation in three dimensions. I. Role of muscle pulleys in determining saccadic trajectory. J Neurophysiol 79:2653–2667.

Raphan T, Cohen B (1980) Integration and its relation to ocular compensatory movement. Mt Sinai J Med 47:410–417.

Raphan T, Cohen B (1986) Multidimensional organization of the vestibulo-ocular reflex (VOR). In: Keller EL, Zee DS (eds) Adaptive Processes in Visual and Oculomotor System. New York: Pergamon Press, pp. 285–292.

Raphan T, Cohen B (1988) Organizational principles of velocity storage in three dimensions: The effect of gravity on cross-coupling of optokinetic after-nystagmus. Ann N Y Acad Sci 545:74–92.

Raphan T, Cohen B (1996) How does the vestibulo-ocular reflex work? In: Baloh RW, Halmagyi GM (eds) Disorders of the Vestibular System. New York: Oxford University Press, pp. 20–47.

Raphan T, Cohen B (2002) The vestibulo-ocular reflex (VOR) in three dimensions. Exp Brain Res 145:1–27.

Raphan T, Sturm D (1991) Modelling the spatiotemporal organization of velocity storage in the vestibuloocular reflex by optokinetic studies. J Neurophysiol 66:1410–1420.

Raphan T, Cohen B, Henn V (1981) Effects of gravity on rotatory nystagmus in monkeys. Ann N Y Acad Sci 374:44–55.

Raphan T, Dai M, Suzuki J-I, Yakushin SB, Cohen B (1992) Semicircular canal input and the spatial orientation of the vestibulo-ocular reflex. In: XVIIth Barany Society Meeting, Czechoslovakia, June 1–5, pp. 97–98.

Raphan T, Matsuo V, Cohen B (1977) A velocity storage mechanism responsible for optokinetic nystagmus (OKN), optokinetic after-nystagmus (OKAN) and vestibular nystagmus. In: Baker R, Berthoz A (eds) Control of Gaze by Brain Stem Neurons. Amsterdam: Elsevier/North Holland, pp. 37–47.

Raphan T, Matsuo V, Cohen B (1979) Velocity storage in the vestibulo-ocular reflexarc (VOR). Exp Brain Res 35:229–248.

Raphan T, Wearne S, Cohen B (1996) Modeling the organization of the linear and angular vestibulo-ocular reflexes. Ann N Y Acad Sci 781:348–363.

Robinson D (1977) Vestibular and optokinetic symbiosis: an example of explaining by modeling. In: Baker R, Berthoz A (eds) Control of Gaze by Brain Stem Neurons. Amsterdam: Elsevier/North Holland, pp. 49–58.

Robinson DA (1964) The mechanics of human saccadic eye movement. J Physiol (Lond) 174:245–264.

Robinson DA (1965) The mechanics of smooth pursuit eye movement. J Physiol (Lond) 180:569–591.

Robinson DA (1968) The oculomotor control system. Proc IEEE 56:1032–1049.

Robinson DA (1971) Models of oculomotor neural organization. In: Bach-Y-Rita P, Collins CC (eds) The Control of Eye Movements. New York: Academic, pp. 519–538.

Robinson DA (1974) Cerebellectomy and vestibulo-ocular reflex arc. Brain Res 71:215–224.

Robinson DA (1975) Oculomotor control signals. In: Lennerstrand G, Bach-Y-Rita P (eds) Basic Mechanisms of Ocular Motility and Their Clinical Implications. Oxford: Pergamon Press, pp. 337–374.

Robinson DA (1976) Adaptive gain control of vestibulo-ocular reflex by the cerebellum. J Neurophysiol 39:954–969.

Robinson DA (1981) Control of eye movements. In: Handbook of Physiology, the Nervous System, Motor Control, Volume II (Chapter 28). Berthesda, MD: American Physiological Society, pp. 1275–1320.

Robinson DA (1982) The use of matrices in analyzing the three dimensional behavior of the vestibulo-ocular reflex. Biol Cybern 46:53–66.

Schiff D, Cohen B, Raphan T (1988) Nystagmus induced by stimulation of the nucleus of the optic tract (NOT) in the monkey. Exp Brain Res 70:1–14.

Schiff D, Cohen B, Buettner-Ennever JA, Matsuo V (1990) Effects of lesions of the nucleus of the optic tract on optokinetic nystagmus and after-nystagmus in the monkey. Exp Brain Res 79:225–239.

Schöne H (1984) Spatial Orientation: The Spatial Control of Behavior in Animals and Man. Princeton, NJ: Princeton University Press.

Schwarz C, Busettini C, Miles FA (1989) Ocular responses to linear motion are inversely proportional to viewing distance. Science 245:1394–1396.

Simpson JI (1984) The accessory optic system. Ann Rev Neurosci 7:13–41.

Simpson JI, Giolli RA, Blanks RHI (1988) The pretectal nuclear complex and the accessory optic system. In: Buettner-Ennever JA (ed) Neuroanatomy of the Oculomotor System, Volume 2. Amsterdam: Elsevier, pp. 335–364.

Simpson JI, Graf W (1981) Eye muscle geometry and compensatory eye movements in lateral and frontal eyed animals. Ann N Y Acad Sci 374:20–30.

Skavenski AA, Robinson DA (1973) Role of abducens neurons in vestibuloocular reflex. J Neurophysiol 36:724–738.

Snyder LH, King WM (1992) Effect of viewing distance and location of the axis of rotation on the monkey's vestibuloocular reflex. I. Eye movement response. J Neurophysiol 67:861–874.

Solomon D, Cohen B (1992a) Stabilization of gaze during circular locomotion in darkness: II. Contribution of velocity storage to compensatory eye and head nystagmus in the running monkey. J Neurophysiol 67:1158–1170.

Solomon D, Cohen B (1992b) Stabilization of gaze during circular locomotion in light: I. Compensatory head and eye nystagmus in the running monkey. J Neurophysiol 67:1146–1157.

Steinhausen W (1933) Über die Beobachtung der Cupula in den Bogengansampullen des Labyrinths des lebenden Hechts. Arch Ges Physiol 232:500–512.

Steinman RM, Collewijn H (1980) Binocular retinal image motion during active head rotation. Vision Res 20:415–429.

Suzuki J, Cohen B (1964) Head, eye, body, and limb movements from semicircular canal nerve. Exp Neurol 10:393–405.

Suzuki J, Cohen B, Bender MB (1964) Compensatory eye movements induced by vertical semicircular canal stimulation. Exp Neurol 9:137–160.

Szentágothai J (1950) The elementary vestibulo-ocular reflex arc. J Neurophysiol 13:395–407.

Tabak S, Collewijn H (1995) Evaluation of the human vestibulo-ocular reflex at high frequencies with a helmet, driven by reactive torque. Acta Otolaryngol 520:4–8.

Telford L, Seidman SH, Paige GD (1996) Canal–otolith interactions driving vertical and horizontal eye movements in the squirrel monkey. Exp Brain Res 109:407–418.

Telford L, Seidman SH, Paige GD (1997) Dynamics of squirrel monkey linear vestibuloocular reflex and interaction with fixation distance. J Neurophysiol 78:1775–1790.

Ter Braak JWG (1936) Untersuchungen ueber optokinetischen nystagmus. Arch Neerl Physiol 21:309–376.

Thurtell MJ, Kunin M, Raphan T (2000) Role of muscle pulleys in producing eye position-dependence in the angular vestibuloocular reflex: a model-based study. J Neurophysiol 84:639–650.

Tweed D (1997) Velocity-to-position transformations in the VOR and the saccadic system. In: Fetter M, Haslwanter T, Misslisch H, Tweed D (eds) Three-dimensional Kinematics of Eye, Head and Limb Movements. Amsterdam: Harwood Academic Publishers, pp. 375–386.

Tweed D, Vilis T (1987) Implications of rotational kinematics for the oculomotor system in three dimensions. J Neurophysiol 58:832–849.

Tweed D, Vilis T (1990) The superior colliculus and spatiotemporal translation in the saccadic system. Neural Networks 3:75–86.

Uchino Y, Sato H, Suwa H (1997) Excitatory and inhibitory inputs from the saccular afferents to single vestibular neurons in the cat. J Neurophysiol 78:2186–2192.

Uchino Y, Sato H, Kushiro K, Zakir M, Imagawa M, Ogawa Y, Katsuta M, Isu N, Suwa H (1999) Cross-striolar and commissural inhibition in the otolith system. Ann N Y Acad Sci 871:162–172.

Virre E, Demer JL (1996) The human vertical vestibulo-ocular reflex during combined linear and angular acceleration with near-target fixation. Exp Brain Res 112:313–324.

Virre E, Tweed D, Milnor K, Vilis T (1986) A reexamination of the gain of the vestibuloocular reflex. J Neurophysiol 56:439–450.

Wade NJ (2000a) Porterfield and Wells on the motions of our eyes. Perception 29:221–239.

Wade NJ (2000b) William Charles Wells (1757–1817) and vestibular research before Purkinje and Flourens. J Vestib Res 10:127–137.

Waespe W, Henn V (1977) Neuronal activity in the vestibular nuclei of the alert monkey during vestibular and optokinetic stimulation. Exp Brain Res 27:523–538.

Waespe W, Henn V (1978) Conflicting visual vestibular stimulation and vestibular nucleus activity in alert monkeys. Exp Brain Res 33:203–211.

Wearne S, Raphan T, Cohen B (1996) Nodulo-uvular control of central vestibular dynamics determines spatial orientation of the angular vestibulo-ocular reflex (aVOR). Ann N Y Acad Sci 781:364–384.

Wearne S, Raphan T, Cohen B (1998) Control of spatial orientation of the angular vestibuloocular reflex by the nodulus and uvula. J Neurophysiol 79:2690–2715.

Wearne S, Raphan T, Cohen B (1999) Effects of tilt of the gravito-inertial acceleration vector on the angular vestibuloocular reflex during centrifugation. J Neurophysiol 81:2175–2190.

Weber KD, Fletcher WA, Jones GM, Block EW (2000) Oculomotor responses to on axis rotational stepping in normal and adaptively altered podokinetic states. Exp Brain Res 135:527–534.

Wersäll J, Bagger-Sjöbäck D (1974) Morphology of the vestibular sense organ. In: Kornhuber HH (ed) Handbook of Sensory Physiology. Volume 6. VI Vestibular System. Part 1. Basic Mechanisms. Berlin: Springer-Verlag, pp. 123–170.

Wilson VJ, Melvill Jones G (1979) Mammalian Vestibular Physiology. New York: Plenum Press.

Yakushin SB, Dai MJ, Suzuki J-I, Raphan T, Cohen B (1995) Semicircular canal contribution to the three-dimensional vestibulo-ocular reflex: a model-based approach. J Neurophysiol 74:2722–2738.

Yakushin SB, Raphan T, Suzuki J-I, Arai Y, Cohen B (1998) Dynamics and kinematics of the angular vestibuloocular reflex in monkey: effects of canal plugging. J Neurophysiol 80:3077–3099.

Yakushin SB, Gizzi M, Reisine H, Raphan T, Buttner-Ennever J, Cohen B (2000a) Functions of the nucleus of the optic tract (NOT). II. Control of ocular pursuit. Exp Brain Res 131:416–432.

Yakushin SB, Reisine H, Buttner-Ennever J, Raphan T, Cohen B (2000b) Functions of the nucleus of the optic tract (NOT). I. Adaptation of the gain of the horizontal vestibulo-ocular reflex. Exp Brain Res 131:416–432.

Yates BJ, Jian BJ, Cotter LA, Cass SP (2000) Responses of vestibular nucleus neurons to tilt following chronic bilateral removal of vestibular inputs. Exp Brain Res 130:151–158.

Young LR, Henn V (1975) Nystagmus produced by pitch and yaw rotation of monkey about non-vertical axes. Fortschr Zool 23:235–246.

Young LR, Henn V, Scherberger HJ (2001) *Fundamentals of the Theory of Movement Perception by Dr. E. Mach.* Translated and annotated by Laurence R. Young, Volker Henn, and Hansjörg Scherberger, *Kluwer Academic/Plenum Publishers,* New York, 191 pages.

Zhu D, Moore ST, Raphan T (1999) Robust pupil center detection using a curvature algorithm. Comput Methods Programs Biomed 59:145–157.

7
Vestibuloautonomic Interactions: A Teleologic Perspective

C.D. BALABAN and B.J. YATES

1. Introduction

"Somatic" and "autonomic" systems are typically treated as being separate and independent sensorimotor compartments. However, coordinated actions of the somatic and autonomic motor systems are required under many circumstances. During exercise, for example, contractions of skeletal muscles are accompanied by stereotyped increases in blood pressure and respiratory muscle activity that provide for the enhanced metabolic needs of the working muscles (Kaufman and Forster 1996; Waldrop et al. 1996). During movement, the effects of gravitoinertial forces on the body may also require active control of distribution of fluid by the cardiovascular system (Berne and Levy 1983). Gravity affects the entire organism, and its actions on body tissues can be detected by both "somatic" receptors (those part of the vestibular, somatosensory, and visual systems) and visceral receptors (Wilson and Melvill Jones 1979; Mittelstaedt 1996). It thus seems practical for somatic and visceral inputs regarding body position in space to be integrated by the central nervous system and for the combined signal to influence both control of movement and autonomic functions.

This review will discuss the challenges imposed by gravitoinertial acceleration on visceral and somatic motor activity and the evidence suggesting that a number of visceral and somatic receptors participate in detecting the position of the body in space. This sensory information, in turn, is used to generate coordinated autonomic and somatic motor activity that is appropriate to prevailing gravitoinertial conditions. A number of recent studies have considered the neural mechanisms through which multiple sensory cues may trigger compensatory changes in blood pressure and respiratory muscle activity during changes in posture. We will thus focus the last portion of this review on discussing these neural pathways and, in particular, connections that may mediate vestibular influences on cardiovascular and respiratory muscle control.

2. Challenges to Somatic and Autonomic Systems Imposed by Gravitoinertial Acceleration: Detection and Compensation

The division of the vertebrate body and nervous system into somatic and autonomic (or visceral) domains is familiar to every student of neuroscience. The somatic motor component innervates striated muscle. The sympathetic and parasympathetic motor divisions of the autonomic nervous system, however, innervate smooth muscle and glands. The existence of somatic and visceral sensory systems is also a familiar concept, the former associated with proprioceptive and exteroceptive functions and the latter with visceral (or interoceptive) sensation. However, it is clear that the contributions of these distinct sensory and motor systems to behavior are integrated at even the most basic levels of segmental reflex organization (Sato 1997). For example, visceral and somatic sensory signals converge directly within both the spinal cord dorsal horn and the dorsal column nuclei (nucleus gracilis and nucleus cuneatus) (Cadden and Morrison 1991; Berkley and Hubscher 1995).

Coordinated activity of somatic and visceral motor pathways is also well-established. For instance, viewing objects close to the face involves the "near triad" of vergence eye movements, lens accommodation, and pupillary diameter regulation (Leigh and Zee 1991). This response is produced by the simultaneous action of striated extraocular muscles and smooth muscle of the ciliary body and iris. Other examples of integrated autonomic and somatic motor responses are provided by the anticipatory change in blood flow to muscle during exercise or stereotyped movements (Waldrop et al. 1996) and by the exercise pressor response to muscle activity (Kaufman and Forster 1996). These example indicate that autonomic and somatic motor and sensory activity act in concert. From this perspective, it is not surprising that brain stem regions that regulate motor and autonomic functions integrate both somatic (vestibular, proprioceptive, somatosensory, visual) and visceral inputs. Furthermore, medial regions of the cerebellar cortex may coordinate autonomic and somatic motor activity (reviewed below and by Balaban 1996).

The recognition of a linkage between vestibular and visceral functions arose in the clinical literature from the cooccurrence of vertigo (defined strictly as an illusory sense of self-motion) and signs of visceral disturbances (defined as nausea, vomiting, epigastric discomfort, and cardiac arrhythmia) in patients with vestibular dysfunction (Goodhill 1979). This association was noted in some of the earliest medical texts and often resulted in confusion between the relative roles of cranial and visceral mechanisms in vertigo and equilibrium. The Greek terms *illingous, dinos*, and *scotodinos* and the Latin equivalent *vertigine* appear to be equivalent to the modern term vertigo, an illusory sense of self-motion. Galen (in the second to third centuries A.D.), citing the earlier work of Archigenes, stated that vertigo "has a two-fold origin, either in the head or the area below the diaphragm (*hypochondria*)" (Siegel 1976). Separate visceral and cephalic (in modern terms, neurologic) mechanisms have been regarded as independent primary etiologic factors in vertigo since these early sources (e.g., Leoni Lunensis' *Ars Medenci Humanos*, 1576). However, the ability of each site to influence the other was recognized, particularly in conjunction with motion sickness (e.g., Thomas Willis' *The London Practice of Physick*, 1692). Even after the recognition in the 1860s that vestibular dysfunction was a major cause of vertigo, abdominal disease remained as an etiologic factor in the disorder termed gastric, stomachic, or intestinal vertigo (e.g., Gowers 1903). More recently, however, the vestibular system has been established as the primary sense organ for linear and angular acceleration of the head (Cyon 1911; Jones 1918). As a result, it has been assumed that the autonomic manifestations of vertigo can be explained strictly by influences of the vestibular system on central autonomic circuits (for a classic review, see Spiegel and Sommer 1944). This review presents a more general perspective: namely, that an integration of vestibular and autonomic inputs and other central signals contributes to both motion perception and coordination of somatic and visceral motor activity during imposed and self-generated changes in gravitoinertial acceleration.

2.1. Gravitoinertial Acceleration Affects Somatic and Visceral Structures

The concept that gravitoinertial sensation is multimodal was recognized prior to the demonstration that the semicircular canal cristae and otolith organ maculae are sensors of head acceleration. For example, the British physiologist Herbert Mayo (1837) stated of the sense of motion that, "It is not easy to say where this sense exists; whether in the muscle only, or in the joints and sinews and integuments, or in the whole frame altogether." He further compared the maintenance of equilibrium to an accomplished musician playing the piano, stating that both skills ". . . are equally the results of long and at first difficult practice in the use of muscles under the guidance of more than one class of sensation." During the late nineteenth century, the general recognition that the semicircular canals and otolith organs sense motion merely added another sensory modality to the resources available for the maintenance of equilibrium. As stated in a widely disseminated essay by Sir Thomas Grainger Stewart (1898), the sensation and control of equilibrium are subserved by ". . . touch, sight, the muscular sense, probably an articular sense, and a visceral sense, along with the great organ of special sense for equilibration—the semicircular canals."

The primacy of the vestibular apparatus for sensation and control of balance has been viewed as dogma since the early twentieth century. This view was summarized rigorously by the French physiologist Élie de Cyon (Cyon 1911), who viewed the semicircular canals as the primary peripheral organ for "the sense of space." This concept of the preeminence of the vestibular apparatus for gravitoinertial sensation and motor responses survives as textbook dogma to this day (Kandel et al. 1991). A secondary role has been ascribed to proprioceptive and visual inputs. However, with the exception of the literature reviewed below, the contributions of other somatic or visceral sensory mechanisms during responses to gravitoinertial acceleration have been relatively neglected.

2.2. Vestibular Mechanisms are Inadequate for Sensing Gravitoinertial Acceleration

Gravitoinertial acceleration affects all tissues. Hence, the body contains many mechanisms that are potential sensors of changes in the relationship of the body to gravity that occur whenever we move. The vestibular end organs appear to serve exclusively as detectors of head motion, with the semicircular canals sensing angular acceleration and the otolith organs sensing linear acceleration (including gravity). However, they do not yield unambiguous information about gravitoinertial stimuli. For example, similar otolith inputs can be induced by head tilt and by linear translation, and semicircular canals cannot detect periods of constant velocity (zero acceleration). Hence, semicircular canal signals cannot be used to distinguish between zero and nonzero constant angular velocity stimulation. One prominent consequence of the ambiguity of information from these sensors is the major individual variation of motion perception during off-vertical axis of "barbecue-spit" rotation in darkness (Guedry 1974).

The psychophysics literature provides numerous other examples of the inadequacy of vestibular input alone for accurate judgments of the orientation of one's body axis with respect to gravitoinertial force (Asch and Witkin 1948; Guedry 1974; Mittelstaedt 1996).Consistent errors in the judgment of spatial orientation have been characterized repeatedly during centrifugation and during water immersion (Schöne 1964; Nelson 1968; Ross et al. 1969). The latter phenomenon is not restricted to underwater or unusual conditions: a study has further reported that subjects consistently overestimate the magnitude of postural tilt when pitched slowly at 1 degree/s (Ito and Gresty 1997). These studies are consistent with the view that multiple sensory modalities (vision, proprioception, mechanoreception), perhaps including those from the viscera (i.e., from nonvestibular graviceptors), are used for perception of orientation relative to gravity. Furthermore, Anastasopoulos et al. (1997) have presented evidence that different combinations of these inputs influence the perception of visual versus postural verticality.

A series of studies by Mittelstaedt (1996) have further established the likelihood that visceral sensations (either visceral movements or blood pooling) can serve as accessory graviceptive signals. These latter findings provide a potential explanation for two conflicting perceptions of astronauts during the early stages of space flight: they perceive that they are either upright or inverted (Mittelstaedt and Glasauer 1993). The visual frame of the space vehicle conveys one spatial framework. The otolith organs convey no information about the gravitoinertial frame. However, there is a cranial shift in blood distribution, produced by a withdrawal of gravitational acceleration acting on the mass of the blood, with vascular tone set to resist the absent force. As a result, blood pooling mimics the condition of standing on one's head. The upright perception reflects the visual framework, whereas the inverted perception is potentially due to the influences of visceral inputs.

During our normal activities, nonvestibular somatic and visceral sensory mechanisms therefore appear to serve a secondary role as accessory gravitoinertial sensors. The visual system provides information regarding motion and the orientation of the spatial vertical. The somatosensory system provides somesthetic information regarding pressure on skin and deep tissues and temperature gradients on the body surface. Other sensory cues that are influenced by gravitoinertial stimuli include proprioceptive information, regional blood distribution, movements of the abdominal viscera within the peritoneal cavity relative to the torso (e.g., traction on mesenteries, contact with the parietal surface of the peritoneum, and stretch due to pressure on the diaphragm) and changes in intraocular and intracranial pressure (see below). Given the ubiquitous exposure to gravity during everyday activities and the multiplicity of gravity-related sensory signals, the direct integration of vestibular, other somatic, and visceral signals in the brain stem may provide a rich neural substrate for accurate determination of the orientation of body segments within a gravitoinertial coordinate frame.

2.3. Sensorimotor Influences of Gravitoinertial Acceleration

We are all aware that gravitoinertial forces present a challenge to maintaining posture during activities such as locomotion and vehicular travel. The necessary muscle activity to counter imposed changes in motion (e.g., change of speed of a bus) is performed unconsciously by a variety of vestibular and proprioceptive reflexes and, if necessary, by "long-loop" reflexes and conscious activity. These compensatory adjustments of muscle activity are responses to neural sensory signals that reflect the actions of gravitoinertial acceleration on tissue mass. The otolith organs transduce total linear acceleration (including gravity) of the head as a function of its action on the mass of the otoconia. The semicircular canal cristae transduce angular acceleration of the head as a function of the inertia of the endolymph and cupula. Proprioceptive and mechanoreceptive signals that reflect the effects of gravitoinertial acceleration on the position of body segments and center of mass are sensed as a function of factors that include pressure on the skin and deep structures, changes in muscle length, and changes in the tension of tendons. The visual system senses the retinal slip velocity that is secondary to movements of the head (and gaze) in space. The combination of signals in these pathways provides contextual information for controlling eye and head movements. As a simple example, passive whole-body rotation in the light activates vestibular receptors and produces a retinal slip (optokinetic) signal but does not affect neck or trunk proprioceptors. The identical passive rotation of the head on a fixed trunk, in contrast, is sensed by changes in vestibular, visual, and neck proprioceptive inputs. The integration of information from these sources is necessary to distinguish which change in body position has occurred and thus which motor response is appropriate (Wilson and Melvill Jones 1979).

2.4. Effects of Gravitoinertial Acceleration on Autonomic Functions and Respiration

Although motor challenges imposed by gravitoinertial acceleration are quite familiar, we are less cognizant of the simultaneous challenges imposed on visceral and respiratory function. As in the case of disturbances to somatic motor activity, these challenges are a direct consequence of the simple fact that every tissue in the body (including blood) has mass. Hence, changes in gravitoinertial acceleration during postural changes, locomotor activity, and vehicular travel can affect the position of internal organs and the distribution of fluid in the body, thereby requiring cardiovascular and other autonomic adjustments. One example of such a challenge is the well-known orthostatic response. The actions of gravitoinertial acceleration on the mass of blood and the fact that veins are distensible (and collapsable) present a challenge to the cardiovascular system for maintaining cardiac output and stroke volume when the orientation of gravitoinertial forces is altered relative to the body axis. When humans change from a supine to a standing posture, it is estimated that venous distension could result in pooling of 300–800 mL of blood in the legs, which may reduce cardiac output by 2 L/min and stroke volume by 40%, thereby eliciting a reduction in blood pressure (Berne and Levy 1983). Although reflexes act quickly to minimize peripheral blood pooling during gravitoinertial acceleration (see below), some redistribution of blood in the body is an inescapable consequence of exposure to a change in gravitoinertial acceleration. Thus, sensory afferents that detect regional changes in blood distribution, including baroreceptors and receptors in limb veins (Yates et al. 1987), are a potential source of gravitoinertial information.

Sensory mechanisms that detect changes of blood distribution are used to elicit well-characterized compensatory changes in the cardiovascular system. For example, if blood pressure drops during changes in posture, the resulting decrease in baroreceptor afferent activity will elicit an acceleration in heart rate, augmented cardiac contractions, and constriction of arterioles and venules to maintain cardiac return (Loewy and Spyer 1990). Although these responses tend to be relatively slow, the net effect of these changes is to restore blood distribution to its status prior to the postural shift. This restoration of blood distribution will then reflect a low-pass filtered representation of the current postural status so that any further changes in posture will produce changes in blood distribution. Hence, from a sensory perspective, these reflexes are an adaptive mechanism to maintain blood distribution at a given "set point" so that new deviations due to gravitoinertial and/or fluid volume changes can be detected.

The control of movements of the torso (including respiration) is also influenced by gravitoinertial acceleration. The mammalian torso consists of an abdominal/pelvic cavity and a thoracic cavity, separated by a muscular septum, the diaphragm. These cavities lie beneath the striated muscles that move the trunk (e.g., intercostal muscles, abdominal muscles, sternocleidomastoideus, and scaleni) and muscles of the pectoral girdle that have the capability to change the configuration of the torso and, hence, the shape of the internal cavities. The thoracic viscera are contained in a median compartment, the mediastinum (e.g., heart and esophagus), and paired lateral pleural compartments that contain the lungs. The abdominal viscera are covered by a visceral peritoneum, which forms both a dorsal mesentery that suspends the viscera from the posterior abdominal wall and a small ventral mesentery that attaches the stomach and proximal duodenum to the anterior abdominal wall. Because the diaphragmatic surface of the liver is firmly attached to the diaphragm by the coronary ligament, the right and left triangular ligaments, and the appendix fibrosa hepatis (an extension of the left triangular ligament), it constitutes a load on the abdominal surface of the diaphragm. As a consequence, actions of gravitoinertial acceleration on the viscera are important determinants of respiratory movements under both static conditions and during locomotion (Campbell et al. 1970). The loosely tethered gastrointestinal tract is a deformable body that produces a variable gravitational load on the diaphragm during attainment of various postures (supine, lateral decubitus, prone, or inverted) such that the Earth-down aspect of the diaphragm is loaded differentially. Agostoni (1977) demonstrated that the abdominal contents are a major determinant of gravity-dependent gradients in transpulmonary pressure. The consequences of this loading have been documented in both physiologic studies of respiratory function (Agostoni 1970) and radiographic studies of human diaphragmatic mechanics (Froese and Bryan 1974) under normal gravitational conditions. The dominant effects of gravitational loading on diaphragm position are also illustrated by the 15% reduction in functional residual respiratory capacity in microgravity during space flight (West et al. 1997) and the effects of hydrostatic loading on respiration during water immersion (Agostoni 1970). For example, an extensive body of literature indicates that the actions of gravitoinertial acceleration on the viscera have prominent effects on factors such as pulmonary blood flow, ventilation, gas exchange, alveolar size and intrapleural pressure, and lung intraparenchymal stress (West 1977). Furthermore, the importance of gravitational loading on respiratory control implies that gravitoinertial accelerations will influence afferent sensory information from pulmonary and other respiratory receptors during movement.

2.5. Multisensory Detection of Gravitoinertial Accelerations

Although the effects of gravitoinertial accelerations can be detected by the vestibular system, visual system, musculoskeletal system, skin, cardiovascular system, and viscera, these different sensory signals are *not* equivalent because they differ in adequate stimuli, sensitivity, and temporal dynamics. This principle is illustrated clearly by the fact that different subsets of inputs have critical roles for specific somatic or visceral motor responses to gravitoinertial challenges. For example, sensory information about head movements (semicircular canal, otolith, cervical proprioceptive, and visual retinal slip inputs) appears to be critical for controlling eye movements during head movements. Compensation for a loss of one of these modalities may be imperfect because each modality contains information about a different aspect of the operating range of head movements (Wilson and Melvill Jones 1979). Similarly, cardiovascular system receptors (baroreceptors, etc.) provide a primary drive for adjustments in cardiovascular tone in response to gravitoinertial challenges. The increased blood pressure lability after baroreflex deafferentation (Schreihofer and Sved 1994) indicates that the information from these afferents regarding blood pressure and/or fluid distribution is not redundant in other sensory signals, such as vestibular inputs. However, it is important to avoid the specious inference from lesion experiments (and biased terminology) that the role of a particular input is limited to one critical motor or sensory function. For example, the term "baroreceptor" conveys only a limited sense of its potential functions. Transduction of arterial stretch may serve as a sensor of changes in vascular tone in one context, blood volume in another context, and gravitoinertial stimulation in yet another context. The functional context is provided by an integration of cognitive information and sensory information from vestibular, visual, proprioceptive, and somesthetic afferents in the central nervous system.

Multisensory integration of vestibular and other signals occurs in several regions of the nervous system, including the vestibular nuclei. For example, vestibulospinal neurons in decerebrate cats (Boyle and Pompeiano 1980, 1981; Pompeiano et al. 1987; Wilson et al. 1990) and a variety of vestibular nucleus units in alert primates (Gdowski and McCrea 2000) respond robustly to neck proprioceptive input. A potential role of nonlabyrinthine influences on vestibular nucleus neuronal activity in vestibular compensation is highlighted in a study in which recordings were made from these cells in decerebrate cats that had undergone a combined bilateral labyrinthectomy and vestibular neurectomy previously and then recovered for over a month (Yates et al. 2000). The firing of 27% of the neurons was modulated by tilt $\leq 15°$ in amplitude despite the fact that labyrinthine inputs were eliminated. The plane of tilt that elicited the maximal response was typically within 25° of pitch, and the response gain was approximately 1 spike/s/° across stimulus frequencies. These findings suggest that limb and trunk inputs may play an important role in graviception and modulating vestibular-elicited reflexes, particularly after compensation for bilateral vestibular loss.

The challenges posed by gravitoinertial acceleration on somatic and visceral structures reveal a complex interaction among multiple sensory signals, multiple effects of motor activity, and the need to "balance" (or juggle) multiple goals for different systems. The striated muscles of the trunk are multifunctional, serving both (1) as postural muscles for movements of the torso (and stabilizing the upper extremity) and (2) as muscles that control the pressure and/or the volume of the abdominal, pelvic, and thoracic cavities. For example, the diaphragm, intercostal muscles, and abdominal muscles are pump muscles for respiration and speech, with the important function of moving air into and out of the lungs by modifying pressure and volume of the thoracic cavity. On the other hand, the diaphragm also has a secondary postural function. Because it is the only muscle that spans the trunk, it appears to stabilize the body cavity during postural changes (Grillner et al. 1978). Trunk muscles also are important for emesis (Grélot and Miller 1996), defecation, micturation, and alterations in blood distribution (Mixter 1953; Moreno et al. 1967; Youmans et al. 1974; Lloyd 1983; Takata and Robotham 1992). Diaphragm contraction and rib cage expansion increase negative intrathoracic pressure, which tends to pull blood into the chest from the extremities. By contrast, contraction of the abdominal muscles pushes blood from the abdominal cavity and toward the heart. Hence, a synthesis of these considerations produces the hypothesis that the instantaneous activity of striated muscles of the torso reflects a balancing of alternative actions to achieve the multiple goals of producing voluntary movement, maintaining posture, and modulating pressures in the abdominal and thoracic cavities to generate respiration and facilitate cardiac return.

3. Visceral Mechanisms as Gravitoinertial Sensors

Gravitoinertial accelerations may have differential effects on a large number of receptors, including those located in the inner ear, muscles, cardiovascular system, and abdominal viscera. It seems reasonable to expect that by integrating multiple sensory inputs, the central nervous system can make a more precise "decision" about the ensuing challenge and thus elicit a precise compensatory response. However, during the twentieth century, relatively little attention was devoted to the earlier hypothesis (Stewart 1898) that visceral sensory inputs contribute to the sense of equilibrium. This hypothesis has been revisited intermittently. Pollack et al. (1955) examined World War II patients with spinal transections and reported the existence of gravity-related postural responses below the level of the lesion, even while the patients were partially submerged in water. Ito and Sanda (1965) further demonstrated that postural limb extension and flexion responses could be reproduced by electrical stimulation of the central stump of splanchnic nerves, pressure on the abdominal wall, or mesenteric traction in decorticate primates. Cutaneous stimulation never produced these effects in either humans or primates, suggesting that activation of visceral receptors and not the occurrence of novel stimulation evoked the responses. More direct experimental evidence for visceral contributions to postural responses was provided by two experimental studies published in 1973 (Biederman-Thorson and Thorson 1973; Delius and Vollrath 1973) that supported the hypothesis that visceral mechanisms contribute to reflex responses to imposed rotations in pigeons with labyrinthectomies or spinal transections. Subsequent clinical studies by Mittelstaedt (1996) are consistent with the hypothesis that visceral sensation influences perception of gravity. This section reviews the properties of visceral receptors that could potentially contribute to detecting gravitoinertial acceleration.

3.1. Cardiovascular Receptors

Change in the orientation of the orthostatic column are one potential source of information regarding changes of the orientation of the long axis of body relative to gravitoinertial acceleration. In addition, gravitoinertial acceleration may affect regional blood flow in organs. For example, gravity is one factor that affects regional pulmonary blood flow such that the dependent (i.e., "downward") part of the upright lung receives more blood than the apex (West et al. 1997). Regional changes in blood pressure and volume are transduced by a variety of visceral receptors, which include carotid baroreceptors, aortic arch baroreceptors, atrial receptors, ventricular pressure receptors, epicardial receptors, pulmonary baroreceptors, renal baroreceptors, and venous stretch receptors (Paintal 1973). Many of these receptors are sensitive to both steady-state blood pressure and to the rate of change (derivative) of blood pressure (Korner 1971; Paintal 1973) Thus, signals from these receptors provide an indication of both the absolute distribution and changes in distribution of blood in the body.

Blood pressure and blood volume can be affected by factors that include postural shifts, neurogenic changes in vascular tone, and hemorrhage. From a cardiovascular control perspective, the local sensory information may be adequate for many reactive adjustments in specific vascular beds that compensate for alterations in blood pressure and fluid distribution *after they have occurred*. However, *predictive* control of blood distribution to minimize changes in blood pressure during challenges such as movement is likely to require additional contextual information regarding the causes of cardiovascular disturbances. From a sensory perspective, however, specific shifts of blood distribution in local vascular beds (e.g., relative blood pooling in the feet when we stand) may provide significant cues regarding gravitoinertial challenges.

In the absence of acute systemic changes in blood volume, the responses of cardiovascular receptors to both mean blood pressure and the rate of change of blood pressure are capable of signaling changes in the orientation of the orthostatic column with respect to gravitoinertial acceleration. One important function of these receptors is to act as the afferent limb for orthostatic responses, which restore a stable mean blood pressure in the face of an orthostatic challenge. However, an additional role emerges when we consider the afferents and reflex from the perspective of a mechanism to detect changes in gravitoinertial acceleration. From this perspective, orthostatic reflexes (feedback control of mean blood pressure) and the phenomenon of "baroreflex resetting" maintain the sensitivity of baroreceptor afferents to rapid changes in body orientation within the context of a gravitoinertial coordinate frame (Guyton and Hall 1996). Just as gamma motoneurons set the operating range of stretch receptors in muscle spindles within the context of the length of extrafusal muscle fibers, one role of baroreflex mechanisms can be conceived as resetting the operating range of the orthostatic column to respond to rapid changes in body orientation relative to gravity.

3.2. Abdominal Viscera, Peritoneum, and Diaphragm

As discussed above, the mammalian torso consists of an abdominal/pelvic cavity and a thoracic cavity, separated by a muscular septum, the diaphragm. Because of the actions of both muscular and gravitoinertial forces, changes in static torso orientation, respiration, posture shifts, and locomotion can all produce movements and compression of the viscera (De Troyer 1989; Bramble and Jenkins 1993). Somatic and visceral receptors may respond to these movements and therefore may relay information about gravitoinertial acceleration. These receptors include those in the abdominal viscera (see below) and in the diaphragm (Road 1990; Revelette et al. 1992), particularly in the central tendon of the diaphragm, which is attached tightly to the liver. For example, slowly adapting mesenteric tension receptors respond to mesenteric traction (Paintal 1973). Compression of the intestines may provide a further signal: distension of the small intestine produces a transient depression of inspiratory discharges of the phrenic nerve and a compensatory enhancement of drive to intercostal muscles (Prabhakar et al. 1985), presumably due to activation of translumenal stretch receptors (Paintal 1973). The upward pressure on the diaphragm produced by the moving visceral mass may be sensed by muscle spindles and Golgi tendon organs in the diaphragm (Revelette et al. 1992) and intercostal muscle spindles. Finally, movements of the abdominal viscera may also stimulate mesenteric Pacinian corpuscles, which respond to pressure pulses up to 500 Hz and may also sense changes in pulse pressure and splanchnic blood flow (Paintal 1973).

One functional response to stimulation of abdominal receptors during movement may be adjustments in fluid distribution in the body. Tactile stimulation of the parietal surface of the peritoneum and viscera (including mesenteric traction) during abdominal surgery produces reflex hypotension in humans (Smith 1953; Rocco and Vandam 1954; Seltzer et al. 1985). Hypotension and bradycardia have also been reported immediately after percutaneous liver biopsy, during gynecologic laparoscopy, and during dilatation and curettage procedures (Doyle and Mark 1990). The primary response to mesenteric traction appears to be a systemic vasodilatation and a compensatory secondary increase in cardiac output (Seltzer et al. 1985). Hence, sensory signals mediating these responses may be an additional source of information regarding gravitoinertial acceleration and its potential effects on fluid distribution. These observations suggest that a visceral graviceptive mechanism may operate in parallel with orthostatic reflexes to maintain stable blood pressure during movement.

3.3. Intracranial and Intraocular Pressure

There are sensations of fullness and pressure associated with blood pooling in the head or extremities that indicate our orientation relative to gravitoinertial acceleration. For example, when we bend over, we sense fullness and warmth in our head and pressure in and around our eyes, which increases when we stand on our head. These apparently somatic sensations are accompanied by increases in both intracranial pressure and intraocular pressure.

Numerous studies indicate that intraocular pressure changes reflect alterations in gravitoinertial acceleration. Rapid postural shifts produce a transient increase in intraocular pressure, which declines exponentially to an elevated pressure plateau (Krieglstein and Langham 1975; Langham 1975; Searles et al. 1996). Human intraocular pressure is lowest in upright stance and increases by only 2–5 mm Hg when moving from an upright or seated posture to a supine posture (Buchanan and Williams 1985; Kothe and Lovasik 1988; Linder et al. 1988; Kothe 1994). When a subject is tilted further from a supine to an inverted posture, however, intraocular pressure increases precipitously to reach approximately three times the pressure when upright. It has been reported that the intraocular pressure in the prone position is approximately 5 mm Hg higher than in the supine position (Lam and Douthwaite 1997). This finding implies (1) that intraocular pressure changes may contain information about the direction of body (or head) movements in the pitch plane and (2) that the intraocular pressure changes cannot be attributed solely to postural effects on episcleral venous pressure. Thus, the adaptation of the transient pressure response to an elevated plateau after head down-tilt (Krieglstein and Langham 1975; Langham 1975; Searles et al. 1996) may reflect autoregulatory changes in retinal circulation (narrowed arterioles and increased central retinal artery pressure) that accompany the increased intraocular pressure during postural changes (Friberg and Weinreb 1985; Baxter et al. 1992). Such autoregulatory effects may be assisted by vestibular nucleus projections to the sources of preganglionic parasympathetic ocular innervation (Balaban 2003) and vestibular influences on sympathetic innervation.

The well-known oculocardiac reflex (Doyle and Mark 1990) suggests that changes in intraocular pressure can contribute to orthostatic responses. Application of pressure to the eye or traction on the orbit during medical procedures produces a reflex vasodepressor response and bradycardia that mimic expected orthostatic responses to body inversion (i.e., standing on one's head). The afferent fibers for the reflex are believed to travel in the ciliary nerves and ophthalmic division of the trigeminal nerve to the brain stem (Doyle and Mark 1990). Because these afferents are sensitive to intraocular pressure (which varies with posture), they are a likely source of information about body orientation relative to gravitoinertial acceleration.

The posture-induced shifts in intraocular pressure also affect both retinal and cortical visual information processing (Linder and Trick 1987; Kothe and Lovasik 1988; Linder et al. 1988). There are no significant effects of postural shift on ocular refraction (i.e., changes in lens shape or position), however, even in inverted subjects (Lovasik and Kothe 1989). Subjects in an inverted posture also show transient visual field deficits (Sanborn et al. 1987) similar to the field defects produced when a suction cup was used to increase intraocular pressure in upright subjects (Drance 1962; Scott and Morris 1967). These visual changes are potential secondary cues for changes in postural orientation with respect to gravity.

4. Processing of Multisensory Gravitoinertial Signals by the Central Vestibular System

Considerable evidence suggests that vestibular and autonomic signals converge in the brain stem. The anatomical organization of vestibuloautonomic pathways (Balaban and Porter 1998) is summarized schematically in Figure 7.1. Four components can be identified heuristically. First, the vestibular nuclei, brain stem autonomic nuclei, and their interconnections are com-

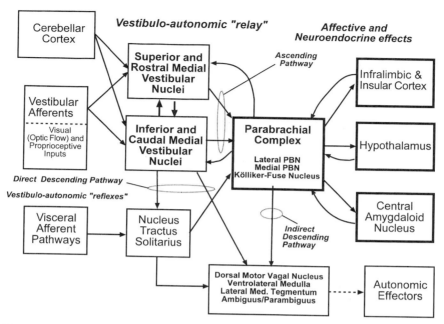

FIGURE 7.1. Schematic diagram of vestibuloautonomic pathways in the central nervous system. The vestibular nuclei project via a direct descending pathway to the nucleus tractus solitarius, dorsal motor vagal nucleus, ventrolateral medulla, lateral medullary tegmentum, and the nucleus ambiguus–parambiguus region. The vestibular nuclei also contribute an ascending projection to the caudal aspect of the lateral parabrachial nucleus (lateral PBN), medial parabrachial nucleus (medial PBN), and the Kölliker–Fuse nucleus. These parabrachial nucleus regions are connected reciprocally with pathways that are related to neuroendocrine and affective responses (e.g., anxiety and conditioned fear), such as the infralimbic and insular cortex, the hypothalamus, and the central amygdaloid nucleus. The parabrachial nucleus also sends projections to brain stem autonomic regions, such as the nucleus tractus solitarius, dorsal motor vagal nucleus, ventrolateral medulla, lateral medullary tegmentum, and the nucleus ambiguus–parambiguus region, which contribute to autonomic effector responses. A direct projection from the vestibular nuclei to ocular preganglionic parasympathetic (choline acetyltransferase-positive) neurons in the anteromedian nucleus and ventral tegmental area has recently been discovered (Balaban 2003).

ponents of a brain stem network that integrates vestibular, somatic, visual, and visceral information. Anatomic evidence indicates that the vestibular nuclear contributions to these pathways originate from the dorsal aspect of the superior vestibular nucleus (SVN), pars alpha (or caudoventral aspect) of the lateral vestibular nucleus (LVN), and the caudal half of the medial vestibular nucleus and the inferior vestibular nucleus (Balaban and Thayer

2001). A caudal vestibuloautonomic region (caudal medial vestibular nucleus and the inferior vestibular nucleus) contributes descending projections to the nucleus of the solitary tract, dorsal motor vagal nucleus, nucleus ambiguus, the ventrolateral medullary reticular formation (rVLM), nucleus raphe magnus, and the lateral medullary tegmentum. These brain stem structures either contain, or are connected directly with, parasympathetic and sympathetic preganglionic neurons (Loewy and Spyer 1990). In addition, both the caudal vestibular nuclei and a rostral vestibuloautonomic region (superior vestibular nucleus and the rostral pole of the medial vestibular nucleus) have ascending projections to the caudal aspect of the parabrachial nucleus (Balaban 1996c; Porter and Balaban 1997; Balaban et al. 2002) that are connected with regions of the medulla that control sympathetic and parasympathetic outflow. Recent electrophysiologic evidence from alert primates has demonstrated that units in this vestibular nucleus recipient region respond in a complex manner to whole-body rotation, with responses that reflect angular velocity, a leaky integration of angular velocity, and position (Balaban et al. 2002). Secondly, specific cerebellar regions (zones) are connected directly with these brain stem circuits that regulate vestibular and autonomic functions. As is the case with other cerebellar zones, these cerebellar pathways may contribute to coordination of activities, learning and adaptive plasticity (including classical conditioning), and predictive behavior of these circuits. Thirdly, descending telencephalic inputs to the vestibular nuclei and brain stem autonomic regions from the cerebral cortex and amygdala may contribute contextual information for interpreting multisensory inputs in terms of graviceptive signals, interoceptive signals, historical information/associations, and affective state. Finally, monoaminergic circuits exert parallel influences on the vestibular nuclei, cerebellum, and other structures related to autonomic and affective functions that may modulate their activity as needed during changes in the environment, external challenges, or alterations in vigilance.

4.1. Vestibular Nuclei, Brain Stem Autonomic Nuclei, and Their Interconnections

Primary vestibular afferents and primary visceral afferents terminate on different populations of neurons in the central nervous system. Primary vestibular afferents terminate heavily within the vestibular nuclei; they also contribute direct projections to cell group y and direct mossy fiber projections to the vermis and vestibulocerebellum. In addition, relatively sparse primary vestibular projections have been reported to nucleus prepositus hypoglossi, cochlear nuclear complex, external cuneate nucleus, abducens nucleus, and restricted regions within the reticular formation (Carleton and Carpenter 1984; Kevetter and Perachio 1989; Burian et al. 1990). Visceral sensory information reaches the brain stem through at least three distinct pathways (Fig. 7.2). Primary visceral afferents that travel in the trigeminal, facial, glossopharyngeal, and vagus nerves terminate in the nucleus tractus solitarius (NTS) (Loewy and Spyer 1990). Many cells in the NTS also receive somatic afferent information, including projections from group II (Aδ) and II (C) spinal afferents (Person 1989). Efferent projections of the NTS to the ventrolateral medulla, parabrachial nucleus, medullary reticular formation, and parasympathetic brain stem nuclei serve an important role in autonomic control. Other visceral afferents (including renal afferents) course through sympathetic nerves and either terminate in laminae I, V, and VII of the thoracolumbar spinal cord or project rostrally in the fasciculus gracilis to terminate in the nucleus gracilis (Kuo et al. 1983, 1984; Kuo and DeGroat 1985; Cadden and Morrison 1991; Kummer et al. 1992;

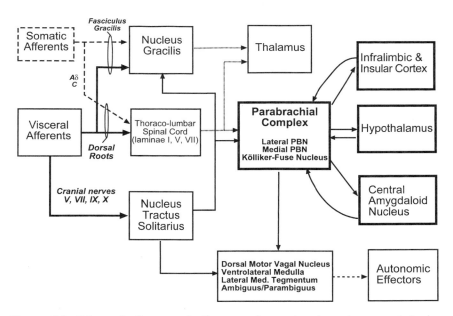

FIGURE 7.2. Schematic diagram of afferent pathways for visceral sensory information. Visceral sensory afferents travel through three main pathways. One group of afferents travels via the trigeminal (V), facial (VII), glossopharyngeal (IX), or vagus (X) nerve to the nucleus tractus solitarius. A second group of afferents travels in dorsal spinal roots to terminate directly in either the dorsal horn or the thoracolumbar spinal cord. The third contingent of fibers enters the spinal cord in dorsal roots before traveling in the fasciculus gracilis to terminate in the nucleus gracilis in the caudal medulla. The termination of the dorsal root fibers in the dorsal horn and nucleus gracilis with sites of termination of nociceptive (Aδ and C) afferents. These afferent pathways project to either the thalamus, parabrachial nucleus pathways related to affective and neuroendocrine responses, or autonomic effector regions (see Fig. 7.1).

Dibona and Kopp 1997). As in the case of nucleus tractus solitarius neurons, these terminal regions in the spinal cord and nucleus gracilis are also likely to receive convergent somatosensory inputs. Solitarioparabrachial and spinoparabrachial projections appear to relay these visceral and somatic signals to the hypothalamus and telencephalon (Ding et al. 1995; Feil and Herbert 1995; Jasmin et al. 1997).

One feature shared by visceral and vestibular sensory pathways is the convergence of information from other sensory modalities even at the earliest levels of processing. Both the nucleus gracilis and the spinal cord gray matter contain neurons that receive convergent visceral and somatic inputs (Ammons 1988; Hubscher and Berkley 1994). The regions of the vestibular nuclei that project to antonomic areas of the brain stem also receive a variety of nonvestibular sensory inputs. The inferior and medial vestibular nuclei (caudal vestibuloautonomic region) and nucleus prepositus hypoglossi receive direct projections from dorsal root ganglion cells (Pfaller and Arvidsson 1988; Neuhuber and Zenker 1989; Bankoul and Neuhuber 1990; Bankoul et al. 1995), whereas afferents from the spinal cord (cervical laminae IV–VIII and lumbar laminae IV, VII, and VIII) terminate in the inferior and medial vestibular nuclei (McKelvey-Briggs et al. 1989). Spinal inputs also reach the vestibular nuclei through multisynaptic pathways (Jian et al. 2002). Accessory optic afferents from the pretectum also terminate in restricted regions of the vestibular nuclei: the dorsal and medial terminal nuclei project contralaterally to the superior and lateral vestibular nuclei (Giolli et al. 1984, 1985, 1988). Thus, in addition to efferent connections, the rostral and caudal vestibuloautonomic regions appear to differ in the relative degree of multimodal sensory inputs, such that spinal inputs terminate extensively in the caudal region and visual inputs in the more rostral region.

A direct convergence of vestibular and visceral information may occur at both brain stem and spinal levels. As discussed above, the vestibular nuclei provide inputs to a number of brain stem nuclei that process visceral signals, including the nucleus tractus solitarius, the lateral medullary reticular formation, the rostral ventrolateral medulla, and the parabrachial nucleus. In addition, lateral and medial vestibulospinal tract projections (Shinoda et al. 1986, 1989) have the potential to influence low cervical, thoracolumbar, and sacral spinal cord interneurons that receive visceral input. Evidence further suggests that the parabrachial nucleus projects caudally to the vestibular nuclei (Balaban 1997). The close interrelationship between vestibular and visceral sensory pathways is consistent with the hypothesis that somatic and autonomic pathways integrate both vestibular (semicircular canal and otolith) and visceral (e.g., blood pooling and visceral proprioception) signals to construct central representations of gravitoinertial forces.

Because the parabrachial nucleus has ascending projections to the hypothalamus, central amygdaloid nucleus, and prefrontal cortex (Fulweiler and

Saper 1984), vestibular inputs to the parabrachial nucleus may be particularly important in autonomic control. These pathways may form a neural substrate for affective, cognitive, and neuroendocrine manifestations of vestibular dysfunction, motion sickness, and responses to altered gravitational environments. In particular, circuitry linking the vestibular and parabrachial nuclei may provide a neural substrate for the interactions between predisposing cognitive and affective factors in the development of both motion sickness and psychiatric disorders (e.g., panic disorder with agoraphobia and height phobia) (Furman et al. 1998; Balaban 1999).

4.2. Cerebellar Modulation of Vestibuloautonomic Circuits

The early experimental studies of Magendie, Flourens, and others in the nineteenth century (reviewed in Ito 1984) established that the cerebellum is not necessary for the *initiation* or *performance* of movement; rather, they noted that the cerebellum was important for *coordination* of movement. Consistent with this finding, more recent studies have shown that cerebellar ablation does not impair the ability of vestibular stimulation to produce alterations in extraocular muscle, postural muscle (Wilson and Melvill Jones 1979; Ito 1984), or sympathetic nerve activity (Yates et al. 1993; Yates and Miller 1994). However, the cerebellum plays an important role in modulating reflexes and thus could be involved in modifying and adjusting vestibuloautonomic responses. Previous studies have identified four regions of the medial aspect of the cerebellar cortex that affect autonomic function (see Fig. 7.3): (1) an intermediolateral site on the border of lobule IX and the nodulus; (2) a caudal posterior lobe region in zone A of lobule IX; (3) a rostral posterior lobe region in zone A of lobules VIIa–VIIIa; and (4) an anterior lobe region within zone A of lobules I–III (for a review, see Balaban 1999b). Cardiovascular responses (pressor or depressor) are elicited by local electrical stimulation in these regions. These responses likely reflect cerebellar Purkinje cell inhibition of brain stem autonomic circuits that influence tonic sympathetic (adrenergic) outflow to blood vessels and/or the heart. The data also suggest that both the anterior lobe and caudal posterior lobe regions can alter the gain (sensitivity) of brain stem cardiovascular reflexes, probably via direct inhibition of vestibuloautonomic regions in the superior, medial, and inferior vestibular nuclei, and, possibly, the parabrachial nucleus. These medial cerebellar regions, particularly the anterior lobe and the uvula–nodulus, also participate in control of posture and locomotion (for reviews, see Ito 1984; Balaban 1996b). Because the cerebellar cortex plays an important role in the coordination of somatic movements, it is possible that these regions coordinate synchronized muscle contraction and changes in autonomic activity during movement. The rostral posterior lobe region (zone A of lobules VIIa–VIIIa), on

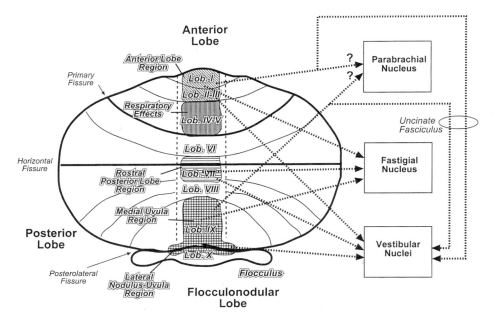

FIGURE 7.3. Schematic diagram of cerebellar projections to regions that integrate vestibular and autonomic information. This illustration shows the general organization of the projections of five cerebellar regions to vestibuloautonomic pathways. These pathways travel to their terminal regions via either the white matter (to the fastigial nucleus and superior vestibular nucleus), juxtarestiform body (medial and inferior vestibular nuclei), or the uncinate fasciculus (the parabrachial nucleus and caudal vestibular nuclei).

the other hand, may contribute to triggering orienting responses and to coordination of the visceral and somatic motor components of orienting responses. In particular, the "autonomic" cerebellar regions may assist in the predictive and context-dependent coordination of autonomic and somatic activity during movement.

4.2.1. Nodulus–Uvula Region

Several convergent lines of evidence suggest that a zone of Purkinje cells in the intermediolateral aspect of the nodulus and ventral uvula directly inhibit superior vestibular nucleus neurons that contribute to vestibular influences on cardiovascular function. First, a region of the intermediolateral aspect of the nodulus–uvual border area (centered approximately 3 mm lateral to the midline) has been termed the "nodulus–uvula depressor site" in rabbits because a transient drop in blood pressure is evoked by electrical microstimulation of this area (Nisimaru and Watanabe 1985; Henry et al. 1989). Second, anatomical evidence from rabbits indicates that the dorsal aspect of the superior vestibular nucleus, which projects to the parabrachial nucleus (Balaban 1996c), receives projections from a sagittal group of Purkinje cells centered approximately 3 mm lateral to the midline (Balaban 1984, 1996; Henry et al. 1989; and unpublished observations). Third, the depressor response elicited from the nodulus–uvula depressor region was blocked by microinjections of bicuculline into the dorsal aspect of the superior vestibular nucleus (Henry et al. 1989). Fourth, the cerebellar-evoked depressor responses reflect inhibition of tonic sympathetic drive to blood vessels because they were abolished by intravenous administration of phentolamine (an α-adrenergic receptor antagonist), attenuated by sympathetic ganglionic blockade, and unaffected by propranolol (a β-adrenergic receptor antagonist) and atropine methyl nitrate (a muscarinic cholinergic blocking agent) (Henry et al. 1989). Finally, the climbing fiber innervation of this zone apparently reflects retinal slip signals in the plane of the vertical semicircular canals (Kano et al. 1990) and possibly and otolithic (utricular) signal (Barmack and Fagerson 1994; Tan et al. 1995). Thus, the directional tuning of the physiologic response properties of the climbing fiber inputs to the nodulus–uvula (near the pitch plane) is similar to the directional tuning of vestibulosympathetic cardiovascular responses (Yates and Miller 1994; Woodring et al. 1997; see below).

4.2.2. Medial Uvula

The medial aspect of the uvula (lobule IX) is another candidate for a cerebellar region that modulates activity in either of the regions of the vestibular nuclei that project to autonomic brain stem regions or in the parabrachial nucleus. Electrical stimulation of the medial uvula produces depressor responses, bradycardia, decreased renal nerve activity, and hindlimb vasodilatation in anesthetized preparations but pressor responses, tachycardia, increased renal nerve activity, and both renal and hindlimb vasoconstriction in decerebrate animals (Rasheed et al. 1970; Bradley et al. 1987a, 1987b). Lesion studies indicated that the depressor/bradycardic responses in anesthetized animals are mediated via ascending Purkinje cell projections (through the superior cerebellar peduncle or the uncinate fasciculus) to the region of the lateral parabrachial nucleus but that pressor/tachycardic responses are mediated via projections through the inferior cerebellar peduncle to either the medial parabrachial nucleus (Paton et al. 1991) or the superior, rostral medial, and inferior vestibular nuclei (Balaban 1984). Although Paton et al. (1991) reported that tritiated amino acid injections into lobule IXb produce anterograde labeling in the caudal aspect of the lateral and medial parabrachial nuclei, it is unclear whether these projections are axon terminals or cerebellovestibular fibers of passage in the uncinate faciculus (Balaban 1996a).

4.2.3. Rostral Posterior Lobe

Zone A of the rostral posterior lobe is a third medial cerebellar region that affects autonomic function. However, unlike the other cerebellar autonomic regions, this area does not project to the vestibular nuclei; rather, it contributes efferent projections to the fastigial nucleus, which in turn makes connections with brain stem nuclei (Ito 1984). Both a depressor response and inhibition of renal sympathetic nerve activity are elicited by electrical pulse train stimulation (10 s duration) of the medial aspect (zone A) of lobules VIIa–VIIIa (Nisimaru 1977; Nisimaru et al. 1984; Bradley et al. 1987a, 1987b).

The mossy fiber inputs to this lobule VII–VIIIa region reflect a combination of vestibular, autonomic, and descending limbic inputs. The region receives input from the superior, medial, and inferior vestibular nuclei and parabrachial nuclei (Batini et al. 1978; Kotchabhakdi and Walberg 1978; Somana and Walberg 1979). However, Purkinje cell responses in this region reflect a wealth of other sensory and motor-related signals. Lobule VII is included in the "auditory" and "visual" areas of the cerebellar cortex (Dow and Moruzzi 1958). Lobule VII was also termed the "oculomotor area" because stimulation of this region can evoke eye movements, and activity of Purkinje cells located there reflects both saccades and smooth pursuit eye movements and stretch receptor information from extraocular muscles (Fuchs and Kornhuber 1969; Noda and Fujikado 1987). Purkinje cells in lobules VI–VIIb respond prominently to electrical stimulation of the fornix (Saint-Cyr and Woodward 1980a, 1980b), indicating a limbic input that is absent in the other cardiovascular regions of the cerebellar cortex. Furthermore, Purkinje cells in this region receive climbing fiber inputs from the medial aspect of the medial accessory olive (Nisimaru et al. 1991), which can produce vagally driven complex spikes (Okahara and Nisimaru 1991) that may reflect a convergence of renal afferent activity (Tong et al. 1993).

The absence of an effect of lobule VI–VII ablation on the cardio-inhibitory component of the baroreflex (La Noce et al. 1991) suggests that the rostral posterior lobe cardiovascular region may serve a different role than other autonomic-related cerebellar regions. This has been confirmed by the finding that ablation of the vermis of lobules IV–VIII in rats severely impairs classical conditioning of bradycardic responses without a significant effect on either the baseline heart rate or responses to the conditioned stimulus alone (Supple and Leaton 1990). The attenuation of conditioned bradycardic responses was confirmed by Sebastiani et al. (1992) in the rabbit, and they further reported that the development of these classically conditioned responses was unaffected by lesions of the uvula (lobule IX) alone. Because the bradycardic component of orienting responses to stimuli was unaffected by lobule VII lesions, these results imply that the lobule VIIa–VIIIa cardiovascular region contributes to modification of the sensory conditions that elicit a bradycardic response (classical conditioning).

The existing data raise the hypothesis that lobule VIIa–VIIIa participates in modifying the criteria for selecting or adjusting stimulus conditions for autonomic components of orienting responses. As indicated above, anatomic and physiologic data indicate that the lobule VIIa–VIIIa region receives mossy fiber afferents that may reflect the magnitude and location of stimuli in exteroceptive space. However, it is important to note that autonomic components of orienting responses are part of a coordinated motor program that includes gaze adjustments, which are mediated by movements of both eyes (saccades, vergence, and smooth pursuit), lens accommodation, pupillary responses, and body movements. A potential role of this region in coordinating somatic and smooth muscle components of the near triad (and, perhaps, linear vestibuloocular reflexes) is consistent with reports that electrical stimulation of lobule VII produces lens accommodation (Hosoba et al. 1978) and pupillary responses (Rasheed et al. 1970) (either mydriasis or miosis). These considerations lead to the hypothesis that the lobule VIIa–VIIIa autonomic region participates in the regulation of orienting responses by (1) contributing to the selection of stimulus criteria for triggering the responses and (2) assisting in coordination of the visceral and somatic motor components of orienting responses.

4.2.4. Medial Anterior Lobe

The medial aspect of the anterior lobe is a fourth potential cerebellar component of vestibuloautonomic circuits. This region was first identified by Moruzzi (Moruzzi 1938, 1940, 1950) and by Wiggers (1943) in electrical stimulation studies as a source of cerebellar modulation of cardiovascular activity. Since their pioneering work, both depressor responses and pressor responses have been elicited from sites within the medial aspect of lobules I–III (Hoffer et al. 1972; Nisimaru et al. 1984). These blood pressure changes are not accompanied by a heart rate response (Moruzzi 1938, 1940), which suggests that activation of these regions may also depress baroreceptor reflexes. Two other lines of evidence suggest that the anterior lobe may contribute to control of blood pressure *lability*. First, anterior lobe stimulation suppresses both spontaneous Meyer waves (fluctuations in mean blood pressure with a period of 10s) and carotid sinus reflexes (Moruzzi 1940, 1950). Second, these effects of cerebellar stimulation on cardiovascular lability are selective because Meyer waves are enhanced by anterior lobe ablation (Ramu and Bergmann 1966) without a concomitant effect on Hering–Traube waves (spontaneous blood pressure variations at respiratory frequency). Nisimaru et al. (1984) reported that stimulation sites that produce cardiovascular effects are located predominantly in zone A, a region that reportedly projects to the rostral fastigial nucleus (Ito 1984), rostral medial vestibular nucleus (Balaban 1984), and parabrachial region (Supple and Kapp 1994). However, more detailed studies are needed to determine whether anterior lobe Purkinje cells influence vestibuloautonomic pathways.

4.3. Telencephalic Contributions to Vestibuloautonomic Integration

It has been noted for more than a half century that vestibular effects on autonomic functions and respiration are extremely sensitive to descending telencephalic influences (Spiegel 1936, 1946). However, the role of telencephalic inputs to brain stem vestibular and autonomic structures in response to gravitoinertial challenges is poorly understood. There are several potential sources of telencephalic influences on these circuits. First, there are direct cortical projections to the vestibular nuclei. Anatomical studies have shown that axons from cells in area 7 of the posterior parietal cortex terminate in discrete fields in all divisions of the vestibular nuclei, including the caudal aspect of the medial vestibular nucleus and the descending vestibular nucleus (Ventre and Faugier-Grimaud 1988; Faugier-Grimaud and Ventre 1989). Area 7 is also connected extensively with the dorsal premotor cortex, the supplemental motor area, and the frontal eye fields (Cavada and Goldman-Rakic 1989). Electrophysiological evidence indicates that area 7 contains populations of neurons that have complex visual receptive properties (including opponent vector organization) (Lynch et al. 1977; Motter and Mountcastle 1981), respond to vestibular inputs (Kawano et al. 1980), and are modulated by directed attention (Lynch et al. 1977; Bushnell et al. 1981). Hence, these connections to the vestibular nuclei may relay contextual information for spatial orientation.

A second potential influence of telencephalic and limbic system information on vestibuloautonomic integration is via direct connections from the cerebral cortex, hypothalamus, and amygdala to the parabrachial nucleus and other brain stem autonomic regions that receive vestibular input. The processing of cortical and limbic inputs by autonomic regions of the cerebellum (see above) provides a third potential influence of emotion, affective, or contextual information for modulating vestibuloautonomic pathways. Finally, descending telencephalic influences on monoaminergic pathways (see below) may contribute to both vestibuloautonomic information processing and adjusting motor responses as a function of requirements for vigilance and emotional status.

4.4. Monoaminergic Influences on Vestibuloautonomic Integration

Noradrenergic and serotonergic axons project extensively within the central nervous system. These axons originate from cell groups in the brain stem and appear to exert a modulatory influence on a wide variety of physiologic and behavioral processes, ranging from changes in arousal and sleep–wakefulness cycles to modulation of sensory processing and motor activity (Moore and Bloom 1979; Jacobs and Azmitia 1992; Jacobs and

Fornal 1993). There is evidence that dysfunction of monoaminergic neuro-transmission may contribute to a variety of behavioral and psychiatric disorders, including anxiety (Gorman et al. 1989; Jacob et al. 1996a, 1996b; Keane and Soubrié 1997). Noradrenergic and serotonergic pathways may influence vestibular and autonomic function at a variety of sites, including the vestibular nuclei, parabrachial nuclei, brain stem areas that regulate sympathetic and parasympathetic outflow, and the cerebellar cortex. Hence, noradrenergic neurons have the potential to modulate sensorimotor integration of vestibular and autonomic inputs and to contribute to behavioral and affective similarities between patients with a history of vestibular dysfunction and patients with psychiatric disorders such as agoraphobia and height vertigo (see Furman and Jacob 2001 for a further discussion of this topic).

4.4.1. Noradrenergic Transmission in the Vestibular Nuclei

The locus coeruleus is a major source of noradrenergic innervation to all levels of the neuraxis (Moore and Bloom 1979). These widespread connections appear to contribute to altering sensorimotor responses in situations requiring increased vigilance or "alertness" (Aston-Jones et al. 1991b), including during stress or anxiety (Bremner et al. 1996). Locus coeruleus activity appears to increase with enhanced alertness and vigilant behavior elicited by exposure to novel or imperative sensory stimuli (Aston-Jones et al. 1991; Foote et al. 1991). The locus coeruleus is also activated during tasks or situations that require the reorienting of attention between stimuli (Aston-Jones et al. 1991b). Anatomic studies (Cedarbaum and Aghajanian 1978; Fung et al. 1987; Balaban 1996c) indicate that the vestibular nuclei project to the locus coeruleus. Because locus coeruleus cells respond to both vestibular stimulation and neck rotation (Barnes et al. 1989; Manzoni et al. 1989a, 1989b), signals related to imperative or novel gravitoinertial accelerations are likely to be reflected in activity of noradrenergic pathways originating from this nucleus. Conversely, locus coeruleus projections to the vestibular nuclei have the potential to influence signal transmission from the vestibular nuclei to the locus coeruleus, possibly to enhance detection of novel or imperative aspects of gravitoinertial stimulation.

The vestibular nuclei receive regionally specialized noradrenergic projections from the locus coeruleus (Schuerger and Balaban 1993, 1999). The postsynaptic actions of norepinephrine appear to be mediated by both α- and β-adrenoreceptors in the vestibular nuclei. Messenger RNAs for α_1-, α_2-, and β_1-adrenoreceptors are expressed by vestibular nucleus neurons (Nicholas et al. 1993a, 1993b; Domyancic and Morilak 1997), and β-adrenoreceptors have been identified immunohistochemically in the same regions (Wanaka et al. 1989). Microiontophoretic application of norepinephrine increased the firing rate of most LVN neurons in the decerebrate cat (Yamamoto 1967; Kirsten and Sharma 1976). More complex results were reported by Licata et al. (1993b), who examined responses of vestibular nucleus neurons with projections to the spinal cord or extraocular motor nuclei in a urethane-anesthetized rat preparation. Their data suggested that some responses of vestibular nucleus neurons to application of norepinephrine are mediated by α_2-adrenoceptors because a norepinephrine-induced reduction of background vestibular neuronal activity was antagonized by yohimbine and mimicked by clonidine. However, many vestibular nucleus neurons were also inhibited through α_1-adrenergic mechanisms. The heterogeneity of responses in the vestibular nuclei is also underscored by the report that bath application of norepinephrine increased the firing rate of MVN neurons in brain slice preparations (Gallagher et al. 1992). The relative density of noradrenergic projections to the superior vestibular nucleus raises the hypothesis that these connections may modulate ascending vestibuloautonomic projections to the parabrachial nucleus.

4.4.2. Serotonergic Transmission in the Vestibular Nuclei

Serotonergic neurons in the raphe nuclei and the adjacent brain stem reticular formation contribute widespread connections to all levels of the neuraxis (Baumgarten and Grozdanovic 1997). Physiologic evidence suggests that most serotonergic neurons in behaving mammals are relatively insensitive to large variations in behavioral, environmental, and physiological conditions (Jacobs and Azmitia 1992). However, there are populations of serotonergic neurons that increase their responses during motor activities such as grooming, chewing, and walking but decrease their activity during orienting responses (Jacobs and Azmitia 1992; Jacobs and Fornal 1993). These data have led to the hypothesis that one function of serotonergic activation is to simultaneously facilitate motor output and inhibit sensory information processing and to coordinate hormonal and neuroendocrine function in parallel with motor programs (Jacobs and Fornal 1993). During conditions when serotonergic neurons are inactive (e.g., orienting responses), these neurons may disfacilitate motor activity and disinhibit sensory processing. However, this is only one aspect of serotonergic influences in the brain: serotonergic transmission has also been implicated as a factor in diverse psychiatric disorders (Keane and Soubrié 1997).

Both anatomical and physiological studies indicate that raphe nuclei receive vestibular information. Anatomical tracer studies have revealed direct connections from the vestibular nuclei to the dorsal raphe nucleus (Kawasaki and Sato 1981; Kalen et al. 1985), nucleus raphe magnus (Carlton et al. 1983), and nucleus raphe pontis (Gerrits et al. 1985). Caudal (medullary) raphe neurons respond to electrical and natural stimulation of vestibular afferents (Yates et al. 1993b), whereas dorsal raphe neurons respond to electrical stimulation of the medial and superior vestibular nuclei (Kawasaki and Sato 1981). However, there is no information regarding responses of raphe neurons to vestibular stimulation in alert, behaving animals.

Serotonergic innervation of the vestibular nuclei originates from two sources, the dorsal raphe nucleus and the nucleus raphe obscurus and nucleus raphe pallidus (Halberstadt and Balaban, 2003). There is a dense subventricular plexus of 5-HT and 5-HT transporter immunoreactive fibers along the medial margins of SVN and MVN that declines in density more laterally (Steinbusch 1991; Halberstadt and Balaban, 2003). Low to very low densities of $5\text{-}HT_1$ (Pazos and Palacios 1985a) and $5\text{-}HT_2$ (Pazos and Palacios 1985b) receptor binding are present in the medial vestibular nucleus. Immunohistochemical data (Balaban 2002) have revealed $5\text{-}HT_{2A}$ receptor immunoreactivity in rat vestibular nuclear neurons, cerebellar Purkinje cells, and central amygdaloid nucleus neurons but not in the parabrachial nucleus (Balaban, 2002). Neuropharmacologic evidence indicates that superior and lateral vestibular nucleus neurons display significant excitatory, inhibitory, or biphasic responses to serotonin application, apparently via a combination of $5\text{-}HT_1$ and $5\text{-}HT_2$ mechanisms (Johnston et al. 1993; Licata et al. 1993a, 1993b). One source of these serotonergic inputs to the vestibular nuclei appears to be the dorsal raphe nucleus (Kishimoto et al. 1991; Licata et al. 1995). However with the exception of the report of Kishimoto et al. (1991) that the lateral vestibular nucleus neurons projecting to the region of the abducens nucleus (but not spinally projecting neurons) tended to be inhibited by these dorsal raphe inputs, there is no information regarding differential influences of serotonergic transmission on functional pools of cells in the vestibular nuclei.

4.4.3. Functional Considerations

The autonomic manifestations of vigilance, alerting, and emotional responses have been the subject of long-standing interest in experimental psychology and physiology (Stürup et al. 1935; Woodworth and Schlosberg 1954; Sokolov 1963). Like other stimuli, a sense of self-motion can elicit an orienting response. On the other hand, sudden unexpected self-motion can be an imperative stimulus that elicits motor responses and a combined emotional and autonomic response akin to panic. Almost everyone has experienced the following scenario: you are stopped at a traffic light in a car on the upward slope of a hill. In anticipation of the traffic signal changing, a large vehicle that is alongside your car moves forward slowly, giving you a wide-field optic flow stimulus that you interpret as a backward drift of your car on the hill. You simultaneously feel a sense of panic and a pounding heart as you vigorously depress the brake pedal. This is followed by a sense of relief as you realize that your sense of movement was merely illusory.

This anecdotal example illustrates three key features of sudden gravitoinertial motion. First, an unexpected sense of rapid self-motion is an imperative or alerting stimulus. The perception of danger is very real: there is potential for injury. Second, the response to our example of sudden perceived self-motion resembles a panic attack. Finally, the response to a single, sudden perturbation has a rapid onset and a relatively rapid resolution, coupled with a realization that the hypothetical danger is nonexistent. Because both the locus coeruleus and the raphe nuclei receive vestibular nuclear input and show altered activity during orienting or alerting stimuli, it seems likely that differential activation of central noradrenergic and serotonergic pathways are involved in the distinct autonomic responses that occur in response to gravitoinertial stimuli. Further, these effects are likely to be manifest in parallel at multiple levels of the neuraxis, including the vestibular nuclei.

The monoaminergic innervation of the vestibular nuclei may also affect sensory processing and motor output of this region as a component of vigilance, alerting, and orientation responses. Although there are no studies that have systematically and explicitly examined the effects of "alerting" or vigilant behavior on the discharges of vestibular nucleus neurons, there is evidence that the discharges of some vestibular nucleus neurons change with drowsiness. Bizzi et al. (1964) reported that the firing rate of vestibular neurons increases with arousal and decreases with drowsiness in cats. Henn et al. (1984), on the other hand, reported that the sleep–wakefulness transition had no effect on the responsiveness of vestibular-only units to rotational stimulation (see also Buettner et al. 1978). However, these investigators showed that drowsiness resulted in: (1) a marked decrease in the spontaneous firing rate in units whose activity is modulated during both vestibular stimulation and eye movements (including TVP neurons); and (2) a selective loss of eye movement modulation on burst–pause units. Evidence is insufficient to attribute these effects to specific actions of monoaminergic pathways.

The terms "vestibular habituation" (Dodge 1922) or vestibular "response decrement" (Hood and Pfaltz 1954) have been used to describe decrements in both the duration and number of quick phases during vestibular-evoked nystagmus with repeated exposures to a constant stimulus (Griffith 1920, 1924; Dodge 1922; Hood and Pfaltz 1954). This decline in responsiveness is persistent. However, the response can be increased transiently by either behavioral arousal or by systemic injections of adrenaline (Hood and Pfaltz 1954). The effects of vigilance or behavioral arousal on vestibuloocular responses are implicit in the clinical admonition to "maintain the alertness" of patients during vestibuloocular reflex testing (Barber and Stockwell 1980). Studies of the effects of "alertness" on vestibular function have typically followed three approaches. First, experiments have documented that instructions to subjects, drowsiness, "alerting" tasks, or "mental set" produce a change in nystagmus during vestibular stimulation (Collins and Guedry 1961; Collins and Poe 1962; Furman et al. 1981). Second, experiments have documented that sleep deprivation affects vestibular-evoked nystagmus

(Wolfe and Brown 1969; Dowd et al. 1975; Collins 1988). Finally, studies have shown that "alerting" pharmacological manipulations, such as administration of amphetamines (a centrally active sympathomimetic), produce an "alerted" pattern of nystagmus (Collins and Poe 1962; Furman et al. 1981; Collins 1988).

A pioneering study by Collins and Guedry (1961) reported that instructions to human subjects can have profound effects on eye movements during vestibular stimulation. When subjects were told to attend to sensations of movements, nystagmus showed fewer quick phases and a reduced slow-phase eye velocity relative to when the same subjects performed mental arithmetic. Subsequent research (Furman et al. 1981) has demonstrated that behavioral alerting increases the gain of the slow phase of vestibuloocular reflexes. A more explicit relationship between nystagmus performance and measures of subject arousal (reaction time to mild electrical shock and electroencephalographic (EEG) activity) was reported by Kasper et al. (1992). These findings therefore suggest that a somatic vestibular response (nystagmus) is affected by "state of arousal," performance of ancillary "alerting" tasks, and administration of centrally active sympathomimetic medications. Because "alerting tasks" such as mental arithmetic are viewed as stressors that increase heart rate and blood pressure, it is of interest to determine whether these concomitant effects of alertness/vigilance facilitate or occlude vestibular-induced transient and sustained autonomic responses in behaving organisms.

5. Existence of Labyrinthine Contributions to Autonomic and Respiratory Muscle Activity: Integrative Approaches to Physiology

In the 1940s, Spiegel presented systematic reviews of contemporary physiological evidence that labyrinthine stimulation evokes pupillary, cardiovascular, gastrointestinal, and respiratory reactions (Spiegel and Sommer 1944; Spiegel 1946). This section reviews more recent physiologic evidence that vestibular stimulation can affect each of these visceral, somatic, or combined somatic and visceral motor systems. Evidence of visceral contributions to the sensation of motion was reviewed earlier (see Section 2 above).

5.1. Evidence that the Vestibular System Affects Pupillary Diameter and Lens Accommodation

The "near triad" of vergence eye movements, lens accommodation, and pupillary diameter regulation (Leigh and Zee 1991) is an example of an integrated somatic and visceral motor response. When we fixate on an object approaching our nose, three motor responses occur simultaneously: a convergent (disconjugate) eye movement mediated by the extraocular (striated) muscles, lens accommodation (relaxation of the ciliary (smooth) muscle leading to lens thickening), and pupillary constriction (smooth muscle response). Although it is well-documented that linear acceleration along the nasooccipital axis elicits vergence eye movements (Paige and Tomko 1991a, 1991b), there is no published evidence regarding changes in either pupillary diameter or lens accommodation as a component of linear vestibuloocular reflexes. However, there are two descriptions of accommodative changes in humans during vestibular stimulation. Clark et al. (1975) reported consistent changes in accommodation during horizontal postrotatory nystagmus, which was prolonged by far-point fixation through a pinhole. Markham et al. (1977) demonstrated monocular accommodative (lens-thickening) changes during slow (1 deg/s) roll rotation that peaked when the ipsilateral ear was down and were absent when the ipsilateral ear was up. The mechanisms mediating these accommodative changes during ocular counterrolling are not understood, and the functional significance of a monocular response is unclear. Furthermore, limited experimental evidence suggests that coordinated actions of smooth and striated muscles of the eye can be elicited by vestibular stimulation. For example, mechanical ("air-puff") stimulation of the utricle in cats and monkeys has been reported to produce conjugate nystagmus, with pupillary constriction during fast phases and dilation during slow phases (DeSantis and Gernandt 1971). Because these responses were unaffected by removing sympathetic input to the eye, they were presumed to be strictly parasympathetic. Electrical stimulation of the utricle in cats also produced lens accommodation (Markham et al. 1973) via parasympathetic mechanisms. Anatomic data (Balaban 2003) indicate that there are direct projections from the vestibular nuclei to putative preganglionic parasympathetic neurons in the anteromedian and parabrachial nuclei that are likely to mediate these effects. The relation of these results to the limited human experimental data, however, needs to be clarified.

5.2. Physiological Evidence that Vestibular Stimulation Affects Blood Pressure

A series of pioneering studies by Spiegel and Démétriades (Spiegel and Démétriades 1922, 1924), reviewed in Spiegel (1946) provided initial evidence that the vestibular system participates in the control of blood pressure. These investigators showed that activation of vestibular afferents by caloric stimulation (perfusion of the ear canal with hot or cold water), by galvanic electrical stimulation, or by natural stimulation produced changes in blood pressure in a number of mammalian species. They also demonstrated that these effects were attenuated or abolished by cocainization of the labyrinths and by lesions of the caudal aspect of the medial vestibular nucleus. Subsequent studies have suggested that a reflexlike component of vestibular–cardiovascular responses is mainly due to labyrinthine influences on the sympathetic nervous system. These responses are probably mediated by descending projections from the vestibular nuclei to medullary structures. Electrical stimulation of vestibular afferents produces changes in activity in sympathetic nerves (Megirian and Manning 1967; Cobbold et al. 1968; Ishikawa and Miyazawa 1980; Kerman and Yates 1998). These vestibulosympathetic reflexes are the result of activation of vestibular afferents (rather than stimulus spread to nontarget nerves) because they are abolished by lesions of the medial and inferior vestibular nuclei (Spiegel and Démétriades 1922; Spiegel 1946; Uchino et al. 1970; Yates and Miller 1994). A study that analyzed the effect of short train electrical vestibular stimulation on the activity of abdominal sympathetic nerves suggested that only sympathetic efferents innervating vascular smooth muscle (and not those involved in regulating motility) received vestibular influences (Kerman and Yates 1998). Furthermore, Tang and Gernandt (1969) demonstrated that electrical vestibular stimulation using short trains has little effect on parasympathetic activity in the vagus nerve; rather, cervical vagus activity appeared to be restricted to the recurrent laryngeal nerve, which innervates all striated muscles of the larynx except the cricothyroid, the constrictor pharyngis inferior muscle (important in deglutition), and smooth muscles of the esophagus and trachea. Thus, cardiovascular effects of vestibular stimulation in decerebrate or anesthetized preparations appear to be restricted to sympathetic innervation of vascular smooth muscle.

The use of natural vestibular stimulation has provided further insights into the organization of vestibulosympathetic reflexes in decerebrate cats. A preparation has been developed that permits the stimulation of vestibular afferents but not other receptors that might affect blood pressure. The head of a decerebrate cat is rotated on a fixed body following a number of denervations to remove nonlabyrinthine inputs that might be produced by the head movement, including transection of the upper cervical dorsal roots and the ninth, tenth, and sometimes the fifth cranial nerves. Use of this preparation has shown that vestibular signals elicited by head movements can affect both sympathetic nerve activity (Yates and Miller 1994) and blood pressure (Woodring et al. 1997). Sympathetic nerve activity is maximal during nose-up head rotations, minimal during nose-down rotations, and is unaffected by ear-down rotations (Yates and Miller 1994). Horizontal head rotations in the denervated preparations also produce no changes in sympathetic outflow (Yates and Miller 1994). In addition, sustained head-up rotations elicit pressor responses; the mean blood pressure increase elicited by a 50° nose-up head movement is approximately

20 mm Hg (Woodring et al. 1997). In contrast, ear-down tilt produces no change in blood pressure (Woodring et al. 1997). The gain of vestibulo-sympathetic responses is flat across the tested frequency range of 0.1–1 Hz, indicating that these responses are predominantly due to activation of otolith afferents (Yates and Miller 1994).

These observations suggest that although baroreflex mechanisms are normally the dominant influence in producing pressor responses to whole-body tilt, otolith signals have the potential to elicit synergistic changes in sympathetic outflow to the vasculature. Furthermore, it is possible that vestibular signals play a different role than baroreceptors in generating responses to reduce orthostasis. As discussed above, changes in baroreceptor activity can be elicited by a number of mechanisms, including an alteration in posture, hemorrhage, or local cardiovascular reflexes that channel blood into particular vascular beds. Each of these cardiovascular challenges requires a particular pattern of sympathetic nervous system responses in order to produce compensation (Loewy and Spyer 1990). Thus, the integration of vestibular, baroreceptor, and perhaps other sensory inputs can produce more accurate compensatory cardiovascular responses than reliance on baroreceptor signals alone. In addition, the vestibular system may be able to detect conditions that might lead to orthostasis prior to the onset of peripheral blood pooling (and a reduction in systemic blood pressure). As a result, vestibular inputs could potentially trigger responses to offset orthostasis at a shorter latency than baroreceptors and therefore provide for more stable blood pressure at the onset of body movement.

Other experiments have explored the effects of vestibular lesions on an animal's ability to maintain stable blood pressure during whole-body rotations. Doba and Reis (1974) compared the blood pressure response of anesthetized and paralyzed cats to 30° and 60° nose-up pitch before and after bilateral transection of the eighth cranial nerves. They showed that removal of vestibular inputs resulted in orthostatic intolerance during pitch rotations. This experiment was more recently repeated in awake animals trained to remain sedentary in a tilting device; the animals were restrained on the tilt table using a screw attached to a bolt mounted to the skull and a harness placed around the torso (Jian et al. 1999). Following vestibular nerve transection, the blood pressure of these animals was less stable during nose-up tilt than when vestibular inputs were present. The deficit was more severe during trials in which the visual environment rotated with the animal, such that no visual cues indicating body position in space were available. Furthermore, the decreased ability to adjust blood pressure during unexpected tilts persisted for only about one week. The subsequent compensation in ability to adjust blood pressure during nose-up tilt may be due to the fact that animals learned to use cues from proprioceptors and other somatosensory receptors to signal the onset of unexpected movements. These observations support the hypothesis that integration of multiple inputs that signal challenges to the cardiovascular system can provide for more stable blood pressure than reliance on baroreceptor mechanisms alone.

5.3. Neural Pathways that Mediate Vestibular–Sympathetic Responses: The Descending Vestibuloautonomic Pathway

The abolition of vestibulocardiovascular responses by lesions of the caudal aspect of the medial vestibular nucleus has been a reproducible finding since the earliest demonstration of these effects (Spiegel 1946; Uchino et al. 1970; Yates and Miller 1994). Anatomical evidence (see above and Figure 7.1) indicates that there are several potential pathways for direct vestibular influences on regions of the brain stem involved in cardiovascular control. A descending pathway provides direct projections from the vestibular nuclei to the nucleus tractus solitarius (Balaban and Beryozkin 1994; Yates et al. 1994; Ruggiero et al. 1996; Porter and Balaban 1997), the dorsal motor vagal nucleus, the lateral medullary tegmental field (Yates et al. 1995a; Porter and Balaban 1997; Stocker et al. 1997), the caudal ventrolateral medulla (Stocker et al. 1997), the nucleus ambiguus/parambiguus (Balaban and Beryozkin 1994), and the rostral ventrolateral medulla (Balaban 1996c; Porter and Balaban 1997; Stocker et al. 1997). The ascending projections to the parabrachial nucleus (Balaban 1996c; Porter and Balaban 1997) also provide a potential indirect link to these brain stem regions, as the parabrachial nucleus makes connections with the medullary cardiovascular-regulatory areas. Of these areas, the nucleus tractus solitarius, the caudal ventrolateral medulla, and the rostral ventrolateral medulla are chiefly involved in producing baroreceptor reflexes. Baroreceptor afferents, which are activated by increases in arterial blood pressure, make connections with neurons in the nucleus tractus solitarius (Loewy and Spyer 1990). Neurons in the nucleus tractus solitarius provide direct excitatory drive to vagal (parasympathetic) preganglionic neurons, located near the nucleus ambiguus in the ventrolateral medulla, which act to slow the heart (Loewy and Spyer 1990). In addition, signals from baroreceptor-sensitive neurons in the nucleus tractus solitarius produce inhibition of sympathetic outflow to the heart and blood vessels, which is necessary to reduce blood pressure during hypertensive episodes (Loewy and Spyer 1990). The inhibitory interneurons in the sympathetic baroreceptor reflex arc are located in the caudal ventrolateral medullary reticular formation; these cells inhibit "vasomotor" neurons in the rostral ventrolateral medulla that control the activity of sympathetic preganglionic neurons in the thoracic spinal cord (Loewy and Spyer 1990).

Despite the presence of vestibular nucleus inputs to several brain stem areas involved in cardiovascular control, physiological and lesion studies have shown that only a subset of these regions are necessary for eliciting vestibulosympathetic responses in decerebrate cats. Neurons in the nucleus tractus solitarius that receive baroreceptor inputs, and presumably participate in regulating blood pressure, receive little vestibular input (Yates et al. 1994). Furthermore, lesions of the nucleus tractus solitarius do not appreciably affect the amplitude of sympathetic nerve responses elicited by electrical stimulation of the vestibular nerve (Steinbacher and Yates 1996a). The two other brain stem regions that mediate baroreceptor reflexes, the caudal ventrolateral medulla and the rostral ventrolateral medulla, appear to be essential for relaying vestibular signals to sympathetic preganglionic neurons in the spinal cord (Yates et al. 1995; Steinbacher and Yates 1996a). However, despite the fact that the caudal ventrolateral medulla contains inhibitory neurons of the baroreceptor reflex arc, cells in this region that receive baroreceptor inputs do not respond to labyrinthine stimulation. Instead, neurons with baroreceptor and vestibular inputs form two overlapping groups located near the medial border of the lateral reticular nucleus in the caudal ventrolateral medulla (Steinbacher and Yates 1996b). These data suggest that although baroreceptor and vestibular inputs are processed by similar regions of the brain stem, these two signals are integrated by different interneurons. However, neurons in the rostral ventrolateral medulla that project to the thoracic spinal cord and presumably make connections with sympathetic preganglionic neurons respond to stimulation of both baroreceptor and vestibular afferents (Yates et al. 1991, 1993a). Thus, even though distinct cardiovascular-regulatory interneurons process labyrinthine and cardiovascular inputs, the same neurons in the rostral ventrolateral medulla relay the baroreceptor and vestibular signals to sympathetic output cells. The functional significance of this pattern of integration of signals remains to be determined.

Figure 7.4 illustrates the minimal circuitry in the brain stem that produces vestibulosympathetic responses and baroreceptor reflexes, as revealed by electrophysiological and lesion studies. As denoted in this figure and as discussed above, the region of the vestibular nuclear complex that mediates vestibulosympathetic reflexes is restricted to the portion of the medial and inferior vestibular nuclei just caudal to Deiters' nucleus. Electrophysiological studies have also shown that the activity of many neurons in the inferior and adjacent medial vestibular nuclei is modulated by the same vestibular stimuli that influence sympathetic outflow: pitch rotations that activate otolith receptors (Schor et al. 1998). Additional critical connections in the vestibulosympathetic "reflex" pathway include projections from the vestibular nuclei to the lateral medullary reticular formation (particularly the caudal ventrolateral medulla) and inputs from the lateral medullary reticular formation to the "pressor" area of the rostral ventrolateral medulla, which relays the signals to sympathetic preganglionic neurons in the spinal cord. As discussed above (see Section 4.2), the cerebellum may also be important in "shaping" the properties of vestibulosympathetic reflexes.

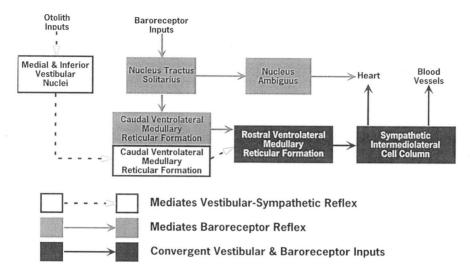

FIGURE 7.4. Schematic illustration of the brain stem and spinal cord pathways that mediate vestibulosympathetic and baroreceptor reflexes in the cat. Different shading patterns of boxes and arrows indicate pathways that convey vestibular or baroreceptor inputs independently or together.

5.4. Evidence that Vestibular Stimulation Affects Gastrointestinal Function

The concept that the vestibular system influences gastrointestinal function was proposed in the late nineteenth century by Irwin (1881) as an explanation for the nausea and vomiting during seasickness. Experiments during the 1920s by Spiegel and Démétriades (reviewed by Spiegel 1946) reported that caloric stimulation increased both resting tonus and "pendular" movements of the small intestine in rabbits, although the response was delayed several seconds from stimulus delivery. Furthermore, Babkin and Bornstein (1943) reported an immediate inhibition of hunger contractions of the stomach in dogs during stimulation on a swing (20 cycles/min, 8 ft radius, 5–6 ft arc) and that this response disappeared after bilateral labyrinthectomy. More recent studies have documented effects of caloric stimulation on gastrointestinal motility and smooth muscle electrical activity in human subjects. Thompson et al. (1982) observed an inhibition of postprandial gastric emptying and motility after caloric stimulation in humans; the duodenum showed aborally migrating contractions, followed by inhibition. Kolev and Altaparmakov (1996) noted that caloric stimulation produces changes in the duration of components of migrating electrical activity of the stomach and duodenum in fasting subjects. Of particular interest was the emergence of "spontaneous nonpropulsive" (i.e., nonpropagating) spike activity in the stomach and duodenum after caloric stimulation, which apparently corresponds to spastic, aperistaltic contractions. Thus, these data suggest that intense, relatively long-duration vestibular stimuli can elicit gastrointestinal reactions.

The duration of vestibular stimulation required to produce gastrointestinal responses is illustrated in the experiments of Ito and Honjo (1990). These investigators reported that pulse train stimulation of one labyrinth (round window electrode, 40–100 Hz, up to 30 s duration) produced contractions of the stomach, which would persist for several minutes after cessation of the stimulus. By contrast, briefer stimuli (single shocks or short trains) have failed to elicit discernible activity in either the abdominal vagus nerve (Tang and Gernandt 1969) or motility-regulating fibers in sympathetic nerves innervating the abdominal viscera (Kerman and Yates 1998). However, single shocks or short stimulus trains are sufficient to evoke responses in sympathetic fibers innervating vascular smooth muscle (Kerman and Yates 1998). The requirement of long-duration, intense vestibular stimulation for producing gastrointestinal effects may reflect the fact that transient postural shifts are not germane to intrinsic or reflex control of gastrointestinal activity. Correction for shifts in blood distribution during movement, on the other hand, must occur very rapidly if blood pressure is to remain constant.

5.5. Physiological Evidence that the Vestibular System Influences "Respiratory Muscle" Activity

The muscles used to move air into and out of the lungs are multifunctional and also participate in postural adjustments, venous return to the heart, and responses such as "straining" while lifting weights or during defecation, deglutition, vocalization, and emesis. Each of these actions will require different coordinated activities of these striated muscles, accompanied by appropriate changes in smooth muscle tone that are regulated by the sympathetic and parasympathetic systems. Because respiratory muscles are used for a number of purposes, the demonstration that their activity is affected by vestibular stimuli does not necessarily imply that the vestibular system can either facilitate blood oxygenation or respiratory rhythm generation. In this sense, the traditional view that vestibular influences on respiratory muscles serve mainly to regulate "respiratory activity" (Spiegel 1946) is a concept of limited utility. Rather, it is of broader utility to focus on elucidating the contributions of vestibular information to both motoneurons innervating these muscles and to central pattern generators that select, regulate, and coordinate various functional movements of the torso, larynx, and pharynx. As recognized by earlier investigators (Spiegel 1936), however, one must be careful to avoid confounding variables such as anxiety, which have a profound impact on respiratory function. Nonetheless, with appropriate caveats, the examination of vestibular influences on the specialized muscles that regulate pressure within the thoracic and abdominal cavities may provide insight into the role of labyrinthine and extralabyrinthine graviceptive inputs in producing integrated motor responses. Furthermore, by contrasting vestibular influences on torso, laryngeal, and pharyngeal muscles with labyrinthine effects on limb and neck muscles, it may be possible to discern basic principles regarding the context-dependent influences of vestibular signals in motor control.

The initial reports of vestibular influences on respiratory function noted that rhythmic linear stimulation in the vertical plane or pitch stimulation produced an increased amplitude and frequency of respiratory movements in experimental animals (Spiegel 1936, 1946). Spiegel (1936) showed that these "respiratory reactions" often persisted after cessation of stimulation and that they were sometimes followed by a period of lower-amplitude, slow respiratory movements. Bilateral labyrinthectomy diminished (but did not abolish) these responses, suggesting that they were only partially due to activation of vestibular receptors. More recent studies, which are described below, have confirmed that stimulation of the vestibular system produces changes in activity of striated muscles of the pharynx, larynx, thorax, and abdomen, although the prolonged respiratory muscle responses described by Spiegel (1946) have not been pursued further.

Electrical stimulation of the vestibular nerve elicits responses in nerves innervating a variety of striated muscles of the oral cavity, pharynx, larynx, and torso, including the diaphragm, abdominal muscles, external intercostal muscles, pharyngeal muscles, laryngeal muscles, and tongue musculature (Megirian 1968; Tang and Gernandt 1969; Yates et al. 1993c; Shiba et al. 1996; Siniaia and Miller 1996; Umezaki et al. 1997; Zheng et al. 1997). These vestibular-evoked responses are abolished by lesions of portions of the medial and inferior vestibular nuclei located just caudal to Deiters' nucleus, demonstrating that they are the result of activation of vestibular afferents (Yates et al. 1993c; Shiba et al. 1996; Siniaia and Miller 1996). Electrical or chemical stimulation of the vestibular nuclei, as well as lesions of the vestibular nuclei, also produce modifications in expiratory and inspiratory spinal nerve discharges (Bassal and Bianchi 1982; Huang et al. 1991, 1993). Selective natural vestibular stimulation (described above in Section 5.2) has also been used to influence activity of the diaphragm, abdominal muscles, and tongue musculature that participate in activities such as respiration, vocalization, deglutition, and defecation. Nose-up vestibular stimulation of cats produces an increase in the activity of nerves innervating these muscles, whereas nose-down vestibular stimulation decreases nerve activity (Rossiter and Yates 1996; Rossiter et al. 1996). Ear-down and horizontal vestibular stimulation typically had no effect on activity of the hypoglossal nerve (which innervates tongue musculature), abdominal nerve, or phrenic nerve (which innervates the diaphragm) in the cat (Rossiter and Yates 1996;

Rossiter et al. 1996). Mameli, Tolu, and co-workers (Mameli and Tolu 1985, 1986) also reported that motoneurons in the ventral aspect of the caudal hypoglossal nucleus respond to vestibular stimulation in rabbits and in addition showed that retinal (light flash) stimulation activated these cells. The vestibular responses in their study were predominantly tonic, occurred at long latency (>20 ms), and differed from those in the cat in that they were elicited by ear-down roll tilt and caloric stimulation (which presumably activates afferents innervating the horizontal semicircular canal) (Mameli and Tolu 1987; Mameli et al. 1988). However, these investigators did not examine responses of hypoglossal motoneurons to pitch rotations, and thus it is unknown whether pitch stimulation affects tongue musculature in the rabbit as it does in the cat.

The effects of postural changes and vestibular lesions on diaphragm and rectus abdominis activity have also been examined in awake cats (Cotter et al. 2001). Spontaneous diaphragm and rectus abdominis activity and modulation of the firing of these muscles during nose-up and ear-down tilts were compared before and after removal of labyrinthine inputs. In vestibular-intact animals, nose-up tilts from the prone position altered rectus abdominis firing, whereas the effects of body rotation on diaphragm activity were not statistically significant. After peripheral vestibular lesions, spontaneous diaphragm and rectus abdominis discharges increased significantly (by ~170%), and augmentation of rectus abdominis activity during nose-up body rotation was diminished. However, spontaneous muscle activity and responses to tilt began to recover after a few days after the lesions, presumably as the central vestibular system began to employ nonlabyrinthine inputs to detect body position in space. These data suggest that the vestibular system provides tonic inhibitory influences on the rectus abdominis and the diaphragm and in addition contributes to eliciting increases in abdominal muscle activity during some changes in body orientation.

5.6. Neural Pathways that Mediate Vestibular Influences on the Diaphragm, Abdominal Muscles, and "Airway" Muscles

Most studies have only considered the neural circuits that impart activity related to respiration on diaphragm, abdominal, and tongue motoneurons. However, it is likely that the complex and diverse functions of these muscles require inputs from many regions. In particular, because these "respiratory muscles" also have a postural role, it seems likely that vestibular signals will be conveyed to diaphragm, abdominal, and "airway" motoneurons through neural circuits that bypass the brain stem respiratory rhythm generator. We will begin by considering whether the medullary respiratory groups may be influenced by vestibular signals and subsequently will discuss other pathways that may mediate vestibular influences on multifunctional muscles that control pressure in the thorax and abdomen.

5.6.1. Vestibular Influences on Circuits that Control the Respiratory Cycle

Experiments conducted since 1936 have shown that neurons located in two columns in the medulla have activity that is correlated with the respiratory cycle (see Feldman 1986; Ezure 1990, 1996 for reviews). One column of respiratory neurons, the dorsal respiratory group, is located in the ventrolateral portion of the nucleus solitarius in the caudal medulla. The dorsal respiratory group almost exclusively contains neurons that fire during the inspiratory phase of the respiratory cycle. The other column of respiratory neurons spans most of the length of the medulla and is located in the ventrolateral reticular formation near the nucleus ambiguus in the caudal medulla and near the retrofacial nucleus in the rostral medulla. This cell column is typically referred to as the ventral respiratory group, although the rostral end of this neuronal group is sometimes called the Bötzinger complex. The ventral respiratory group is comprised of both enpiratory and inspiratory neurons and appears to contain the pattern generator that produces the respiratory rhythm (Ezure 1990). Many neurons in the dorsal and ventral respiratory groups have axons that project to the spinal cord and make connections with spinal respiratory motoneurons. The ventral respiratory group also contains motoneurons that innervate laryngeal and pharyngeal muscles (Feldman 1986; Ezure 1996).

There may be species differences in the organization of vestibular nucleus projections to the dorsal and ventral respiratory groups. Anterograde tracing studies have indicated that neurons in the medial and inferior vestibular nuclei send projections to the ventrolateral medullary reticular formation, near neurons of the ventral respiratory group, in rats (Porter and Balaban 1997), rabbits (Balaban 1996c), and cats (Yates et al. 1995a). Although rats (Porter and Balaban 1997) and rabbits (Balaban 1996c) possess an appreciable vestibular nuclear projection to the vicinity of dorsal respiratory group neurons in the ventrolateral portion of the nucleus solitarius, the equivalent projections in cats are extremely sparse (Yates et al. 1994).

Electrophysiological studies have confirmed the anatomical observation that dorsal respiratory group neurons receive sparse vestibular input in cats. Electrical stimulation of the vestibular nerve has little effect on the firing rate of dorsal respiratory group neurons, despite the fact that this stimulus activates other neurons in the nucleus solitarius (Yates et al. 1994). In contrast, almost 50% of inspiratory bulbospinal neurons in the ventral respiratory group (Miller et al. 1995), and over 80% of bulbospinal expiratory ventral respiratory group neurons (Shiba et al. 1996), responded to electrical stimulation of the vestibular nerve. The activity of respiratory neurons whose axons did not exit from the medulla, and that presumably participate in generating the respiratory rhythm, was also influenced by electrical vestibular stimulation (Nakazawa et al. 1997; Umezaki et al. 1997; Zheng et al. 1997).

Responses were also recorded from ventral respiratory group neurons during natural vestibular stimulation in multiple transverse planes (Woodring and Yates 1997). Only a subpopulation of ventral respiratory group neurons with inspiratory (20/80 cells) or expiratory (11/59 cells) activity responded to tilts up to 15° in amplitude, delivered at frequencies between 0.02 Hz and 2 Hz. In particular, responses were infrequent in ventral respiratory group neurons with projections to the spinal cord (2/34 inspiratory and expiratory cells), despite the fact that the tilts employed produced robust modulation of activity of spinal respiratory nerves. Furthermore, the characteristics of the responses to tilt of ventral respiratory group neurons with vestibular inputs did not match those of respiratory muscles. These data suggest that despite the fact that ventral respiratory group neurons respond to powerful electrical vestibular stimulation, they are unlikely to be major participants in generating vestibular-evoked responses of abdominal, phrenic, and hypoglossal nerves. This conclusion is reinforced by two additional observations. First, large chemical lesions that inactivate the dorsal and ventral respiratory groups have little effect on the amplitude of responses of phrenic and abdominal nerves to vestibular stimulation (Yates et al. 1995b). Second, although midsagittal lesions of the ventral caudal medulla eliminate transmission of respiratory-related signals to abdominal motoneurons (Merrill 1974), they do not abolish responses of abdominal muscles to electrical (Shiba et al. 1996) or natural (Rossiter et al. 1996) vestibular stimulation.

5.6.2. Medial Reticulospinal Pathways

The locations of neurons outside of the dorsal and ventral respiratory groups that influence phrenic and abdominal motoneurons have recently been determined in the ferret using the transneuronal transport of pseudorabies virus. Injection of pseudorabies virus into the diaphragm labels a large number of cell bodies in the medial medullary reticular formation (Billig et al. 2000), which is known to receive a considerable input from otolith receptors (Manzoni et al. 1983; Pompeiano et al. 1984; Bolton et al. 1992). In addition, the dorsal and ventral respiratory groups and the raphe nuclei contained labeled neurons following virus injections into the diaphragm (Billig et al. 2000). However, no direct vestibular nucleus projections to diaphragm motoneurons were detected in this study. Injections of pseudorabies virus into the rectus abdominis muscle of the ferret also labeled a considerable number of cells in the medial medullary reticular formation as well as neurons in the ventral respiratory group and raphe nuclei (Billig et al. 2000). Thus, these anatomical studies indicate that the medullary reticulospinal pathway is likely to be involved in mediating vestibular influences on the muscles that regulate pressure in the thorax and abdomen.

The anatomical demonstration of medial reticulospinal inputs to diaphragm and abdominal motoneurons is not surprising. A previous study showed that chemical lesions of the medial medullary reticular formation abolished abdominal and diaphragm contractions during the emetic response to injection of toxins (Miller et al. 1996). However, direct electrophysiological experiments testing the hypothesis that medial medullary reticulospinal projections are prominently involved in relaying vestibular signals to abdominal and diaphragm motoneurons are yet to be conducted.

5.6.3. Vestibular Influences on Hypoglossal and Pharyngeal Motoneurons

In general, motoneuron pools innervating laryngeal, pharyngeal, and tongue muscles are small and are often interspersed with neurons that have other functions. Thus, the determination of pathways that provide inputs to these motoneurons has been very difficult. Only inputs to tongue motoneurons have been considered in any detail. Injection of pseudorabies virus into the tongue musculature produces labeled cell bodies in the hypoglossal nucleus, nucleus subcoeruleus, trigeminal sensory areas, the parvicellular reticular formation, and the dorsal medullary reticular fields (Fay and Norgren 1997). Portions of the parvicellular reticular formation, including the lateral tegmental field and the caudal ventrolateral medulla, have previously been shown to receive vestibular signals (Steinbacher and Yates, 1996b; Yates et al. 1995b). In particular, the caudal ventrolateral medulla contributes a prominent projection to the ventral half of the caudal aspect of the hypoglossal nucleus, which corresponds to the location of hypoglossal motoneurons that respond to vestibular stimulation (Mameli and Tolu 1986, 1987; Mameli et al. 1988). Retrograde tracing studies have shown that this region of the hypoglossal nucleus contains motoneurons that innervate the thyrohyoid, geniohyoid, genioglossus, and intrinsic tongue muscles (Uemura et al. 1979; Miyazaki et al. 1981). However, further studies will be required to determine the role of the caudal ventrolateral medulla in relaying labyrinthine signals to hypoglossal motoneurons.

6. Coordination of Vestibular Influences on Somatic and Autonomic Systems

As discussed above, considerable evidence shows that the vestibular system influences both somatic musculature (including neck, limb, trunk, pharyngeal, and tongue muscles) and autonomic systems (including those regulating blood pressure, pupillary shape, lens shape, and perhaps gastrointestinal activity). Presumably, these influences are important in producing coordinated muscle contractions and autonomic adjustments during movement and changes in posture. Furthermore, it seems likely that particular regions of the central nervous system are involved in programming, triggering, and executing these coordinated somatic and autonomic responses. For example, portions of the medial cerebellum have appropriate connections to simultaneously modulate motor activity and blood pressure (see above). A second example is the caudal ventrolateral medullary reticular formation, which appears to be essential for eliciting vestibulosympathetic responses (Steinbacher and Yates 1996a). This caudal brain stem region also projects to regions containing tongue, pharyngeal, facial, and abdominal motoneurons as well as motoneurons innervating middle ear muscles and muscles that move the pinna (Holstege 1989; Stocker et al. 1997). Stimulation of this area has previously been shown to elicit both vocalization and pressor responses in cats (Zhang et al. 1992). Nonetheless, it seems likely that "higher" regions of the central nervous system, including the cerebral cortex, are involved in simultaneously adjusting autonomic functions and coordinating muscle contractions.

The concept of "central commands" that trigger simultaneous excitation of neuronal circuits that control the locomotor and cardiorespiratory systems is familiar in the exercise literature (For a review, see Waldrop et al. 1996). However, it is likely that this concept can be extended and applied to a large number of movements that do not necessarily involve "exercise." It is possible that "central motor programs" that are used to coordinate the sequence of muscle contractions during particular movements also give rise to simultaneous autonomic responses that are essential for the movements to take place. This hypothesis remains to be tested.

7. Summary and Conclusions

A number of lines of evidence suggest that multiple sensory inputs that are influenced by gravitoinertial accelerations, including those from the vestibular labyrinth, retina, cutaneous and muscle receptors, baroreceptors, and abdominal visceral receptors, are integrated to produce an accurate perception of the location of the body in space, particularly during unusual conditions (e.g., swimming underwater). Such a processing of multiple sensory inputs appears to be important in coordinating appropriate motor and autonomic responses during movement and changes in posture. Although particular effector systems are most strongly influenced by particular sensory inputs (e.g., extraocular muscle contractions are driven powerfully be signals from semicircular canals, whereas components of the sympathetic nervous system that innervate vascular smooth muscle are strongly regulated by baroreceptor inputs), the "secondary inputs" and cerebellar contributions can potentially shape the responses, reduce their latency, and improve their accuracy. Because movement frequently requires coordinated changes in limb, axial, and respiratory muscle activity, accompanied by stereotyped adjustments in the cardiovascular system, it is important to explore the organization of components of "central motor programs" or "central pattern generators" that coordinate both motor and autonomic responses during the execution of movement. In addition to motor control, it is important to acknowledge the implications of the extensive convergence of vestibular and autonomic afferent information in the brain stem and cerebellum for spatial perception and affective changes associated with motion sickness and vestibular dysfunction. It appears that responses of vestibular and nonvestibular receptors to gravitoinertial challenges are integrated centrally to generate perceptual representations of gravitoinertial challenges. In addition, the visceral manifestations of motion sickness and vestibular dysfunction may be regarded as referred visceral discomfort related to gravitoinertial stimulation in the same sense that angina pectoris is a referred somatic pain related to cardiac dysfunction (Balaban 1999). Because these referred complaints are unpleasant and are not readily attributed to balance function, they may serve as eliciting or reinforcing stimuli for conditioned avoidance of situations that evoke discomfort. This conditioned avoidance may be one aspect of the linkage between balance disorders, height vertigo, and agoraphobia (Balaban 1999).

Acknowledgments. Dr. Joseph M. Furman provided helpful comments on an earlier version of this manuscript. Dr. Balaban's research is supported by National Institutes of Health grants R01 DC00739 and P01 DC03417 and NASA grant NAG2-1186; Dr. Yates' research is supported by National Institutes of Health grants R01 DC00693, R01 DC03732, and P01 DC03417.

References

Agostoni E (1970) Statics. In: Campbell EJM, Agostoni E, and Davis JN (eds) The Respiratory Muscles: Mechanics and Neural Control, Second edition. Philadelphia: W.B. Saunders Company, pp. 48–79.

Agostoni E (1977) Transpulmonary pressure. In: West JB (ed) Regional Differences in the Lung. New York: Academic Press, pp. 245–280.

Ammons WS (1988) Renal and somatic input to spinal neurons antidromically activated from the ventrolateral medulla. J Neurophysiol 60:1967–1981.

Anastasopoulos D, Haslwanter T, Bronstein A, Fetter M, Dichgans J (1997) Dissociation between the perception of body verticality and visual vertical in acute peripheral vestibular disorder in humans. Neurosci Lett 233:151–153.

Asch SE, Witkin HA (1948) Studies in space orientation. II. Perception of the upright with displaced visual fields and with body tilted. J Exp Psychol 38:455–477.

Aston-Jones G, Shipley MT, Chouvet GEM, van Bockstaele E, Pieribone V, Shiekhatter R, Akaoka H, Drolet G, Atier B, Charléty P, Valentino RJ, Williams JT (1991a) Afferent regulation of locus coeruleus neurons: anatomy, physiology and pharmacology. In: Barnes CD, Pompeiano O (eds) Neurobiology of the Locus Coeruleus. Amsterdam: Elsevier, pp. 47–75.

Aston-Jones G, Chiang C, Alexinsky T (1991b) Discharge of noradrenergic locus coeruleus neurons in behaving rats and monkeys suggests a role in vigilance. In: Barnes CD, Pompeiano O (eds) Neurobiology of the Locus Coeruleus. Amsterdam: Elsevier, pp. 501–520.

Babkin BP, Bornstein MB (1943) Effect of swinging and binaural galvanic stimulation on the motility of the stomach in dogs. Rev Can Biol 2:336–349.

Balaban CD (1984) Olivovestibular and cerebellovestibular connections in albino rabbits. Neuroscience 12:129–149.

Balaban CD (1996a) Efferent projections from the cerebellar nodulus and uvula in rabbits: potential substrates for cerebellar modulation of vestibulo-autonomic interactions. Abstracts, 1996 Midwinter Research Meeting of the Association for Research in Otolaryngology, p. 176. ARO, Mount Royal NJ.

Balaban CD (1996b) The role of the cerebellum in vestibular autonomic function. In: Yates BJ, Miller AD (eds) Vestibular Autonomic Regulation. Boca Raton: CRC Press, pp. 127–144.

Balaban CD (1996c) Vestibular nucleus projections to the parabrachial nucleus in rabbits: implications for vestibular influences on autonomic function. Exp Brain Res 108:367–381.

Balaban CD (1997) Projections from the parabrachial nucleus to the vestibular nuclei in rabbits: a visceral relay to vestibular circuits. Abstracts, Midwinter Meeting of Association for Research in Otolaryngology, p. 69. ARO, Mount Royal NJ.

Balaban CD (1999) Vestibular autonomic regulation. Curr Opin Neurol 12:29–33.

Balaban CD (2002) Neural substrates linking balance control and anxiety. Physiol Behav 77:469–475.

Balaban CD (2003) Vestibular nucleus projections to the Edinger–Westphal and anteromedian nuclei of rabbits. Brain Res., 963:121–131.

Balaban CD, Beryozkin G (1994) Vestibular nucleus projections to nucleus tractus solitarius and the dorsal motor nucleus of the vagus nerve: potential substrates for vestibulo-autonomic interactions. Exp Brain Res 98:200–212.

Balaban CD, Porter JD (1998) Neuroanatomical substrates for vestibulo-autonomic interactions. J Vestib Res 8:7–16.

Balaban CD, Thayer JF (2001) Neurological bases for balance-anxiety links. J Anxiety Disord 15:53–79.

Balaban CD, McGee DM, Zhou J, Scudder CA (2002) Responses of primate caudal parabrachial nucleus and Kölliker–Fuse nucleus neurons to whole body rotation. J Neurophysiol 88:3175–3193.

Bankoul S, Neuhuber WL (1990) A cervical primary afferent input to vestibular nuclei as demonstrated by retrograde transport of wheat germ agglutinin-horseradish peroxidase in the rat. Exp Brain Res 79:405–411.

Bankoul S, Goto T, Yates B, Wilson VJ (1995) Cervical primary afferent input to vestibulospinal neurons projecting to the cervical dorsal horn: an anterograde and retrograde tracing study in the cat. J Comp Neurol 353:529–538.

Barber HO, Stockwell CW (1980) Manual of Electronystagmography, Second edition. St. Louis: C.V. Mosby Company.

Barmack NH, Fagerson MH (1994) Vestibularly evoked activity of single units in the dorsomedial cell column of the inferior olive of the rabbit. Neurosci Abstr 20:1190.

Barnes CD, Manzoni D, Pompeiano O, Stampacchia G, d'Ascanio P (1989) Responses of locus coeruleus and subcoeruleus neurons to sinusoidal neck rotation in decerebrate cat. Neuroscience 31:317–392.

Bassal M, Bianchi AL (1982) Inspiratory onset or termination induced by electrical stimulation of the brain. Respir Physiol 50:23–40.

Batini C, Corvisier J, Hardy O, Jassik-Gerschenfeld D (1978) Brain stem nuclei giving fibers to lobules VI and VII of the cerebellar vermis. Brain Res 152: 241–261.

Baumgarten HG, Grozdanovic Z (1997) Anatomy of central serotonergic projection systems. In: Baumgarten HG, Gothert M (eds) Serotonergic neurons and 5-HT receptors in the CNS. Berlin: Springer, pp. 41–89.

Baxter GM, Williamson TH, McKillop G, Dutton GN (1992) Color Doppler ultrasound of orbital and optic nerve blood flow: effects of posture and timolol 0.5%. Invest Ophthalmol Vis Sci 33:604–610.

Berkley KJ, Hubscher CH (1995) Are there separate central nervous system pathways for touch and pain? Nat Med 1:766–773.

Berne RM, Levy MN (1983) Physiology. St. Louis: Mosby.

Biederman-Thorson M, Thorson J (1973) Rotation-compensating reflexes independent of the labyrinth and eye. J Comp Physiol 83:103–122.

Billig I, Foris JM, Enquist LW, Card JP, Yates BJ (2000) Definition of neuronal circuitry controlling the activity of phrenic and abdominal motoneurons in the ferret using recombinant strains of pseudorabies virus. J Neurosci 20:7446–7454.

Bizzi EO, Pompeiano O, Somogyi I (1964) Vestibular nuclei: activity of single neurons during natural sleep and wakefulness. Science 145:414–415.

Bolton PS, Goto T, Schor RH, Wilson VJ, Yamagata Y, Yates BJ (1992) Response of pontomedullary reticulospinal neurons to vestibular stimuli in vertical planes. Role in vertical vestibulospinal reflexes of the decerebrate cat. J Neurophysiol 67:639–647.

Boyle R, Pompeiano O (1980) Responses of vestibulospinal neurons to sinusoidal rotation of the neck. J Neurophysiol 44:633–649.

Boyle R, Pompeiano O (1981) Convergence and interaction of neck and macular vestibular inputs on vestibulospinal neurons. J Neurophysiol 45:852–868.

Bradley DJ, Gherlarducci B, Paton JFR, Spyer KM (1987a) The cardiovascular responses elicited from the posterior cerebellar cortex in the anaesthetized and decerebrate rabbit. J Physiol 383:537–550.

Bradley DJ, Pascoe JP, Paton JFR, Spyer KM (1987b) Cardiovascular responses and respiratory responses evoked from the posterior cerebellar cortex and fastigial nucleus in the cat. J Physiol 393:107–121.

Bramble DM, Jenkins FA Jr (1993) Mammalian locomotor–respiratory integration: implications for diaphragmatic and pulmonary design. Science 262:235–240.

Bremner JD, Krystal JH, Southwick SM, Charney DS (1996) Noradrenergic mechanisms in stress and anxiety: I. Preclinical studies. Synapse 23:28–38.

Buchanan RA, Williams TD (1985) Intraocular pressure, ocular pulse pressure, and body position. Am J Optom Physiol Opt 62:59–62.

Buettner UW, Büttner U, Henn V (1978) Transfer characteristics of neurons in vestibular nuclei of the alert monkey. J Neurophysiol 41:1614–1628.

Burian M, Gstoettner W, Mayr R (1990) Brainstem projection of the vestibular nerve in the guinea pig: an HRP (horseradish peroxidase) and WGA-HRP (wheat germ agglutinin-HRP) study. J Comp Neurol 293:165–177.

Bushnell MC, Goldberg ME, Robinson DL (1981) Behavioral enhancement of visual responses in monkey cerebral cortex. I. Modulation in posterior parietal cortex related to selective visual attention. J Neurophysiol 46:755–771.

Cadden SW, Morrison JF (1991) Effects of visceral distension on the activities of neurones receiving cutaneous inputs in the rat lumber dorsal horn: comparison with effects of remote noxious stimuli. Brain Res 558:63–74.

Campbell EJM, Agostoni E, Davis JN (1970) The Respiratory Muscles: Mechanics and Neural Control, Second edition. Philadelphia: Saunders.

Carleton SC, Carpenter MB (1984) Distribution of primary vestibular fibers in the brainstem and cerebellum of the monkey. Brain Res 294:281–298.

Carlton SM, Leichnetz GR, Young EG, Mayer DJ (1983) Supramedullary afferents of the nucleus rephe magnus in the rat: a study using transcannula HRP gel and autoradiographic techniques. J Comp Neurol 214:43–58.

Cavada C, Goldman-Rakic PS (1989) Posterior parietal cortex in rhesus monkeys: II. Evidence for segregated corticocortical networks linking sensory and limbic areas with the frontal lobe. J Comp Neurol 287:422–445.

Cedarbaum JM, Aghajanian GK (1978) Afferent projections of the rat locus coeruleus as determined by a retrograde tracing technique. J Comp Neurol 178:1–16.

Clark B, Randle RJ, Stewart JD (1975) Vestibulo-ocular accommodation reflex in man. Aviat Space Environ Med 46:1336–1339.

Cobbold AF, Meghirian D, Sherrey JH (1968) Vestibular evoked activity in autonomic motor outflows. Arch Ital Biol 106:113–123.

Collins WE (1988) Some effects of sleep loss on vestibular responses. Aviat Space Environ Med 59:523–529.

Collins WE, Guedry FE Jr (1961) Arousal effects and nystagmus during prolonged constant acceleration. Acta Otolaryngol (Stockh) 54:349–362.

Collins WE, Poe RH (1962) Amphetamine arousal and human vestibular nystagmus. J Pharmacol Exp Ther 138:120–125.

Cotter LA, Arendt HE, Jasko JG, Sprando C, Cass SP, Yates BJ (2001) Effects of postural changes and vestibular lesions on diaphragm and vectus abdominis activity in awake cats. J Appl Physiol 91:137–144.

Cyon É de (1911) L'oreille: Organe d'Orientation dans le Temps et dans l'Espace. Paris: Librairie Félix Alcan.

Delius JD, Vollrath FW (1973) Rotation compensating reflexes independent of the labyrinth: neurosensory correlates in pigeons. J Comp Physiol 83:123–134.

DeSantis M, Gernandt BE (1971) Effect of vestibular stimulation on pupillary size. Exp Neurol 30:66–77.

De Troyer A (1989) The mechanism of the inspiratory expansion of the rib cage. J Lab Clin Med 114:97–104.

Dibona GF, Kopp UC (1997) Neural control of renal function. Physiol Rev 77: 75–197.

Ding YQ, Takada M, Shigemoto R, Mizuno N (1995) Spinoparabrachial tract neurons showing substance P receptor-like immunoreactivity in the lumbar spinal cord of the rat. Brain Res 674:336–340.

Doba N, Reis DJ (1974) Role of cerebellum and vestibular apparatus in regulation of orthostatic reflexes in the cat. Circ Res 34:9–18.

Dodge R (1922) Habituation to rotation. J Exp Psychol 6:1–35.

Domyancic AV, Morilak DA (1997) Distributution of α_{1A}-adrenergic receptor mRNA in the rat brain visualized by in situ hybridization. J Comp Neurol 386:358–378.

Dow RS, Moruzzi G (1958) The Physiology and Pathology of the Cerebellum. Minneapolis: University of Minnesota Press.

Dowd PJ, Moore EW, Cramer RL (1975) Relationships of fatigue and motion sickness to vestibulo-ocular responses to Coriolis stimulation. Hum Factors 17:98–105.

Doyle DJ, Mark PWS (1990) Reflex bradycardia during surgery. Can J Anaesth 37:219–222.

Drance SM (1962) Studies in the susceptibility of the eye to raised intraocular pressure. Arch Ophthalmol 68:478–485.

Ezure K (1990) Synaptic connections between medullary respiratory neurons and considerations on the generation of respiratory rhythm. Prog Neurobiol 35: 429–450.

Ezure K (1996) Respiratory control. In: Yates BJ, Miller AD (eds) Vestibular Autonomic Regulation. Boca Raton: CRC Press, pp. 53–84.

Faugier-Grimaud S, Ventre J (1989) Anatomic connections of inferior parietal cortex (area 7) with subcortical structures related to vestibulo-ocular function in a monkey (*Macaca fascicularis*). J Comp Neurol 280:1–14.

Fay RA, Norgren R (1997) Identification of rat brainstem multisynaptic connections to the oral motor nuclei using pseudorabies virus. III. Lingual muscle motor systems. Brain Res Rev 25:291–311.

Feil K, Herbert H (1995) Topographic organization of spinal and trigeminal somatosensory pathways to the rat parabrachial and Kölliker–Fuse nuclei. J Comp Neurol 353:506–528.

Feldman JL (1986) Neurophysiology of breathing in mammals. In: Bloom FE (ed) Handbook of Physiology. The Nervous System. IV. Intrinsic Regulatory Systems of the Brain. Bethesda, MD: American Physiological Society, pp. 463–524.

Foote SL, Berridge CW, Adams LM, Pineda JA (1991) Electrophysiological evidence for the involvement of the locus coeruleus in alerting, orienting and attending. In: Barnes CD, Pompeiano O (eds) Neurobiology of the Locus Coeruleus. Amsterdam: Elsevier, pp. 521–532.

Friberg TR, Weinreb RN (1985) Ocular manifestations of gravity inversion. JAMA 253:1755–1757.

Froese AB, Bryan AC (1974) Effects of anesthesia and diaphragmatic mechanics in man. Anesthesiology 41:242–255.

Fuchs A, Kornhuber HH (1969) Extraocular muscle afferents to the cerebellum of the cat. J Physiol 200:713–722.

Fulweiler CE, Saper C (1984) Subnuclear organization of the efferent connections of the parabrachial nucleus in the rat. Brain Res Rev 7:229–259.

Fung SJ, Reddy RM, Barnes CD (1987) Differential labeling of the vestibular complex following unilateral injections of horseradish peroxidase into the cat and rat locus coeruleus. Brain Res 401:347–352.

Furman JM, Jacob RG (2001). A clinical taxonomy of dizziness and anxiety in the otoneurologic setting. J Anxiety Disord 15:9–26.

Furman JM, O'Leary DM, Wolfe JW (1981) Changes in the horizontal vestibulo-ocular reflex of the rhesus monkey with behavioral and pharmacologic alerting. Brain Res 206:490–494.

Furman JM, Jacob RG, Redfern MS (1998) Clinical evidence that the vestibular system participates in autonomic control. J Vestib Res 8:27–34.

Gallagher JP, Phelan KD, Shinnick-Gallagher P (1992) Modulation of excitatory transmission at the rat medial vestibular nucleus synapse. Ann N Y Acad Sci 656:630–644.

Gdowski GT, McCrea RA (2000) Neck proprioceptive inputs to primate vestibular nucleus neurons. Exp Brain Res 135:511–526.

Gerrits NM, Voogd J, Magras IN (1985) Vestibular nuclear efferents to the nucleus raphe pontis, the nucleus reticularis tegmenti pontis and the nucleus pontis in the cat. Neurosci Lett 54:357–362.

Giolli RA, Blanks RHI, Torigoe Y (1984) Pretectal and brain stem projections of the medial terminal nucleus of the accessory optic system of the rabbit and rat as studied by anterograde and retrograde neuronal tracing methods. J Comp Neurol 227:228–251.

Giolli RA, Blanks RHI, Torigoe Y, Williams DD (1985) Projections of the medial terminal accessory optic nucleus, ventral tegmental nuclei, and substantia nigra of rabbit and rat as studied by retrograde transport of horseradish peroxidase. J Comp Neurol 232:99–116.

Giolli RA, Torigoe Y, Blanks RHI, McDonald HM (1988) Projections of the dorsal and lateral terminal accessory optic nuclei and of the interstitial nucleus of the superior fasciculus (posterior fibers) in the rabbit and rat. J Comp Neurol 277: 608–620.

Goodhill V (1979) Ear Diseases, Deafness and Dizziness. Hagerstown, MD: Harper & Row.

Gorman JM, Liebowitz MR, Fyer AJ, Stein J (1989) A neuroanatomical hypothesis for panic disorder. Am J Psychiatry 146:148–161.

Gowers WR (1903) A Manual of Diseases of the Nervous System, Second edition. Philadelphia: P. Blakiston's Son & Co.

Grélot L, Miller AD (1996) Neural control of respiratory muscle activity during vomiting. In: Miller AD, Bianchi AL, Bishop BP (eds) Neural Control of the Respiratory Muscles. Boca Raton: CRC Press, pp. 239–248.

Griffith CR (1920) The organic effects of complete body rotation. J Exp Psychol 3:15–47.

Griffith CR (1924) A note on the persistence of the "practice effect" in rotation experiments. J Comp Psychol 4:137–149.

Grillner S, Nilsson J, Thorstensson (1978) Intra-abdomimal pressure changes during natural movements in man. Acta Physiol Scand 103:275–283.

Guedry, FE Jr (1974) Psychophysics of vestibular sensation. In: Kornhuber HH (ed) Handbook of Sensory Physiology: Vestibular System. Psychophysics, Applied Aspects, and General Interpretation. Berlin: Springer-Verlag, Volume VI, Chapter 2, pp. 4–190.

Guyton AC, Hall JE (1996) Textbook of Medical Physiology, Ninth edition. Philadelphia: Saunders.

Halberstadt AL, Balaban CD (2003) Organization of projections from the raphe nuclei to the vestibular nuclei in rats. Neuroscience 120:571–592.

Henn V, Baloh RW, Hepp K (1984) The sleep–wake transition in the oculomotor system. Exp Brain Res 54:166–176.

Henry RT, Connor JD, Balaban CD (1989) Nodulus–uvula depressor response: central GABA-mediated inhibition of α-adrenergic outflow. Am J Physiol 256: H1601–H1608.

Hokffer BJ, Mitra J, Snider RS (1972) Cerebellar influences on the cardiovascular system. In: Hockman CH (ed) Limbic System Mechanisms and Autonomic Function. Springfield, IL: Charles C. Thomas, pp. 91–112.

Holstege G (1989) Anatomical study of the final common pathway for vocalization in the cat. J Comp Neurol 284:242–252.

Hood JD, Pfaltz CR (1954) Observations upon the effects of repeated stimulation upon rotational and caloric nystagmus. J Physiol (Lond) 124:130–144.

Hosoba M, Bando T, Tsukahara N (1978) The cerebellar control of accommodation of the eye in the cat. Brain Res 153:495–505.

Huang Q, Zhou D, St John WM (1991) Vestibular and cerebellar modulation of expiratory motor activities in the cat. J Physiol (Lond) 436:385–404.

Huang Q, Zhou D, St John WM (1993) Cerebellar control of expiratory activities of medullary neurons and spinal nerves. J Appl Physiol 74:1934–1940.

Hubscher CH, Berkley KJ (1994) Responses of neurons in caudal solitary nucleus of female rats to stimulation of the vagina, uterine horn and colon. Brain Res 664:1–8.

Irwin JA (1881) The pathology of sea-sickness. Lancet 2:907–909.

Ishikawa T, Miyazawa T (1980) Sympathetic responses evoked by vestibular stimulation and their interactions with somato-sympathetic reflexes. J Auton Res 1:243–254.

Ito J, Honjo I (1990) Central fiber connections of the vestibulo-autonomic reflex arc in cats. Acta Otolaryngol (Stockh) 110:379–385.

Ito M (1984) The Cerebellum and Neural Control. New York: Raven Press.

Ito T, Sanda Y (1965) Location of receptors for righting reflexes acting upon the body in primates. Jpn J Physiol 15:235–242.

Ito Y, Gresty MA (1997) Subjective postural orientation and visual vertical during slow pitch tilt for the seated human subject. Aviat Space Environ Med 68:3–12.

Jacob RG, Furman JM, Balaban CD (1996a) Psychiatric aspects of vestibular disorders. In: Baloh RW, Halmagyi GM (eds) Handbook of Neurotology/Vestibular System: New York: Oxford University Press, pp. 509–528.

Jacob RG, Furman JM, Perel JM (1996b) Panic, phobia, and vestibular dysfunction. In: Yates BJ, Miller AD (eds) Vestibular Autonomic Regulation. Boca Raton: CRC Press, pp. 197–227.

Jacobs BL, Azmitia E (1992) Structure and function of the brain serotonin system. Physiol Rev 72:165–229.

Jacobs BL, Fornal CA (1993) 5-HT and motor control: a hypothesis. Trends Neurosci 16:346–352.

Jasmin L, Burkey AR, Card JP, Basbaum AI (1997) Transneuronal labeling of a nociceptive pathway, the spino-(trigemino-)parabrachio-amygdaloid, in the rat. J Neurosci 17:3751–3765.

Jian BJ, Cotter LA, Emanuel BA, Cass SP, Yates BJ (1999) Effects of bilateral vestibular lesions on orthostatic tolerance in awake cats. J Appl Physiol 86: 1552–1560.

Jian BJ, Shintani T, Emanuel BA, Yates BJ (2002) Convergence of limb, visceral and vertical semicircular canal or otolith in puts onto vestibular nucleus neurons. Eyp Brain Res 144:247–257.

Johnston AR, Murnion B, McQueen DS, Dutia MB (1993) Excitation and inhibition of rat medial vestibular nucleus neurons by 5-hydroxytryptamine. Exp Brain Res 93:293–298.

Jones IH (1918) Equilibrium and Vertigo. Philadelphia: J.B. Lippincott Co.

Kalen P, Karlson M, Wiklund L (1985) Possible excitatory amino acid afferents to nucleus raphe dorsalis of the rat investigated with retrograde wheat germ agglutinin and [^{3}H]aspartate tracing. Brain Res 360:285–297.

Kandel ER, Schwartz JH, Jessell TM (1991) Principles of Neural Science, Third edition. New York: Elsevier.

Kano M, Kano M-S, Kusonoki M, Maekawa K (1990) Nature of optokinetic response and zonal organization of climbing fiber afferents in the vestibulocerebellum of the rabbit. II. The nodulus. Exp Brain Res 80:238–251.

Kasper J, Diefenhardt A, Mackert A, Thoden U (1992) The vestibulo-ocular response during transient arousal shifts in man. Acta Otolaryngol (Stockh) 112:1–6.

Kaufman MP, Forster HV (1996) Reflexes controlling circulatory, ventilatory and airway responses to exercise. In: Rowell L, Shepherd J (eds) Handbook of Physiology, Section 12, Exercise: Regulation and Integration of Multiple Systems. Bethesda, MD: American Physiological Society/Oxford University Press, pp. 381–447.

Kawano K, Sasaki M, Yamashita M (1980) Vestibular input to visual tracking neurons in the posterior parietal association cortex of the monkey. Neurosci Lett 17:55–60.

Kawasaki T, Sato Y (1981) Afferent projections to the caudal part of the dorsal nucleus of the raphe in cats. Brain Res 211:439–444.

Keane PE, Soubrié P (1997) Animal models of integrated serotonergic functions: their predictive value for the clinical applicability of drugs interfering with serotonergic transmission. In: Baumgarten HG, Gothert M (eds) Serotonergic neurons and 5-HT receptors in the CNS. Berlin: Springer, pp. 707–725.

Kerman IA, Yates BJ (1998) Regional and functional differences in the distribution of vestibular-sympathetic reflexes. Am J Physiol Reg Integ Comp Physiol 44:R828–R835.

Kevetter GA, Perachio AA (1989) Projections from the sacculus to the cochlear nuclei in the mongolian gerbil. Brain Behav Evol 34:193–200.

Kirsten EB, Sharma JA (1976) Characteristics and response differences to iontophoretically applied norepinephrine, D-amphetamine and acetylcholine on neurons in the medial and lateral vestibular nuclei of the cat. Brain Res 112:77–90.

Kishimoto T, Sasa M, Takaori S (1991) Inhibition of lateral vestibular nucleus neurons by 5-hydroxytryptamine derived from the dorsal raphe nucleus. Brain Res 553:229–237.

Kolev OI, Altaparmakov IA (1996) Changes in the gastrointestinal electric pattern to motion sickness in susceptibles and insusceptibles during fasting. J Vestib Res 6:15–21.

Korner PI (1971) Integrative neural cardiovascular control. Physiol Rev 51:312–367.

Kotchabhakdi N, Walberg F (1978) Cerebellar afferent projections from the vestibular nuclei in the cat: an experimental study with the method of retrograde axonal transport of horseradish peroxidase. Exp Brain Res 31:591–604.

Kothe AC (1994) The effect of posture on intraocular pressure and pulsatile ocular blood flow in normal and glaucomatous eyes. Surv Ophthalmol Suppl 38:S191–S197.

Kothe AC, Lovasik JV (1988) Neural effects of body inversion: photopic oscillatory potentials. Curr Eye Res 7:1221–1229.

Krieglstein GK, Langham ME (1975) Influence of body position on the intraocular pressure of normal and glaucomatous eyes. Ophthalmologica (Basel) 171:132–145.

Kummer W, Fischer A, Kurkowski R, Heym C (1992) The sensory and sympathetic innervation of the guinea-pig lung and trachea as studied by retrograde neuronal tracing and double-labelling immunohistochemistry. Neuroscience 49:715–737.

Kuo DC, DeGroat WC (1985) Primary afferent projections of the major splanchnic nerve to the spinal cord and nucleus gracilis of the cat. J Comp Neurol 231:421–434.

Kuo DC, Nadelhaft I, Hisamitsu T, DeGroat WC (1983) Segmental distribution and central projections of renal afferent fibers in the cat studied by transganglionic transport of horseradish peroxidase. J Comp Neurol 216:162–174.

Kuo DC, Oravitz JJ, DeGroat WC (1984) Tracing of afferent and efferent pathways of the left inferior cardiac nerve of the cat using retrograde and transganglionic transport of horseradish peroxidase. Brain Res 321:111–118.

La Noce A, Bradley DJ, Goring MA, Spyer KM (1991) The influence of lobule IX of the cerebellar posterior vermis on the baroreceptor reflex in the decerebrate rabbit. J Auton Nerv Syst 32:31–36.

Lam AKC, Douthwaite WA (1997) Does the change of anterior chamber depth or/and episcleral venous pressure cause intraocular pressure change in postural variation? Optom Vis Sci 74:664–667.

Langham ME (1975) Vascular pathology of the ocular postural response. A pneumotonographic study. Trans Ophthalmol Soc U K 95:281–287.

Leigh RJ, Zee DS (1991) The Neurology of Eye Movements, second edition Contemporary Neurology Series No. 35. Philadelphia: Davis.

Leoni Lunensis D (1576) Ars medendi humanos, particuaresque; morbos à capite, usque; ad pedes. Bononiae: Apud Io. Rossium.

Licata F, LiVolsi G, Maugeri G, Ciranna L, Santangelo F (1993a) Serotonin-evoked modifications of the neuronal firing rate in the superior vestibular nucleus. Neuroscience 52:941–949.

Licata F, LiVolsi G, Maugeri G, Ciranna L, Santangelo F (1993b) Effects of noradrenaline on the firing rate of vestibular neurons. Neuroscience 53:149–158.

Licata F, LiVolsi G, Maugeri G, Ciranna L, Santangelo F (1993c) Effects of 5-hydroxytryptamine on the firing rates of neurons in the lateral vestibular nucleus of rats. Exp Brain Res 79:293–298.

Licata F, LiVolsi G, Maugeri G, Santangelo F (1995) Neuronal responses in vestibular nuclei to dorsal raphe electrical stimulation. J Vestib Res 5:137–145.

Linder BJ, Trick GL (1987) Simulation of spaceflight with whole-body head-down tilt: influence on intraocular pressure and retinocortical processing. Aviat Space Environ Med Suppl 58:A139–A142.

Linder BJ, Trick GL, Wolf ML (1988) Altering body position affects intraocular pressure and visual function. Invest Ophthalmol Vis Sci 29:1492–1497.

Lloyd TC (1983) Effect of inspiration on inferior caval blood flow in dogs. J Appl Physiol 55:1701–1708.

Loewy AD, Spyer KM (1990) Central Regulation of Autonomic Functions. New York: Oxford University Press.

Lovasik JV, Kothe AC (1989) Ocular refraction with body orientation. Aviat Space Environ Med 60:321–328.

Lynch JC, Mountcastle VB, Talbot WH, Yin TCT (1977) Parietal lobe mechanisms for directed visual attention. J Neurophysiol 40:362–389.

Mameli O, Tolu E (1985) Visual input to the hypoglossal nucleus. Exp Neurol 90:341–349.

Mameli O, Tolu E (1986) Vestibular ampullary modulation of hypoglossal neurons. Physiol Behav 37:773–775.

Mameli O, Tolu E (1987) Hypoglossal responses to macular stimulation in the rabbit. Physiol Behav 39:273–275.

Mameli O, Tolu E, Melis F, Caria MA (1988) Labyrinthine projection to the hypoglossal nucleus. Brain Res Bull 20:83–88.

Manzoni D, Pompeiano O, Stampacchia G, Srivastava UC (1983) Responses of medullary reticulospinal cells to sinusoidal stimulation of labyrinth receptors in decerebrate cat. J Neurophysiol 50:1059–1079.

Manzoni D, Pompeiano O, Barnes CD, Stampacchia G, D'Ascanio P (1989a) Convergence and interaction of neck and macular vestibular inputs on locus coeruleus and subcoeruleus neurons. Pflügers Arch 413:580–598.

Manzoni D, Pompeiano O, Barnes CD, Stampacchia G, D'Ascanio P (1989b) Responses of locus coeruleus neurons to convergent neck and vestibular inputs. Acta Otolaryngol (Stockh) 468:129–135.

Markham CH, Estes MS, Blanks RHI (1973) Vestibular influences on ocular accommodation in cats. Int J Equilibrium Res 3:102–115.

Markham CH, Diamond SG, Simpson NE (1977) Ocular accommodative changes in humans induced by positional changes with respect to gravity. Electroencephalog Clin Neurophysiol 42:332–340.

Mayo H (1837) Outlines of Human Physiology, Fourth edition. London: Henry Renshaw and J. Churchill.

McKelvey-Briggs DK, Saint-Cyr JA, Spence SJ, Partlow GD (1989) A reinvestigation of the spinovestibular projection in the cat using axonal transport techniques. Anat Embryol 180:281–291.

Megirian D (1968) Vestibular control of laryngeal and phrenic motoneurons of cat. Arch Ital Biol 106:333–342.

Megirian D, Manning JW (1967) Input–output relations in the vestibular system. Arch Ital Biol 105:15–30.

Merrill EG (1974) Finding a respiratory function for the medullary respiratory neurons. In: Bellairs R, Gray EG (eds) Essays on the Nervous System. Oxford: Clarendon, pp. 451–486.

Miller AD, Yamaguchi T, Siniaia MS, Yates BJ (1995) Ventral respiratory group bulbospinal inspiratory neurons participate in vestibular-respiratory reflexes. J Neurophysiol 73:1303–1307.

Miller AD, Nonaka S, Jakus J, Yates BJ (1996) Modulation of vomiting by the medullary midline. Brain Res 737:51–58.

Mittelstaedt H (1996) Somatic graviception. Biol Psychol 42:53–74.

Mittelstaedt H, Glasauer S (1993) Illusions of verticality in weightlessness. Clin Invest 71:732–739.

Mixter G (1953) Respiratory augmentation of inferior venal caval flow demonstrated by a low-resistance phasic flowmeter. Am J Physiol 172:446–456.

Miyazaki T, Yoshida YHMST, Kanaseki T (1981) Central location of the motoneurons supplying the thyrohyoid and the geniohyoid muscles as demonstrated by horseradish peroxidase method. Brain Res 219:423–427.

Moore RY, Bloom FE (1979) Central catecholamine neuron systems: anatomy and physiology of the norepinephrine and epinephrine systems. Annu Rev Neurosci 2:113–168.

Moreno AH, Burchell AR, Burke JH (1967) Respiratory regulation of splanchnic and systemic venous return. Am J Physiol 213:455–465.

Moruzzi G (1938) Action inhibitrice du paléocervelet sur les réflexes circulatoires et respiratoires d'origine sino-carotidienne. C R Soc Belge Biol 128: 533–538.

Moruzzi G (1940) Paleocerebellar inhibition of vasomotor and respiratory carotid sinus reflexes. J Neurophysiol 3:20–32.

Moruzzi G (1950) Problems in Cerebellar Physiology. Springfield, IL: Charles C. Thomas.

Motter B, Mountcastle VB (1981) The functional properties of the light-sensitive neurons of the posterior parietal cortex studied in waking monkeys: foveal sparring and opponent vector organization. J Neurosci 1:3–26.

Nakazawa K, Zheng Y, Umezaki T, Miller AD (1997) Vestibular inputs to bulbar respiratory interneurons in the cat. Neuroreport 8:3395–3398.

Nelson JG (1968) Effect of water immersion and body position upon perception of the gravitational vertical. Aerosp Med 39:806–811.

Neuhuber WL, Zenker W (1989) Central distribution of cervical primary afferents in the rat, with emphasis on proprioceptive projections to vestibular, perihypoglossal, and upper thoracic spinal nuclei. J Comp Neurol 280:231–253.

Nicholas AP, Pieribone VA, Hökfelt T (1993a) Cellular localization of messenger RNA for beta-1 and beta-2 adrenergic receptors in rat brain: an *in situ* hybridization study. Neuroscience 56:1024–1039.

Nicholas AP, Pieribone VA, Hökfelt T (1993b) Distribution of mRNAs for alpha-2 adrenergic receptor subtypes in rat brain: an *in situ* hybridization study. J Comp Neurol 328:575–594.

Nisimaru N (1977) Depressant action of the posterior lobe of the cerebellum upon renal sympathetic nerve activity. Brain Res 133:371–375.

Nisimaru N, Watanabe Y (1985) A depressant area in the lateral nodulus–uvula of the cerebellum for renal sympathetic nerve activity and systemic blood pressure in the rabbit. Neurosci Res 3:177–181.

Nisimaru N, Yamamoto M, Shimoyama I (1984) Inhibitory effects of cerebellar cortical stimulation on sympathetic nerve activity in rabbits. Jpn J Physiol 34:539–551.

Nisimaru N, Okahara K, Nagao S (1991) Olivocerebellar projection to the cardiovascular zone of rabbit cerebellum. Neurosci Res 12:240–250.

Noda H, Fujikado T (1987) Topography of the oculomotor area of the cerebellar vermis in macaques as determined by microstimulation. J Neurophysiol 58: 359–378.

Okahara K, Nisimaru N (1991) Climbing fiber responses evoked in lobule VII of the posterior cerebellum from the vagal nerve in rabbits. Neurosci Res 12:232–239.

Paige GD, Tomko DL (1991a) Eye movement responses to linear head motion in the squirrel monkey. I. Basic characteristics. J Neurophysiol 65:1170–1182.

Paige GD, Tomko DL (1991b) Eye movement responses to linear head motion in the squirrel monkey. II. Visual–vestibular interactions and kinematic considerations. J Neurophysiol 65:1183–1196.

Paintal AS (1973) Vagal sensory receptors and their reflex effects. Physiol Rev 53:159–227.

Paton JFR, LaNoce A, Sykes RM, Sebastiani L, Bagnoli P, Gherlarducci B, Bradley DJ (1991) Efferent connections of lobule IX of the posterior cerebellar cortex in the rabbit—some functional considerations. J Auton Nerv Syst 36:209–224.

Pazos A, Palacios J (1985a) Quantitative autoradiographic mapping of serotonin receptors in the rat brain. I. Serotonin-1 receptors. Brain Res 346:205–230.

Pazos A, Palacios J (1985b) Quantitative autoradiographic mapping of serotonin receptors in the rat brain. II. Serotonin-2 receptors. Brain Res 346:231–249.

Person RJ (1989) Somatic and vagal afferent convergence on solitary tract neurons in cat: electrophysiological characteristics. Neuroscience 30:283–295.

Pfaller K, Arvidsson J (1988) Central distribution of trigeminal as upper cervical primary afferents in the rat studied by anterograde transport of horseradish peroxidate conjugated to wheat germ agglutinin. J Comp Nerol 268:91–108.

Pollack LL, Boshes B, Zivin I, Pyzik SW, Finkle JR, Tigay ELKBH, Arieff AJ, Finkelman I, Brown M, Dobin NB (1955) Body reflexes acting on the body. AMA Arch Neurol Psychiatry 74:527–533.

Pompeiano O, Manzoni D, Srivastava UC, Stampacchia G (1984) Convergence and interaction of neck and macular vestibular inputs on reticulospinal cells. Neuroscience 12:111–128.

Pompeiano O, Mazoni S, Marchand AR, Stampacchia G (1987) Effects of roll tilt and neck rotation on different size vestibulospinal neurons in decerebrate cats with the cerebellum intact. Pflügers Arch 409:24–38.

Porter JD, Balaban CD (1997) Connections between the vestibular nuclei and regions that mediate autonomic function in the rat. J Vestib Res 7:63–76.

Prabhakar NR, Marek W, Lowschcke HH (1985) Altered breathing pattern elicited by stimulation of abdominal visceral afferents. J Appl Physiol 58:1755–1760.

Ramu A, Bergmann F (1966) The role of the cerebellum in blood pressure regulation. Experientia 23:383–384.

Rasheed BMA, Manchanda SK, Anand BK (1970) Effects of stimulation of paleocerebellum on certain vegetative functions in the cat. Brain Res 20:293–308.

Revelette R, Reynolds S, Brown D, Taylor R (1992) Effect of abdominal compression on diaphragmatic tendon organ activity. J Appl Physiol 72:288–292.

Road JD (1990) Phrenic afferents and ventilatory control. Lung 168:137–149.

Rocco AG, Vandam LD (1954) Changes in circulation consequent to manipulation during abdominal surgery. JAMA 164:14–18.

Ross HE, Crickmar SD, Sills NV, Owen EP (1969) Orientation to the vertical in free divers. Aerosp Med 40:728–732.

Rossiter CD, Yates BJ (1996) Vestibular influences on hypoglossal nerve activity in the cat. Neurosci Lett 211:25–28.

Rossiter CD, Hayden NL, Stocker SD, Yates BJ (1996) Changes in outflow to respiratory muscles produced by natural vestibular stimulation. J Neurophysiol 76:3274–3284.

Ruggiero DA, Mtui EP, Otake K, Anwar M (1996) Vestibular afferents to the dorsal vagal complex: substrate for vestibulo-autonomic interactions in the rat. Brain Res 743:294–302.

Saint-Cyr JA, Woodward DJ (1980a) Activation of mossy and climbing fiber pathways to the cerebellar cortex by stimulation of the fornix in the rat. Exp Brain Res 40:1–12.

Saint-Cyr JA, Woodward DJ (1980b) A topographic analysis of limbic and somatic inputs to the cerebellar cortex in the rat. Exp Brain Res 40:13–22.

Sanborn GE, Friberg TR, Allen R (1987) Optic nerve dysfunction during gravity inversion. Arch Ophthalmol 105:774–776.

Sato A (1997) Neural mechanisms of autonomic responses elicited by somatic sensory information. Neurosci Behav Physiol 27:610–621.

Schöne H (1964) On the role of gravity in human spatial orientation. Aerosp Med 35:764–772.

Schor RH, Steinbacher BC Jr, Yates BJ (1998) Horizontal linear and angular responses of neurons in the medial vestibular nucleus of the decerebrate cat. J Vestib Res 8(1):107–116.

Schreihofer AM, Sved AF (1994) The use of sinoaortic denervation to study the role of baroreceptors in cardiovascular regulation–special communication. Am J Physiol 266:R1705–R1710.

Schuerger RJ, Balaban CD (1993) Noradrenergic projections to the vestibular nuclei in monkey. Soc Neurosci Abstr 19:136.

Schuerger RJ, Balaban CD (1999) Organization of the coeruleo-vestibular pathway in rats, rabbits and monkeys. Brain Res Rev 30:189–217.

Scott AB, Morris A (1967) Visual field changes produced by artificially elevated intraocular pressure. Am J Ophthalmol 63:308–312.

Searles RV, Balaban CD, Severs WB (1996) Interaction between head-down tilt and anterior chamber infusions on intraocular pressure of anesthetized rats. Exp Eye Res 62:621–626.

Sebastiani L, La Noce A, Paton JFR, Gherlarducci B (1992) Influence of the cerebellar posterior vermis on the acquisition of the classically conditioned bradycardic response in the rabbit. Exp Brain Res 88:193–198.

Seltzer JL, Ritter DE, Starsnic MA, Marr AT (1985) The hemodynamic response to traction on the abdominal mesentery. Anesthesiology 63:96–99.

Shiba K, Siniaia MS, Miller AD (1996) Role of ventral respiratory group bulbospinal expiratory neurons in vestibular-respiratory reflexes. J Neurophysiol 76:2271–2279.

Shinoda Y, Ohgaki T, Futami T (1986) The morphology of single lateral vestibulospinal tract axons in the lower cervical spinal cord of the cat. J Comp Neurol 249:226–241.

Shinoda Y, Ohgaki T, Sugiuchi Y, Futami T (1989) Comparison of the branching patterns of lateral and medial vestibulospinal tract axons in the cervical spinal cord. Prog Brain Res 80:137–147.

Siegel RE (1976) Galen On the Affected Parts. Translation of De locis affectus. Basel: S. Karger, pp. 98–99.

Siniaia MS, Miller AD (1996) Vestibular effects on upper airway musculature. Brain Res 736:160–164.

Smith BH (1953) Nature and treatment of the celiac-plexus reflex in man. Lancet 2:223–227.

Sokolov YN (1963) Perception and the Conditioned Reflex. New York: Pergamon Press/MacMillan Company.

Somana R, Walberg F (1979) The cerebellar projection from the parabrachial nucleus in the cat. Brain Res 172:144–149.

Spiegel EA (1936) Respiratory reactions upon vertical movements. Am J Physiol 117:349–354.

Spiegel EA (1946) Effect of labyrinthine reflexes on the vegetative nervous system. Arch Otolaryngol 44:61–72.

Spiegel EA, Démétriades TD (1922) Beiträge zum Studium des vegetativen Nervensystems. III. Metteilung. Der einfluß des Vestibularapparates auf das Gefäßsystem. Pflügers Arch 196:185–199.

Spiegel EA, Démétriades TD (1924) Beiträge zum Studium des vegetativen Nervensystems. VII. Metteilung. Der zentrale Mechanismus der vestibulären Blutdrucksenkung und ihre Bedeutung für die Entstehung des Labyrinthschwindels. Pflügers Arch 205:328–337.

Spiegel EA, Sommer I (1944) Vestibular mechanisms. In: Glasser O (ed) Medical Physics. Chicago: Year Book Publishers, Volume 1, pp. 1638–1653.

Steinbacher BC, Yates BJ (1996a) Brainstem neurons necessary for vestibular influences on sympathetic outflow. Brain Res 720:204–210.

Steinbacher BC, Yates BJ (1996b) Processing of vestibular and other inputs by the caudal ventrolateral reticular formation. Am J Physiol Regul Integ Comp Physiol 40:R1070–R1077.

Steinbusch HWM (1991) Distribution of histaminergic neurons and fibers in rat brain. Comparison with noradrenergic and serotonergic innervation of the vestibular system. Acta Otolaryngol (Stockh) Suppl 479:12–23.

Stewart TG (1898) Lectures on Giddiness and Hysteria in the Male, second edition. Edinburgh and London: Young J. Pentland.

Stocker SD, Steinbacher BC, Balaban CD, Yates BJ (1997) Connections of the caudal ventrolateral medullary reticular formation in the cat brainstem. Exp Brain Res 116:270–282.

Stürup G, Bolton B, Williams DJ, Carmichael EA (1935) Vasomotor responses in hemiplegic patients. Brain 58:456–469.

Supple WF Jr, Kapp BS (1994) Anatomical and physiological relationships between the anterior cerebellar vermis and the pontine parabrachial nucleus in the rabbit. Brain Res Bull 33:561–574.

Supple WF Jr, Leaton RN (1990) Cerebellar vermis: essential for classically conditioned bradycardia in the rat. Brain Res 509:17–23.

Takata M, Robotham JL (1992) Effects of diaphragm descent on inferior vena caval venous return. J Appl Physiol 72:597–607.

Tan J, Gerrits NM, Nanhoe R, Simpson JI, Voogd J (1995) Zonal organization of the climbing fiber projection to the flocculus and nodulus of the rabbit: a combined axonal tracing and acetylcholinesterase histochemistry tracing study. J Comp Neurol 356:23–50.

Tang PC, Gernandt BE (1969) Autonomic responses to vestibular stimulation. Exp Neurol 24:558–578.

Thompson DG, Richelson E, Malagelada J-R (1982) Perturbation of gastric emptying and duodenal motility through the central nervous system. Gastroenterology 83:1200–1206.

Tong G, Robertson LT, Brons J (1993) Climbing fiber representation of the renal afferent nerve in the vermal cortex of the cat cerebellum. Brain Res 601:65–75.

Uchino Y, Kudo N, Tsuda K, Iwamura Y (1970) Vestibular inhibition of sympathetic nerve activities. Brain Res 22:195–206.

Uemura M, Matsuda K, Kume M, Takeuchi Y, Matsushima R, Mizuno N (1979) Topographical arrangement of hypoglossal motoneurons: an HRP study in the cat. Neurosci Lett 13:99–104.

Umezaki T, Zheng Y, Shiba K, Miller AD (1997) Role of nucleus retroambigualis in respiratory reflexes evoked by superior laryngeal and vestibular nerve afferents in emesis. Brain Res 769:347–356.

Ventre J, Faugier-Grimaud S (1988) Projections of the temporo-parietal cortex on vestibular complex in the macaque monkey (*Macaca fascicularis*). Exp Brain Res 72:653–658.

Waldrop TG, Eldridge FL, Iwamoto GA, Mitchell JH (1996) Central neural control of respiration and circulation during exercise. In: Rowell LB, Shepherd JT (eds) Handbook of Physiology, Section 12: Exercise: Regulation and Integration of Multiple Systems. New York: American Physiological Society/Oxford University Press, pp. 333–380.

Wanaka A, Kiyama H, Murakami T, Matsumoto M, Kamada T, Malbon CC, Tohyama M (1989) Immunohistochemical localization of β-adrenergic receptors in rat brain. Brain Res 485:125–140.

West JB (1977) Regional Differences in the Lung. New York: Academic Press.

West JB, Elliott AR, Guy HJB, Prisk GK (1997) Pulmonary function in space. JAMA 277:1957–1961.

Wiggers K (1943) The influence of the cerebellum on the heart and circulation of the blood. II. Arch Physiol (Neerl) 27:301–303.

Willis T (1692) The London Practice of Physick, being the Practical Part of Physick Contain'd in the Works of the Famous Dr. Willis. London: T. Basset, T. Dring, C. Harper and W. Crook.

Wilson VJ, Melvill Jones G (1979) Mammalian Vestibular Physiology. New York: Plenum.

Wilson VJ, Yamagata Y, Yates BJ, Schor RH, Nonaka S (1990) Response of vestibular neurons to head rotations in vertical planes. III. Responses of vestibulocollic neurons to vestibular and neck stimulation. J Neurophysiol 64:1695–1703.

Wolfe JW, Brown JH (1969) Effects of sleep deprivation on the vestibulo-ocular reflex. Aerosp Med 39:947–949.

Woodring SF, Yates BJ (1997) Responses of ventral respiratory group neurons of the cat to natural vestibular stimulation. Am J Physiol Regul Integr Comp Physiol 42:R1946–R1956.

Woodring SF, Rossiter CD, Yates BJ (1997) Pressor response elicited by nose-up vestibular stimulation in cats. Exp Brain Res 113:165–168.

Woodworth RS, Schlosberg H (1954) Experimental Psychology (Revised edition). New York: Henry Holt and Company.

Yamamoto C (1967) Pharmacologic studies of norepinephrine, acetylcholine and related compounds on neurons in Deiters' nucleus and the cerebellum. J Pharmacol Exp Ther 156:39–47.

Yates BJ, Miller AD (1994) Properties of sympathetic reflexes elicited by natural vestibular stimulation: implications for cardiovascular control. J Neurophysiol 71:2087–2092.

Yates BJ, Mickle JP, Hedden WJ, Thompson FJ (1987) Tracing of afferent pathways from the femoral-saphenous vein to dorsal root ganglia using transport of horseradish peroxidase. J Auton Nerv Syst 20:1–11.

Yates BJ, Yamagata Y, Bolton PS (1991) The ventrolateral medulla of the cat mediates vestibulosympathetic reflexes. Brain Res 552:265–272.

Yates BJ, Goto T, Bolton PS (1993a) Responses of neurons in rostral ventrolateral medulla of the cat to natural stimulation. Brain Res 601:255–264.

Yates BJ, Goto T, Kerman I, Bolton PS (1993b) Responses of caudal medullary raphe neurons to natural vestibular stimulation. J Neurophysiol 70:938–946.

Yates BJ, Jakus J, Miller AD (1993c) Vestibular effects on respiratory outflow in the decerebrate cat. Brain Res 629:209–217.

Yates BJ, Grelot L, Kerman IA, Balaban CD, Jakus J, Miller AD (1994) The organization of vestibular inputs to nucleus tractus solitarius (NTS) and adjacent structures in the cat brainstem. Am J Physiol 267:R974–R983.

Yates BJ, Balaban CD, Miller AD, Endo K, Yamaguchi Y (1995a) Vestibular inputs to the lateral tegmental field of the cat: potential role in autonomic control. Brain Res 689:197–206.

Yates BJ, Siniaia MS, Miller AD (1995b) Descending pathways necessary for vestibular influences on sympathetic and inspiratory outflow. Am J Physiol Regul Integr Comp Physiol 37:R1381–R1385.

Yates BJ, Jian BJ, Cotter LA, Cass SP (2000) Responses of vestibular nucleus neurons to tilt following chronic bilateral removal of vestibular inputs. Exp Brain Res 130:151–158.

Youmans WB, Tjioe DT, Tong EY (1974) Control of involuntary activity of abdominal muscles. Am J Phys Med 53:57–74.

Zhang SP, Davis PJ, Carrive P, Bandler R (1992) Vocalization and marked pressor effect evoked from the region of nucleus retroambigualis in the caudal ventral medulla of the cat. Neurosci Lett 140:103–107.

Zheng Y, Umezaki T, Nakazawa K, Miller AD (1997) Role of pre-inspiratory neurons in vestibular and laryngeal reflexes and in swallowing and vomiting. Neurosci Lett 225:161–164.

8
Vestibulocollic Reflexes

Barry W. Peterson and Richard D. Boyle

1. Introduction

The vestibulocollic reflexes (VCRs) are a set of relatively short-latency (less than 100 ms), automatic responses of the neck muscles to activation of the receptors of the vestibular labyrinth. They are often supplemented by longer-latency voluntary responses to vestibular input. Although they are distinct from voluntary head movements, the VCRs may be modified by voluntary processes and by the context in which the vestibular input occurs. As will be described below, for instance, vestibular inputs to vestibulospinal neurons involved in the VCR are largely suppressed during voluntary head movements. The VCRs face a formidable challenge in stabilizing the massive head, which must be held in an unstable inverted pendulum posture under most conditions. They are assisted in this by the voluntary responses mentioned earlier, by proprioceptive mechanisms, including the stretch-reflexlike cervicocollic reflex, and by passive mechanical properties of the head–neck system. The second section of this chapter considers how these mechanisms work together to stabilize the head and hold it upright against the force of gravity. The third section will describe the anatomy of VCR pathways and properties of neurons within them.

2. Spatial and Dynamic Properties of Vestibulocollic Reflexes

2.1. Studies of the VCR in Reduced Preparations

Early studies of the VCR examined the modulation of neck muscle electromyographic (EMG) activity during whole-body rotations in decerebrate animals (Schor and Miller 1981; Bilotto et al. 1982; Baker et al. 1985; Dutia 1988; Banovetz et al. 1995). Figure 8.1A illustrates the modulation of EMG activity of the complexus muscle produced by sinusoidal rotations about a number of different axes. When activation was plotted against orientation

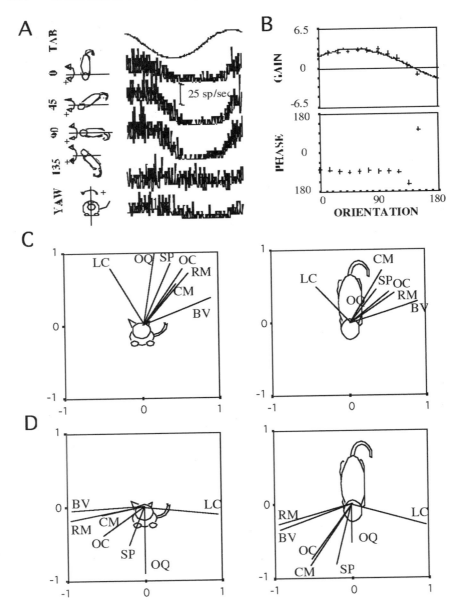

of the body with respect to the Earth-horizontal rotation axis (Fig. 8.1B), response sensitivity or gain varied as the cosine of the angle between the stimulus axis and a preferred direction, referred to as the muscle's maximal activation direction (MAD). The third component of this MAD vector was determined by the muscle's response to yaw rotation (bottom trace in Fig. 8.1A). The resultant vector was very similar to the mean vector for complexus (CM) in Figure 8.1C. Examination of the six mean muscle MAD vectors in Figure 8.1C reveals another challenge for the central vestibular pathways—converting the three nearly orthogonal semicircular canal activation signals into a diverse set of nonorthogonal motor commands. This is not done by simply activating each muscle maximally for motions in the plane of its maximum pulling direction (MPD). The MPDs of the same six muscles, plotted in Figure 8.1D, do not exhibit the mirror-image alignment with the MADs in Figure 8.1C that such a simple scheme requires.

When analyzed over a range of frequencies, the EMG modulation produced by the VCR exhibits complex dynamic behavior, which has been modeled as a second-order lead–lag system containing two pole-zero pairs (Boyle and Pompeiano 1979; Schor and Miller 1981; Bilotto et al. 1982). Recordings from second-order vestibular neurons suggested that one pole-zero pair represented the input from irregular vestibular afferents and the second pole-zero pair the CNS mechanisms operating between the vestibular nuclei and neck motor neurons (Wilson et al. 1979). Parallel studies of the cervicocollic reflex revealed similar second-order dynamics (Peterson et al. 1985). This similarity will allow the two reflexes to work together, providing an angular-position-related output at low frequencies and an angular-acceleration-related output at high frequencies appropriate to stabilize the inertially dominated head–neck system (Goldberg and Peterson 1986; Dutia 1988). Later studies showed that these reflex properties are also observed in alert animals with the head either fixed or free to move in

---

FIGURE 8.1. Activation of neck muscles by rotations in multiple directions. (A) Response of complexus muscle to rotations about four Earth-horizontal axes (orientation angle of 0° is pitch, 90° is roll) and about the Earth-vertical axis (Yaw). (B) Plot of gain and phase of EMG modulation as a function of rotation orientation angle. Complexus responds best at angles between 45° and 90°. (C) Average maximal activation direction vectors for six neck muscles (BV, biventer cerviciis; CM, complexus; LC, longus capitis; OC, occipitoscapularis; OQ, Obliquus; RM, rectus capitiis major; SP, splenius). (D) Maximal pulling directions of the same muscles (from Wickland et al. 1991). Note that MADs and MPDs do not exhibit the simple mirror-image alignment that would occur if each muscle were maximally activated by the rotation that evokes a VCR aligned with its pulling direction. For instance, LC is a pure head flexor (horizontal MPD in D) but is strongly activated during yaw rotations (upward angle of MAD in C). It may be activated in yaw to counter the extension forces produced by the dorsal neck muscles.

response to the applied rotation of the body (Goldberg and Peterson 1986; Lacour and Borel 1989; Keshner et al. 1992, 1997; Banovetz et al. 1995; Gdowski and McCrea 1999).

2.2. Studies of the Human VCR and Head–Neck System Models

There have also been a number of studies of the VCR in human subjects. One theme in these studies has been the influence of instructions given to the subject upon head-stabilizing behavior (Keshner and Peterson 1995; Keshner et al. 1995). When subjects are instructed to maintain their head stable in space, good stabilization is observed at lower frequencies between 0.1 and 1.0 Hz. Response latencies suggest that such stabilization is the result of voluntary mechanisms having large central delays. When subjects are distracted by performing a mental task, the voluntary contribution disappears and stabilization gain (the ratio of stabilizing rotation of the head on the trunk to the applied rotation of the trunk) falls to ~0.1 at low frequencies. At higher frequencies, behavior is similar in both conditions and appears to reflect the combined action of reflexes and head–neck mechanics. The plots in Figure 8.2 illustrate this behavior.

Another emphasis in human studies has been on the interaction of forces produced by neural activation of neck muscles with the passive properties of the head–neck motor plant. This interaction can best be understood by comparing data to behavior of a dynamical model of the system such as that illustrated in Figure 8.3 (Peterson et al. 2001). The model includes four neural control loops that converge on neck muscles, which are modeled as a low-pass torque converter. Muscle-generated torques combine with externally applied torques and with the torque produced by the head's inertial response to head acceleration with respect to space. The net torque then acts via the head–neck biomechanical plant to produce head rotations on the trunk. The VCR controller receives the head acceleration with respect to space signal (labeled 1 in Fig. 8.3) as its input and acts in negative feedback fashion on the plant. The cervicocollic reflex (CCR) controller receives signal 2—head acceleration with respect to trunk—and also acts in negative feedback. As shown by the inset, the combined action of these two negative feedback pathways (dashed traces) is to damp out head oscillations that would otherwise follow an externally applied rotation of the body in space (solid traces). They also produce a small static reduction in head excursion, but this is less important than the damping effect.

As shown by the solid lines in Figure 8.2, the model does a good job of predicting and explaining the data for yaw rotations (Peng et al. 1996). Biomechanical forces are simplest during such horizontal rotations about the axis of the erect spinal column, where the system can be treated as having a single rotating joint. The system is dominated by the inertia of the massive

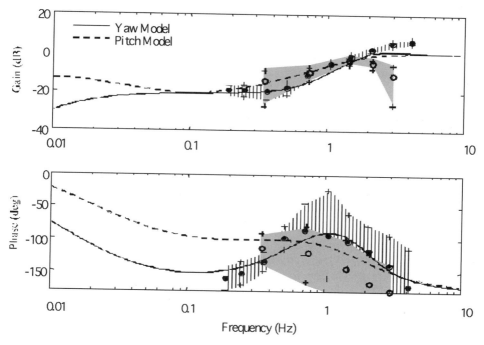

FIGURE 8.2. Dynamic properties of head movements evoked by yaw (filled circles with standard error indicated by vertical hatched area) and pitch (open circles with standard error indicated by gray shading) rotations while subjects performed a mental task. Solid and dashed lines are simulations of data by model of the type shown in Figure 8.3.

head and the springlike elastic properties of neck muscles. It therefore exhibits a second-order resonant behavior analogous to a ball oscillating up and down on a spring. At its resonant frequency, the head will rotate with an amplitude several times the amplitude of the applied body motion. At still higher frequencies, the head's inertia will cause it to remain stable in space, resulting in perfect stabilization. At very low frequencies, on the other hand, the spring force of the muscles will cause the head to move with the body, resulting in a stabilization gain near zero. Actual behavior (from experiments of Keshner et al. 1995) departs from this passive behavior in two ways. At low frequencies, there is the low-gain stabilization described above. More importantly, at frequencies near the resonance point, the head is much more stable than the passive model predicts. This appears to be primarily due to muscle activation produced by the VCR, which helps damp out the resonant oscillations that are a major sign of acute vestibular dysfunction. The cervicocollic reflex plays a subtler role, helping to damp excessive head movements at frequencies above the resonant point.

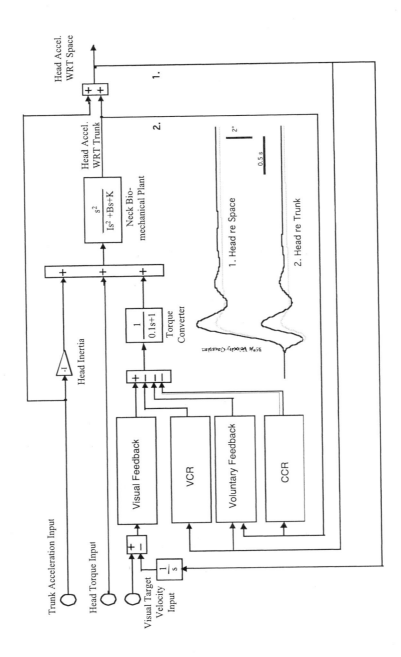

In its current version, the VCR/CCR model for pitch rotations is less accurate because it is more difficult to account for all of the forces acting about the two centers of rotation in this situation where gravitational forces and kinematic interaction torques come into play. Another problem is that relatively little is known about activation of neck muscles by the otolith organs. As described below, signals from these linear acceleration sensors converge on vestibulocollic neurons. Lacour et al. (1987) reported activation of neck muscles by vertical linear accelerations. Activation was in phase with downward acceleration and increased in amplitude with increasing stimulus frequency. During locomotion, downward and interaural (side-to-side) linear accelerations are accompanied by rotations of the head, which act to stabilize gaze (Imai et al. 2001). It is not clear, however, whether such motions are produced by a VCR as opposed to mechanical forces or longer-latency predictive motor commands.

A final issue is how the head stabilization system functions during active, voluntary head movements. If the VCR were to function unmodified, it would oppose such voluntary movements. It might seem ideal, therefore, to have no VCR during a voluntary movement. Yet observations in animals shortly after inactivation of semicircular canals by plugging reveals large head oscillations whenever an attempt is made to shift head position (Schor et al. 1984; Newlands et al. 1999). Clearly, sudden total loss of vestibular input is not desirable. Evidence will be presented below that feedforward voluntary signals null out the portion of vestibular neuron response that would be produced by a voluntary head movement, leaving the system able to respond to additional unexpected perturbations. Presumably, the latter are not only additional motions imposed from outside but hypermetric head motions induced by the voluntary activation of the neck musculature. If this is indeed the case, the VCR can be thought of as a follow-up dynamic stabilizing servo system that shapes head motions to correspond to the CNS' intentions.

FIGURE 8.3. Model of head movement system used to investigate the roles of VCR and CCR in head stabilization. The head responds to passive torques related to its inertial response to acceleration with respect to space and to directly applied forces and to active torques generated by the VCR, CCR, visual feedback system, and voluntary feedback system. During voluntary head movements, there is an additional input to the neck muscles, and a signal representing the predicted input from vestibular end organs is subtracted from the input driving the VCR (not shown). All torques act on the head via the neck biomechanical plant, which is modeled by the second-order transfer function shown. The inset shows model simulations of head motion in space (1) and head-on-trunk motions (2) in response to a 36°/s Gaussian velocity input. Darker traces show the response of the system without the VCR and CCR, lighter traces the response of the intact system.

3. Electrophysiology of Vestibulocollic Reflexes

3.1. Input–Output Studies of VCR

Accurate stabilization of the head requires mechanisms that respond both to angular accelerations sensed by semicircular canals and to linear accelerations sensed by otolith organs of the utriculus and sacculus. Recordings of electromyographic activity of neck muscles following electrical stimulation of these receptors have provided evidence for relatively direct vestibulocollic pathways from both types of receptors.

Most neck motoneurons show excitatory of inhibitory responses following electrical stimulation of all six individual semicircular canals, and four clear patterns of connectivity to the extensor, ventral and lateral flexor, and rotator muscles are evident (Wilson and Maeda, 1974; Shinoda et al. 1992b, 1994, 1996, 1997). The extensor muscles (rectus capitis posterior, cervical multifidus, biventer cervicis, and complexus) typically were excited by stimulation of the ipsilateral and contralateral anterior canal nerves (ACN) and of the contralateral horizontal canal nerve (HCN); they were inhibited by stimulation of the ipsilateral and contralateral posterior canal nerves (PCN) and the ipsilateral HCN. Ventral flexor motoneurons innervating the longus capitis muscle were excited by ipsilateral and contralateral PCN stimulation and by the contralateral HCN and inhibited in the opposite sense. Motoneurons of the obliquus capitis superior, splenius, and longissimus lateral flexor muscles and the sternocleidomastoid rotator muscles were excited by stimulating each of the three contralateral canal nerves and inhibited by stimulating each of the three ipsilateral canal nerves. Lastly, motoneurons of the obliquus capitis inferior rotator muscle were excited by stimulating the ipsilateral ACN, PCN, and contralateral HCN and inhibited by stimulation of the contralateral ACN, PCN, and ipsilateral HCN. These patterns of connectivity match the angular head movements elicited by stimulation of the same canal nerve (Suzuki and Cohen 1966).

Although less studied, powerful otolith inputs from the utriculus and sacculus reach the cervical circuits controlling head movement and posture via the medial and lateral vestibulospinal tracts (Bolton et al. 1992; Ikegami et al. 1994). Stimulating the utricular nerve evokes short-latency excitation of motoneurons innervating ipsilateral extensor and ventral flexor muscles and long-latency inhibition of the corresponding muscles on the other side. Stimulating the saccular nerve produces bilateral excitation of the motoneurons innervating the biventer cervicis and complexus extensor muscle and bilateral inhibition of motoneurons innervating the longus capitis muscle (Uchino 2001). In the alert squirrel monkey (*Saimiri sciureus*), identified second-order vestibulospinal neurons respond to translational accelerations with sensitivities up to 500 impulses/s/g and modulation tightly tuned to the stimulus direction (Boyle 1997). Despite the 360° range of directional selectivity of otolith afferents in each macula, these otolith–neck connections are organized to produce outputs consistent with the reflex responses evoked by translational accelerations (Lacour et al. 1987).

The convergence of both short- and long-latency inputs from selective stimulation of separate end organs onto vestibulospinal neurons in the cat has been investigated by Uchino and colleagues (see Uchino 2001 for a review). Approximately 30% of vestibulospinal neurons received convergent input from a semicircular canal (posterior or horizontal canal) and an otolith organ (saccule or utricle), with the vertical canals and saccule showing higher convergence than that from the horizontal canal and utricle. Of the convergent cells, about 50% received short-latency inputs from both; the remaining were longer latency (i.e., polysynaptic). The majority of the convergent cells are reported to travel in the lateral vestibulospinal tract. These findings suggest that the convergent vestibulospinal cells are particularly sensitive to head tilts and thus serve to stabilize the head when the head is inclined near the upright position. Nonconvergent cells comprise the majority (70%) of vestibulospinal neurons. Because neck motoneurons receive direct (inhibitory or excitatory) inputs from the different canals and otoliths, they do so largely through separate vestibulospinal pathways. Perlmutter et al. (1999) observed static tilt responses of vestibulospinal neurons that may indicate the presence of otolith inputs related to linear VCRs on these neurons.

3.2. Morphology of Vestibulospinal Neurons

The simplest VCR pathways are three-neuron arcs interconnecting vestibular-nerve afferents to cervical motoneurons through neurons of both the medial (MVST) and lateral (LVST) tracts (Wilson and Peterson 1988). MVST axons arise from neighboring neurons in the vestibular nuclei that project to regions of the ventral horn extending from the upper cervical segments to the cervicothoracic junction (Wilson and Peterson 1988; Boyle et al. 1992). LVST axons supply all levels of the spinal cord with a topographic arrangement of somata of origin according to level of descent (Wilson and Peterson 1988; Boyle et al. 1992). LVST axons that target the neck arise from neurons in the ventral regions of the lateral nucleus and adjacent rostroventral inferior nucleus. Sections that follow will describe the termination morphology of vestibulospinal axons in the cervical segments of the spinal cord to lay the structural basis of vestibulospinal control of head/neck posture and movement. We will then turn to the head movement signal content carried by the same class of secondary vestibulospinal neurons during the actual execution of the VCR and during self-generated, or active, rapid head movements.

3.2.1. Soma Location

Spinal-projecting vestibular neurons are conventionally typed by the location of the parent axon in the cervical white matter as medial (MVST) or lateral (LVST) vestibulospinal tract neurons and by the side and the site of termination in the spinal neuraxis. LVST neurons project ipsilaterally with respect to the cell body to their spinal target sites, and separate populations terminate in the cervical and lumbosacral segments of the spinal cord. MVST neurons innervate the cervical spinal segments, and separate populations project to either side of the spinal cord. Another class of vestibulospinal neuron exists: the vestibulooculocollic (VOC) neuron. Away from the soma, the axon of the VOC neuron projects rostromedially through the ipsilateral abducens nucleus and crosses the midline, where it bifurcates into an ascending and descending branch (Minor et al. 1990; Uchino and Hirai 1984); the descending branch typically has a larger diameter than its ascending counterpart. Ipsilaterally projecting VOC neurons have been described in the cat (Isu et al. 1991), but their counterpart in the primate has not yet been observed. The soma location of secondary neurons, identified as cervical- and lumbar-projecting LVST, ipsilaterally and contralaterally projecting MVST, VOC and vestibuloocular (VOR) neurons, has been mapped using orthodromic and antidromic stimulation and intracellular recording techniques in the squirrel monkey (*S. sciureus*) (Boyle et al. 1992). To a large extent, the results obtained in the squirrel monkey (*S. sciureus*) are consistent with those obtained in the rabbit (Akaike et al. 1973) and cat (Wilson et al. 1967; Rapoport et al. 1977a, 1977b; Akaike 1983). LVST neurons in the monkey are concentrated in lateral (LV) and descending vestibular (DV) nuclei and show a somatotopic organization. Those terminating in the cervical cord have their cell bodies in the ventral LV, the transition zone between the ventral LV and DV, and to a lesser extent the dorsal LV. Lumbar-projecting neurons are located more dorsally in the dorsal LV (dorsal Deiters' nucleus) and more caudally in the DV. MVST and VOC neurons lie intermingled together with cervical-projecting LVST and VOR neurons, primarily in the central region of vestibular nuclei in the ventrolateral MV, in the adjacent part of the ventral LV, and in the ventral LV–DV transition area at the level of the eighth nerve point of entry to the brain stem. The idealized representations of the soma location of an LVST (filled circle) and an MVST (open circle) neuron are given in the brain stem sketch of Figure 8.4A.

3.2.2. Axon Trajectory

The axon of an LVST neuron projects caudally through the brain stem, often coursing in fascicles below the cuneate nucleus (Fig. 8.4B, filled area) and enters the spinal cord from the lateral to the ventral funiculi. Most LVST axons enter the ventral–ventrolateral funiculus by the cervical enlargement. MVST axons take a more varied course through the caudal

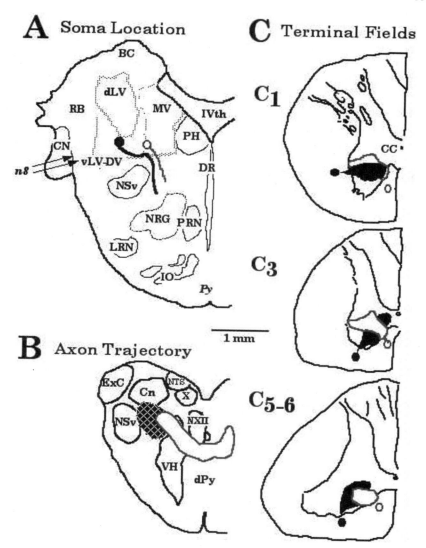

FIGURE 8.4. Schematic representation of secondary vestibulospinal morphology in a primate. (**A**) Soma location. Two idealized vestibulospinal neurons are shown, a cervical-projecting LVST (filled circle) and an MVST (open circle) neuron. The cell bodies of both neurons are intermingled in the ventromedial portions of the vestibular nuclei at the entry zone of the eighth nerve. (**B**) Axon trajectory. The averaged course through the caudal brain stem of axons of each neuron depicted in A is shown. MVST fibers project both ipsilaterally and contralaterally to the spinal cord. (**C**) Terminal fields. The averaged terminal field of each neuron is shown at three segments in the cervical cord.

brain stem to reach the ventromedial funiculus of the cervical cord, with considerable variation in the location at which the contralaterally projecting MVST neuron crosses the midline before the first cervical dorsal root (Fig. 8.4B, open area). Some MVST neurons follow the same course as VOC neurons, but, after crossing the midline, the axon turns caudally without issuing an ascending branch. Collateralization of MVST and VOC neurons in the ipsilateral vestibular nuclei is sparse (Boyle et al. 1992; Boyle and Moschovakis 1993).

3.2.3. Terminations in the Cervical Segments

Terminal synaptic fields of squirrel monkey (*S. sciureus*) MVST neurons (including ipsilaterally and contralaterally projecting MVST neurons and contralaterally projecting VOC neurons) have a characteristic distribution pattern in laminae VII, VIII, and IX in the cervical spinal cord. Individual MVST and VOC neurons may differ in the details, such as the axon course within and outside the ventromedial funiculus, the number of collaterals issued from the parent axon, the intercollateral distance, the cross-sectional area of the terminal field, and the density of boutons within each terminal field. However, the most striking variability observed among these neurons is the extent to which their axons either selectively terminate in specific cervical segments or issue relatively evenly spaced collaterals supplying each cervical segment along the entire length of the axon.

An example of a biocytin-labeled secondary MVST neuron that provided relatively uniform input to ventral horn targets is shown in the reconstruction of Figure 8.5. The cell's axon traveled in the contralateral ventromedial funiculus, issuing 19 collaterals from C1 to its terminus at C6–7. Nearly 3000 synaptic terminals were observed within the outline of the cell's terminal fields, principally the medial wall of lamina VIII (splenis motor pool) and the spinal accessory motor pools of the trapezius and sternocleidomastoid. The pattern of innervation of this cell is characteristic of other MVST neurons. Figure 8.4C shows an idealized representation of the averaged terminal field of MVST neurons (open fields) at three cervical segments.

Figure 8.6 shows the reconstructions in the horizontal plane of three neurons, an ipsilateral and contralateral MVST and a VOC neuron, and their innervation of the cervical segments; the three cells were individually labeled in the same experiment. The VOC neuron issued ten collaterals between C1 and C4, at which point the label faded. The pattern of innervation was similar to that of the MVST cell shown in Figure 8.5, except that the VOC cell projected more dorsally within lamina VII to target the central cervical nucleus. The ipsilateral MVST cell bypassed the upper cervical segments of C1 and C2 and issued its first collateral at C3, provided five separated terminal areas within C3–C4, skipped the next segment, and the label faded at C5. The contralateral MVST cell bypassed the entire upper cervi-

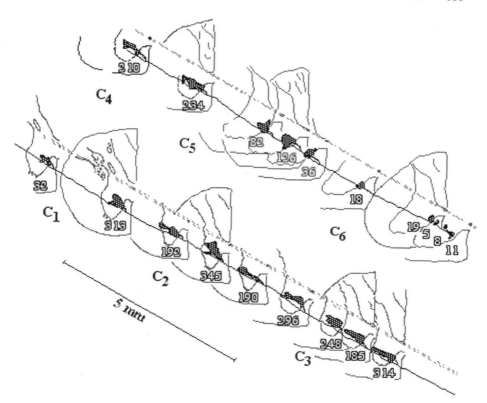

FIGURE 8.5. Reconstruction of the innervation of a contralateral MVST cell to the cervical spinal cord. The cell issued 19 intraspinal collaterals with a total of 2864 observed synaptic specializations, indicated separately for each terminal field, and supplied the medial wall of laminae VII and VIII and spinal accessory nuclei; collaterals were distributed along C1–C6 to the axon's terminus at C6.

cal segments from C1 to C5 and issued its first collateral mid-C5. The three axons preferentially targeted the medial wall of lamina VIII. Unlike MVST neurons that target the entire cervical spinal cord (example shown in Fig. 8.5), neurons such as those of Figure 8.6 preferentially innervate more selected regions, such as C1–C2 and C3–C4, or bypass the upper segments entirely.

Cervical-projecting LVST neurons also have a characteristic pattern of termination. Idealized representations of the averaged terminal field at three cervical segments (darkened fields) are shown in the sketches of Figure 8.4C. LVST neurons coursing in the lateral funiculus at C1 issue collaterals that innervate the spinal accessory nucleus (cranial motor XI), overlapping with the terminal fields of the MVST and VOC cells. LVST neurons

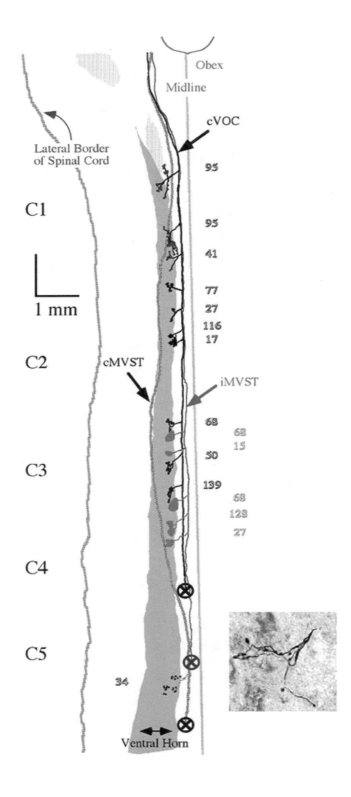

Obex

Midline

cVOC

Lateral Border
of Spinal Cord

95

C1

95

41

1 mm

77

27

116

17

C2

cMVST

iMVST

68

68

15

50

C3

139

68

128

27

C4

C5

34

Ventral Horn

coursing in the ventrolateral–ventral funiculus issue collaterals that innervate the ventromedial motoneurons of lamina IX and central and medial lamina VIII, including the medial wall, and extend into lamina VII and often into lamina X (see Fig. 8.7).

In a recent intracellular labeling study (Boyle 2000), it was found that secondary, lumbar-projecting LVST neurons bypass the cervical spinal cord and thus do not contribute to head/neck posture and stabilization. In essence, lumbar-projecting neurons provide a private, and mostly rapid, communication pathway between the dorsal Deiters' nucleus and the lower limb motor circuits.

3.3. Conclusions from Morphophysiological Studies

Morphological properties of MVST and LVST neurons suggest several general conclusions. The observation that some MVST neurons can innervate the cervical spinal cord along its entire length points to a more generalized, postural control of reflex head movement for these cells. This control would include a widely distributive excitatory input into motor circuits to maintain motoneuron excitability and consequently head/neck stability. Further, the multisegmental termination pattern of these cells suggests that they distribute and coordinate reflex head movements about multiple degrees of freedom. As suggested by Shinoda et al. (1992a), these neurons might activate muscles that "compose a functional synergy" of head movement. That other MVST neurons can target selective spinal segments, and bypass other segments, points to a more specific role in head movement. This subpopulation of vestibulospinal neurons could provide a pathway between canal-specific vestibular nerve afferents and motoneurons that initiate head movements about a limited number of vertebrae.

Evidence that some MVST neurons control broad synergistic groups of motoneurons while others target more specific subsets of motoneurons was provided by studies of Perlmutter et al. (1998a, 1998b), who combined

---

FIGURE 8.6. Collateral input patterns and termination fields of three vestibulospinal axons, individually identified as receiving a short-latency (monosynaptic) input from the eighth nerve and labeled with biocytin, in the cervical spinal cord of the squirrel monkey (S. sciureus). Reconstruction is shown in the horizontal plane from the obex to mid-C5; the labeling of each cell faded (marked by the X within a circle) before reaching the axon terminus. The cell labeled cVOC (contralateral vestibuloocular collic) issued ten collaterals from C1 to C4, with 725 combined synaptic boutons. The cell labeled iMVST (ipsilateral MvST) issued its first collateral at C3, with five separate terminal fields with 306 synaptic specializations. The cell labeled cMVST (contralateral MVST) bypassed the entire upper cervical segments and issued its first collateral at C5; note that the axon trajectory changed along its course. The photomicrograph shows the terminal specializations from the cell.

Vestibulospinal Tract Cells

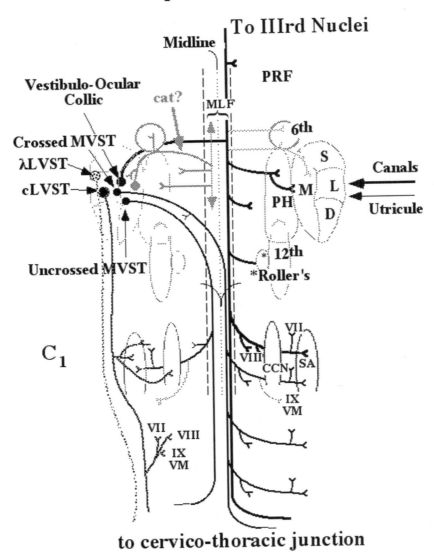

antidromic and morphological identification of the projection of MVST neurons with observation of their responses to rotation of the head. As shown in Figure 8.8, VOC neurons typically had their maximum activation directions (the direction of rotation that gave the greatest modulation of their discharge) aligned with the maximum activation direction of one of the three semicircular canals (dashed vectors). On the other hand, neurons projecting to the cervical spinal cord but not to the oculomotor nucleus (VC neurons) were more likely to have a maximum activation direction indicating input from more than one canal or from a contralateral canal. Furthermore, neurons with convergent canal input were found to have more restricted patterns of termination than those carrying a simple signal arising from a single canal pair. This leads to the conclusion that VOC neurons carry an unprocessed signal from a single canal pair to many motor pools,

◄——

FIGURE 8.7. Schematic representation of the major projections of crossed and uncrossed MVST VCR pathways drawn without respect to vestibular input specificity (Isu and Yokota 1983; McCrea et al. 1987a, 1987b; Isu et al. 1988, 1991; Shinoda et al. 1988a, 1988b, 1992a, 1992b; Minor et al. 1990; Boyle et al. 1992; Boyle and Moschovakis 1993). Some vestibulospinal axons may provide bilateral input to cervical cell groups (Shinoda et al. 1992a); they appear to be few and were not included in the figure. Primary sites of termination of both crossed and uncrossed MVST axons are (1) the medial wall of lamina VIII, which contains both dorsal neck motoneurons such as the splenius capitis (Richmond et al. 1978) and interneurons including commissural cells (Szentágothai 1951; Bolton et al. 1991); (2) lamina VII, which contains interneuron circuits including propriospinal neurons and scattered motoneurons supplying ventral as well as dorsal neck muscles; and (3) the lateral cell groups of lamina IX (labeled SA, spinal accessory motor pool) that include motoneurons of cranial nerve XI, which supply the sternocleidomastoid and trapezius muscles. A major feature of crossed axons is the high degree of collateralization in the cervical segments, which contrasts with the fewer collaterals issued from uncrossed MVST fibers. VOC neurons are particularly noteworthy. They resemble other crossed MVST fibers in the extent of their branching and terminations in the ventral horn except that some VOC cells also target the precerebellar Roller's nucleus and the central cervical nucleus, which receive neck proprioceptive inputs (Hongo et al. 1988). Crossed MVST cells provide fewer contacts to these proprioceptive nuclei. This finding indicates that signals carried by VOC cells are widely distributed to include extraocular motor nuclei, medially located cervical motoneurons, interneuronal segmental circuits, and likely propriospinal neurons as well as proprioceptive pathways to the cerebellum. Cervical LVST neurons (labeled cLVST) provide a dense input to the cervical ventral horn, often providing 20 or more collateral inputs along the cervical neuraxis. The primary sites of termination of cLVST are (1) the spinal accessory and lateral cell groups of lamina IX; (2) ventromedial motoneuron groups (vm IX) of lamina IX; (3) the medial wall of lamina VIII; and (4) partial innervation of lamina X and commissural cell groups. Lumbar LVST cells (λLVST) bypass the cervical segments.

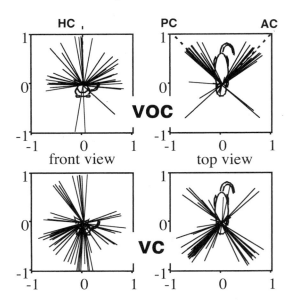

FIGURE 8.8. Maximum activation directions of MVST neurons. Unit vectors indicate the direction of rotation that maximally activated each neuron as determined by cosine fits to responses to rotations in multiple directions. The preferred direction of rotation can be visualized by the right-hand rule: when the right thumb is aligned with the vector, the direction that the fingers curl is the direction of rotation that would elicit the strongest response. Activation directions of the ipsilateral horizontal (HC), anterior (AC), and posterior (PC) canals are shown by dashed lines for comparison. Upper and lower panels show behavior of VOC neurons and neurons projecting only to the neck (VC).

where such inputs are combined in the proper ratio to produce the signal required by a particular muscle. VC neurons, on the other hand, carry a convergent signal that is preprocessed to match the requirements of a particular muscle or muscle group.

Viewed in the context of the different vestibular signals they carry, it is not surprising that individual vestibulospinal neurons target a wide range of possible segmental targets—one preferentially targets a segmental cell group bypassed by another vestibulospinal neuron. Based on morphological properties of vestibulospinal axons in the squirrel monkey (*S. sciureus*), two broad groups of cells emerge: one involved with postural control and synergistic movements of the head and neck and the other involved in generating reflex movements more compensatory to the external perturbation. VOC cells likely play a prominent action in coordinated reflex movements of both the eyes and the head. Interestingly, individual VOC neurons have morphological characteristics of both groups of MVST axons. The observation that the VOC cells also heavily target precerebellar nuclei suggests that the signal they carry might contribute a motor component to an extra-vestibular control pathway. Lastly, cervical-projecting LVST neurons provide both a separate and overlapping input to the ventral horn. These axons preferentially target the spinal accessory motor nuclei, as do MVST axons, and, in contrast to MVST cells, terminate extensively in ventrome-dial lamina IX.

3.4. Discharge Behavior of Vestibulospinal Neurons During Active and Passive Head Movements

To understand the control signals used by the central vestibular pathways in the generation of reflex head stabilization, such as the vestibulocollic reflex (VCR), and the maintenance of head posture, it is necessary to observe behavior of identified vestibulospinal neurons projecting to the cervical spinal segments in the alert animal. Secondary canal-related vestibu-lospinal neurons respond to an externally applied (passive) movement of the head in the form of a firing rate modulation that encodes the angular velocity of the movement and reflects in large part the input "head veloc-ity in space" signal carried by the semicircular canal afferents. In addition to this head velocity signal, vestibulospinal neurons can carry a more processed signal that includes eye position, eye velocity, or both (Boyle 1993). The head movement signal content carried by the same class of sec-ondary vestibulospinal neurons during the actual execution of the VCR and during self-generated, or active, rapid head movements was recorded in the alert squirrel monkey (*S. sciureus*) (Boyle et al. 1996; McCrea et al. 1999). To identify secondary vestibulospinal neurons, shocks were applied to wires implanted into the ventromedial funiculi at C1 to antidromically excite them and into the middle ear space to determine the same neuron's ortho-dromic relationship to the eighth nerve. In this study, the animal could vol-untarily make ±45° head movements about the vertical axis and ±10° head movements about the interaural axis while single-unit and eye movement recordings were acquired. The general conclusion is that vestibulospinal neurons do not encode the actual head velocity in space during self-generated head movements but instead encode the velocity of *externally applied* head movements not only as the animal is performing the VCR but also when the animal is executing rapid, orienting head movements.

Figure 8.9 shows the firing rate of a secondary vestibulospinal neuron during three paradigms: *passive* movement of the head in space (Fig. 8.9A) and passive movement of the head on shoulders (Fig. 8.9B), self-generated, *active* movements of the head on shoulders (Fig. 8.9C,D), and *combined* passive movement of the head in space and active movement of the head on shoulders (Fig. 8.9E–G). Figure 8.9A shows the standard test of whole-body rotation with the head restrained for vestibular input identification. The firing rate response (lower histogram) to a 1 Hz whole-body rotation

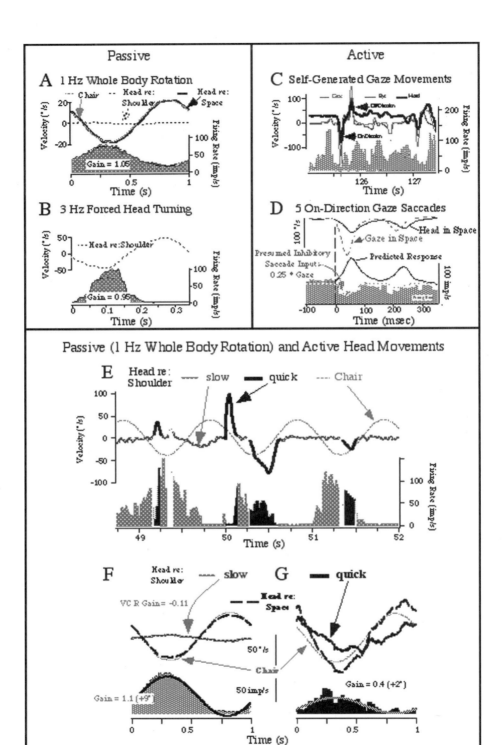

with the head restrained was averaged over 42 cycles of chair rotation (gray line, "Chair"). The other velocity traces are "Head re: Shoulder" (dashed line), representing the differentiated head position signal detected by the head coil, and the "Head re: Space" (solid and thick line), representing the sum of the chair velocity and the head velocity re: shoulder. In the head restrained condition, the "head re: shoulder" is near $0°$, and the "head re: space" closely corresponds to the applied chair velocity. The averaged

◄━━

FIGURE 8.9. Response of a secondary vestibulospinal neuron to passive or active and passive and active head movements. (**A**) Response to 1 Hz whole-body rotation; and (**B**) 3 Hz forced head turning. Both stimuli represent an external (passive) head perturbation. In A, the head is stationary with respect to the shoulders and the head in space is equivalent to the underlying body rotation; in B, the body is stationary and the head on shoulders is equivalent to the head in space. The cell's response is given in the form of a binned histogram in each panel, and the measured or calculated velocity traces are given in the upper curves (see key). (**C**) Cell firing rate during self-generated horizontal head movements in both the cell's on direction and off direction (with respect to the whole-body rotation response). Upper traces in panel C are the velocity records of horizontal eye (thin dashed line) and head (thick line) and their sum or gaze (solid and lighter line); in panel C, only the head and gaze traces are given (see key). The lower histogram in each panel is the neuron's firing rate over the indicated time segment. The record shows the firing behavior during an on-direction or leftward (downward or negative deflection of the velocity trace) gaze saccade followed by an off-direction or rightward (upward or positive deflection of the velocity trace) gaze saccade. (**D**) Averaged response for eight on-direction gaze saccades having similar amplitudes and time courses. Two models of the cell's firing rate are given: one based on the predicted response based on the cell's response to a 3 Hz whole-body rotation and the other modeled as an inhibitory input to the cell proportional to the gaze velocity. (**E–G**) Combined passive and active head movement records of the firing rate during 1 Hz sinusoidal chair rotation in the head-free paradigm. In E, the head velocity record is divided into two epochs, a slow one (stippled line) and a quick one (thick, solid line) (see the text for explanation), to illustrate the execution of the vestibulocollic reflex and the periods of rapid head movements. The cell's firing rate, given in the rate histograms, is shaded to correspond to the two epochs of head movement. F and G show the averaged cell responses to applied chair rotation. Head velocity records were divided into slow (F) and quick (G) epochs of head movement as shown in E. In F, the vestibulocollic reflex is counterrotating the head at a gain of about 0.1; as a result, the head velocity in space is correspondingly reduced by ~10%. The cell's response during the execution of the VCR corresponds closely to that obtained during whole-body rotation. In G, the cell's firing rate again was not correlated with the velocity of voluntary head movements. Importantly, the cell continued to respond, although at a reduced firing rate and sensitivity, to the applied whole-body (chair) rotation during the performance of active, rapid head movements. Also note that rapid head movements were more frequently made in the same direction as the chair rotation. (Modified from Boyle et al. 1996.)

response to applied chair rotation was characterized as having a sensitivity of 1.05 impulses/s per degree/s of chair or head re: space. In the head-free condition, however, when the animal makes a self-generated head movement, head velocity on shoulder and head velocity in space are identical in the absence of an external stimulus. This situation can be mimicked by forced turning of the head relative to a stationary trunk. Figure 8.9B shows the same cell's response to such a forced head turning at around 3 Hz. The neuron's response (gain −0.95) was similar to that during whole-body rotation. It thus responds in a comparable fashion to passive displacements of the head in space either *en bloc* or with respect to the body.

Self-generated head movements can reach velocities up to 350–400°/s for extremely rapid gaze saccades. However, normal exploratory gaze saccades are made in the range <100°/s. Figure 8.9C shows a secondary vestibulospinal neuron's firing rate (lower histograms) during self-generated horizontal eye (thin trace in upper curves) and head (thick solid trace) movements; the sum of eye in head and head on shoulder, or horizontal gaze velocity (gray solid trace), is also given. Without regard to the velocity traces, the cell's firing rate showed considerable variation with periods of increased or decreased activity. Linear regression analysis found no significant relationship between the cell's firing rate and the head velocity during self-generated head movements over the entire or limited ranges of head velocity. Figure 8.9D shows the same cell's firing behavior averaged over five comparable gaze saccades made in the cell's on direction (left and ipsilateral to the recorded cell). The upper curves are the horizontal gaze (lighter and dashed trace) and head (darker and solid trace) in space. The lower portion of the panel gives the cell's firing rate: histograms are the averaged binned results, the dark solid trace is the predicted response (again based on the cell's passive response to the head in space), and the striped trace is the presumed inhibitory input to the neuron during gaze saccades. This secondary vestibulospinal neuron responded to a passive head in space stimulus but not to head velocity in space when actively generated by the animal. This particular vestibulospinal neuron exhibited a decrease in discharge during voluntary head movements that might reflect an inhibitory input proportional to the gaze velocity. About half of the vestibulospinal neurons studied so far exhibited either a reduction or inhibition of the firing rate during a portion of the head movement. The source of this inhibition is as yet undetermined, but a local inhibitory gaze or head saccade neuron, such as the inhibitory burst neuron (Strassman et al. 1986), is a likely candidate. Discharge of the remaining secondary vestibulospinal neurons was completely unaffected by self-generated eye, head, or combined gaze movements. The main finding from the first two paradigms is that horizontal canal-related vestibulospinal neurons that receive monosynaptic excitation from vestibular nerve afferents do not faithfully encode velocity of the head in space during self-generated head saccades. This holds for head saccades of all amplitudes. Some show a decrease in discharge associated with a head saccade (Fig. 8.9C,D), whereas others remain unaffected.

One of the primary findings is that, although the secondary vestibulospinal neuron does not encode the velocity of the head in space for voluntary movements, the cell nevertheless can register the passive components of an applied perturbation of the head, even during large voluntary gaze saccades. To test the cell's ability to encode an external perturbation, whole-body rotations were delivered in the head-free paradigm. Here the animal could perform any number of behaviors, such as executing the vestibulocollic reflex, a compensatory response with head movements made in the direction opposite to the applied perturbation, or generate exploratory or other orienting head movements. Figure 8.9E–G shows the results obtained from one secondary vestibulospinal neuron. A brief ~3s segment of raw data during a 1 Hz applied chair rotation (solid and lighter trace in upper velocity curves) is given in Figure 8.9E. The head velocity re: shoulder, together with the corresponding firing rate, is parsed into two epochs: one is termed "slow" (lighter trace) and represents head movements that were either stationary or made within a window of an ideal sine wave that modeled the vestibulocollic reflex (mean −0.5 ± 5°/s), and the other is termed "quick" (darker trace) and represents rapid head movements. Lighter and darker filled histograms correspond to the slow and quick epochs of head velocity, respectively. Small discontinuities (open bins in the histogram) existed between the two epochs and were discarded. Figure 8.9F,G plot the cell's averaged firing rate (histograms), chair (solid and lighter trace), head re: shoulder (lighter trace in Fig. 8.9F, darker trace in Fig. 8.9G), and head re: space (dashed trace) velocity traces for the two separate epochs of head velocity; more than 150 cycles of rotation at 1 Hz were averaged for each plot. In Figure 8.9F, the head re: shoulder trace (marked "slow", thicker gray trace) shows the low-amplitude compensatory vestibulocollic reflex (head velocity/chair velocity, or gain = 0.11). During these periods of the record, the cell responded to applied chair rotation (1.1 impulses/s per degree/s and a phase lead of 9°) with sensitivity comparable to its whole-body response to passive rotation. Interestingly, vestibulospinal neurons do not respond equally during execution of the VCR: some reflect the head velocity in space input and have a corresponding reduced gain (with respect to the stimulus), whereas others have an unexpectedly higher gain (with respect to the head velocity in space input) and mirror the whole-body input (Gdowski et al. 2000). In Figure 8.9G, a reduced but persistent modulation related to the applied perturbation is still observed during epochs when the animal made quick head movements (sensitivity of 0.4 impulses/s per degree/s and a phase lead of 2°). Thus, although the cells' firing rate was independent of head velocity for self-generated head movements, the applied head velocity was still monitored by the cell.

The most striking finding is that the firing rate output of vestibulospinal neurons, also shown to receive a direct, short-latency synaptic input from eighth cranial nerve afferents, is dynamically modified by the behavioral context in which a movement is made. This, the central vestibular representation of movement is dynamically controlled by volitional head movements, beginning even at the initial stages of processing of afferent sensory signals in the brain stem. The head velocity in space signal produced by self-generated, active movements of the head is effectively canceled. As a result, vestibulospinal neurons selectively detect *external* perturbations of the head and translate only those passive components of the overall head movement into control signals to facilitate reflex behaviors and postural stability. To effectively and rapidly encode an applied head perturbation, a baseline firing rate of the vestibulospinal neuron is required. The results show that, even for very large gaze saccades, the vestibulospinal neuron's baseline firing rate is either unaffected or marginally decreased. Redundant signals created by voluntary head movement are canceled from the cell's firing, leaving the cell able to detect unexpected passive movement signals. The functional significance of this finding is clear: individuals can actively explore the environs using rapid, orienting head (and gaze) movements and maintain head and body stability in the event of an unexpected and externally induced head perturbation. In vestibular patients, the separation between active and passive components of a head displacement in space is impaired or lost. Viewed in this context, it is not surprising to see these patients use more *en bloc* adjustments of the head and body and minimize large head on shoulders orientation.

Figure 8.10 shows a simple conceptual model of the voluntary and reflex control of head movement. The VCR pathways are activated by external perturbations of the head. Premotor generators provide the command to make voluntary head movements. Following the reafference principle of von Holst and Mittelstaedt (1950), an "efference copy" of the command is constructed and sent to the VCR pathways to cancel the expected (or redundant) head acceleration signals transmitted by the labyrinth during self-generated head movement. Sensory reafference signals, arising from the neck muscles and joints (Boyle and Pompeiano 1981; Wilson et al. 1990), may also participate directly in extracting the active components of the head movement from the VCR pathways or indirectly by constructing a head posture reference on which both the active and passive head movements act. Viewed in this context, it is also not surprising to see why some patients suffering from cervical trauma mimic the behavior of vestibular patients.

3.5. Other Pathways Involved in the VCR

In addition to the direct VCR pathways described above, there are other direct and indirect pathways that influence head/neck stability and posture.

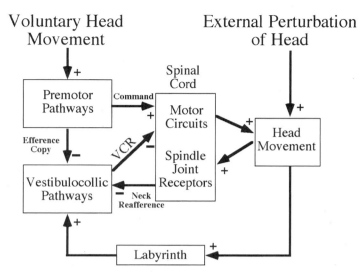

FIGURE 8.10. Simple model of active and passive contributions to the vestibulospinal control of reflex head movement and posture. The mechanism by which resposes to active components of a head movement are suppressed is as yet undefined but is modeled here as an efference copy signal, as postulated by von Holst and Mittelstaedt (1950), that effectively cancels the resultant input from the labyrinth onto the secondary vestibulospinal neuron, thus removing the redundant information and permitting throughput of external perturbations. The role of spinal afferents, such as those arising from muscle spindles and joints, which are activated by both active and passive head movements, remains unresolved.

One of these is the vestibuloreticulospinal pathway. Rapidly conducting reticulospinal neurons in the medial pontine and medullary brain stem excite motoneurons at all levels of the neuroaxis, with particularly strong actions on neck motoneurons, and direct inhibition of these motoneurons is also observed following activation of medullary reticulospinal pathways (Peterson et al. 1978). Work of Peterson et al. (1980), illustrated in Figure 8.11, revealed that such neurons may respond both to activation of the superior colliculus (thus participating in gaze shifts) and to sinusoidal modulation of activity of vestibular afferent fibers (thus participating in the VCR). Interestingly, discharge of reticulospinal neurons during the VCR includes both a head velocity and a head position component, as does the discharge of neck motoneurons. Vestibulospinal neurons, on the other hand, carry only the head velocity component (Wilson et al. 1979). The latter study also reported that transection of the medial longitudinal fasciculus (MLF), which contains most MVST fibers in the cat, fails to abolish the VCR or shift its phase. This suggests an important role for reticulospinal pathways that run outside the MLF. In their studies of reticulospinal neurons in the

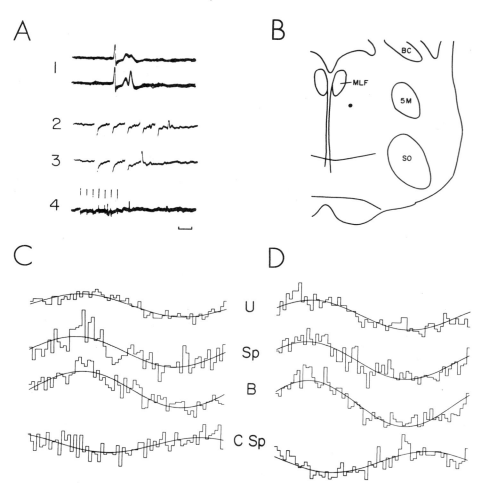

FIGURE 8.11. Reticulospinal neuron located in medial pons (dot in B). Neuron was antidromically activated from C1 (A1) and orthodromically activated when a train of shocks was applied to the motor cortex (A2–3) and superior colliculus (A4). It was also modulated (trace labeled U) when vestibular nerve fibers were activated sinusoidally at 0.5 Hz (C) and 2 Hz (D). Its activation was in phase with activation of the ipsilateral splenius (Sp) and biventer cervicis (B) muscles and out of phase with activation of the contralateral splenius (CSp) during this electrically evoked VCR.

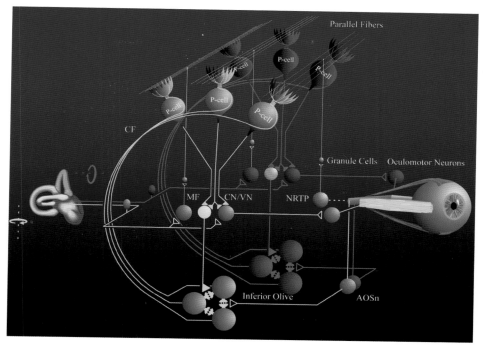

FIGURE 9.1. Schematic drawing of the vestibulocerebellar circuitry and its major afferent and efferent pathways that are involved in the control of the vestibulo-ocular reflex (VOR) and optokinetic reflex (OKR). The major pathway that mediates the VOR is the three-neuron arc of Lorente de Nó. This open loop is formed by the vestibular ganglion cells that receive input from the vestibular apparatus, the second-order vestibular neurons that are innervated by the primary afferents from these ganglion neurons, and the oculomotor neurons that innervate the oculomotor muscles. The major pathway that mediates the OKR is the accessory optic system (AOS). The AOS is embedded in a closed loop that is formed by the retinal ganglion cells; several mesencephalic nuclei (AOSn) such as the nucleus of the optic tract; the medial tegmental nucleus and the visual tegmental relay zone, which transmit optokinetic signals to the inferior olive; the vestibulocerebellum, which receives its climbing fibers (CF) from the inferior olive; and the complex of cerebellar and vestibular nuclei (CN/VN), which in turn innervate the oculomotor neurons. The granule cells in the vestibulocerebellar cortex receive mossy fiber (MF) signals that can carry vestibular, optokinetic, and/or eye movement signals. The Purkinje cells of the vestibulocerebellum control both the vestibular and optokinetic pathways by innervating the neurons in the CN/VN (dark yellow and red) that innervate the oculomotor neurons as well as the GABAergic neurons in the CN/VN (light yellow and red) that innervate the dendrites in the inferior olive that are coupled by gap junctions. Interneurons of the vestibulocerebellum are omitted for simplicity. P-cell and NRTP indicate Purkinje cell and nucleus reticularis tegmenti pontis, respectively.

alert cat, Kitama et al. (1995) reported that vestibular inputs were particularly strong on reticulospinal neurons that were not strongly activated during gaze shifts. Perhaps these neurons are responsible for the maintenance of VCR activity following transection of the MLF.

Another source of direct excitatory projections to neck motoneurons is the interstitial nucleus of Cajal (Fukushima et al. 1978), which provides terminal fields in the upper cervical segments that overlap those of vestibulospinal fibers (Kokkoroyannis et al. 1996). Lesions of this nucleus result in head tilts, but analysis of vestibular inputs to interstitiospinal neurons and interstitial neurons projecting to the vestibular nuclei suggests that any role of the interstitial nucleus in the VCR is likely mediated by its interconnections with the vestibular nuclei (Fukushima 1987). Other pathways that could carry signals related to the VCR are mesencephalic reticulospinal projections (Fukushima et al. 1981) and cerebellar fastigiospinal fibers (Batton et al. 1977; Wilson et al. 1978). Further studies are required to determine whether these neurons carry the appropriate signals to participate in the VCR.

4. Summary

This chapter examined the functional properties and neural substrates of a key vestibulospinal reflex—the vestibular-neck or vestibulocollic reflex (VCR). Functionally, the VCR is a dynamic stabilizing system that shapes head movements, particularly in the critical 1–3 Hz range where the head–neck system exhibits resonant instability. VCR circuits are designed to suppress reflex responses to active, intended head motions, leaving the reflex free to oppose unwanted oscillations or motions induced by external perturbations. A great deal is known about the anatomy and physiology of neurons that comprise the central leg of the important three-neuron VCR reflex arcs. These include both lateral and medial vestibulospinal tract neurons. The latter can be functionally divided into two classes. One carries unmodified signals from semicircular canals to a wide array of motoneurons. Included in this class are the branching vestibulooculocollic neurons. The second class selectively targets small groups of motoneurons and carries a processed canal signal that is presumably designed to control a small synergistic group of muscles. During active gaze shifts, many vestibulooculocollic neurons cease discharging. Vestibulocollic neurons, on the other hand, continue discharging and responding to externally applied head motions while ignoring the canal activation produced by the active head movements. The mechanisms responsible for this behavior are under active investigation. It also remains to be seen how these neurons participate in reflexes generated by activation of otolith receptors.

References

Akaike T (1983) Neuronal organization of the vestibulospinal system in the cat. Brain Res 259:217–227.

Akaike T, Fanardjian VV, Ito M, Kumada M, Nakajima H (1973) Electrophysiological analysis of the vestibulospinal reflex pathway of rabbit. I. Classification of tract cells. Exp Brain Res 17:477–496.

Baker J, Goldberg J, Peterson B (1985) Spatial and temporal response properties of the vestibulocollic reflex in decerebrate cats. J Neurophysiol 54:735–756.

Banovetz JM, Peterson BW, Baker JF (1995) Spatial coordination by descending vestibular signals. 1. Reflex excitation of neck muscles in alert and decerebrate cats. Exp Brain Res 105:345–362.

Batton RR III, Jayaraman A, Ruggiero D, Carpenter MB (1977) Fastigial efferent projections in the monkey, an autoradiographic study. J Comp Neurol 174: 281–305.

Bilotto G, Goldberg J, Peterson BW, Wilson VJ (1982) Dynamic properties of vestibular reflexes in the decerebrate cat. Exp Brain Res 47(3):343–352.

Bolton PS, Endo K, Goto T, Imagawa M, Sasaki M, Uchino Y, Wilson VJ (1992) Connections between utricular nerve and dorsal neck motoneurons of the decerebrate cat. J Neurophysiol 67(6):1695–1697.

Boyle R (1993) Activity of medial vestibulospinal tract cells during rotation and ocular movement in the alert squirrel monkey. J Neurophysiol 70:2176–2180.

Boyle R (1997) Activity of lateral vestibulospinal neurons during applied linear and angular head acceleration in the alert squirrel monkey. Soc Neurosci Abstr 23: 753.

Boyle R (2000) Morphology of lumbar-projecting lateral vestibulospinal neurons in the brainstem and cervical spinal cord in the squirrel monkey. Arch Ital Biol 138:107–122.

Boyle R, Pompeiano O (1979) Frequency response characteristics of vestibulospinal neurons during sinusoidal neck rotation. Brain Res 173:344–349.

Boyle R, Pompeiano O (1981) Convergence and interaction of neck and macular vestibular inputs on vestibulospinal neurons. J Neurophysiol 45:852–868.

Boyle R, Moschovakis AK (1993) Vestibular control of head movement in squirrel monkey: morphology of individual vestibulospinal axons. Soc Neurosci Abstr 19:138.

Boyle R, Goldberg JM, Highstein SM (1992) Inputs from regularly and irregularly discharging vestibular nerve afferents to secondary neurons in the vestibular nuclei of the squirrel monkey. III. Correlation with vestibulospinal and vestibuloocular output pathways. J Neurophysiol 68:471–484.

Boyle R, Belton T, McCrea RA (1996) Responses of identified vestibulospinal neurons to voluntary and reflex eye and head movements in the alert squirrel monkey. Am NY Acad Sci 781:244–263.

Dutia MB (1988) Interaction between vestibulocollic and cervicocollic reflexes: automatic compensation of reflex gain by muscle afferents. Prog Brain Res 76:173–180.

Fukushima K (1987) The intersitial nucleus of Cajal and its role in the control of movements of head and eyes. Prog Neurobiol 29:107–192.

Fukushima K, Peterson BW, Wilson VJ (1978) Vestibulospinal, reticulospinal and interstitiospinal pathways in the cat. Prog Brain Res 50:121–136.

Fukushima K, Ohno M, Kato M (1981) Responses of cat mesencephalic reticulospinal neurons to stimulation of superior colliculus, pericruciate cortex, and neck muscle afferents. Exp Brain Res 44:441–444.

Gdowski GT, McCrea RA (1999) Integration of vestibular and head movement signals in the vestibular nuclei during whole-body rotation. J Neurophysiol 82:436–449.

Gdowski G, Boyle R, McCrea RA (2000) Sensory processing in the vestibular nuclei during active head movements. Arch Ital Biol 138:15–28.

Goldberg J, Peterson BW (1986) Reflex and mechanical contributions to head stabilization in alert cats. J Neurophysiol 56:857–875.

Hongo T, Kitami T, Yoshida K (1988) Integration of vestibular and neck signals in the central cervical nucleus. Prog Brain Res 76:155–162.

Ikegami H, Sasaki M, Uchino Y (1994) Connections between utricular nerve and neck flexor motoneurons of decerebrate cats. Exp Brain Res 98:373–378.

Imai T, Moore ST, Raphan T, Cohen B (2001) Interaction of the body, head, and eyes during walking and turning. Exp Brain Res 136(1):1–18.

Isu N, Yokota J (1983) Morphophysiological study on the divergent projection of axon collaterals of medial vestibular neurons in the cat. Exp Brain Res 53:151–162.

Isu N, Uchino Y, Nakashima H, Satoh S, Ichikawa T, Watanabe S (1988) Axonal trajectories of posterior canal-activated secondary vestibular neurons and their coactivation of extraocular and neck flexor motoneurons in the cat. Exp Brain Res 70:181–191.

Isu N, Sakuma A, Hiranuma K, Uchino H, Sasaki S-I, Imagawa M, Uchino Y (1991) The neuronal organization of horizontal semicircular canal-activated inhibitory vestibulocollic neurons in the cat. Exp Brain Res 86:9–17.

Keshner EA, Peterson BW (1995) Mechanisms controlling human head stabilization. I. Head–neck dynamics during random rotations in the horizontal plane. J Neurophysiol 73:2293–2301.

Keshner EA, Baker JF, Banovetz J, Peterson BW (1992) Patterns of neck muscle activation in cats during reflex and voluntary head movements. Exp Brain Res 88:361–374.

Keshner EA, Cromwell RL, Peterson BW (1995) Mechanisms controlling human head stabilization. II. Head–neck characteristics during random rotations in the vertical plane. J Neurophysiol 73:2302–2312.

Keshner EA, Statler KD, Delp S (1997) Kinematics of the freely moving head and neck in the alert cat. Exp Brain Res 115:257–266.

Kitama T, Grantyn A, Berthoz A (1995) Orienting-related eye–neck neurons of the medial ponto-bulbar reticular formation do not participate in horizontal canal-dependent vestibular reflexes of alert cats. Brain Res Bull 38:337–347.

Kokkoroyannis T, Scudder CA, Balaban CD, Highstein SM, Moschovakis AK (1996) Anatomy and physiology of the primate intersitial nucleus of Cajal. I. Efferent projections. J Neurophysiol 75:725–739.

Lacour M, Borel L (1989) Functional coupling of the stabilizing gaze reflexes during vertical linear motion in the alert cat. Prog Brain Res 80:385–394.

Lacour M, Borel L, Barthelemy J, Harlay F, Xerri C (1987) Dynamic properties of the vertical otolith neck reflexes in the alert cat. Exp Brain Res 65:559–568.

McCrea RA, Strassman A, May E, Highstein SM (1987a) Anatomical and physiological characteristics of vestibular neurons mediating the horizontal vestibulo-ocular reflexes of the squirrel monkey. J Comp Neurol 264:547–570.

McCrea RA, Strassman A, Highstein SM (1987b) Anatomical and physiological characteristics of vestibular neurons mediating the vertical vestibulo-ocular reflexes of the squirrel monkey. J Comp Neurol 264:571–592.

McCrea RA, Gdowski G, Boyle R, Belton T (1999) Firing behavior of vestibular nucleus neurons during active and passive head movements. II. Vestibulo-spinal and other non-eye-movement related neurons. J Neurophysiol 82:416–428.

Minor LB, McCrea RA, Goldberg JM (1990) Dual projections of secondary vestibular axons in the medial longitudinal fasciculus to extraocular motor nuclei and the spinal cord of the squirrel monkey. Exp Brain Res 83:9–21.

Newlands SD, Ling L, Phillips JO, Siebold C, Duckert L, Fuchs AF (1999) Short- and long-term consequences of canal plugging on gaze shifts in the rhesus monkey. I. Effects on gaze stabilization. J Neurophysiol 81:2119–2130.

Peng GC, Hain TC, Peterson BW (1996) A dynamical model for reflex activated head movements in the horizontal plane. Biol Cybern 75:309–319.

Perlmutter SI, Iwamoto Y, Baker JF, Peterson BW (1998a) Interdependence of spatial properties and projection patterns of medial vestibulospinal tract neurons in the cat. J Neurophysiol 79:270–284.

Perlmutter SI, Iwamoto Y, Barke LF, Baker JF, Peterson BW (1998b) Relation between axon morphology in C1 spinal cord and spatial properties of medial vestibulospinal tract neurons in the cat. J Neurophysiol 79:285–303.

Perlmutter SI, Iwamoto Y, Baker JF, Peterson BW (1999) Spatial alignment of rotational and static tilt responses of vestibulospinal neurons in the cat. J Neurophysiol 82:855–862.

Peterson BW, Choi H, Hain TC, Keshner E, Peng GC (2001) Dynamic and kinematic strategies for head movement control. Ann N Y Acad Sci 942:381–393.

Peterson BW, Fukushima K, Hirai N, Schor RH, Wilson VJ (1980) Responses of vestibulospinal and reticulospinal neurons to sinusoidal vestibular stimulation. J Neurophysiol 43:1236–1250.

Peterson BW, Goldberg J, Bilotto G, Fuller JH (1985) Cervicocollic reflex: its dynamic properties and interaction with vestibular reflexes. J Neurophysiol 54:90–109.

Peterson BW, Pitts NG, Fukushima K, Mackel R (1978) Reticulospinal excitation and inhibition of neck motoneurons. Exp Brain Res 32:471–489.

Rapoport S, Susswein A, Uchino Y, Wilson VJ (1977a) Properties of vestibular neurones projecting to neck segments of the cat spinal cord. J Physiol (Lond) 268:493–510.

Rapoport S, Susswein A, Uchino Y, Wilson VJ (1977b) Synaptic actions of individual vestibular neurones on cat neck motoneurones. J Physiol (Lond) 272:367–382.

Richmond FJR, Scott DA, Abrahams VC (1978) Distribution of motoneurones to the neck muscles, biventer cervicis, splenius and complexus in the cat. J Comp Neurol 181:451–464.

Schor RH, Miller AD (1981) Vestibular reflexes in neck and forelimb muscles evoked by roll tilt. J Neurophysiol 46:167–178.

Schor RH, Miller AD, Tomko DL (1984) Responses to head tilt in cat central vestibular neurons. I. Direction of maximum sensitivity. J Neurophysiol 51:136–146.

Shinoda Y, Ohgaki T, Futami T, Sugiuchi Y (1988a) Vestibular projections to the spinal cord: the morphology of single vestibulospinal axons. Prog Brain Res 76:17–27.

Shinoda Y, Ohgaki T, Sugiuchi Y, Futami T (1988b) Structural basis for three-dimensional coding in the vestibulospinal reflex: morphology of single vestibulospinal axons in the cervical cord. Ann N Y Acad Sci 545:216–227.

Shinoda Y, Ohgaki T, Sugiuchi Y, Futami T (1992a) Morphology of single medial vestibulospinal axons in the upper cervical spinal cord of the cat. J Comp Neurol 316:151–172.

Shinoda Y, Ohgaki T, Sugiuchi Y, Futami T, Kakei S (1992b) Functional synergies of neck muscles innervated by single medial vestibulospinal axons. Ann N Y Acad Sci 656:507–518.

Shinoda Y, Sugiuchi Y, Futami T, Ando N, Kawasaki T (1994) Input patterns and pathways from six semicircular canals to motoneurons of neck muscles. I. The multifidus muscle group. J Neurophysiol 72:2691–2702.

Shinoda Y, Sugiuchi Y, Futami T, Kakei S, Izawa Y, Na J (1996) Four convergent patterns of input from the six semicircular canals to motoneurons of different neck muscles in the upper cervical cord. Ann N Y Acad Sci 781:264–275.

Shinoda Y, Sugiuchi Y, Futami T, Ando N, Yagi J (1997) Input patterns and pathways from six semicircular canals to motoneurons of neck muscles. II. The longissimus and semispinalis muscle groups. J Neurophysiol 77:1234–1258.

Strassman A, Highstein SM, McCrea RA (1986) Anatomy and physiology of saccadic burst neurons in the alert squirrel monkey. II. Inhibitory burst neurons. J Comp Neurol 249:358–380.

Suzuki JI, Cohen B (1966) Integration of semicircular canal activity. J Neurophysiol 29(6):981–995.

Szentágothai J (1951) Short propriospinal neurons and intrinsic connections of the spinal grey matter. Acta Morphol Acad Sci Hung 1:81–94.

Uchino Y (2001) Otolith and semicircular canal inputs to single vestibular neurons in cat. Jpn Biol Sci Space 15:375–381.

Uchino Y, Hirai N (1984) Axon collaterals of anterior semicircular canal-activated vestibular neurons and their coactivation of extraocular and neck motoneurons in the cat. Neurosci Res 1:309–325.

von Holst E, Mittelstaedt H (1950) Das Reafferenzprinzip. Naturwissenschaften 37:464–476.

Wickland CR, Baker JF, Peterson BW (1991) Torque vectors of neck muscles in the cat. Exp Brain Res 84:649–659.

Wilson VJ, Maeda M (1974) Connections between semicircular canals and neck motoneurons in the cat. J Neurophysiol 37:346–357.

Wilson VJ, Peterson BW (1988) Vestibular and reticular projections to the neck. In: Peterson BW, Richmond FJ (eds) Control of Head Movement. New York: Oxford University Press, pp. 129–140.

Wilson VJ, Kato M, Peterson BW, Wylie RM (1967) A single-unit analysis of the organization of Deiters' nucleus. J Neurophysiol 30:603–619.

Wilson VJ, Uchino Y, Maunz RA, Susswein A, Fukushima K (1978) Properties and connections of cat fastigiospinal neurons. Exp Brain Res 32:1–17.

Wilson VJ, Peterson BW, Fukushima K, Hirai N, Uchino Y (1979) Analysis of vestibulocollic reflexes by sinusoidal polarization of vestibular afferent fibers. J Neurophysiol 42:331–346.

Wilson VJ, Yamagata Y, Yates BJ, Schor RH, Nonaka S (1990) Response of vestibular neurons to head rotations in vertical planes. III. Response of vestibulocollic neurons to vestibular and neck stimulation. J Neurophysiol 64:1695–1703.

Wilson VJ, Boyle R, Fukushima K, Rose PK, Shinoda Y, Sugiuchi Y (1995) The vestibulocollic reflex. J Vestib Res 5:147–170.

9
Gain and Phase Control of Compensatory Eye Movements by the Flocculus of the Vestibulocerebellum

CHRIS I. DE ZEEUW, SEBASTIAAN K.E. KOEKKOEK,
ARJAN M. VAN ALPHEN, CHONGDE LUO, FREEK HOEBEEK,
JOHANNES VAN DER STEEN, MAARTEN A. FRENS, JOHN SUN,
HIERONYMUS H.L.M. GOOSSENS, DICK JAARSMA,
MICHIEL P.H. COESMANS, MATTHEW T. SCHMOLESKY,
MARCEL T.G. DE JEU, and NIELS GALJART

1. Introduction

The cerebellum controls the amplitude and timing of movements. Neurons in the cerebellar nuclei are thought to coordinate the management of commands, corrections, and feedback through coactivation of paths descending to alpha-motoneurons for generation of muscle power as well as to gamma-motoneurons for adjustment of muscular position sensors and velocity sensors (Brooks and Thach 1981). In this way, the cerebellum probably exerts two mutually supportive actions to govern posture and motor control, namely phasic triggering of programs and tonic support of reflexes (Bloedel and Courville 1981). Lesions of the cerebellum can cause general dysfunctions such as ataxia and hypotonia and, in addition, rather specific defects can be attributed to lesions of particular parts of the cerebellum. For example, lesions of the spinocerebellum can produce an abnormality in the sequence of muscle contractions during rapid movements and dysmetria and tremor during slow movements, whereas lesions of the cerebrocerebellum are more characterized by delays in movement initiation and in coordination of distal limb movements (Kennedy et al. 1982; Keele and Ivry 1990). Lesions of the nodulus and flocculus of the vestibulocerebellum, on the other hand, predominantly evoke disturbances in balance and eye movement control, respectively (Ito 1984). This chapter focuses on the role of the flocculus in the control of the amplitude and the timing of compensatory eye movements in that it is divided into different sections pertaining to the gain (i.e., amplitude of eye movement/amplitude of stimulus movement) and the phase (i.e., time difference between eye and stimulus

× frequency × 360) of these movements. Special emphasis will be put on cellular and molecular factors that directly influence these parameters. Some aspects of the plastic processes that underlie adaptive processes of eye movements will be mentioned, but more details on models of motor learning in the vestibulocerebellum will be presented in Chapter 10 by Green et al. Apart from some supplementary reports of experiments in rats, cats, and monkeys, the bulk of the material reviewed here has been obtained from studies in rabbits and mice.

2. Floccular Pathways Involved in the Control of Compensatory Eye Movements

The flocculus of the cerebellum is involved in the control of the gain and phase dynamics of the optokinetic reflex (OKR) and the vestibuloocular reflex (VOR), adaptation of the VOR, and in the case of primates in the control of smooth pursuit eye movements (for a review, see Ito 1984). To fulfill these functions, the flocculus receives many different afferents conveying different signals related to eye movements, and, in turn, it projects to various nuclei in the complex of vestibular and cerebellar nuclei to influence ultimately the activity of oculomotor neurons (Fig. 9.1). The afferents include the climbing fibers, mossy fibers, and monoaminergic fibers, and the efferent pathway of the flocculus is formed by the Purkinje cell projections.

2.1. Floccular Input

The climbing fiber input to the flocculus is, similar to that of other parts of the cerebellum, strongly topographically organized. The climbing fiber input to Purkinje cells in the rabbit flocculus can be divided into five zones: four visual zones and one nonvisual zone (De Zeeuw et al. 1994b; Tan et al. 1995). The visual Purkinje cell zones (zones 1–4) receive their climbing fiber input from the dorsal cap and/or ventrolateral outgrowth of the inferior olive, whereas the Purkinje cell zone of the nonvisual zone (zone C2) receives its climbing fibers from the rostral tip of the rostral medial accessory olive. The C2 zone is presumably involved in the control of neck movements because electrical stimulation of this zone elicits short-latency neck movements (De Zeeuw and Koekkoek 1997). The climbing fiber activities (i.e., complex spikes) of the visual Purkinje cells modulate optimally to an optokinetic stimulus rotating about the vertical axis (zones 2 and 4) or about the horizontal axis perpendicular to the ipsilateral anterior semicircular canal (zones 1 and 3) (see Fig. 9.2) (De Zeeuw et al. 1994b). These different slip signals are conveyed from the retina through differents sets of nuclei of the accessory optic system and inferior olive (for a review, see Simpson et al. 1996). The optokinetic input to the horizonal axis zones is

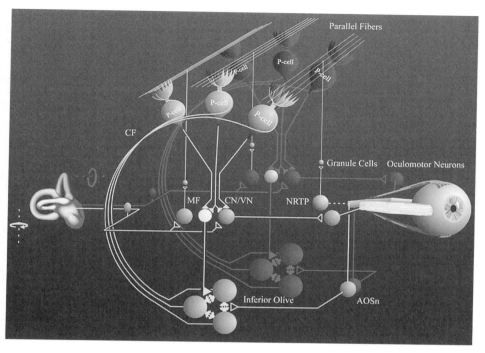

FIGURE 9.1. Schematic drawing of the vestibulocerebellar circuitry and its major afferent and efferent pathways that are involved in the control of the vestibulo-ocular reflex (VOR) and optokinetic reflex (OKR). The major pathway that mediates the VOR is the three-neuron arc of Lorente de Nó. This open loop is formed by the vestibular ganglion cells that receive input from the vestibular apparatus, the second-order vestibular neurons that are innervated by the primary afferents from these ganglion neurons, and the oculomotor neurons that innervate the oculomotor muscles. The major pathway that mediates the OKR is the accessory optic system (AOS). The AOS is embedded in a closed loop that is formed by the retinal ganglion cells; several mesencephalic nuclei (AOSn) such as the nucleus of the optic tract; the medial tegmental nucleus and the visual tegmental relay zone, which transmit optokinetic signals to the inferior olive; the vestibulocerebellum, which receives its climbing fibers (CF) from the inferior olive; and the complex of cerebellar and vestibular nuclei (CN/VN), which in turn innervate the oculomotor neurons. The granule cells in the vestibulocerebellar cortex receive mossy fiber (MF) signals that can carry vestibular, optokinetic, and/or eye movement signals. The Purkinje cells of the vestibulocerebellum control both the vestibular and optokinetic pathways by innervating the neurons in the CN/VN (dark yellow and red) that innervate the oculomotor neurons as well as the GABAergic neurons in the CN/VN (light yellow and red) that innervate the dendrites in the inferior olive that are coupled by gap junctions. Interneurons of the vestibulocerebellum are omitted for simplicity. P-cell and NRTP indicate Purkinje cell and nucleus reticularis tegmenti pontis, respectively. (See color insert.)

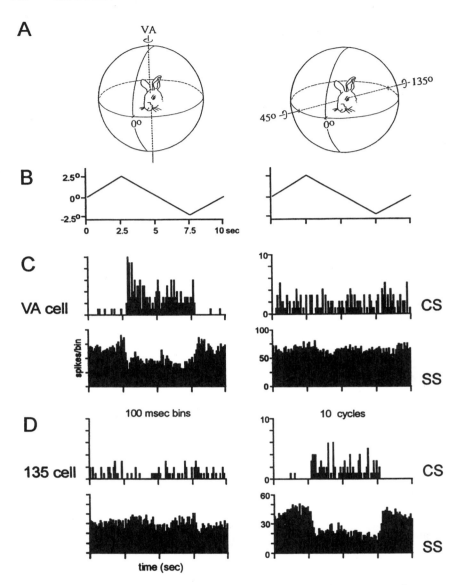

FIGURE 9.2. Perstimulus time histograms of the CS and SS responses of a representative vertical axis Purkinje cell (VA cell; **C**) and horizontal axis Purkinje cell (135 cell; **D**) in the rabbit flocculus to optokinetic stimulation (see **A** and **B**) about the vertical axis (left-hand column) and horizontal axis perpendicular to the ipsilateral anterior canal (right-hand column). The horizontal axis Purkinje cell is referred to as a 135 cell because the horizontal axis is oriented at 135° ipsilateral azimuth. Note the reciprocity of the CS and SS modulations at the preferred axis of stimulation (adapted from De Zeeuw and Simpson; see De Zeeuw et al. 1995b).

mediated via the contralateral medial terminal nucleus and the ipsilateral visual tegmental relay zone of the accessory optic system and via either the rostral dorsal cap or the ventrolateral outgrowth of the inferior olive, whereas the visual signals to the vertical axis zones are mediated via the ipsilateral dorsal terminal nucleus and pretectal nucleus of the optic tract and via the caudal dorsal cap (Simpson et al. 1988; Soodak and Simpson 1988). Thus, to a large extent, the complex spike modulation of Purkinje cells in the different zones is already encoded in the descending projections from the accessory optic system and the neurons in the different olivary subnuclei. Apart from the visual signals, evidence has been gathered that the climbing fiber input to the visual zones of the flocculus may also relay nonvisual, vestibularlike signals (De Zeeuw et al. 1995b; Simpson et al. 2002). In the rabbit, the majority of the climbing fibers in these zones modulate when the rabbit is oscillated in the dark at frequencies and amplitudes higher than 0.8 Hz and 5°, respectively. This complex spike modulation is approximately in phase with the concomitant simple spike modulation, and the activity increase consistently occurs when the head rotates contralaterally. Thereby, this nonreciprocal relation between the complex spike and simple spike modulation during vestibular stimulation stands in marked contrast to the reciprocal relation found when the rabbit is afforded vision. Possibly, the complex spike signals that arise with vestibular stimulation in the dark reflect the inhibitory input to the caudal dorsal cap from the nucleus prepositus hypoglossi (De Zeeuw et al. 1993). The third possible signal carried by climbing fibers to the flocculus may be more directly related to the actual movement (Frens et al. 2001). Similar to the vestibularlike signals, it remains to be determined how and where this motor-related signal is generated, but it can become apparent by presenting transparently moving optokinetic stimuli. Applying this stimulation paradigm to alert rabbits has demonstrated that identical retinal slip patterns can result in different complex spike modulation depending on eye movement behavior (Fig. 9.3).

The mossy fiber input to the flocculus, which is mostly bilateral, is much more diverse than the climbing fiber input and is derived from many different sources. The sources of the mossy fiber projections to the flocculus include the medial vestibular nucleus, descending vestibular nucleus, and superior vestibular nucleus (rabbits, Alley et al. 1975; Yamamoto 1979; cats, Kotchabhakdi and Walberg 1978; Sato et al. 1983; rats, Blanks and Precht 1983), group y (rhesus macaque (*Macaca mulatta*), Langer et al. 1985a), nucleus prepositus hypoglossi (cats, Kotchabhakdi et al. 1978; rabbits, Yamamoto 1979; Barmack et al. 1992a, 1992b), nucleus abducens (cats, Kotchabhakdi and Walberg 1977; rabbits, Yamamoto 1979), nucleus reticularis tegmenti pontis (cats, Hoddevik 1978; rabbits, Yamamoto 1979), and, via the output of the brush cells, the flocculus itself (Mugnaini et al. 1997). Based on electrophysiological recordings that showed short latencies, which may be due to axosomatic gap junctions at the terminals of the vestibular

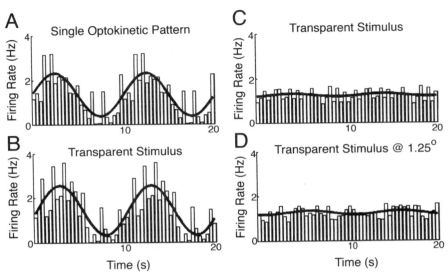

FIGURE 9.3. CS modulation as a result of transparent stimulation. (**A**) and (**B**): CS modulations are virtually identical while the rabbit makes virtually identical movements (gain values are 0.60 and 0.58, respectively), despite considerably different slip patterns on the retina. (**C**): The eyes did not move, and no significant CS modulation was present. (**D**) shows the response to transparent stimulation with an amplitude of 1.25°. Here, there is no modulation, despite a retinal slip that is very similar to the slip in B. Thus, these panels show the CS activities of a Purkinje cell that can correlate with motor behavior rather than retinal slip (from Frens et al. 2001).

afferents innervating the second-order vestibular neurons (De Zeeuw and Berrebi 1995), it was originally assumed that the primary afferents from the vestibular ganglion also directly innervate the flocculus (Shinoda and Yoshida 1975; Ito 1984). However, using sensitive autoradiography, Gerrits and colleagues (Gerrits et al. 1989) demonstrated that primary vestibular afferents are virtually absent in the flocculus. Instead, most, if not all, of the vestibular mossy fiber input to the flocculus is probably derived from second-order neurons in the vestibular nuclei. The information transmitted from the nucleus reticularis tegmenti pontis to the flocculus is probably mainly visual, whereas the nucleus prepositus hypoglossi may relay multiple types of information, including vestibular, visual, and eye-movement-related signals, and the abducens nucleus may relay eye movement signals only (Maekawa and Takeda 1977; Maekawa and Kimura 1980; Miyashita et al. 1980).

The preferred axis of modulation for floccular simple spike responses evoked by the mossy fiber/parallel fiber pathways during optokinetic stimulation is mostly the same as those of complex spike response caused by climbing fiber activation (Fig. 9.2), despite the patchy distribution of the mossy fiber projection and despite the facts that the mossy fiber projection is bilateral and that the parallel fibers cross the climbing fiber zones orthogonally (Brand et al. 1976; Mugnaini 1983). This spatial alignment may be explained in part by the possibility that the ascending axons of granule cells serve as the main, mossy-fiber mediated input to the Purkinje cells (Llinás 1982; Bower 1997) and in part by the possibility that a subpopulation of the Golgi cell axons exert their inhibitory actions preferentially in neighboring (micro)zones (De Zeeuw et al. 1995a). In particular, the glycinergic Lugaro cells may be relevant candidates in this respect as they traverse perpendicularly through the sagittal zones (Laine and Axelrad 2002; De Zeeuw, unpublished observations).

Similar to the climbing fibers, most of the mossy fibers in the flocculus use glutamate as their neurotransmitter (for a review, see Voogd et al. 1996). Phosphate-activated glutaminase and/or conjugates of glutamate are found in many neurons of the medial vestibular nucleus, superior vestibular nucleus, group y, nucleus prepositus hypoglossi, and nucleus reticularis tegmenti pontis. However, there is considerable evidence for more heterogeneity in the mossy fibers of the flocculus because many have been shown to contain neuroactive peptides or to be cholinergic. For example, in the flocculus of the opossum, the octapeptide cholecystokinin occurs in mossy fibers and enkephalin and corticotropin-releasing factor are colocalized in both climbing fibers and mossy fibers (Cummings and King 1990; Madtes and King 1994). In rabbits and presumably also in mice, these mossy fibers originate from the nucleus prepositus hypoglossi, vestibular nuclei, reticular nucleus, nucleus reticularis gigantocellularis, and raphe nuclei (Errico and Barmack 1993; Overbeck and King 1999). The cholinergic mossy fiber input to the flocculus originates predominantly in the caudal medial vestibular nucleus and the nucleus prepositus hypoglossi (Ojima et al. 1989; Barmack et al. 1992a; Jaarsma et al. 1995, 1996) and innervates both granule cells and unipolar brush cells (Jaarsma et al. 1996). Interestingly, although glutamate and corticotropin-releasing factor can be colocalized and coreleased at some mossy fiber terminals in the flocculus, it is unlikely that such colocalization and release occur with acetylcholine and peptides (Ikeda et al. 1992). It should be noted, however, that there is apart from a cholinergic mossy fiber projection also another population of cholinergic fibers in the flocculus (Ojima et al. 1989; Barmack et al. 1992a; Jaarsma et al. 1996). This input is derived from the lateral paragigantocellular nucleus and projects via a sparse diffuse network of thin, beaded fibers predominantly to dendrites of cortical interneurons in both the granular and molecular layers of the floccular cortex (Jaarsma et al. 1997).

The distribution of acetylcholine receptors in the cortex of the flocculus appears more prominent and widespread than that of the cholinergic fibers, but there is some variability among species (Jaarsma et al. 1997). For example, in most species examined (i.e., mouse, rat, cat, rabbit, and monkey), muscarinic m2-receptors are localized on the dendrites of Golgi cells and a subset of mossy fibers in the flocculus, but in the rabbit they are also prominent on a subset of parallel fibers and on most of the Purkinje cells in zone 1 (Fig. 9.4; Jaarsma et al. 1995, 1997). Nicotinic acetylcholine receptors are generally sparse in the cerebellar cortex, occurring at low levels in granule cells (Jaarsma et al. 1997). However, a special class of nicotinic receptors, which can be labeled by $[^{125}I]$ α-bungarotoxin and is putatively composed of α7-nicotine receptor subunits, has been identified in the flocculonodular lobe (Hunt and Schmidt 1978). Their distribution suggests a specific association with unipolar brush cells.

The monoaminergic inputs to the flocculus are mainly formed by serotonergic and noradrenergic fibers; the dopaminergic input to the flocculus is virtually absent (Panagopoulos et al. 1991; Ikai et al. 1992). The serotonergic input to the flocculus, which is present but also relatively weak, is mainly derived from the nucleus pontis oralis and the nucleus reticularis

FIGURE 9.4. Origin of cholinergic innervation of the flocculus in the rabbit (**A**) and its distribution of muscarinic receptors (**B,C**). Part A shows an injection site of horseradish peroxidase in the left flocculus (left) and the distribution of the retrogradely labeled neurons and/or immunocytochemically labeled neurons in the vestibular nuclei following this injection and immunocytochemistry against choline acetyltransferase (right). The filled diamonds, open circles, and filled circles indicate horseradish peroxidase labeled, choline acetyltransferase labeled, and double choline acetyltransferase and horseradish peroxidase labeled neurons, respectively (from Barmack et al. 1992b). Note that double choline acetyltransferase and horseradish peroxidase labeled neurons are present in the nucleus prepositus hypoglossi (NPH) and to a lesser extent in the medial vestibular nuclei (MVN), indicating that cholinergic projections to the flocculus originate in these nuclei (from Barmack et al. 1992b). Parts B and C show the autoradiographic and immunocytochemical distributions of muscarinic receptors in the flocculus labeled with the muscarinic antagonist [3H]QNB and an antibody specific to the m2 muscarinic receptor, respectively. Muscarinic receptors in the rabbit flocculus are of the m2 type and are predominantly located on Purkinje cell dendrites of zone 1 (large arrow in B) as identified by acetylcholine-esterase histochemistry. In addition, muscarinic receptors are present on Golgi cells (arrows in C) and on a subset of parallel fibers that are concentrated in a thin stratum immediately adjacent to the Purkinje cell layer (PCl; arrowheads in C). Note in C that Golgi cells in the rabbit cerebellum are localized both in the molecular (ml) and the granular (gl) cell layers. Fl, dPf, vPf, and arrows in the left panel of A indicate flocculus, dorsal paraflocculus, ventral paraflocculus, and their borders, respectively. VII, DVN, CN, ICP, Tr V, N V, DMN X, CE, Pyr, N VII, and ION in the right panel of A indicate the facial nerve, descending vestibular nucleus, cochlear nucleus, inferior cerebellar peduncle, tractus of the trigeminal nucleus, the trigeminal nucleus, vagus nucleus, external cuneate nucleus, pyramidal tract, facial nucleus, and the inferior olivary nucleus, respectively. Numerals 2, 3, and 4 in B indicate zones 2, 3, and 4, respectively.

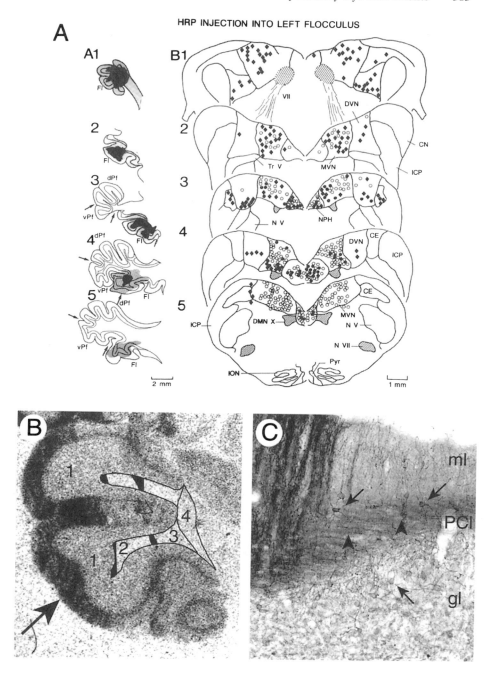

(para)gigantocellularis (Bishop et al. 1985). The fibers consist mainly of long axons containing many boutons de passage terminating in all three cortical layers (i.e., the granular cell layer, Purkinje cell layer, and molecular layer) (Bishop et al. 1985). Ligand-binding autoradiography shows a low density of serotonin receptors in the cerebellar cortex, with the highest density in the molecular layer consisting primarily of $5HT_{1b}$ receptors (Pazos and Palacios 1985). The noradrenergic projection to the flocculus originates from the locus coeruleus (for a review, see Van Neerven 1990). These fibers also terminate in both the granular cell layer and molecular layer of the flocculus (Kimoto et al. 1981). Autoradiographic studies have shown that the noradrenergic receptors are mainly of the β type and that they are predominantly concentrated in irregular patches in the Purkinje cell layer (Sutin and Minneman 1985).

2.2. Floccular Output

Despite earlier reports on an efferent floccular pathway formed by the axons of unipolar brush cells, the Purkinje cell axons form the one and only output of the flocculus (Mugnaini and Floris 1994; Mugnaini et al. 1997). They project to target cells in the cerebellar and vestibular nuclei that themselves do not project to the flocculus. The projections of Purkinje cells from the visual floccular zones to the complex of cerebellar and vestibular nuclei in the rabbit are, similar to the climbing fiber projections, strongly topographically organized (Fig. 9.5). Collectively, Purkinje cells of horizontal axis zone 1 project to the ipsilateral ventral dentate nucleus, dorsal group y, and superior vestibular nucleus; Purkinje cells of vertical axis zones 2 and 4 project to the ipsilateral magnocellular medial vestibular nucleus and parvocellular medial vestibular nucleus; and Purkinje cells of horizontal axis zone 3 project to the ipsilateral dorsal and ventral group y and the superior vestibular nucleus (for a review on rabbits, see De Zeeuw et al. 1994b; for the monkey, Langer et al. 1985b; for the cat, Sato et al. 1988; for the rat, Umetani 1992). Individual Purkinje cell axons originating in the flocculus can branch and innervate different nuclei (De Zeeuw et al. 1994b). Branching axons from zone 1 either innervate both the ventral dentate nucleus and superior vestibular nucleus or both dorsal group y and the superior vestibular nucleus; branching axons from zones 2 and 4 innervate both the magnocellular medial vestibular nucleus and parvocellular medial vestibular nucleus (and the nucleus prepositus hypoglossi, if folium p is included); and branching axons from zone 3 innervate both dorsal group y and the superior vestibular nucleus or both ventral group y and the superior vestibular nucleus. The terminal varicosities of an individual floccular Purkinje cell axon can contact both smaller inhibitory and larger excitatory neuron cells in their target nuclei (De Zeeuw and Berrebi 1995). For example, individual Purkinje cell axons of the horizontal axis zones give off terminals in both the dorsolateral and central parts of the superior vestibular nucleus,

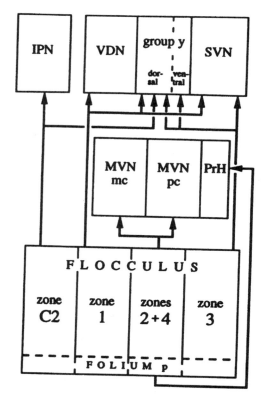

FIGURE 9.5. Summary of the projections from the Purkinje cell zones in the flocculus and folium p to the cerebellar and vestibular nuclei in the rabbit (from De Zeeuw et al. 1994b). Note that the zones involved in vertical (zones 1 and 3) and horizontal (zones 2 and 4) compensatory eye movements project to different sets of nuclei. IPN, VDN, SVN, MVN mc, MVN pc, and PrH indicate posterior interposed cerebellar nucleus, ventral dentate nucleus, superior vestibular nucleus, magnocellular medial vestibular nucleus, parvocellular medial vestibular nucleus, and nucleus prepositus hypoglossi, respectively.

which predominantly contain, respectively, the excitatory and inhibitory second-order vestibular neurons that are monosynaptically innervated by afferents from the ipsilateral anterior semicircular canal (Highstein and Reisine 1979; Sato and Kawasaki 1990; De Zeeuw et al. 1994b).

2.3. Open and Closed Floccular Pathways

An individual Purkinje cell of the flocculus can be involved in two pathways: an open (via predominantly excitatory neurons) and a closed (via inhibitory neurons) anatomical pathway involved in compensatory eye movements (Fig. 9.6) (De Zeeuw et al. 1994b). All nuclei that receive input from the visual floccular zones (i.e., the medial vestibular nucleus, the superior vestibular nucleus, the ventral dentate nucleus, dorsal group y, and the nucleus prepositus hypoglossi) project to the oculomotor complex (Highstein et al. 1971; Graybiel and Hartwieg 1974; Yamamoto et al. 1986; Evinger 1988; Büttner-Ennever 1992). The ventral dentate nucleus, dorsal group y, and nucleus prepositus hypoglossi, which project to the dorsal cap and ventrolateral outgrowth, are part of a circuit linking the inferior olive, the cerebellar cortex, and the cerebellar or vestibular nuclei (Voogd and Bigaré 1980; De Zeeuw et al. 1993, 1994a). This circuit is referred to as the closed olivofloccular pathway (Fig. 9.6A,B) because a specific olivary subnucleus provides the climbing fibers to a particular zone of Purkinje cells that innervates a specific cerebellar nucleus that, in turn, projects to the corresponding olivary subnucleus (Voogd and Bigaré 1980; De Zeeuw et al. 1997). In contrast, the medial vestibular nucleus and superior vestibular nucleus do not project to the dorsal cap and ventrolateral outgrowth, and thus this arrangement is referred to as the open olivofloccular pathway (Fig. 9.6C,D). Interestingly, the flocculus-receiving neurons in the open pathway (medial vestibular nucleus and superior vestibular nucleus) receive a direct input from the semicircular canals and are second-order neurons in the three-neuron arc that underlies the vestibuloocular reflex (for a review, see Büttner and Büttner-Ennever 1988), whereas the flocculus-receiving neurons in the closed pathway (ventral dentate nucleus, dorsal y, and nucleus prepositus hypoglossi) do not receive a direct input from the primary afferents of the semicircular canals. Thus, because an axon of an individual Purkinje cell can branch and innervate two different nuclei, one branch can be part of the open pathway while the other is part of the closed pathway, and thereby the Purkinje cell can simultaneously influence loops that are intimately connected with either the primary vestibular afferents or the inferior olive, establishing a site for cross coupling.

The open and closed pathways each contribute to the control of both horizontal (Fig. 9.6A,C) and vertical (Fig. 9.6B,D) compensatory eye movements. With regard to the open pathway, the medial vestibular nucleus neurons are excited by primary afferents from all ipsilateral semicircular canals (Büttner and Büttner-Ennever 1988), but the flocculus inhibits only those neurons that are monosynaptically innervated by the primary afferents from the horizontal canals (Ito et al. 1973; Sato et al. 1988). These findings are consistent with the fact that only floccular zones 2 and 4, which are involved in the horizontal compensatory eye movements, project to the medial vestibular nucleus. In the rabbit, the flocculus-receiving neurons in the medial vestibular nucleus can be either excitatory or inhibitory (Kawaguchi 1985; Sato et al. 1988). The excitatory flocculus-receiving neurons in the medial vestibular nucleus project to the ipsilateral medial rectus motoneurons, whereas the inhibitory flocculus-receiving neurons in the medial vestibular nucleus project to the ipsilateral abducens nucleus

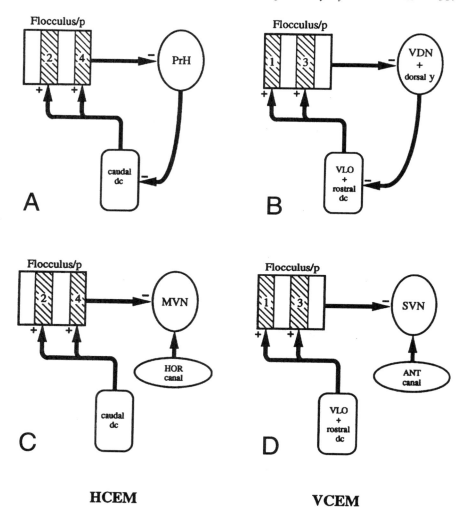

HCEM   **VCEM**

FIGURE 9.6. Summary of closed (**A,B**) and open (**C,D**) loops between the flocculus including folium p (p), the cerebellar and vestibular nuclei, and corresponding olivary subnuclei involved in horizontal (A,C) and vertical (B,D) compensatory eye movements (HCEM and VCEM) (from De Zeeuw et al. 1994b). It should be noted that individual Purkinje cell axons can collateralize into two branches, one of which is involved in the open-loop system the other and involved in the closed-loop system. PrH, dc, VLO, VDN, dorsal y, MVN, SVN, HOR canal, and ANT canal indicate nucleus prepositus hypoglossi, dorsal cap, ventrolateral outgrowth, ventral dentate nucleus, dorsal group y, medial vestibular nucleus, superior vestibular nucleus, horizontal canal, and anterior canal, respectively.

(rabbit, Highstein 1973; cat, Highstein and Reisine 1979; Reisine and Highstein 1979; Reisine et al. 1981; Uchino et al. 1982; Uchino and Suzuki 1983; monkey, McCrea et al. 1987). The flocculus-receiving neurons in the superior vestibular nucleus receive vestibular primary afferents only from the anterior canal (Ito et al. 1973, 1982). These findings are consistent with the fact that only floccular zones 1 and 3, which are involved in the vertical compensatory eye movements (i.e., eye movements dominated by the vertical recti and oblique muscles), project to the superior vestibular nucleus (see also Van Der Steen et al. 1994). The excitatory second-order flocculus-receiving neurons in the superior vestibular nucleus innervate contralateral superior rectus and inferior oblique motoneurons, whereas the inhibitory ones innervate ipsilateral inferior rectus and superior oblique motoneurons (Highstein 1973; Ito et al. 1976; Yamamoto et al. 1978; Highstein and Reisine 1979; Sato and Kawasaki 1990).

With regard to the closed floccular pathway, the ventral dentate nucleus, dorsal group y, and nucleus prepositus hypoglossi are part of the loop linking the inferior olive, the cerebellar cortex, and the cerebellar or vestibular nuclei. The ventral dentate nucleus and dorsal group y provide a GABAergic input to the rostral dorsal cap and ventrolateral outgrowth of the inferior olive (De Zeeuw et al. 1994a), the neurons of which give rise to the climbing fibers innervating Purkinje cells in zones 1 and 3, which, in turn, innervate the ventral dentate nucleus and dorsal group y. The nucleus prepositus hypoglossi provides the major GABAergic input to the caudal dorsal cap (De Zeeuw et al. 1993), which projects to zones 2 and 4 of the flocculus and the adjacent folium p (Ruigrok et al. 1992; Tan et al. 1995). Folium p in turn innervates the nucleus prepositus hypoglossi (Yamamoto 1978). Therefore, the ventral dentate nucleus and dorsal y are part of the closed olivofloccular pathway mainly involved in the vertical compensatory eye movements, and the nucleus prepositus hypoglossi is part of an analogous pathway mainly involved in horizontal compensatory eye movements (see also Blanks and Bok 1977; Chubb and Fuchs 1982; McCrea and Baker 1985; Delgado-Garcia et al. 1989). Thus, taken together, we can conclude that in both the open and closed floccular pathways, the nuclei involved in horizontal and vertical eye movements are segregated and follow the same functional organization as that of the floccular zones.

3. Gain Control

3.1. Effects of Floccular Lesions on Gain of Compensatory Eye Movements

The possible roles of the flocculus in gain control have been extensively studied (in the rabbit, Ito et al. 1982; Nagao 1983, 1989a, 1989b; in primates, Takemori and Cohen 1974; Zee et al. 1981; Waespe et al. 1983; Lisberger et al. 1984; Lisberger and Pavelko 1988; in the chinchilla, Daniels et al. 1978; in the cat, Robinson 1976). Ito initially proposed that the flocculus augments OKR gain and mediates enhancement of VOR gain by vision (Ito et al. 1974, 1982; Ito 1982). Lesions of the flocculus (and the ventral paraflocculus) can change the gain of the OKR and to a lesser extent that of the VOR (Takemori and Cohen 1974; Robinson 1976; Daniels et al. 1978; Zee et al. 1981; Ito et al. 1982; Nagao 1983, 1989b; Waespe et al. 1983; Lisberger et al. 1984). For example, bilateral lesions of the rabbit flocculus result in an initial spontaneous conjugate nystagmus with the slow phase to the contralateral side when the animal is placed in the dark and a more permanent decrease in the OKR gain of up to 50% and in the VOR gain of up to 20% (Nagao 1983; for unilateral lesions, see Barmack and Pettorossi 1985). Bilateral microinjections of the GABAergic agonist muscimol or baclofen can cause effectively a functional ablation of the rabbit flocculus and reversibly induce the same behavioral gain changes (Van Neerven et al. 1989).

3.2. Impact of Floccular Afferents on Gain of Compensatory Eye Movements

A unilateral lesion of the inferior olive, the source of the climbing fibers, evokes in the rabbit: (1) an immediate, spontaneous, conjugate drift of the eyes to the side contralateral to the lesion; (2) a reduction of the OKR gain of the contralateral eye during optokinetic stimulation in the posterior to anterior direction; and (3) a VOR velocity bias to the contralateral side at low stimulus frequencies (0.02–0.05 Hz) but no effect on the VOR gain (Barmack and Simpson 1980). Disturbances of the mossy fiber inputs to the flocculus can also influence the gain of the OKR and VOR. Unilateral lesions of the nucleus reticularis tegmentis pontis in rabbits reduce the OKR gain of the contralateral eye but not the VOR gain either before or after adaptation (Miyashita et al. 1980). In rats, unilateral lesions of the nucleus reticularis tegmentis pontis result in a reduction of the OKR gain, whereas the VOR gain is maintained during velocity step stimulation and reduced during sinusoidal stimulation (Hess et al. 1989).

Another approach to investigate the impact of the mossy fiber input to the flocculus is to influence its neurotransmission by pharmacological interference. This approach was used by Tan and colleagues to investigate the function of the cholinergic input to the flocculus of the rabbit (Tan and Collewijn 1991; Tan et al. 1992, 1993; for a review, see Van der Steen and Tan 1997). Their studies on the effects of intrafloccular injections of cholinergic agonists and antagonists on compensatory eye movements in the rabbit suggest that acetylcholine may play a modulatory role in the rabbit cerebellar circuitry. The drugs tested by Tan and Collewijn (1991, 1992a, 1992b) included carbachol (aspecific cholinergic agonist—i.e., both muscarinic and nicotinic), eserine/physostigmine (acetylcholinesterase inhibitor), bethanechol (selective muscarinic agonist), mecamulamine (nicotinic antagonist), and atropine (muscarinic antagonist). Of all the drugs tested, the effects of carbachol were most pronounced. This cholinergic agonist strongly enhanced the visually evoked eye movement responses (OKR) while it moderately enhanced the vestibular responses (VOR) (see also Fig. 9.7); for the OKR, the mean gain increase was 0.46, compared to 0.14 for VOR in the dark. Due to the relatively high baseline gain values during VOR in the light, the gain changes during this paradigm were not significant. Application of the acetylcholinesterase inhibitor physostigmine caused a similar, though less pronounced, temporary increase in the gain of both the OKR and VOR. Selective cholinergic agonists (bethanechol) and antagonists (atropine) of the muscarinic type resulted in a significant OKR gain increase and decrease, respectively, whereas their nicotinic counterparts had no effect (Tan 1992). In conclusion, the results of these behavioral studies suggest a role of the muscarinic cholinergic system in signal processing in the flocculus. In line with the general prominent control of the flocculus in the OKR, the cholinergic

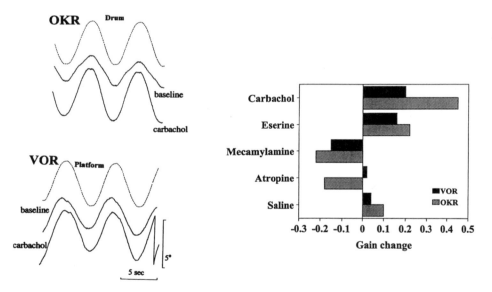

FIGURE 9.7. The effects of cholinergic agonists and antagonists on the gain of the OKR and VOR. Left panel: the top panel of this figure shows the OKR performance under sinusoidal optokinetic drum stimulation (dashed line) before (middle trace) and 6 minutes after injection of 1 μL of carbachol (lower solid trace). The bottom panel shows the same effect, but less prominently, for the VOR. For the OKR, the mean gain increase ($n = 6$) was 0.46, compared with 0.14 for VOR in the dark. Right panel: gain changes induced by floccular injection of cholinergic agonists carbachol and eserine and antagonists mecamylamine and atropine.

system of the flocculus has a particularly positive effect on the gain characteristics of the OKR.

In contrast to cholinergic agents, serotonergic and noradrenergic agonists and antagonists do not have a strong direct effect on the eye movement performance of rabbits (Van Neerven 1990; Van Neerven et al. 1990; Tan et al. 1991). The β noradrenergic agonist isoproterenol and antagonist sotalol can facilitate and inhibit adaptation of the VOR, respectively, but they do not prominently influence the gain of the OKR or VOR without any visuovestibular training paradigm. On the other hand, conjoint injection of carbachol and isoproterenol increased the gain of the OKR to an even greater extent than carbachol alone (Tan and Collewijn 1992a, 1992b). Moreover, Nagao and colleagues (Nagao et al. 1997) reported that the gain of the horizontal VOR in the rabbit is increased by 0.1–0.2 following systemic application of (6R)-5,5,6,7-tetrahydro-L-biopterin (R-THBP). Because this effect on the gain is prevented by lesions of the flocculus, and because R-THBP is a cofactor for monoamine hydroxylases and therefore involved in the biosynthesis of catecholamines and indolamines, we cannot rule out the possibility that the noradrenergic and serotonergic afferents to the flocculus do have some effect on gain dynamics.

3.3. Gain of Compensatory Eye Movements in Mouse Mutants

In general, full-field optokinetic stimulation of pigmented wild-type mice (e.g., C57BL/6J +/+, WCB6F1 +/+, WCB6F1 Sl/Sl^d, CE/J c^e/c^e, SEC/ReJ c^{ch}/c^{ch}, and 129/Sv) produces a clear in-phase eye movement response with gain values that can vary somewhat among different backgrounds (Mitchiner et al. 1976; Balkema et al. 1984; Mangini et al. 1985; Katoh et al. 1998). When measured with the minicoil method or video method, they can reach levels as high as 0.8 and 0.9, respectively (Fig. 9.8) (Stahl et al. 2000; Van Alphen et al. 2001). For comparison, hypopigmented mutants with a reduced density of melanosomes in their retinal pigment epithelium (e.g., pale ears, ep/ep; underwhite, uw/uw; maroon, ru-$2^{mr}/ru$-2^{mr}; albino, c^{2J}/c^{2J}; and pearl, pe/pe) respond very poorly to optokinetic stimuli and reach OKR gain values not higher than 0.3–0.4 (Mangini et al. 1985). During VOR in the dark, the gain values in normal pigmented wild-type mice (e.g., C57BL/6J +/+) can reach levels as high as 0.6–0.7 (stimulus parameters: 10° amplitude, 1.0 Hz) (Van Alphen et al. 2001). For this paradigm, however, the gain values of mice with a 129/Sv background are significantly lower than those of C57BL/6 mice (Katoh et al. 1998). Thus, because ES cells derived from the 129/Sv background are still the most common cells used for homologous recombination, particular attention should be paid to sufficient backcrossing before the knockout mice are subjected to compensatory eye movement tests.

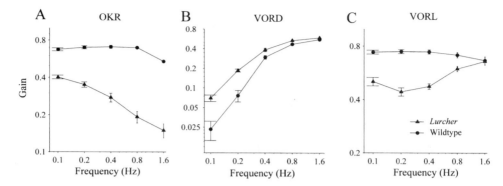

FIGURE 9.8. The contribution of Purkinje cells to the gain of OKR, VOR in the dark (VORD), and VOR in the light (VORL). The OKR (**A**), VORD (**B**), and VORL (**C**) of *lurchers* (triangles) and control animals (circles) are presented in Bode plots in which the gain of the eye movements is related to stimulus frequency (for details, see also Van Alphen et al. 2002).

The first spontaneous cerebellar mutant of which the eye movements were investigated was the *Weaver*. This animal loses its granule cells in the first postnatal weeks due to a mutation in the G-protein-linked inwardly rectifying K^+ channel type 2 gene (Hess 1996). As far as can be deduced from electrooculograms, these mutants generally show a reduced number of nystagmus beats per second during the slow phase of their optokinetic responses and a decreased OKR gain (Grüsser-Cornehls and Bohn 1988). Eye movement recordings from spontaneous mutants such as the *lurcher* are more extensive. These mutants lack their Purkinje cells due to a mutation in their δ2-glutamate receptor (Zuo et al. 1997), and they show a significant reduction in their gain values for both their OKR and VOR in the light (Fig. 9.8) (De Zeeuw et al. 1998; Van Alphen et al. 2002). This phenotype is most likely mainly due to the lack of Purkinje cells in the flocculus, as a flocculectomy in mice produces the same behavioral deficits in this respect (Koekkoek et al. 1997). Yet, the gain during VOR in the dark is enhanced in the *lurcher*. Possibly, the elevated VOR gain level in the *lurcher* reflects secondary developmental compensations in their vestibular and cerebellar nuclei (Heckroth and Eisenman 1991) because we did not observe such a difference following acute flocculectomy (Koekkoek et al. 1997). The observation that the VOR, but not the OKR, is enhanced in the *lurcher* suggests that the vestibular signals driving oculomotor activity can largely circumvent the flocculus, whereas the visual signals cannot (see also Fig. 9.1). In fact, the enhanced gain during VOR in the dark may primarily reflect efforts of the vestibular nuclei to compensate for a reduced VOR in the light due to impairment of the visual pathway through the flocculus.

Another recent study on a spontaneous mutant, the *Shaker-1* mutant, shows that the opposite pattern can also occur (Sun et al. 2001). In this mutant, the gain of the OKR is not significantly reduced, and the VOR is absent (Fig. 9.9). The genetic defect responsible for this abnormality is a mutation in the *myosin VIIA* gene (Hasson et al. 1997), which is very similar

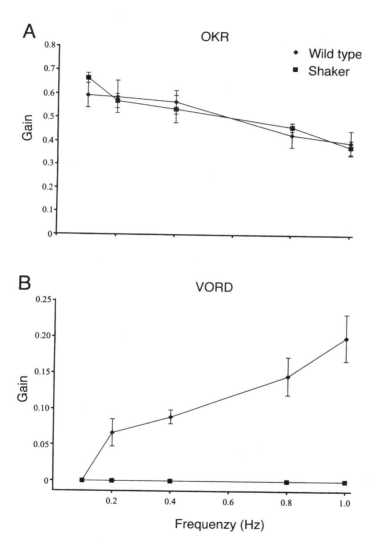

FIGURE 9.9. Eye movement performance in the *shaker-1* mutant mouse. The *shaker-1* mutant, which has a mutation in its *myosin VIIa* gene similar to patients with Usher syndrome type 1B, shows normal OKR gain values (**A**) but an absent VOR (**B**).

to the most common mutation in the *myosin VIIA* gene that occurs in patients with Usher syndrome type 1B (USH1B) (Weil et al. 1995; Weston et al. 1996). Although cerebellar deformations have been found in USH1B patients (Tamayo et al. 1996), the vestibular deficits in *Shaker-1* mutants are probably purely peripheral because Myosin VIIa is present in the stereocilia, cuticular plates, and cell bodies of both type I and type II hair cells in the semicircular canals and utriculus but not in neurons of the cerebellum (Hasson et al. 1997; Sun et al. 2001). This notion is also in line with the observation that electrical stimulation of the semicircular canals in *shaker* mice can circumvent the impaired transmission between the hair cells and their afferent fibers and thereby evoke eye movements at normal latencies (Sun et al. 2001). Moreover, the fact that gain and phase values of the OKR are normal in *Shaker-1* mutants also suggests that the function of the flocculus of the cerebellum is sufficiently intact.

Apart from studies in spontaneous mutants, eye movements have also been recorded in knockout mice and transgenic mice. For example, both homozygous and heterozygous knockout mice of the panneuronal transcription factor ZFP37 show a significantly reduced gain of the OKR and visual contribution to the VOR (i.e., gain of VOR in the light minus gain of VOR in the dark) but not of the VOR in the dark (Fig. 9.10) (Van Alphen et al., in preparation). ZFP37 knockout animals in which the expression of ZFP37 is genetically rescued do not show this phenotype, indicating that the phenotype is a result of a lack of ZFP37. Because ZFP37 is prominent in granule cells, Purkinje cells, and brush cells of the vestibulocerebellum, and because the phenotype of the ZFP37 knockouts corresponds to the gain deficits of animals with lesions of the flocculus, it appears possible that the phenotype of ZFP37 knockout mice is in fact due to a partial dysfunction of the flocculus. The exact mechanism by which their OKR gain is affected is not clear, but it may be related to a general deficit in their protein synthesis. ZFP37 is particularly prominent in nucleoli and therefore most likely relevant for construction of the ribosomal apparatus (Payen et al. 1998). The fact that the gain reductions in the homozygous and heterozygous ZFP37 mice show an effect dependent on gene dosage is in line with such a general effect on the level of protein synthesis. It appears unlikely that the expression of ZFP37 in brush cells is very important of the phenotype because transgenic L7-ras mice in which the number of brush cells in the flocculus is reduced to approximately 20% of the normal level show a normal OKR (De Zeeuw, Berrebi, and Oberdick, unpublished observations). Possibly, a lack of ZFP37 causes a similar deficit in Purkinje cells as does a lack of calbindin. Mice in which calbindin is specifically knocked out in Purkinje cells show the same deficits as ZFP37 knockouts (i.e., decreased gains of the OKR and a reduced visual contribution to the VOR without any phase defects) (Barski et al. 2003). Interestingly, these deficits can occur without any impact on either VOR adaptation or the induction of long-term depression (LTD) (De Zeeuw and Luo, unpublished observations). Thus,

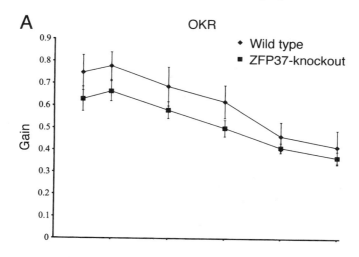

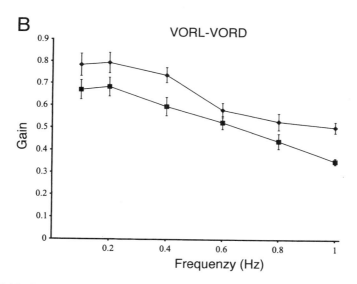

FIGURE 9.10. Eye movement deficits in the mouse knockout of transcription factor ZFP37. The knockout of ZFP37 shows a specific impairment of the amplitude of its visual reflexes; the gain values of both the OKR (**A**) and the visual contribution to the VOR (i.e., VORL – VORD) (**B**) are reduced, but its VORD gain values as well as all of its phase values, including those of the visual responses, are normal (data not shown).

performance deficits can occur to a certain extent without impairing the ability to adapt.

The opposite can also occur in that mutant mice in which LTD of the parallel fiber/Purkinje cell synapse is impaired can show a relatively normal eye movement performance. For example, mice in which the expression of glial fibrillary acid protein (GFAP) is blocked (Shibuki et al. 1996) do not show any adaptation of their OKR gain after 1 hour of optokinetic training, but their OKR gain values before the training period are normal and their gain values during VOR are even somewhat higher (Katoh et al. 1998). The blockage of LTD induction in these mutants may be an indirect effect of a lack of GFAP in Bergmann glia fibers that normally may interact with Purkinje cells via the release of some diffusible factors, such as homocysteate, an endogenous transmitterlike substance present in Bergmann glia (Cuénod et al. 1990; Shibuki et al. 1996). The caveats in this interpretation of the phenotype are the same as those of many other global knockouts: (1) there may be developmental aberrations are the protein is expressed from early on; (2) there may be a contribution to the phenotype from dysfunctions elsewhere in the brain, as there are various cell types in the brain that express GFAP; and (3) there may be an up-regulation of and compensation by related proteins. These problems are overcome in a new transgenic approach in which multiple isoforms of the same protein are inhibited in a specific cell type in the brain (De Zeeuw et al. 1998). In this mutant (*L7-PKCI*), various isoforms of Protein Kinase C are effectively inhibited by overexpressing an inhibitory peptide, PKC (19-31), specifically in Purkinje cells, with the use of the L7 promoter. In contrast to the global knockout of a single PKC isoform, such as PKCγ (Chen et al. 1995), LTD induction is blocked in *L7-PKCI* animals (Fig. 9.11). Similar to the *GFAP* mutants described above, these animals show a normal OKR and VOR motor performance, even though the blockage of LTD impairs motor learning (Fig. 9.12). Initially, the caveat of misinterpretations due to developmental aberrations mentioned above also appeared to hold true for this mutant because about half of all Purkinje cells in this mutant showed a persistent multiple climbing fiber innervation until the age of 3 months (De Zeeuw et al. 1998). Yet, Purkinje cells of *L7-PKCI* animals older than 6 months do not show any form of multiple climbing fiber innervation and still have impaired LTD induction and VOR adaptation (Goossens et al. 2001; De Zeeuw, unpublished observations). Thus, it seems that none of the three potential caveats of global knockouts apply to the *L7-PKCI* mutant. Interestingly, the Purkinje cell specific knockout of PK*G* shows qualitatively the same phenotype as the *L7-PKCI* mutant except in a milder form (Feil et al. 2003). These findings indicate that indeed the NO-PKG pathway and the PKC pathway can both play a similar role in LTD induction in Purkinje cells but apparently cannot fully compensate for each other. The fact that all three LTD-deficient mutants described above (i.e., the GFAP knockout, the L7-PKCI transgenic, and the L7-PKG knockout) all show rel-

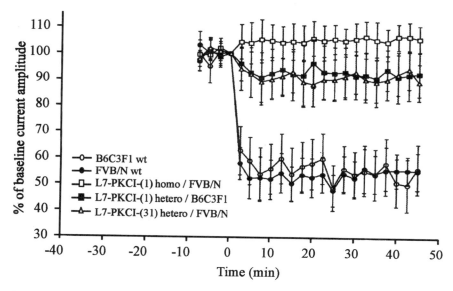

FIGURE 9.11. LTD induction is deficient in PKC-I mice. Both homozygous and heterozygous animals that express an inhibitory peptide of various isoforms of PKC have a blockade of LTD induction. Note that different backgrounds and different transgenic lines are tested (data by D. Linden; for details, see De Zeeuw et al. 1998).

atively normal gain values during their eye movement performance indicates that other sites for motor plasticity, such as the vestibular nuclei (Khater et al. 1993; Highstein et al. 1997; Lisberger 1998), are apparently sufficient for the elevation of OKR and VOR gains to normal levels if prolonged periods of training are available, as occurs in daily life (De Zeeuw et al. 1998). Indeed, if one trains LTD-deficient *L7-PKCI* mutants for prolonged periods of time, they do adapt their VOR, although at a much slower rate than wild-type animals (Van Alphen and De Zeeuw 2002).

3.4. Relations Between Gain of Compensatory Eye Movements and Purkinje Cell Activity in the Flocculus

The relation between the complex spike and simple spike activities of Purkinje cells in the flocculus and the short-term dynamics and long-term adaptive processes of compensatory eye movements has been the subject of intense research over the past decades and still provides ground for many fervent debates in the field of vestibular research (e.g., Lisberger 1998). The possible causes and consequences of climbing fiber (i.e., complex spike)

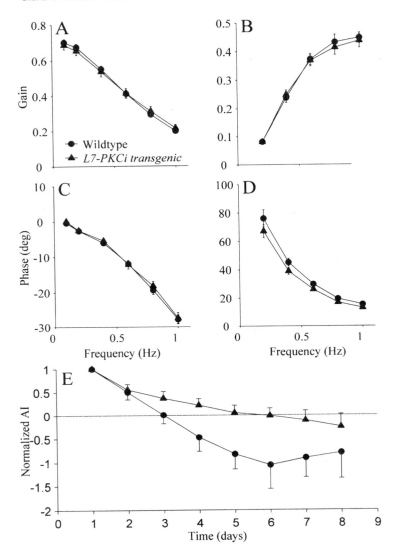

activity of floccular Purkinje cells are briefly discussed above and have been reviewed extensively (Simpson et al. 1996). The ins and outs of the relations between the simple spike activity and eye movement performance, however, are at least as controversial. As mentioned above, the simple spike activities of floccular Purkinje cells probably result from a variety of signals that converge onto these neurons, such as vestibular signals, retinal slip signals, eye position and eye velocity signals, and other proprioceptive and eye command signals (Miles et al. 1980; Ito 1984; Shidara et al. 1993; De Zeeuw et al. 1995b). Although it is difficult, if not impossible, to measure the exact contribution of each of these individual, partly even nonlinearly related, signal components to the overall simple spike response during a particular eye movement response, one can still describe a general trend in the relation between gain and simple spike response. In the rabbit, the modulation amplitude (amplitude of simple spike modulation/mean discharge rate × 100%) generally increases as the OKR or VOR gain increases (Nagao 1989a, 1989b; De Zeeuw et al. 1995b). For example, 180° out-of-phase sinusoidal visual and vestibular stimulation for one hour (0.1 Hz, 5° screen and 5° turntable) increases the VOR gain on average by 0.16 while the modulation amplitude increases by 4.7% (Nagao 1989a, 1989b). Similarly, the modulation amplitude during VOR in the light (high gain) is significantly higher than that during VOR in the dark (low gain) at a broad range of frequencies (De Zeeuw et al. 1995b). During these paradigms, the modulation amplitude of the simple spike activity of vertical axis Purkinje cells in the flocculus increases as the maximum eye velocity increases, with

FIGURE 9.12. LTD-deficient PKC-I mice show normal eye movement performance even though adaptation of the VOR is impaired. Normal performances of OKR (**A,C**) and VOR (**B,D**) in untrained L7-PKCi transgenic (triangles) and control animals (circles) are shown. Gain and phase values of the eye movement relative to stimulus movement were plotted against stimulus frequency. Neither OKR nor VOR values were statistically different between the groups. When L7-PKCi transgenic and control animals were subjected to a VOR adaptation paradigm in which they were trained to reverse direction of the VOR, L7-PKCi transgenic animals were not as able as wild-type animals to reverse direction of their VOR (**E**). The amount of adaptation in each animal was quantified by calculating an index of adaptation (AI) according to the equation $AI = G * \cos(\phi)$. In this equation, G denotes VOR and ϕ represents the VOR phase. The normalized adaptation index is shown at 0.4 Hz to outline two distinct features. First, the VOR reverses direction as the phase of the eye movement surpasses 90° lead, which causes the AI to change sign. Second, normalizing AI to each animal's initial performance shows that the L7-PKCi transgenic animals can learn but that they are limited in their ability to express this form of motor learning. Error bars indicate SEM. (Modified from De Zeeuw et al. 1998; Van Alphen and De Zeeuw 2002).

correlation coefficients as high as 0.98 (VOR dark) and 0.87 (VOR light) (De Zeeuw et al. 1995b). It is difficult to determine to what extent these changes in simple spike activity are a cause or a consequence of a gain change in the eye movement performance. In at least some cases, it is likely that they can cause a gain change. For example, the studies by Van der Steen and colleagues on the effects of cholinergic agents on eye movement performance and simple spike activity in rabbits strongly suggest than an increase in modulation amplitude of simple spike activity can cause a higher gain of the optokinetic response. Application of acetylcholine to the flocculus evokes an enhanced simple spike modulation during optokinetic stimulation, whereas the baseline level of the simple spike frequency is not affected (Van der Steen and Tan 1997). Figure 9.13 shows an increase in simple spike firing rate of a vertical axis Purkinje cell before and after

FIGURE 9.13. The impact of acetylcholine on simple spike firing frequency in the flocculus. These plots demonstrate an example of the simple spike modulation of a vertical axis cell in response to optokinetic stimulation (10° peak-to-peak at 0.2 Hz) before (**A**) and after (**B**) iontophoretic application of acetylcholine to the flocculus (current: 90 nA). The left panels show the peristimulus histograms based on cell responses averaged over ten stimulus cycles. The right panels show the simple spike frequency versus stimulus velocity. Positive values represent a movement of the optokinetic stimulus in the nasal-temporal direction of the ipsilateral (= left) eye. Note that the increase in modulation depth after acetylcholine application is due to the increase in spike frequency during the excitatory phase (modified from Van der Steen and Tan 1997). At this point, we can only speculate about the exact mechanisms of how this action takes place. Tan and Collewijn (1991) suggested that it could be an excitatory action via muscarinic receptors at the Purkinje cell level. André and colleagues (André et al. 1993, 1994) indeed showed a long-lasting enhancement of Purkinje cell responses in rats to glutamate after activation of muscarinic receptors by bethanechol. However, the type of muscarinic receptors found in the rabbit cerebellum is mainly m2 (see above). In neuronal tissue, the effect of m2 receptor stimulation is usually an increased potassium conductance, and consequently a hyperpolarization of the cell, which causes an inhibition (Bonner 1992; Caulfield 1993). This inhibitory action makes an explanation of cholinergic agonists acting directly on Purkinje cell activity via m2 receptors less likely. Another argument against the action via the m2 receptors is that they are distributed mainly in zone 1 (see Fig. 9.4), a horizontal axis zone, whereas the effect of application of acetylcholine on vertical axis Purkinje cells is significantly greater than on horizontal axis cells. Another possibility is that the cholinergic system exerts its modulatory action more indirectly via an inhibition of the Golgi cells, which contain muscarinic receptors of the m2 type (Fig. 9.4) (Jaarsma et al. 1995). Because Golgi cell axons project to the glomeruli, where mossy fibers make synaptic contacts with granule cells, the activation of m2 receptors on Golgi cells could lead to less inhibitory action on granule cells. This effect in turn would have a positive modulatory effect on the simple spike activity.

A Baseline

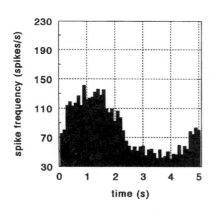

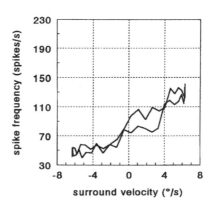

B Acetylcholine

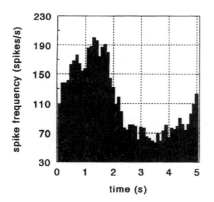

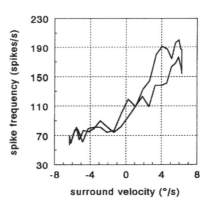

iontophoretic application of acetylcholine during continuous sinusoidal optokinetic stimulation. This increase occurs gradually in a period of about 100 s after iontophoretic application of acetylcholine and gradually recovers toward baseline values after cessation of the application. Because, as described above, local application of the cholinergic agonist carbachol to the flocculus increases the OKR gain (Tan et al. 1992), we suggest that a stronger simple spike modulation can in principle underlie a gain increase of compensatory eye movements. These effects are also in line with stimulation studies of the floccular white matter (Van der Steen et al. 1994) and with the output connections of the floccular Purkinje cells (see above). For example, an increase in the simple spike activity in vertical axis Purkinje cells during contralateral head movements is expected to exert an inhibitory effect on the excitatory and inhibitory flocculus-receiving neurons in the medial vestibular nucleus, which in turn innervate the ipsilateral medial rectus and ipsilateral lateral rectus motoneurons, respectively. Thus, the medial rectus muscle will be inhibited by this increase in simple spike activity during contralateral head movements, and the lateral rectus muscle will be disinhibited and thereby cause larger ipsiversive eye movements.

The notion that an increase in simple spike modulation correlates with an increase in the gain of the eye movement performance raises the question of how induction of LTD of the efficacy of the parallel fiber/Purkinje cell synapse could possibly lead to a gain increase (Ito 1982, 1998). Nagao (1989a, 1989b) reported that two types of vertical axis Purkinje cells exist in the flocculus of the rabbit—those that are excited during contralateral head movement (out-of-phase cells) and others that are excited during ipsilateral head movements (in-phase cells). The flocculus hypothesis predicts that LTD occurs specifically in in-phase cells during a gain-enhancing visuovestibular training paradigm and that, thereby, the out-of-phase cells will carry relatively more weight after the training process so that the gain will effectively increase (Ito 1982, 1998). More recent recordings, however, have indicated that in-phase cells are virtually absent in the flocculus of the rabbit and mouse (De Zeeuw et al. 1995b; Hoebeek and De Zeeuw, unpublished observations). Presumably, most of the previously reported in-phase cells were in fact either horizontal axis Purkinje cells or interneurons (for a detailed explanation, see De Zeeuw et al. 1995b). Thus, the short-term dynamic responses of Purkinje cells in the rabbit are not in line with the idea that a shift in dominance between the activities of two types of Purkinje cells underlies a change in VOR gain. If the phase-shift mechanism plays a role, it can only occur via a shift within the out-of-phase group (as proposed below). Recordings of Purkinje cell activity during VOR adaptation in monkeys are also difficult to reconcile with the typeI/II-shift mechanisms proposed in the classical LTD hypothesis (Ito 1998). Measured changes in the vestibular sensitivity of vertical axis Purkinje cells (i.e., horizontal gaze velocity cells; see also below) accompanying VOR adaptation are in the opposite direction from that predicted by LTD (Lisberger et al.

1994; Raymond and Lisberger 1998). Furthermore, multiple linear regression analysis of simple spike activity of horizontal axis Purkinje cells during visual following in squirrel monkeys (*Saimiri sciureus*) shows that the transfer function between flocculus and eye does not change consistently with a change in VOR gain (Hirata et al. 1998). Thus, single-unit recordings from Purkinje cells in the flocculus of various mammalian species suggest that, if LTD indeed plays a role in VOR adaptation (e.g., De Zeeuw et al. 1998), the mechanism via which it may evoke its effects on the simple spike frequency during modulation may be different from that originally proposed by Ito and Nagao (Ito 1982; Nagao 1989a, 1989b).

It is therefore possible that the increase in modulation amplitude of the simple spike activities after visuovestibular training, as observed by Nagao (1989a, 1989b), could reflect predominantly an increase in input signals carried by mossy fibers such as eye velocity signals rather than a long-term change in the excitability of Purkinje cells. The latter is expected to also affect the baseline activity, whereas no change in baseline activity might be expected when changes in modulation amplitude merely reflect differences in eye velocity. Thus, it is possible that LTD induction does not directly lead to (or at least not only lead to) a change in the amplitude of the modulation but also evokes its effects by reducing the baseline level of the simple spike response. This effect, which is not contradicted by recordings of the simple spike activity of floccular Purkinje cells and is in fact in line with the effect of climbing fiber activity on the simple spike firing rate in the flocculus during optokinetic stimulation (Leonard and Simpson 1986; Simpson et al. 1996), may increase the signal-to-noise ratio and thereby increase the gain. Because many floccular Purkinje cells receive two types of vestibular parallel fiber inputs (i.e., those that relay signals from ipsilateral second-order vestibular neurons and those that relay signals from contralateral second-order vestibular neurons) (Tan and Gerrits 1992), the ipsilateral and contralateral vestibular signals are probably always in counterphase so that the balance between their contributions to the simple spike activity of a single Purkinje cell would affect not only its modulation amplitude but also its baseline activity. LTD could cause a shift in the contributions of parallel fiber inputs carrying ipsilateral and contralateral vestibular signals by a differential modification of the synaptic weights. For example, during a gain-increase training paradigm, the complex spikes of Purkinje cells in zones 2 and 4 of the rabbit flocculus are elicited during ipsilateral head rotation (De Zeeuw et al. 1994b). Because neurons in the ipsilateral medial vestibular nucleus are excited during ipsilateral head rotation, complex spikes will tend to coincide with parallel fiber inputs relaying in-phase ipsilateral vestibular signals so that LTD would occur predominantly in the subpopulation of parallel fiber synapses that transmit ipsilateral vestibular signals. Thus, in this concept, a Purkinje cell will show an increase in modulation amplitude around a decreased baseline activity after the training procedure, and it may still hold true that the net simple spike activity of all Purkinje cells modulates out of phase with their complex spikes. Theoretically, plastic gain changes of eye movements might also be caused by plastic changes in the exact timing of simple spikes at a scale of milliseconds or tens of milliseconds. We have recently observed a spontaneous mutant in which the absolute firing frequency and modulation amplitude are normal, whereas the coefficient of variation of the simple spike activities and the gain of their compensatory eye movements are increased and decreased, respectively. These phenomena have been observed in the *tottering* mutant (Hoebeek, Stahl, and De Zeeuw, unpublished observations), which suffers from a mutation in the P/Q type calcium channel alpha 1A subunit gene and shows a reduction of the amplitude of its parallel fiber-mediated excitatory postsynaptic currents (Terwindt et al. 1998; Matsushita et al. 2002). Thus, the fact that the decrease of the gain in this mutant may be specifically due to an enhanced irregularity of simple spike activities raises the possibility that particular cellular plastic processes mediate behavioral gain changes by altering regularity of firing rather than by modulating firing frequency.

Independent from the issue of whether gain changes are conveyed via changes in modulation amplitude, baseline levels, or regularity of simple spike activities, the question remains how long these changes hold. Because we have not detected any permanent changes in the simple spike and complex spike activities in the LTD-deficient *L7-PKCI* mutants (Fig. 9.14) (Goossens, Hoebeek, Frens, and De Zeeuw, unpublished observations), it may well be that changes in modulation amplitude, baseline levels, and/or regularities of simple spike activities due to LTD induction only occur transiently and are somehow stored downstream at the Purkinje cell synapse onto the cerebellar and vestibular nuclei neurons or in these neurons themselves. Similar mechanisms have been proposed for memory processing in other brain regions such as the hippocampus and cerebral cortex (Otten and Rugg 2002).

FIGURE 9.14. Simple spike and complex spike activities of Purkinje cells in the flocculus of L7-PKCi mutants do not differ from those in wild-type mice during OKR. (**A**) Example of complex spike (bottom panel) and simple spike (middle panel) activity of a Purkinje cell in the flocculus of an awake mouse during optokinetic stimulation (top panel) about the vertical axis (0.1 Hz, 3°/s peak velocity); these per-stimulus histograms were obtained by averaging 12 cycles. (**B**) The average of simple spike modulation amplitude over different frequencies. (**C**) Then average of the simple spike phase with regard to eye velocity. (**D**) The average of complex spike modulation amplitude over different frequencies. (**E**) The average of the complex spike phase with regard to eye velocity. Note that there are no significant differences between activities in mutant and wild-type mice (from Goossens, Hoebeek, Stahl, Frens, and De Zeeuw, unpublished observations).

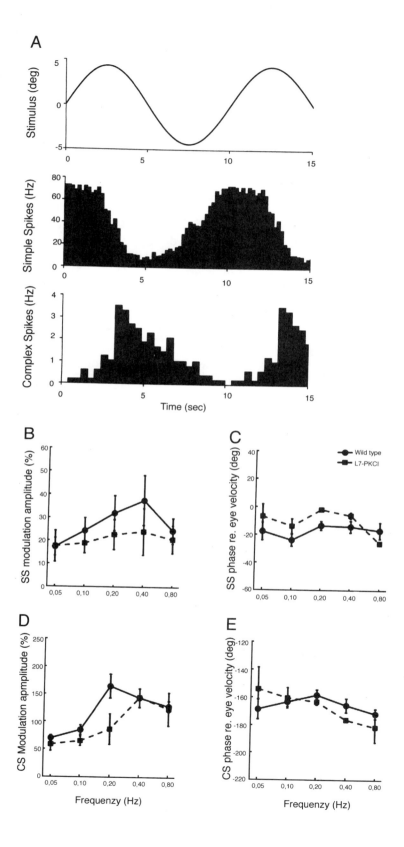

4. Phase Control

4.1. Effects of Floccular Lesions on Phase Dynamics of Compensatory Eye Movements

The role of the flocculus in control of the phase of compensatory eye movements is not as extensively documented as that of the gain. In general, eye movements in flocculectomized animals lag those of normal animals during OKR and, to a lesser degree, during VOR (rabbits, Ito et al. 1982; Nagao 1983; cats, Robinson 1976). In pigmented rabbits, bilateral flocculectomy produces a significant phase lag during OKR at a broad range of frequencies and during VOR in the dark at the higher frequencies (Nagao 1983, 1989b). During VOR in the light at 0.01–0.5 Hz, flocculectomy in pigmented rabbits induces a phase lag varying from 8° to 15° (Kimura et al. 1991). In addition, a bilateral lesion of the flocculus with kainic acid can abolish a significant adaptive phase lead in the VOR that is induced by optokinetic stimulation in phase with whole-body rotation but at twice the amplitude (Nagao 1983). In albino rabbits, Ito et al. (1982) found that unilateral flocculectomy produces, in comparison to normal animals, a significant phase lag (13°–20°) of the ipsilateral eye during rotation in the dark (at 0.1 Hz and 0.5 Hz) and during OKR (at 0.033 Hz and 0.05 Hz). In the cat, flocculectomy can decrease the phase lead by 7° during VOR in the dark at 0.05 Hz (Robinson 1976; cf. Keller and Precht 1979). In the monkey, on the other hand, lesions of the ventral paraflocculus and/or flocculus do not produce consistent phase shifts of compensatory eye movements (Takemori and Cohen 1974; Zee et al. 1980, 1981). This difference may be due to the fact that a large portion of the primate's Purkinje cell population in the flocculus consists of gaze velocity cells (Miles et al. 1980), which are absent in the flocculus of the rabbit (Leonard 1986). The gaze velocity cells are characterized by an increase of their simple spike firing frequency in phase with ipsiversive eye velocity during smooth pursuit eye movements and by an increase of their simple spike firing frequency in phase with ipsiversive head velocity during cancellation of the VOR (Raymond and Lisberger 1998). In mice, a flocculectomy also induces a considerable phase lag during OKR (Koekkoek et al. 1997). The optokinetic response in *lurcher* mice generally also lagged stimulus movement more than that of control wild types, while their phase lead during VOR in the dark was significantly greater than that of control animals at the lower frequencies of 0.1 and 0.2 Hz (Fig. 9.15) (Van Alphen et al. 2002). During VOR in the light, the phase values of *lurcher* and wild-type mice basically showed a pattern that can be predicted by an addition of their OKR and VOR values (i.e., dominated by vision at the lower frequencies and by vestibular signals at the higher frequencies). The variability of the phase values in *lurcher* mice was significantly greater than that of controls (see also Fig. 9.15). Similar effects on the phase of the OKR and VOR can be observed in mice in which the granule cells degenerate

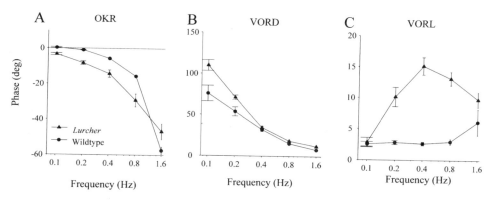

FIGURE 9.15. Contribution of Purkinje cells to the phase of the OKR and VOR in the dark (VORD) and in the light (VORL). The OKR (**A**), VORD (**B**), and VORL (**C**) of *lurchers* (triangles) and control animals (circles) are presented in Bode plots in which the phase of the eye movements is related to stimulus frequency (for details, see Van Alphen et al. 2002).

(Grüsser-Cornehls and Bohm 1988; Koekkoek and De Zeeuw, unpublished observations). For example, in *weavers*, OKR lags that of normal animals, and variation of the phase of the VOR increases dramatically. A phase lag in OKR can also be induced by a lesion of the visual mossy fiber pathway further upstream of the flocculus. This possibility is exemplified by experiments of Hess et al. (1989), who showed that unilateral lesions of the nucleus reticularis tegmentis pontis in rats not only result in a decrease of OKR gain but also in an OKR phase lag (for the position of the NRTP in pathways, see Fig. 9.1). Thus, at least some of the visual inputs to the flocculus are essential to advance the phase signals of the OKR.

Yet, not all lesions that affect OKR gain also affect the phase dynamics of the compensatory eye movements. For example, the ZFP37 and L7-calbindin knockouts described above show gain deficits, but their phase values during OKR and VOR are normal (Barski et al. 2003). Similarly, the OKR and VOR gain changes that can be induced following application of cholinergic antagonists and agonists are more prominent than the phase shifts that can be evoked by these drugs (Frens and van der Steen, unpublished observations). This holds true for both phase changes in eye movement performance and in simple spike activity (Fig. 9.16). Thus, in general, the effects of a mechanical or genetic flocculectomy on the phase of eye movements in rabbits, cats, and mice suggest that the flocculus advances the phase of the net preoculomotor signal for the OKR and stabilizes that of the VOR. The data obtained from mouse mutants so far suggest that the phase regulation of the flocculus is implemented in the hardware circuitry of its cortical layers or their afferents rather than by molecular factors that do not have prominent secondary effects on the wiring.

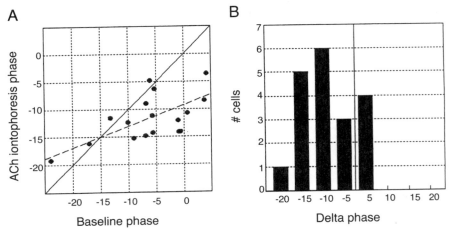

FIGURE 9.16. The impact of acetylcholine on the phase of the simple spike modulation during optokinetic stimulation. (**A**) Plot of phase of simple spike modulation during iontophoretic injection of acetylcholine (dashed line) versus baseline response (n = 19 cells). (**B**) Distribution of the phase change induced by acetylcholine compared to the baseline response of each cell. Both figures show a phase advance after injection of acetylcholine. Note that application of acetylcholine into the flocculus produces a phase lead of the simple spike activity. Dashed line: linear fit, slope 0.39, r = 0.64; the mean phase change equals $-7.9°$.

4.2. Phase Relations of Activities of Floccular Purkinje Cells and Flocculus-Receiving Neurons in the Vestibular Nuclei

If, as described above, eye movements in flocculectomized animals lag those of normal animals, one would expect that in normal animals the phase of the neurons in vestibular nuclei that project to the oculomotor nucleus and receive an input from the flocculus lead the neurons that also project to the oculomotor nucleus but do not receive this floccular input. This phase relation indeed exists. Stahl and Simpson (1995) demonstrated in awake, pigmented rabbits that flocculus-receiving neurons in the medial vestibular nucleus have a phase lead with respect to head position that is significantly greater than that of other oculomotor-projecting neurons in the medial vestibular nucleus that do not receive an input from the flocculus. This phase difference is present throughout a wide range of frequencies (0.05–0.8 Hz) for sinusoidal rotation about the vertical axis in both dark and light as well as for optokinetic stimulation about this axis. This finding suggests that the phase lead of the flocculus-receiving neurons over that of the nonflocculus-receiving neurons could be produced by the signal contributed by floccular Purkinje cells. If so, the Purkinje cells should functionally lead the flocculus-receiving neurons. As described above, the flocculus-receiving neurons in the rabbit medial vestibular nucleus are innervated by those floccular Purkinje cells whose climbing fibers respond best to optokinetic stimulation about the vertical axis (zones 2 and 4) but not by those Purkinje cells whose climbing fibers respond best to stimulation about a horizontal axis (De Zeeuw et al. 1994b). Thus, the phase of flocculus-receiving neurons in the medial vestibular nucleus should lag that of simple spike activities of vertical axis Purkinje cells; this relation indeed holds true (De Zeeuw et al. 1995b). Vertical axis Purkinje cells functionally lead the flocculus-receiving neurons in the medial vestibular nucleus at all frequencies (0.05–0.8 Hz) during VOR both in the light and in the dark (Fig. 9.17). Thus, the floccular signals in the rabbit can account for a phase advance of the flocculus-receiving neurons and thereby of the net premotor signal (defined as the summed activity of all cells synapsing on extraocular motoneuron pools). Similar phase relations presumably occur in mice (Fig. 9.14) (Grüsser-Cornehls 1995; Grüsser-Cornehls et al. 1995; Goossens, Hoebeek, Frens, and De Zeeuw, unpublished observations). Although no direct comparisons were made between flocculus-receiving neurons and nonflocculus-receiving neurons that project to the oculomotor complex, Grüsser-Cornehls and colleagues were able to show that the phase of the simple spike activity of floccular Purkinje cells in mice generally leads those of the vestibular nuclei neurons. Moreover, it was shown that the phase values of floccular Purkinje cells and vestibular nuclei neurons in *Weaver* mutant mice (B6CBA *wv/wv*) show very high standard deviations, indicating that a proper organization of the granule cell layer is necessary for a steady phase control (Grüsser-Cornehls 1995). Whether the same phase relations also hold true in the flocculus of monkeys is unclear. The primate Purkinje cells comparable to those of the rabbit and mice may be the "eye movement only" group, which makes up approximately a quarter of the recorded population in the monkey (Miles et al. 1980). It is unknown whether these "eye movement only" Purkinje cells lead their medial vestibular targets (Miles et al. 1980; Lisberger et al. 1994).

Because the flocculus-receiving neurons in the magnocellular and parvicellular medial vestibular nuclei lead the neurons that do not receive an input from the flocculus, the signals of the flocculus-receiving neurons can be imagined to be synthesized by imposing the vertical axis Purkinje cell activity on a neuronal group whose properties are, in the absence of the floccular input, identical to those of the neurons that do not receive an input from the flocculus (De Zeeuw et al. 1995b; Stahl and Simpson 1995). A consequence of this synthesis would be an increase in the phase lead of the net premotor signal. The phase-leading property of the floccular signal may indicate that one role of the flocculus is to compensate for excessive integration by the brain stem neural circuitry. If the brain stem integrator network has limitations in the precision of the phase lag that it produces, a separate structure may be required for fine-tuning. The flocculus could

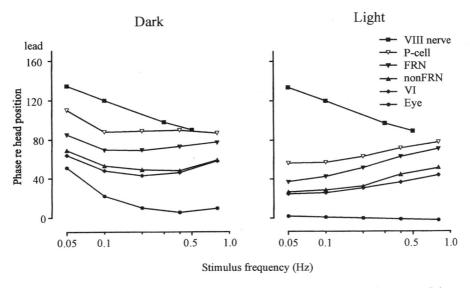

FIGURE 9.17. Summary of the phase relationships among different elements of the VOR circuit. The Purkinje cells lead the FRNs (flocculus-receiving neurons) at all frequencies in both the dark and the light, while the FRNs in turn lead the non-FRNs at all frequencies in both the dark and the light. Phases (medians) are referenced to contralateral head position (+ = lead). For clarity of presentation, the phases of the medial vestibular nucleus neurons (FRNs and non-FRNs) and primary afferents (VIIIth nerve) have been shifted by 180° to reflect, respectively, their direct and indirect inhibitory actions on the ipsilateral sixth nucleus neurons. Thus, the simple spike response of vertical axis Purkinje cells (P cells) and the firing rate of primary vestibular afferents increases and decreases with contralateral head movements, respectively. Therefore, the simple spike activity of these Purkinje cells cannot be solely the result of an excitatory mossy fiber input derived from ipsilateral second-order vestibular neurons. Instead, these phase relations indicate that other sources such as eye velocity signals or vestibular inputs from the contralateral side must contribute. Sixth nucleus data were obtained in the light, and the sixth nucleus phase values in the dark were obtained by referencing to eye position (Eye) in the dark. (Modified from Stahl and Simpson 1995; De Zeeuw et al. 1995b).

perform this role by taking the overintegrated premotor signal from the brain stem, emphasizing the component in phase with eye velocity relative to the component in phase with eye position, and then injecting that signal into the vestibular nuclei. As such, during sinusoidal stimulation, the flocculus creates a signal that leads the original input signal. This scenario predicts the existence of a cell group producing an overintegrated signal whose phase lies between the motoneurons and eye position (see also Fig. 9.17). The weighted sum of the flocculus-receiving neurons, the nonflocculus-receiving neurons, and the phase-lagged premotor cell group would be in phase with the abducens neurons. The proposed lagging cell group could be located in the prepositus hypoglossi, which projects to the flocculus, the oculomotor nucleus, and the abducens nucleus (for a review, see McCrea et al. 1979; cat, McCrea and Baker 1985; McCrea et al. 1980; rabbit, Barmack et al. 1992a, 1992b), and which contains some cells with a phase close to eye position (cat, Lopez-Barneo et al. 1982; Escudero et al. 1992; monkey, McFarland and Fuchs 1992). Another possible location for the phase-lagged premotor cell group is the caudal medial vestibular nucleus, which also projects to the flocculus (Kotchabhakdi and Walberg 1978; Sato et al. 1983) and the oculomotor complex (Thunnissen 1990) and receives a prominent input from nodular (Wylie et al. 1994) but not floccular vertical axis Purkinje cells (De Zeeuw et al. 1994a).

If, as described above, the phase lag of the compensatory eye movement performance in flocculectomized animals is particularly prominent during OKR, even more so than during VOR, one would expect that the flocculus-receiving neurons would have a stronger visual signal than the nonflocculus-receiving neurons in the medial vestibular nucleus (Stahl and Simpson 1995). As shown in Figure 9.18, the difference between the phase during VOR in the light and in the dark is greater for flocculus-receiving neurons than for nonflocculus-receiving neurons at all frequencies, indicating that vision makes a larger contribution to the phase of flocculus-receiving neurons. If this difference is attributable to the vertical axis Purkinje cell input to the flocculus-receiving neurons, then the vertical axis Purkinje cells should exhibit an even larger dark versus light phase difference than flocculus-receiving neurons. This prediction is upheld (De Zeeuw et al. 1995b). Thus, the visual signals of the floccular Purkinje cells not only evoke a prominent effect in gain control but also in phase control of compensatory eye movements.

5. General Conclusions

The input and output connections of the flocculus of the vestibulocerebellum are strongly topographically organized in line with the organization of the semicircular canals, and the simple spike and complex spike activities of its Purkinje cells are optimally modulated following optokinetic and/or vestibular stimulation about the axes that run through these canals. The flocculus plays a role in controling both OKR and VOR, but it acts to a greater degree on the OKR than on the VOR. Usually, the flocculus exerts a gain-enhancing and a phase-leading effect on these reflexes. Recent "genetic lesion" studies in mouse mutants indicate that OKR and VOR performances can be separately altered and that their gain and phase parameters can also be partially separately influenced. In general, the gain values can be influenced by merely changing the expression of genes that may influence the synaptic activity of neurons but that do not necessarily affect

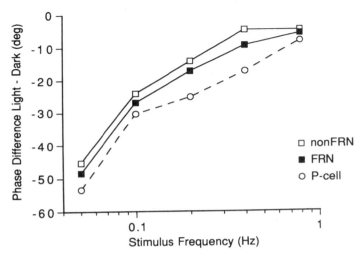

FIGURE 9.18. Contribution of vision to neuronal phase of floccular Purkinje cells (P cells) and flocculus-receiving neurons (FRNs) and nonflocculus-receiving neurons (FRNs) in the medial vestibular nucleus of the rabbit. The firing rate phase for the VOR in the dark was subtracted from the phase for the VOR in the light and plotted versus the stimulus frequency. The presence of vision had large effects on the phase of all three cell groups, particularly at the lowest frequencies. However, the Purkinje cells were most affected, indicating that the stronger visual signal of the FRNs (in comparison with the signal of non-FRNs) is probably attributable to the vertical axis Purkinje cell input to the FRNs. All phases are in reference to contralateral head position. (From De Zeeuw et al. 1995b.)

the cytoarchitecture of the vestibulocerebellum; phase changes, on the other hand, usually require more robust genetic effects that do affect the cytoarchitecture and the hardware wiring.

Acknowledgments. The authors would like to acknowledge Eddie Dalm, Edith Klink, Erika Goedknegt, and Ton Verkerk for excellent technical support. The present study is supported by HERSENSTICHTING NOW-SLW/ALW and NWO-MW (Netherlands Organization for Scientific Research).

References

Alley K, Baker R, Simpson JI (1975) Afferents to the vestibulo-cerebellum and the origin of the visual climbing fibers in the rabbit. Brain Res 98:582–589.
André P, Pompeiano O, White SR (1993) Activation of muscarinic receptors induces a long-lasting enhancement of Purkinje cell responses to glutamate. Brain Res 617:28–36.

André P, Fascetti F, Pompeiano O, White SR (1994) The muscarinic agonist, bethane-chol, enhances GABA-induced inhibition of Purkinje cells in the cerebellar cortex. Brain Res 637:1–9.

Balkema GW, Mangini NJ, Pinto LH, Vanable JW Jr (1984) Visually evoked eye movements in mouse mutants and inbred strains. A screening report. Invest Ophthalmol Vis Sci 25:795–800.

Barmack NH, Pettorossi VE (1985) Effects of unilateral lesions of the flocculus on optokinetic and vestibuloocular reflexes of the rabbit. J Neurophysiol 53:481–496.

Barmack NH, Simpson JI (1980) Effects of microlesions of dorsal cap of inferior olive of rabbits on optokinetic and vestibuloocular reflexes. J Neurophysiol 43:182–206.

Barmack NH, Baughman RW, Eckenstein FP (1992a) Cholinergic innervation of the cerebellum of rat, rabbit, cat, and monkey as revealed by choline acetyl-transferase activity and immunohistochemistry. J Comp Neurol 317:233–249.

Barmack NH, Baughman RW, Eckenstein FP, Shojaku H (1992b) Secondary vestibular cholinergic projection to the cerebellum of rabbit and rat as revealed by choline acetyltransferase immunohistochemistry, retrograde and orthograde tracers. J Comp Neurol 317:250–270.

Barski JJ, Hartmann J, Rose CR, Hoebeek F, Morl K, Noll-Hussong M, Related Articles, Links De Zeeuw CI, Konnerth A, Meyer M (2003) Calbindin in cerebellar Purkinje cells is a critical determinant of the precision of motor coordination. J Neurosci Apr 15;23(8):3469–3477.

Bishop G, Ho RH, King JS (1985) Localization of serotonin immunoreactivity in the opossum cerebellum. J Comp Neurol 15:301–321.

Blanks JC, Bok D (1977) An autoradiographic analysis of postnatal cell proliferation in the normal and degenerative mouse retina. J Comp Neurol 174:317–327.

Blanks RH, Precht W (1983) Responses of units in the rat cerebellar flocculus during optokinetic and vestibular stimulation. Exp Brain Res 53(1):1–15.

Bloedel JR, Courville J (1981) A review of cerebellar afferent systems. In: Brooks VB (ed) Handbook of Physiology, Volume II. Motor Control. Baltimore: Williams & Wilkins, pp. 725–730.

Bonner TI (1992) Domains of muscarinic acetylcholine receptors that confer specificity of G protein coupling. Trends Pharmacol Sci 13:48–50.

Bower JM (1997) Is the cerebellum sensory for motor's sake, or motor for sensory's sake: the view from the whiskers of a rat? Prog Brain Res 114:463–496.

Brand S, Dahl AL, Mugnaini E (1976) The length of parallel fibers in the cat cerebellar cortex. An experimental light and electron microscopic study. Exp Brain Res 26(1):39–58.

Brooks VB, Thach WT (1981) Cerebellar control of posture and movement. In: Brookhart JM, Mountcastle VB (eds) Handbook of Physiology, Section I: The Nervous System, Volume II: Motor Control. Bethesda, MD: American Physiological Society, pp. 877–946.

Büttner-Ennever JA (1992) Patterns of connectivity in the vestibular nuclei. Ann N Y Acad Sci 656:363–378.

Büttner U, Büttner-Ennever JA (1988) Neuroanatomy of the oculomotor system. Present concepts of oculomotor organization. Rev Oculomot Res 2:3–33.

Caulfield MP (1993) Muscarinic receptors—characterization, coupling and function. Pharmacol Ther 58:319–379.

Chen C, Kano M, Chen L, Bao S, Kim JJ, Hashimoto K, Thompson RF, Tonegawa S (1995) Impaired motor coordination correlates with persistent multiple climbing fiber innervation in PKCγ mutant mice. Cell 83:1233–1242.

Chubb MC, Fuchs AF (1982) Contribution of y group of vestibular nuclei and dentate nucleus of cerebellum to generation of vertical smooth eye movements. J Neurophysiol 48:75–99.

Cuénod M, Audinat E, Do KQ, Gahwiler BH, Grandes P, Herrling P, Knopfel T, Perschak H, Streit P, Vollenweider F (1990) Homocysteic acid as transmitter candidate in the mammalian brain and excitatory amino acids in epilepsy. Adv Exp Med Biol 268:57–63.

Cummings S, King JS (1990) Coexistence of corticotropin releasing factor and enkephalin in cerebellar afferent systems. Synapse 5:167–174.

Daniels PD, Hassul M, Kimm J (1978) Dynamic analysis of the vestibulo-ocular reflex in the normal and flocculectomized chinchilla. Exp Neurol 58:32–45.

Delgado-Garcia JM, Vidal PP, Gomez C, Berthoz A (1989) A neurophysiological study of prepositus hypoglossi neurons projecting to oculomotor and preoculomotor nuclei in the alert cat. Neuroscience 29:291–307.

De Zeeuw CI, Berrebi AS (1995) Postsynaptic targets of Purkinje cell terminals in the cerebellar and vestibular nuclei of the rat. Eur J Neurosci 7:2322–2333.

De Zeeuw CI, Koekkoek SKE (1997) Signal processing in the C2-module of the flocculus and its role in head movement control. Prog Brain Res 114:299–321.

De Zeeuw CI, Wentzel PR, Mugnaini E (1993) Fine structure of the dorsal cap of the inferior olive and its GABAergic and non-GABAergic input from the nucleus prepositus hypoglossi in rat and rabbit. J Comp Neurol 327:63–82.

De Zeeuw CI, Gerrits NM, Voogd J, Leonard CS, Simpson JI (1994a) The rostral dorsal cap and ventrolateral outgrowth of the rabbit inferior olive receive a GABAergic input from dorsal group y and the ventral dentate nucleus. J Comp Neurol 341:420–432.

De Zeeuw CI, Wylie DR, DiGiorgi PL, Simpson JI (1994b) Projections of individual Purkinje cells of identified zones in the flocculus to the vestibular and cerebellar nuclei in the rabbit. J Comp Neurol 349:428–448.

De Zeeuw CI, Van den Burg J, Wylie DR, DiGiorgi PL, Ruigrok TJH, Teune T, Simpson JI (1995a) Morphological evidence for interzonal inhibition by Golgi cells in the rabbit vestibulo-cerebellum. Eur J Morphol 33:328–329.

De Zeeuw CI, Wylie DR, Stahl JS, Simpson JI (1995b) Phase relations of Purkinje cells in the rabbit flocculus during compensatory eye movements. J Neurophysiol 74:2051–2064.

De Zeeuw CI, Ruigrok TJH, Hawkins R, van Alphen AM (1997) Climbing fiber collaterals contact neurons in the cerebellar nuclei that provide a GABAergic feedback to the inferior olive. Neuroscience 80:981–987.

De Zeeuw CI, Hansel C, Bian F, Koekkoek SKE, van Alphen A, Linden DJ, Oberdick J (1998) Expression of a protein kinase C inhibitor in Purkinje cells blocks cerebellar long term depression and adaptation of the vestibulo-ocular reflex. Neuron 20:495–508.

Errico P, Barmack NH (1993) Origins of cerebellar mossy and climbing fibers immunoreactive for corticotropin-releasing factor in the rabbit. J Comp Neurol 336(2):307–320.

Escudero M, de la Cruz RR, Delgado-Garcia JM (1992) A physiological study of vestibular and prepositus hypoglossi neurones projecting to the abducens nucleus in the alert cat. J Physiol 458:539–560.

Evinger C (1988) Extraocular motor nuclei: location, morphology and afferents. Rev Oculomot Res 2:81–117.

Feil R, Harkmann J, Luo C, Wolfsgruber W, Schilling K, Feil S, Barski J, Meyer M, Konnerkh A, De Zeeuw CI, Hofmann F (2003) Impairment of LTD and cerebellar learning by Puskinge cell-specific ablazion of cGMP-dependent protein kinas I.P. J. Cell Biology, in press.

Frens MA, Mathoera AL, van der Steen J (2001) Floccular complex spike response to transparent retinal slip. Neuron 30:795–801.

Gerrits NM, Epema AH, van Linge A, Dalm E (1989) The primary vestibulocerebellar projection in the rabbit: absence of primary afferents in the flocculus. Neurosci Lett 105:27–33.

Goossens J, Daniel H, Rancillac A, van der Steen J, Oberdick J, Crepel F, De Zeeuw CI, Frens MA (2001) Expression of protein kinase C inhibitor blocks cerebellar long-term depression without affecting Purkinje cell excitability in alert mice. J Neurosci 21(15):5813–5823.

Graybiel AM, Hartwieg EA (1974) Afferent connections of the oculomotor complex in the cat: an experimental study with tracer techniques. Brain Res 68:167–173.

Grüsser-Cornehls U (1995) Responses of flocculus and vestibular nuclei neurons in Weaver mutant mice (B6CBA wv/wv) to combined head and body rotation. Exp Brain Res 107:26–33.

Grüsser-Cornehls U, Bohm P (1988) Horizontal optokinetic ocular nystagmus in wildtype (B6CBA+/+) and weaver mutant mice. Exp Brain Res 72:29–36.

Grüsser-Cornehls U, Niemschynski A, Plassman W (1995) Vestibular responses of flocculus and vestibular nuclei neurons in mice (B6cBA). Exp Brain Res 107:17–25.

Hasson T, Gillespie PG, Garcia JA, MacDonald RB, Zhao Y, Yee AG, Mooseker MS, Corey DP (1997) Unconventional myosins in inner-ear sensory epithelia. J Cell Biol 137:1287–1307.

Heckroth JA, Eisenman LM (1991) Olivary morphology and olivocerebellar topography in adult lurcher mutant mice. J Comp Neurol 312:641–651.

Hess EJ (1996) Identification of the weaver mouse mutation: the end of the beginning. Neuron 16:1073–1076.

Hess BJ, Blanks RH, Lannou J, Precht W (1989) Effects of kainic acid lesions of the nucleus reticularis tegmenti pontis on fast and slow phases of vestibulo-ocular and optokinetic reflexes in the pigmented rat. Exp Brain Res 74(1):63–79.

Highstein SM (1973) Synaptic linkage in the vestibulo-ocular and cerebello-vestibular pathways to the VIth nucleus in the rabbit. Exp Brain Res 17(3): 301–314.

Highstein SM, Ito M, Tsuchiya T (1971) Synaptic linkage in the vestibulo-ocular reflex pathway of rabbit. Exp Brain Res 113(3):306–326.

Highstein SM, Reisine H (1979) Synaptic and functional organization of vestibulo-ocular reflex pathways. Prog Brain Res 50:431–442.

Highstein SM, Partsalis A, Arikan R (1997) Role of the Y-group of the vestibular nuclei and flocculus of the cerebellum in motor learning of the vestibulo-ocular reflex. Prog Brain Res 114:383–401.

Hirata Y, Arikin R, Highstein SM (1998) Multiple linear regression analysis of floccular Purkinje cell simple spike activity during vertical visual following in squirrel monkeys. Soc Neurosci Abstr 554.8.

Hoddevik GH (1978) The projection from nucleus reticularis tegmenti pontis onto the cerebellum in the cat. A study using the methods of anterograde degeneration and retrograde axonal transport of horseradish peroxidase. Anat Embryol 153:227–242.

Hunt S, Schmidt J (1978) Some observations on the binding patterns of α-bungarotoxin in the central nervous system of the rat. Brain Res 157:213–232.

Ikai Y, Takada M, Shinonaga Y, Mizuno N (1992) Dopaminergic and non-dopaminergic neurons in the ventral tegmental area of the rat project respectively to the cerebellar cortex and deep cerebellar nuclei. Neuroscience 51:719–728.

Ikeda M, Houtani T, Ueyama T, Sugimoto T (1992) Distribution and cerebellar projections of cholinergic and corticotropin-releasing factor-containing neurons in the caudal vestibular nuclear complex and adjacent brainstem structures. Neuroscience 49(3):635–651.

Ito M (1982) Cerebellar control of the vestibulo-ocular reflex—Around the flocculus hypothesis. Annu Rev Neurosci 301:275–296.

Ito M (1984) The Cerebellum and Neural Control. Raven Press, New York.

Ito M (1998) Cerebellar learning in the vestibulo-ocular reflex. Trends Cognit Sci 2:305–371.

Ito M, Nisimaru N, Yamamoto M (1973) Specific neural connections for the cerebellar control of vestibulo-ocular reflexes. Brain Res 60(1):238–243.

Ito M, Shida T, Yagi N, Yamamoto M (1974) The cerebellar modification of rabbit's horizontal vestibulo-ocular reflex induced by sustained head rotation combined with visual stimulation. Proc Jpn Acad 50:85–89.

Ito M, Nisimaru N, Yamamoto M (1976) Pathways for the vestibulo-ocular reflex excitation arising from semicircular canals of rabbits. Exp Brain Res 24:257–271.

Ito M, Jastreboff PJ, Miyashita Y (1982) Specific effects of unilateral lesions in the flocculus upon eye movements in albino rabbits. Exp Brain Res 45:233–242.

Jaarsma D, Levey AI, Frostholm A, Rotter A, Voogd J (1995) Light-microscopic distribution and parasagittal organisation of muscarinic receptors in rabbit cerebellar cortex. J Chem Neuroanat 9:241–259.

Jaarsma D, Dino MR, Cozzari C, Mugnaini E (1996) Cerebellar choline acetyltransferase positive mossy fibres and their granule and unipolar brush cell targets: a model for central cholinergic nicotinic neurotransmission. J Neurocytol 25(12):829–842.

Jaarsma D, Ruigrok TJ, Caffe R, Cozzari C, Levey AI, Mugnaini E, Voogd J (1997) Cholinergic innervation and receptors in the cerebellum. Prog Brain Res 114:67–96.

Katoh A, Kitazawa H, Itohara S, Nagao S (1998) Dynamic characteristics and adaptibility of mouse vestibulo-ocular and optokinetic response eye movements and the role of the flocculo-olivary system revealed by chemical lesions. Proc Natl Acad Sci USA 95:7705–7710.

Kawaguchi Y (1985) Two groups of secondary vestibular neurons mediating horizontal canal signals, probably to the ipsilateral medial rectus muscle, under inhibitory influences from the cerebellar flocculus in rabbits. Neurosci Res 2:434–446.

Keele SW, Ivry R (1990) Does the cerebellum provide a common computation for diverse tasks? A timing hypothesis (Review). Ann NY Acad Sci 608:179–211.

Keller EL, Precht W (1979) Visual-vestibular responses in vestibular nuclear neurons in the intact and cerebellectomized, alert cat. Neuroscience 4:1599–1613.

Kennedy H, Courjon JH, Flandrin JM (1982) Vestibulo-ocular reflex and optokinetic nystagmus in adult cats reared in stroboscopic illumination. Exp Brain Res 48(2):279–287.

Khater TT, Quinn KJ, Pena, J, Baker JF, Peterson BW (1993) The latency of the cat vestibulo-ocular reflex before and after short- and long-term adaptation. Exp Brain Res 94:16–32.

Kimoto Y, Tohyama M, Satoh K, Sakumoto T, Takahashi Y, Shimizu N (1981) Fine structure of rat cerebellar noradrenaline terminals as visualized by potassium permanganate "in situ perfusion" fixation method. Neuroscience 6:47–58.

Kimura M, Takeda T, Maekawa K (1991) Contribution of eye muscle proprioception to velocity-response characteristics of eye movements: involvement of the cerebellar flocculus. Neurosci Res 12:160–168.

Koekkoek SKE, Van Alphen AM, Van den Burg J, Grosveld F, Galjart N, Zeeuw CI (1997) Gain adaptation and phase dynamics of compensatory eye movements in mice. Genes Function 1:175–190.

Kotchabhakdi N, Walberg F (1977) Cerebeller afferents from neurons in motor nuclei of cranial nerves demonstrated by retrograde axonal transport of horseradish peroxidase. Brain Res 137:158–163.

Kotchabhakdi N, Walberg F (1978) Cerebeller afferents from the vestibular nuclei in the cat: an experimental study with the method of retrograde axonal transport of horseradish peroxidase. Exp Brain Res 31:591–604.

Kotchabhakdi N, Hoddevik GH, Walberg F (1978) Cerebellar afferent projections from the perihypoglossal nuclei: an experimental study with the method of retrograde axonal transport of horseradish peroxidase. Exp Brain Res 31(1): 13–29.

Laine J, Axelrad H (2002) Extending the cerebellar Lugaro cell class. Neuroscience 115:363–374.

Langer T, Fuchs AF, Chubb MC, Scudder CA, Lisberger SG (1985a) Floccular efferents in the rhesus macaque as revealed by autoradiography and horseradish peroxidase. J Comp Neurol 235:26–37.

Langer T, Fuchs AF, Scudder CA, Chubb MC (1985b) Afferents to the flocculus of the cerebellum in the rhesus macaque as revealed by retrograde transport of horseradish peroxidase. J Comp Neurol 235:1–25.

Leonard CS (1986) Signal characteristics of cerebellar Purkinje cells in the rabbit flocculus during compensatory eye movements. Ph.D. Thesis, New York University, New York.

Leonard CS, Simpson JI (1986) Simple spike modulation of floccular Purkinje cells during the reversible blockade of their climbing fiber afferents. In: Keller EL, Zee DS (eds) Adaptive Processes in Visual and Oculomotor Systems. Oxford: Pergamon, pp. 429–434.

Lisberger SG (1998) Cerebellar LTD: a molecular mechanism of behavioral learning? Cell 92:701–704.

Lisberger SG, Pavelko TA (1988) Brain stem neurons in modified pathways for motor learning in the primate vestibulo-ocular reflex. Science 242:771–773.

Lisberger SG, Miles FA, Zee DS (1984) Signals used to compute errors in monkey vestibule-ocular reflex: possible role of flocculus. J Neurophysiol 52: 1140–1153.

Lisberger SG, Pavelko TA, Broussard DM (1994) Responses during eye movements of brainstem neurons that receive monosynaptic inhibition from the flocculus and ventral paraflocculus in monkeys. J Neurophysiol 72:909–927.

Llinás RR (1982) Radial connectivity in the cerebellar cortex: a novel view regarding the functional organization of the molecular layer. Exp Brain Res Suppl 6:189–194.

Lopez-Barneo J, Darlot C, Berthoz A, Baker R (1982) Neuronal activity in prepositus nucleus correlated with eye movement in the alert cat. J Neurophysiol 47:329–352.

Madtes PC Jr, King JS (1994) Distribution of cholecystokinin binding sites in the North American opossum cerebellum. J Chem Neuroanat 7(1–2):105–112.

Maekawa K, Takeda T (1977) Afferent pathways from the visual system to the cerebellar flocculus in the rabbit. In: Baker R, Berthoz A (eds) Control by Gaze of Brain Stem Neurons. Amsterdam: Elsevier, pp. 187–195.

Maekawa K, Kimura M (1980) Mossy fiber projection to the cerebellar flocculus from the extraocular muscle afferents. Brain Res 191:313–325.

Mangini NJ, Vanable JW Jr, Williams MA, Pinto LH (1985) The optokinetic nystagmus and ocular pigmentation of hypopigmented mouse mutants. J Comp Neurol 241(2):191–209.

Matsushita K, Wakamori M, Rhyu IJ, Arii T, Oda S, Mori Y, Imoto K (2002) Bidirectional alterations in cerebellar synaptic transmission of tottering and rolling Ca^{2+} channel mutant mice. J Neurosci 22(11):4388–4398.

McCrea RA, Baker R (1985) Anatomical connections of the nucleus prepositus of the cat. J Comp Neurol 237:377–407.

McCrea RA, Baker R, Delgado-Garcia J (1979) Afferent and efferent organization of the prepositus hypoglossi nucleus. Prog Brain Res 50:653–665.

McCrea RA, Yoshida K, Berthoz A, Baker R (1980) Eye movement related activity and morphology of second order vestibular neurons terminating in the cat abducens nucleus. Exp Brain Res 40:468–473.

McCrea RA, Strassman A, May E, Highstein SM (1987) Anatomical and physiological characteristics of vestibular neurons mediating the horizontal vestibulo-ocular reflex of the squirrel monkey. J Comp Neurol 264:547–570.

McFarland JL, Fuchs AF (1992) Discharge patterns in nucleus prepositus hypoglossi and adjacent medial vestibular nucleus during horizontal eye movement in behaving macaques. J Neurophysiol 68:319–332.

Miles FA, Fuller JH, Braitman DJ, Dow BM (1980) Long-term adaptive changes in primate vestibuloocular reflex. III. Electrophysiological observations in flocculus of normal monkeys. J Neurophysiol 43:1437–1476.

Mitchiner JC, Pinto LH, Vanable JW Jr (1976) Evoked eye movements in the mouse (*Mus musculus*). Vision Res 16:1169–1171.

Miyashita Y, Ito M, Jastreboff PJ, Maekawa K, Nagao S (1980) Effect upon eye movements of rabbits induced by severance of mossy fiber visual pathway to the cerebellar flocculus. Brain Res 198:210–215.

Mugnaini E (1983) The length of cerebellar parallel fibers in chicken and rhesus monkey. J Comp Neurol 220:7–15.

Mugnaini E, Floris A (1994) The unipolar brush cell: a neglected neuron of the mammalian cerebellar cortex. J Comp Neurol 339:174–180.

Mugnaini E, Dino M, Jaarsma D (1997) The unipolar brush cells of the mammalian cerebellum and cochlear nucleus: cytology and microcircuitry. Prog Brain Res 114:131–151.

Nagao S (1983) Effects of vestibulocerebellar lesions upon dynamic characteristics and adaptation of vestibulo-ocular and optokinetic responses in pigmented rabbits. Exp Brain Res 53:36–46.

Nagao S (1989a) Behavior of floccular Purkinje cells correlated with adaptation of vestibulo-ocular reflex in pigmented rabbits. Exp Brain Res 77:531–540.

Nagao S (1989b) Role of cerebellar flocculus in adaptive interaction between optokinetic eye movement response and vestibulo-ocular reflex in pigmented rabbits. Exp Brain Res 77:541–551.

Nagao S, Kitazawa H, Osanai R, Hiramatsu T (1997) Acute effects of tetrahydrobiopterin on the dynamic characteristics and adaptability of the vestibulo-ocular reflex in normal and flocculus lesioned rabbits. Neurosci Lett 231:41–44.

Ojima H, Kawajiri SI, Yamasaki T (1989) Cholinergic innervation of the rat cerebellum: qualitative and quantitative analyses of elements immunoreactive to a monoclonal antibody against choline acetyltransferase. J Comp Neurol 290:41–52.

Otten LJ, Rugg MD (2002) The birth of a memory. Trends Neurosci 25:279–282.

Overbeck TL, King JS (1999) Developmental expression of corticotropin-releasing factor in the postnatal murine cerebellum. Brain Res Dev Brain Res 115:145–159.

Panagopoulos NT, Papadopoulos GC, Matsokis NA (1991) Dopaminergic innervation and binding in the rat cerebellum. Neurosci Lett 130:208–212.

Payen ET, Verkerk D, Michalovich SD, Dreyer A, Winterpacht A, Lee B, De Zeeuw CI, Grosveld F, Galjart N (1998) The centromeric/nucleolar chromatin protein ZFP-37 may function to specify neuronal nuclear domains. J Biol Chem 273:9099–9109.

Pazos A, Palacios JM (1985) Quantitative autoradiographic mapping of serotonin receptors in the rat brain. I. Serotonin-1 receptors. Brain Res 346:205–230.

Raymond JL, Lisberger SG (1998) Neural learning rules for the vestibulo-ocular reflex. J Neurosci 18:9112–9129.

Reisine H, Highstein SM (1979) The ascending tract of Deiters conveys a head velocity signal to medial rectus motoneurons. Brain Res 170:172–176.

Reisine H, Strassman A, Highstein SM (1981) Eye position and head velocity signals are conveyed to medial rectus motoneurons in the alert cat by the ascending tract of Deiters. Brain Res 211:153–157.

Robinson DA (1976) Adaptive gain control of vestibuloocular reflex by the cerebellum. J Neurophysiol 39:954–968.

Ruigrok TJH, Osse RJ, Voogd J (1992) Organization of inferior olivary projections to the flocculus and ventral paraflocculus of the rat cerebellum. J Comp Neurol 316:129–150.

Sato Y, Kawasaki T (1990) Operational unit responsible for plane-specific control of eye movement by cerebellar flocculus in cat. J Neurophysiol 64:551–564.

Sato Y, Kawasaki T, Ikarashi K (1983) Afferent projections from the brainstem to the floccular three zones in cats. II. Mossy fiber projections. Brain Res 272:37–48.

Sato Y, Kanda K, Kawasaki T (1988) Target neurons of floccular middle zone inhibition in medial vestibular nucleus. Brain Res 446:225–235.

Shibuki K, Gomi H, Chen L, Bao S, Kim JJ, Wakatsuki H, Fujisaki T, Fujimoto K, Katoh A, Ikeda T, Chen C, Thompson RF, Itohara S (1996) Deficient cerebellar long-term depression, impaired eyeblink conditioning, and normal motor coordination in GFAP mutant mice. Neuron 16:587–599.

Shidara M, Kawano K, Gomi H, Kawato M (1993) Inverse-dynamics model eye movement control by Purkinje cells in the cerebellum. Nature 365:50–52.

Shinoda Y, Yoshida K (1975) Neural pathways from the vestibular labyrinths to the flocculus in the cat. Exp Brain Res 22:97–111.

Simpson JI, Leonard CS, Soodak RE (1988) The accessory optic system of rabbit. II. Spatial organization of direction selectivity. J Neurophysiol 60:2055–2072.

Simpson JI, Wylie DR, De Zeeuw CI (1996) On climbing fiber signals and their consequences. Behav Brain Sci 19:380–394.

Simpson JI, Belton T, Suh M, Winkelman BHJ (2002) Complex spike activity in the flocculus signals more than the eye can see. Annals of NYAS 978:232–237.

Soodak RE, Simpson JI (1988) The accessory optic system of rabbit. I. Basic visual response properties. J Neurophysiol 60:2037–2054.

Stahl JS, Simpson JI (1995) Dynamics of rabbit vestibular nucleus neurons and the influence of the flocculus. J Neurophysiol 73:1396–1413.

Stahl JS, van Alphen AM, De Zeeuw CI (2000) A comparison of video and magnetic search coil recordings of mouse eye movements. J Neurosci Methods 99:101–110.

Sun JC, van Alphen AM, Huygens P, Wagenaar M, Hoogenraad CC, Hasson T, Koekkoek SKE, Bohne BA, De Zeeuw CI (2001) Origin of vestibular dysfunction in Usher syndrome type 1B/shaker-1 mice. Neurobiol Dis 8:69–77.

Sutin J, Minneman KP (1985) Adrenergic beta receptors are not uniformly distributed in the cerebellar cortex. J Comp Neurol 236:547–554.

Takemori S, Cohen B (1974) Loss of visual suppression of vestibular nystagmus after flocculus lesions. Brain Res 72:213–224.

Tamayo ML, Maldonado C, Plaza SL, Alvira GM, Tamayo GE, Zambrano M, Frias JL, Bernal JE (1996) Neuroradiology and clinical aspects of Usher syndrome. Clin Genet 50:126–132.

Tan HS (1992) Gaze stabilization in the rabbit. Thesis, Erasmus University, Rotterdam.

Tan HS, van Neerven J, Collewijn H, Pompeiano O (1991) Effects of alpha-noradrenergic substances on the optokinetic and vestibulo-ocular responses in the rabbit: a study with systemic and intrafloccular injections. Brain Res 562(2):207–215.

Tan HS, Collewijn H (1991) Cholinergic modulation of optokinetic and vestibulo-ocular responses: a study with microinjections in the flocculus of the rabbit. Exp Brain Res 85(3):475–481.

Tan HS, Collewijn H (1992a) Cholinergic and noradrenergic stimulation in the rabbit flocculus has synergistic facilitatory effects on optokinetic responses. Brain Res 586:130–134.

Tan HS, Collewijn H (1992b) Muscarinic nature of cholinergic receptors in the cerebellar flocculus involved in the enhancement of the rabbit's optokinetic response. Brain Res 591:337–340.

Tan HS, Gerrits NM (1992) Laterality in the vestibulo-cerebellar mossy fiber projection to flocculus and caudal vermis in the rabbit: a retrograde fluorescent double-labeling study. Neuroscience 47:909–919.

Tan HS, Collewijn H, Van der Steen J (1992) Optokinetic nystagmus in the rabbit and its modulation by bilateral microinjection of carbachol in the cerebellar flocculus. Exp Brain Res 90:456–468.

Tan HS, Collewijn H, Van der Steen J (1993) Shortening of vestibular nystagmus in response to velocity steps by microinjection of carbachol in the rabbit's cerebellar flocculus. Exp Brain Res 92:385–390.

Tan J, Gerrits NM, Nanhoe RS, Simpson JI, Voogd J (1995) Zonal organization of the climbing fiber projection to the flocculus and nodulus of the rabbit. A combined axonal tracing and acetylcholinesterase histochemical study. J Comp Neurol 356:23–50.

Terwindt GM, Ophoff RA, Haan J, Sandkuijl LA, Frants RR, Ferrari MD (1998) Migraine, ataxia and epilepsy: a challenging spectrum of genetically determined calcium channelopathies. Eur J Hum Genet 6:297–307.

Thunnissen IE (1990) The vestibulo-cerebellar and vestibulo-oculomotor projections in the rabbit. Ph.D. Thesis, Erasmus University, Rotterdam.

Uchino Y, Suzuki S (1983) Axon collaterals to the extraocular motoneuron pools of inhibitory vestibuloocular neurons activated from the anterior, posterior, and horizontal semicircular canals in the cat. Neurosci Lett 37:129–135.

Uchino Y, Hirai N, Suzuki S (1982) Branching pattern and properties of vertical- and horizontal-related excitatory vestibulo-ocular neurons in the cat. J Neurophysiol 48:891–903.

Umetani T (1992) Efferent projections from the flocculus in the albino rat as revealed by an autoradiographic orthograde tracing method. Brain Res 586: 91–103.

Van Alphen AM, De Zeeuw CI (2002) Cerebellar LTD facilitates but is not essential for longterm adaptation of the vestibulo-ocular reflex. Eur J Neurosci 16:486–490.

Van Alphen AM, Stahl J, De Zeeuw CI (2001) The dynamic characteristics of the mouse horizontal vestibulo-ocular and optokinetic response. Brain Res 890: 296–305.

Van Alphen AM, Schepers T, Luo C, De Zeeuw CI (2002) Motor performance and motor learning in Lurcher mice. Proc N Y Acad Sci 978:413–425.

Van Der Steen J, Tan HS (1997) Cholinergic control in the floccular cerebellum of the rabbit. Prog Brain Res 114:335–345.

Van Der Steen J, Simpson JI, Tan J (1994) Functional and anatomic organization of three-dimensional eye movements in rabbit cerebellar flocculus. J Neurophysiol 72:31–46.

Van Neerven J (1990) Visuo-vestibular interactions in the rabbit: the role of the flocculus and its mono-aminergic inputs. Thesis, Erasmus University, Rotterdam.

Van Neerven J, Pomepeiano O, Collewijn H (1989) Depression of the vestibulo-ocular and optokinetic responses by intrafloccular microinjection of GABA-A and GABA-B agonists in the rabbit. Arch Ital Biol 127:243–263.

Van Neerven J, Pompeiano O, Collewijn H, Van der Steen J (1990) Injections of beta-noradrenergic substances in the flocculus of rabbits affect adaptation of the VOR gain. Exp Brain Res 79:249–260.

Voogd J, Bigaré F (1980) Topographical distribution of olivary and corticonuclear fibers in the cerebellum. The inferior olivary nucleus. New York: Raven Press, pp. 207–305.

Voogd J, Gerrits NM, Ruigrok TJH (1996) Organization of the vestibulocerebellum. Ann N Y Acad Sci 781:553–579.

Waespe W, Cohen B, Raphan T (1983) Role of the flocculus and paraflocculus in optokinetic nystagmus and visual-vestibular interactions: effects of lesions. Exp Brain Res 50:9–33.

Weil D, Blanchard S, Kaplan J, Guilford P, Gibson F, Walsh J, Mburu P, Varela A, Levilliers J, Weston MD (1995) Defective myosin VIIA gene responsible for Usher syndrome type 1B. Nature 374:60–61.

Weston MD, Kelley PM, Overbeck LD, Wagenaar M, Orten DJ, Hasson T, Chen ZY, Corey P, Mooseker M, Sumegi J, Cremers C, Moller C, Jacobsen SG, Gorin MB, Kimberling WJ (1996) Myosin VIIA mutation screening in 189 Usher syndrome type 1 patients. Am J Hum Genet 59:1074–1083.

Wylie DR, De Zeeuw CI, DiGiorgi PL, Simpson JI (1994) Projections of individual Purkinje cells of identified zones in the ventral nodulus to the vestibular and cerebellar nuclei in the rabbit. J Comp Neurol 349:448–464.

Yamamoto F, Sato Y, Kawasaki T (1986) The neuronal pathway from the flocculus to the oculomotor nucleus: an electrophysiological study of group y nucleus in cats. Brain Res 371:350–354.

Yamamoto M (1978) Localization of rabbit's flocculus Purkinje cells projecting to the lateral cerebellar nucleus and the nucleus prepositus hypoglossi investigated by means of the horseradish peroxidase retrograde axonal transport. Neurosci Lett 7:197–202.

Yamamoto M (1979) Vestibulo-ocular reflex pathways of rabbits and their representation in the cerebellar flocculus. Prog Brain Res 50:451–457.

Yamamoto M, Shimoyama I, Highstein SM (1978) Vestibular nucleus neurons relaying excitation from the anterior canal to the oculomotor nucleus. Brain Res 148:31–42.

Zee DS, Leigh RJ, Mathieu-Millaire F (1980) Cerebellar control of ocular gaze stability. Ann Neurol 7:37–40.

Zee DS, Yamazaki A, Butler PH, Gucer G (1981) Effects of ablation of flocculus and para-flocculus on eye movements in primate. J Neurophysiol 46:878–899.

Zuo J, De Jager PL, Takahashi KA, Jiang W, Linden DJ, Heintz N (1997) Neurodegeneration in Lurcher mice caused by mutation in delta2 glutamate receptor gene. Nature 388:769–773.

10
Localizing Sites for Plasticity in the Vestibular System

A.M. GREEN, Y. HIRATA, H.L. GALIANA, and S.M. HIGHSTEIN

1. Introduction

Motor skills are not innate but are acquired and perfected by a repetitive process of trial and error on the basis of sensory feedback. They can be considered a simple form of learning, generally referred to as "motor learning." How animals learn and store memories has been the focus of a great deal of research ranging from studies of learned motor behavior (du Lac et al. 1995; Raymond et al. 1996; Kim and Thompson 1997; Mauk 1997; Thach 1998) and spatial learning in the hippocampus (Bures et al. 1997; McNaughton 1998; Silva et al. 1998) to studies of language acquisition (Seidenberg 1997; Neville and Bavelier 1998). Significant progress has been made in understanding mechanisms of plasticity such as long-term potentiation (LTP) and long-term depression (LTD) at the cellular level (Ito 1989; Artola and Singer 1993; Bliss and Collingridge 1993; Bear and Malenka 1994; Daniel et al. 1998). Ultimately, however, an understanding of the neural basis for learning and memory requires that such cellular mechanisms be considered in a context where changes in neuronal responses may be directly associated with well-controlled behaviors. The vestibuloocular reflex (VOR) has been considered an excellent model for the study of motor learning due to its relative simplicity compared with other motor control systems and the precision with which the behavioral performance of the reflex may be measured. Furthermore, the VOR has been well-studied in alert, behaving animals because of the accessibility of its neural networks (see du Lac et al. 1995 for a review).

1.1. Adaptation of the VOR Performance

Because the VOR is required to compensate for high-frequency head movements during natural activities such as locomotion, it must operate at latencies much shorter than those associated with visual feedback delays; thus, even under viewing conditions, the reflex operates open-loop during high-frequency head movements. Nevertheless, visual feedback is required to continuously recalibrate or adapt the gain of the reflex for optimal performance in the face of changes in head size and eye growth during development, neural loss, and potential injury to the extraocular muscles or vestibular labyrinths. The ability to substantially modify reflex performance was first demonstrated by Gonshor and Melvill Jones (1971) when the VOR was altered after subjects wore reversing prisms. Subsequent investigations have revealed the ability to adapt the VOR gain in a variety of species, including humans (Gauthier and Robinson 1975; Gonshor and Melvill Jones 1976a, 1976b; Collewijn et al. 1983; Paige and Sargent 1991; Tiliket et al. 1994), monkeys (Miles and Fuller 1974; Miles and Eighmy 1980; Lisberger et al. 1983; Bello et al. 1991), rabbits (Ito et al. 1974; Collewijn and Grootendorst 1979), cats (Melvill Jones and Davies 1976; Robinson 1976; Godaux et al. 1983; Snyder and King 1988), and goldfish (*Carassius auratus*) (Schairer and Bennett 1981; Pastor et al. 1992). In these experiments, VOR gain (eye velocity/head velocity) was trained up or down by exposing the subject to conflicting visual–vestibular sensory information. For long-term training, this typically involves fitting an animal with magnifying lenses to increase the gain, miniaturizing or fixed-field lenses to reduce the gain, or reversing prisms to achieve a reversal in the direction of compensatory responses. The reflex then adapts over several hours, days, or even weeks in the course of attempting to perform normal daily behaviors. Alternatively, the gain may be adjusted over the short term by exposing the animal to a rotary stimulus that is paired with some "abnormal" behavior of the visual environment. For example, head rotation paired with rotation of an optokinetic or small target visual stimulus at the same speed but in the opposite direction is an appropriate stimulus to double the gain of the reflex, whereas rotating the subject and visual surround in tandem may be used to train the reflex gain toward zero.

When such visual–vestibular stimuli are used to adapt the VOR at a single temporal frequency, changes in reflex gain have been demonstrated to be frequency-selective. Specifically, observed increases or decreases in gain are largest at the adapting frequency with appropriate phase, but changes in gain are less robust at other frequencies and typically accompanied by phase lags or leads relative to the ideal compensatory response (Godaux et al. 1983; Lisberger et al. 1983; Powell et al. 1991; Raymond and Lisberger 1996). In addition, it has been shown that is possible to adapt the vertical VOR gain in opposite directions at low versus high frequencies (Hirata et al. 2002). Although most studies have focused on the ability to adapt reflex gain, appropriate manipulations of both the phase and speed of visual surround movement relative to head velocity have been used to selectively train the phase of the VOR, although in practice both the gain and phase of the VOR have been shown to be modified by such an adaptation protocol (Kramer et al. 1995, 1998). Furthermore, the ability to alter the direction of compensatory responses has been demonstrated following exposure to conflicting visual and vestibular rotational stimuli about mutually orthogonal axes (Schultheis and Robinson 1981; Baker et al. 1986, 1987; Harrison et al. 1986; Angelaki and Hess 1998). More recently, both cross-axis adaptation and modification of the viewing-location-dependent behavior of reflexive ocular responses to head *translation* (translational VOR) have also been demonstrated (Seidman et al. 1999; Wei and Angelaki 2001). In addition, many studies have considered the ability to adapt to injured states (e.g., unilateral or bilateral damage to the vestibular organs) in an effort to restore functional visual–vestibular oculomotor performance (Paige 1983b; Curthoys and Halmagyi 1995; Dieringer 1995).

1.2. Neural Basis for Gain Adaptation

To date, the neural basis for adaptation of the VOR has generally been studied by correlating the *average* responses of neural populations with behavior both before and following reflex gain adaptation. Hence, these observations do not necessarily reflect the process of motor learning per se but rather convey information about the final adapted state of the system (i.e., the "memory" of the learning that took place). Notable exceptions include the study of Partsalis et al. (1995a), who recorded the activity of Y group cells (vertical system vestibular neurons that receive floccular projections), as well as investigations of floccular Purkinje cell responses in primates (Watanabe 1984; Hirata and Highstein 2001), in the rabbit (Dufosse et al. 1978), and in goldfish (*C. auratus*) (Pastor et al. 1997). In these investigations, cell responses were recorded *during* the process of learning over the short term.

In attempts to localize central sites of plasticity, it has generally been assumed that the neurons at these locations should exhibit sensitivity changes that at least qualitatively parallel changes in behavioral performance. However, such deductions are complicated by the fact that in a system with multiple, recursive feedback loops, local changes must be distinguished from activity transmitted from a modified site elsewhere. As a result of this complexity, several approaches have been used to aid in the interpretation of experimental observations. These include: (1) consideration of the latencies at which neural versus behavioral adaptation-related changes are apparent (Lisberger 1984; Broussard et al. 1992; Pastor et al. 1992, 1994; Khater et al. 1993; Lisberger et al. 1994b, 1994c; Clendaniel et al. 2001); (2) lesion/inactivation studies (Ito et al. 1974; Robinson 1976; Nagao 1983; Lisberger et al. 1984; Luebke and Robinson 1994; Pastor et al. 1994; Partsalis et al. 1995b; McElligott et al. 1998); (3) quantitative modeling approaches (Miles et al. 1980b; Fujita 1982; Galiana 1986; Gomi and Kawato 1992; Lisberger and Sejnowski 1992; Quinn et al. 1992a, 1992b; Lisberger

1994; Galiana and Green 1998; Green 2000; Clendaniel et al. 2001; Hirata and Highstein 2001; Tabata et al. 2002).

1.2.1. The Cerebellar Hypothesis

Early investigations on plasticity in the VOR pathways led to two main hypotheses concerning potential sites for motor learning. The first suggested the cerebellar flocculus (the most ancient part of the cerebellum) as the main site for plasticity (Ito 1972) and has its origins in the quest to understand the function of the cerebellum. Despite much continued debate, the idea that a key role of the cerebellum is to learn motor skills has become almost common "knowledge." This hypothesis follows from the theoretical works of Brindley (1964), Marr (1969), and Albus (1971), who proposed that climbing fibers conveying sensory signals from the inferior olive to the cerebellar cortex play a "teacher" role in modifying the strength of synapses between parallel fibers and Purkinje cells. As a result, the pattern of outputs from Purkinje cells would be modified to reflect learned motor behavior in different sensory contexts. Investigation of the neural circuitry underlying the VOR played a key role in reinforcing this hypothesis.

Ito (1972) proposed that the flocculus provides a "side path" to the main brain stem VOR circuitry that is used specifically to adjust reflex performance. Early demonstrations in the rabbit (Ito et al. 1974) and cat (Robinson 1976) that the ability to adapt the gain of the VOR was abolished following flocculectomy supported this view. Subsequently, long-term depression (LTD) was demonstrated *in vitro* at the parallel fiber to purkinje cell synapse following paired climbing fiber and parallel fiber stimulation (Ito et al. 1981, 1982; Ito 1989). Thus, evidence for the necessity of an intact cerebellum to initiate motor adaptation as well as the demonstration of a mechanism by which it could be achieved provided strong reinforcement of the Brindley–Marr–Albus hypothesis. The suggestion of the cerebellum as a general site for motor learning has since been supported by numerous lesion studies, extracellular recordings, and stimulation studies in different movement systems. However, there remains considerable debate over the issue, in particular with respect to the role of LTD under natural learning conditions (Llinás et al. 1997; Mauk et al. 1998).

In the specific case of the VOR, single-unit recordings in alert, behaving animals have revealed changes in the modulation of Purkinje cells following or during reflex adaptation that are in the appropriate direction to support reflex adaptation during head rotation in the dark (Dufosse et al. 1978; Miles et al. 1980b; Watanabe 1984; Nagao 1989; Lisberger et al. 1994c; Pastor et al. 1997; Hirata and Highstein 2001). Furthermore, lesion studies demonstrate the requirement of an intact vestibulocerebellum for motor learning (Ito et al. 1974; Robinson 1976; Lisberger et al. 1984; Luebke and Robinson 1994; McElligott et al. 1998) and provide evidence that the cerebellum plays an important and necessary role in VOR adaptation. However, such observations do not provide conclusive evidence of a cerebellar *site* for plasticity.

1.2.2. The Brain Stem Hypothesis

The "cerebellar hypothesis" was questioned by the observations of Miles et al. (1980b), who investigated the effect of motor learning on the responses of floccular Purkinje cells, focusing on so-called gaze velocity Purkinje (GVP) cells (Miles et al. 1980a). During rotation in the dark, changes in GVP cell responses were observed that were in the appropriate direction to support motor learning after adaptation using reversing prisms to lower the VOR gain. However, following high-gain adaptation, changes sufficient to make a substantial contribution to the increase in behavioral VOR gain were not evident.[1] Furthermore, it was recognized that because GVP cells carry signals related to both head rotation and eye movement (Lisberger and Fuchs 1978; Miles et al. 1980a), the changes observed in cell responses following motor learning in the dark might largely reflect feedback from brain stem sites carrying efference copies of eye movement signals. To isolate the vestibular components of GVP cell activity from such signals related to eye movement, Miles et al. (1980b) also recorded cell responses following motor learning during head-fixed target stabilization (VOR cancellation).[2] Significant changes in cell responses were again apparent, but this time they were in the wrong direction to support the changes in reflex performance. It was concluded that the observed changes in the "head velocity sensitivity" of these Purkinje cells might be necessary to ensure appropriate system function following gain adaptation; however, they were unlikely to represent the main site for plasticity subserving VOR gain changes. Furthermore, it was noted that the earliest behavioral gain changes were observed at latencies too short to be mediated by GVP cells.

These observations led Miles and Lisberger (1981) to propose a new hypothesis (second main hypothesis), which suggested that the primary modifiable elements underlying gain changes in the VOR are located in the brain stem and are likely to be found in disynaptic or trisynaptic VOR pathways. The simple spike output of floccular Purkinje cells was proposed to provide the error or "teacher" signal to guide recalibration of the reflex.

---

[1] Significant changes in GVP cell activities following high-gain adaptation during head rotation in the dark, however, have subsequently been observed in other primate studies (Watanabe 1984; Lisberger et al. 1994c; Hirata and Highstein 2001).
[2] During a head-fixed target stabilization paradigm, the visual surround or target is moved in tandem with the head so that no reflexive counterrotation of the eyes in response to head movement is necessary to maintain stable viewing. Under these circumstances, the VOR is said to be cancelled or suppressed. This protocol has often been used to isolate cell sensitivities to head movement in the absence of eye movement (e.g., Scudder and Fuchs 1992).

Thus, floccular lesions would eliminate the ability to adapt reflex gain not because memory of the learning was stored in the flocculus but rather because signals conveyed from the flocculus to the brain stem were required to *induce* learning. Nevertheless, an extensive survey of primate medial vestibular nucleus (MVN) cell responses, including those of numerous position-vestibular-pause (PVP) type neurons (at the time considered the main interneurons in the "three-neuron arc" of the VOR), failed to demonstrate changes in average neuronal responses following motor learning that would be sufficient to account for the learned changes in VOR performance (Lisberger and Miles 1980); the actual sites for VOR plasticity thus remained in question. An earlier investigation in the cat, however, had demonstrated large changes in a subpopulation of vestibular nucleus (VN) cells (Keller and Precht 1979), suggesting that either this cell type had simply not been recorded in the primate study or that the failure to observe comparable VN cell behavior in the primate could be attributed to a species difference.

For the "brain stem hypothesis" of Miles and Lisberger (1981) to be valid, brain stem neurons that constitute sites for plasticity in the primary VN pathways should receive direct input from the cerebellar flocculus. An extensive reinvestigation of primate MVN cells identified a subset that receive monosynaptic inhibitory projections from the flocculus and thus have been called floccular target neurons (FTN) (Lisberger and Pavelko 1988; Lisberger et al. 1994a). These cells receive monosynaptic inputs from the ipsilateral vestibular nerve (Broussard and Lisberger 1992). Based on their similarity to the premotor eye-head-velocity (EHV) cells recorded by Scudder and Fuchs (1992) as well as demonstrations that floccular-receiving vestibular neurons make projections to the motor nuclei in other animals (Baker et al. 1972; Highstein 1973; Ito et al. 1977; Sato et al. 1988), primate FTN cells are also likely to project directly to motoneurons. Hence, these cells are likely to be interneurons in a direct disynaptic VOR pathway. Furthermore, FTN cells were shown on average to exhibit large changes in modulation following motor learning that were in the appropriate direction to support the adapted gain state when observed during both head rotation in the dark and head-fixed target stabilization paradigms (Lisberger 1988; Lisberger and Pavelko 1988; Lisberger et al. 1994b; Partsalis et al. 1995a).

Additional support for FTN cells as a site for plasticity was provided by the latency of their responses. Lisberger (1984) used rapid changes in head velocity to estimate the shortest latency at which modification in the behavioral gain was evident following VOR adaptation. At the shortest latencies (14 ms), no modification in the VOR was observed, whereas the evidence of gain adaptation became apparent after an average of 19 ms. Using similar stimuli, FTN cells were shown to respond with a median latency of 11 ms, from which it was estimated that they would influence eye movements after about 19 ms, in agreement with the latency for the modified component of the VOR (Lisberger et al. 1994b). Hence, there appears to be strong evidence that this cell type constitutes a site for VOR plasticity, although this does not preclude a modified site in other disynaptic or polysynaptic pathways. It should be noted that, using very rapid head accelerations, Khater et al. (1993) actually demonstrated learning-related changes in the earliest portion of compensatory responses, as has also been demonstrated when eye movements were elicited by electrical stimulation of the vestibular apparatus (Broussard and Lisberger 1992; Bronté-Stewart and Lisberger 1994). Clendaniel et al. (2001) have confirmed that indeed the latency of onset of the adapted response is not fixed but varies with the dynamic properties of the stimulus. With the lower head accelerations employed by Lisberger (1984), primary afferents were shown to respond with a range of latencies (Lisberger and Pavelko 1986). It has been suggested that the difference in observations with respect to the latency at which adapted behavioral responses are apparent might be explained if afferents with shorter latencies project into unmodified pathways, while those with longer latencies project into modified pathways (du Lac et al. 1995). As a result, learning might not be evident at the shortest latencies for lower head accelerations, while gain changes would be apparent at the onset of eye movement in response either to electrical stimulation, where afferents respond synchronously (Bronté-Stewart and Lisberger 1994), or to higher levels of acceleration, where synchronous afferent excitation is also assumed. Alternatively, it has been proposed that a frequency-dependent adaptation process may explain the stimulus-dependent latency of the adapted response (Clendaniel et al. 2001).

In the vertical VOR system, changes in neural modulation associated with VOR adaptation similar to those in the horizontal system have been observed on Y group cells. Like horizontal system FTNs, these cells receive monosynaptic inhibition from the flocculus; unlike FTNs, however, they receive only disynaptic or polysynaptic vestibular canal innervation (Sato and Kawasaki 1987). Y group cells have been shown to exhibit changes in their modulations both during and following motor learning that are in the appropriate direction to support the adaptation observed at the behavioral level (Partsalis et al. 1995a; Highstein et al. 1997).

In contrast to the large changes observed in the responses of FTN cells, on average, horizontal system PVP cells exhibited quite small differences in responses when the VOR gain was high as compared to when it was low, and only during contraversive head turns (Lisberger et al. 1994b). These changes were attributed to the feedback of eye movement signals onto this cell type because they were largely eliminated after correcting for the eye velocity sensitivities of these neurons estimated from their responses during pursuit. Furthermore, no changes in the head velocity sensitivities of PVP neurons were evident when cell responses were measured during head-fixed target viewing (Lisberger et al. 1994b). Based on these observations, it has been concluded that PVP cells do not constitute a primary site for memory following motor learning (Lisberger et al. 1994b; du Lac et al. 1995; Lisberger 1998).

1.2.3. Current Evidence for the Flocculus as a Site for Plasticity

Although there is ample evidence that the flocculus is required for VOR adaptation, a floccular site for plasticity is less well supported. The behavior of GVP cells following motor learning was reinvestigated by Lisberger et al. (1994c), who reconfirmed previous observations that GVP cells modulate in the correct direction to support motor learning during head rotation in the dark but demonstrate changes in an inappropriate direction to support motor learning when observed during VOR cancellation. Latency considerations also were consistent with previous observations. Because GVP cells respond with an average latency of 23 ms to ramps in head velocity and therefore would influence eye movements only after approximately 32 ms (Lisberger et al. 1994c; du Lac et al. 1995), they are unlikely to contribute to the earliest modified components of the VOR. However, despite observations more or less similar to those made in the earlier study of Miles et al. (1980a), it has now been concluded that GVP cells represent an important second site for plasticity (Lisberger et al. 1994c; du Lac et al. 1995; Raymond et al. 1996; Lisberger 1998). The basis for this assertion stems from: (1) observed changes in the activity of these cells following motor learning (Miles et al. 1980b; Watanabe 1984; Lisberger et al. 1994c; Hirata and Highstein 2001); (2) the conclusions of theoretical investigations that demonstrated that parallel changes in the strength of vestibular inputs to FTN and GVP cells were required to replicate experimental observations yet prevent unstable runaway behavior in a proposed VOR model structure (Lisberger and Sejnowski 1992; Lisberger 1994).

The flocculus has also been reemphasized as a likely site for plasticity based on studies in the vertical VOR system (Partsalis et al. 1995b). However, in this case, the evidence stems from observations made when the flocculus was temporarily pharmacologically inactivated following motor learning. After high or low gain adaptation, flocculus inactivation resulted in a partial loss of the observed change in Y cell behavior that was accompanied by a partial loss in the adapted gain at the behavioral level. These observations were taken to imply that the changes in Y cell modulation following gain adaptation were associated in part with an adaptive change at the brain stem level while another component of their observed change in modulation was likely due to signals conveyed from a second modified site through the flocculus (Partsalis et al. 1995b). It has been noted in a review of ongoing work in this lab, however, that although the observations described above suggest that some learning is transmitted to the Y group through the flocculus, "it has not been conclusively demonstrated that gain changes actually occur on floccular P cells" (Highstein 1998).

On the other hand, a behavioral study involving reversible inactivation of the flocculus in the cat led to different conclusions (Luebke and Robinson 1994). In this study, the VOR gain was adapted up or down over a period of several days using magnifying or fixed-field goggles. It was demonstrated that the gain could be quickly "deadapted" to the normal gain value over a period of approximately 30 minutes by visual–vestibular mismatch training. However, during "floccular shutdown", achieved by using 7 Hz stimulation of climbing fibers to temporarily eliminate the simple spike firing of floccular Purkinje cells, no readaptation occurred. The gain remained at its adapted value, suggesting that the modifiable synapses associated with motor learning are not in the flocculus and furthermore that an intact flocculus is not required to *maintain* the adapted state (as opposed to guiding adaptation).

Conflicting observations have also been made in goldfish (*C. auratus*) studies. Pastor et al. (1994) demonstrated that following VOR adaptation, changes in gain were apparent from the onset of the compensatory response to a step in head velocity and thus at a latency too short to be attributed solely to a site for plasticity in the flocculus. Furthermore, the behavioral expression of adaptation was partially retained following acute cerebellectomy. These observations are similar to those that have been reported in primates. However, unlike the observations made in primates, a detailed study of Purkinje cell responses during motor learning revealed no change in the head velocity sensitivity of these cells (Pastor et al. 1997). In addition, several months after complete cerebellar removal, a limited ability to adapt the steady-state component of the compensatory VOR response was demonstrated (Pastor et al. 1994). In contrast, McElligott et al. (1998) reported no change in goldfish (*C. auratus*) VOR gain in the dark or for various combinations of visual–vestibular stimulation prior to gain adaptation when the cerebellum was reversibly inactivated with lidocaine. However, following motor learning, cerebellar inactivation resulted in a complete loss of adaptation, which was restored after the effect of the lidocaine had worn off, suggesting that either the sites of gain modification do indeed reside in the cerebellum or that such sites are directly influenced by cerebellar signals. In addition, no ability to adapt the reflex was demonstrated during cerebellar inactivation.

1.3. Summary

In summary, there is general agreement that at least one site for plasticity in the VOR pathways resides in the brain stem. In primates, of the two main VN cell types that have been identified as premotor (i.e., FTN/EHV cells and PVP cells), only FTN cells (Y group cells in the vertical system) have been proposed as a modifiable site (Lisberger et al. 1994b; du Lac et al. 1995; Partsalis et al. 1995b; Highstein et al. 1997). The evidence supporting a site for plasticity in the cerebellum is less robust. Nevertheless, current opinion among primate investigators suggests either that GVP cells of the flocculus constitute a second modifiable site or that the flocculus transmits to the brain stem some portion of the learning occurring elsewhere.

2. Localizing Potential Sites for Plasticity in the VOR

As reviewed above, the ability to adapt VOR performance has been examined extensively, and several *potential* neural correlates for motor learning in this reflex have now been identified. To date, the majority of studies have focused on localizing central sites at which there is evidence for changes in sensitivity to sensory head velocity signals appropriate to drive changes in high-frequency reflex performance. Such changes in "head velocity sensitivity" have been identified by examining populations of neural responses to sinusoidal rotations at a particular stimulus frequency under different visual–vestibular stimulus conditions and/or to transient head velocity stimuli in the dark. Average neural responses prior to and following motor learning have then been compared and correlated with changes in reflex behavior. Nevertheless, because of the highly interconnected nature of the premotor circuitry, observed changes in firing activity need not be directly linked to a *local* change in sensitivity. Identification of the neural populations in which plastic changes actually occur thus is not trivial, and quantitative modeling approaches have been used extensively to aid in the interpretation of experimental observations.

The goal in the following sections will be to illustrate how conclusions with respect to potential sites for plasticity in the VOR can be significantly influenced by presumptions concerning the structure of the neural circuitry that underlies the reflex. By examining the predictions of several simple models, it will be shown that structural assumptions can affect the interpretation of experimental observations and influence perceptions with respect to the locations at which plastic changes are expected to take place. Limitations in knowledge of premotor cell interconnectivity and in the frequency range over which neural responses have been explored can also undermine the validity of the data analysis techniques commonly used to extract central sensitivities to particular signals. In the context of learning in the VOR, these issues will be investigated to explore where current knowledge limits the ability to localize sites for plasticity. Experimental paradigms and theoretical approaches that may aid in identifying the neural correlates for VOR adaptation will be proposed.

2.1. Simple Static Models of the VOR

2.1.1. No Local Interconnections Between cells

We may begin by considering the very simple schematics in Figures 10.1 and 10.2. These represent static VOR models, which are only appropriate to explore high-frequency changes in VOR *gain*, associated with sites for plasticity in the shortest-latency VOR pathways. For the moment, a single side of the brain is represented (i.e., the influence of the bilateral aspects of the premotor circuitry are not taken into account), and only modulations in neural activity about an assumed resting firing rate are considered.

In Figure 10.1, the circles labeled C_A and C_B represent interneurons involved in the shortest-latency disynaptic or "three-neuron-arc" VOR pathways (e.g., involving the PVP and/or FTN/EHV cells described above). Because neither visual pathways nor cerebellar inputs are included in the schematics of Figure 10.1, we will simply assume that C_A and C_B represent two distinct populations of premotor vestibular neurons. Each cell type receives angular head velocity signals, $\dot{H}$, from the semicircular canals via primary afferent projections with synaptic strengths $s_{(A,B)}$ and each makes a direct projection with synaptic strength $m_{(A,B)}$ onto extraocular motoneurons (MN). Motor projections are labeled as inhibitory to indicate that each cell is assumed to make either an excitatory projection to the contralateral MN or an inhibitory projection to the ipsilateral MN (McCrea et al. 1987; Scudder and Fuchs 1992) appropriate to drive conjugate eye velocity, $\dot{E}$, in the direction opposite to head motion. The same conventions are used in all of the examples below.

In this first example, notice that only feedforward pathways are represented and that there are no local interconnections between cells C_A and C_B. The VOR gain is then simply described by the expression (VOR$_{gain}$ = $\dot{E}/\dot{H} = -(s_A m_A + s_B m_B)$, and cell modulations are described by $C_A = s_A \dot{H}$ and $C_B = s_B \dot{H}$. Because premotor cells in the VOR pathways are known to participate in all types of eye movements (e.g., Scudder and Fuchs 1992; Cullen and McCrea 1993; Cullen et al. 1993; Lisberger et al. 1994a), we will assume that only synapses s_A and s_B represent appropriate potential sites for selective plasticity in the VOR. Under "normal" gain conditions (i.e., prior to any adaptation), we may define the baseline strengths of synapses s_A and s_B as s_{A0} and s_{B0}, respectively, and assume for the moment that these are equal. To simulate adapted gain states, we may define $s_{(A,B)} = s_{(A,B)0} + s_{(A,B)a}$, where s_{Aa} and s_{Ba} represent the changes in the strengths of synapses s_A and s_B, respectively, associated with motor learning. Figures 10.1B through 10.1G illustrate the sensitivity of a given cell to head rotation (i.e., modulation depth normalized for a 1 deg/s rotation) plotted as a function of the VOR gain when different assumptions are made about sites for plasticity and the relative strengths of motor projections m_A and m_B.

In Figures 10.1B and 10.1C, motor projection strengths m_A and m_B are both equal to 1, and baseline synaptic strengths s_{A0} and s_{B0} were both set to 0.425 to produce a normal VOR gain of 0.85 (as observed for monkeys in the dark; e.g., Paige 1983a). Figure 10.1B illustrates the case in which a site for plasticity exists only on one cell type (synapse s_B of cell C_B). Modifications in VOR gain are then associated with large changes in the modulation depth of C_B (solid trace in Fig. 10.1B) but no change in the activity of C_A (dashed trace in Fig. 10.2B). The changes in modulation on C_B are significantly larger than the reflex gain change (e.g., change in modulation

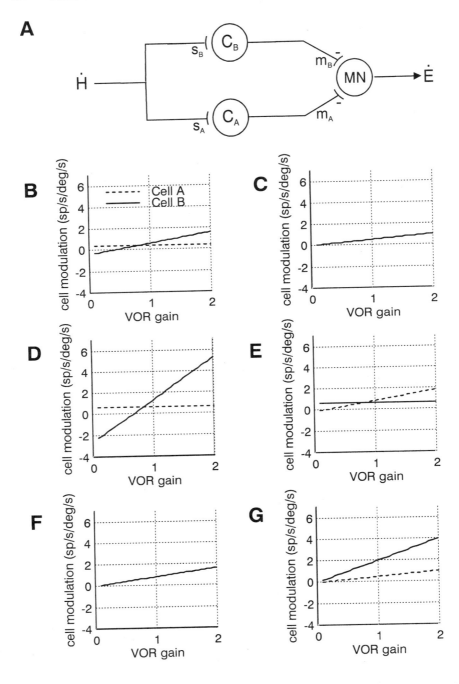

depth of approximately 3 for a doubling of the VOR gain) because under normal conditions both cell types contribute to reflex performance whereas only C_B makes a contribution to the *changes* in VOR gain above its set-point value. In Figure 10.1C, both cells are assumed to contribute to reflex adaptation to the same degree. Thus, a change in modulation is apparent on both cells, although it is now much smaller and comparable to the change in reflex gain. Figures 10.1D through 10.1G illustrate similar examples; however, the possibility of different motor projection strengths for each cell is now considered. For illustration purposes, it is now assumed that the motor projection strength $m_B = m_A/4$. If only s_B represents a site for plasticity (Fig. 10.1D), then the change in modulation depth on C_B must now be significantly larger than in Figure 10.1B (i.e., change in modulation depth of ~6 for a doubling of the VOR gain) to compensate for the cell's weaker motor projection. Alternatively, if only s_A represents a site for plasticity (Fig. 10.1E), then the change in modulation depth on C_A is much smaller given the cell's much stronger relative contribution to driving the reflex. Finally, Figures 10.1F and 10.1G illustrate cases where s_A and s_B are both assumed to be sites for plasticity. The changes in strength of the synapses are assumed to be equal in Figure 10.1F, whereas in Figure 10.1G the synaptic changes are unequal, so that each cell contributes equally to the new gain state despite different motor projection strengths (i.e., $(s_{A0} + s_{Aa})m_A = (s_{B0} + s_{Ba})m_B$).

FIGURE 10.1. Static VOR model with unconnected premotor cell types. (**A**) Schematic of the shortest-latency disynaptic rotational VOR pathways. Two distinct interneuron cell populations, C_A and C_B, in the vestibular nuclei receive head velocity signals, $\dot{H}$, via primary afferent projections from the semicircular canals (associated with synaptic weights s_A and s_B). Each cell type makes a direct projection onto motor neurons, MN, to drive a compensatory eye velocity response, $\dot{E}$. Motor projections associated with synaptic weights m_A and m_B are illustrated as inhibitory to indicate either a contralateral excitatory or an ipsilateral inhibitory projection to abducens MN appropriate to drive $\dot{E}$ in the opposite direction of $\dot{H}$ during rotation in the horizontal plane. Plots B–G illustrate the predicted modulation of each cell as a function of VOR gain ($\dot{E}/\dot{H}$) when adaptive gain changes are associated with different assumed sites for plasticity and different relative motor projection strengths. VOR gain adaptation about a default value of 0.85 is simulated by assuming: (**B**) changes in the strength of synapse s_B only for $m_B = m_A$; (**C**) equal changes in the strengths of synapses s_A and s_B for $m_B = m_A$; (**D**) changes in the strength of synapse s_B only for $m_B = m_A/4$; (**E**) changes in the strength of synapse s_A only for $m_B = m_A/4$; (**F**) equal changes in the strengths of synapses s_A and s_B for $m_B = m_A/4$; (**G**) changes in the strengths of synapses s_A and s_B for $m_B = m_A/4$ such that each cell contributes equally to the motor drive (see the text). In B and C, synaptic weights s_A and/or s_B modulate about baseline strengths $s_{A0} = s_{B0} = 0.425$ and $m_A = m_B = 1$. In D–G, s_A and/or s_B modulate about baseline strengths $s_{A0} = s_{B0} = 0.68$ for $m_A = 1$ and $m_B = 0.25$.

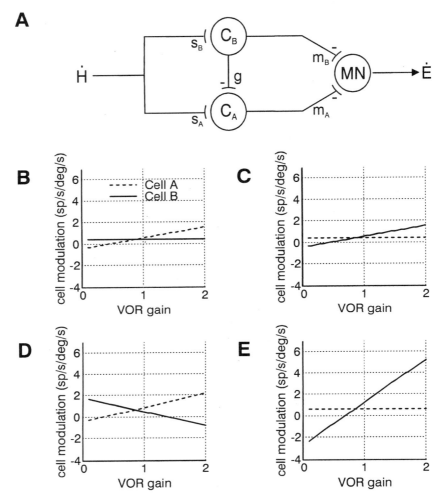

FIGURE 10.2. Static VOR model with interconnected premotor cell types. (**A**) Schematic of the shortest-latency disynaptic VOR pathways in which premotor neuron C_B is now assumed to make an efferent projection (associated with weight g) onto cell C_A. All other labels are as described in the legend of Figure 10.1. Plots B–D illustrate the predicted modulation of each cell as a function of VOR gain when adaptive gain changes are associated with different assumed sites for plasticity and relative motor projection strengths. VOR gain adaptation about a default value of 0.85 is simulated by assuming: (**B**) changes in the strength of synapse s_A only for $m_B = m_A$; (**C**) equal changes in the strength of synapses s_A and s_B for $m_B = m_A$; (**D**) changes in the strength of synapse s_B only for $m_B = m_A/4$; (**E**) equal changes in the strengths of synapses s_A and s_B for $m_B = m_A/4$. In B and C, synaptic weights s_A and/or s_B modulate about baseline strengths $s_{A0} = 0.85$ and $s_{B0} = 0.425$ for $m_A = m_B = 1$. In C and D, synaptic weights s_A and/or s_B modulate about baseline strengths $s_{A0} = 1.36$ and $s_{B0} = 0.68$ for $m_A = 1$ and $m_B = 0.25$. In all cases, weight $g = 1$.

The simplistic examples in Figure 10.1 illustrate that the *degree* of change observed in a premotor cell's response, compared with the behavioral gain change, depends on how many sites for plasticity are involved and the contribution of each cell to the motor response. However, in all cases, the existence of a synaptic site for plasticity is directly correlated with an *observable* change in the response of the associated cell. Hence, if such a simple feedforward model for the VOR were valid, then the observation of changes in a cell response should be sufficient to localize a potential site for plasticity. Under this assumption, the experimental observation that reflex adaptation involves large changes in the modulation depth of FTN cells accompanied by at most small or negligible changes in PVP cell responses would imply that primary afferent synapses onto FTN cells constitute the major site for plasticity. Furthermore, if only FTN cells were to contribute to reflex adaptation, the disproportionately large changes observed in their modulation depths could point to a significantly weaker motor contribution than that of PVP cells. Indeed, EHV cells (at least a subset of which are FTN cells) so far appear to make fewer motor projections than do PVP cells (Scudder and Fuchs 1992).[3] *The question to be addressed, however, is whether this interpretation remains valid if in fact the premotor VOR circuitry is not accurately represented by independent feedforward pathways.*

2.1.2. Feedforward Interconnections Between Different Cell Types

The schematic in Figure 10.2A is identical to that in Figure 10.1A, with the exception that now the possibility of a feedforward interconnection between the cells is considered. Notice that this projection from cell C_B onto C_A is labeled with weight g and is inhibitory to indicate either a contralateral excitatory or ipsilateral inhibitory projection, compatible with known anatomy (McCrea et al. 1981, 1987).

The VOR gain and cell modulations are now described by the following expressions:

$$\text{VOR}_{\text{gain}} = \dot{E}/\dot{H} = -(s_A - gs_B)m_A - s_B m_B, \tag{10.1}$$

$$C_A = (s_A - gs_B)\dot{H}, \tag{10.2}$$

$$C_B = s_B \dot{H}, \tag{10.3}$$

where $s_A = s_{A0} + s_{Aa}$ and $s_B = s_{B0} + s_{Ba}$. As in the previous example, the implications of choosing different sites for plasticity can be considered. We may begin by assuming for simplicity that weights m_A, m_B, and g are equal to 1.

---

[3] A caveat with regard to this statement is that the numbers of excitatory projections from PVP cells and of inhibitory projections from EHV cells to the abducens nucleus were estimated using the technique of spike-triggered averaging. This technique is likely to reveal excitatory connections with greater probability than inhibitory ones (Scudder and Fuchs 1992). Furthermore, it cannot be assumed that the cells in this study were encountered in their natural proportions.

Weights s_{A0} and s_{B0} were then again set to satisfy the following conditions: (1) VOR gain in darkness of 0.85; (2) equal sensitivities to head rotation on both cell types under normal gain conditions (see Figure 10.2 legend for parameters).

In Figure 10.2B, as in the previous example, when only s_A represents a site for plasticity, changes in reflex gain are accompanied by observed changes in modulation only on cell C_A. More interesting, however, is the case illustrated in Figure 10.2C, where both weights s_A and s_B are postulated to adapt such that $s_{Aa} = s_{Ba}$. Despite the fact that sites for plasticity exist on both cells, evidence for this plasticity is only apparent on C_B. Without *a priori* knowledge of the interconnectivity between the two cell types, it might therefore be concluded that cell C_B is wholly responsible for adaptive changes in VOR gain whereas cell C_A plays no role. However, in fact, sites for plasticity were simulated on both cells. The reason for the absence of a change in modulation on cell C_A is clear from Eq. (10.2) above. Because $g = 1$, $s_{Aa} = gs_{Ba}$ and the plasticity associated with s_A is masked in the overall modulation of cell C_A.

At this point, it might be concluded that C_B plays the more important "role" in reflex adaptation because in the absence of any change in the modulation of cell C_A, the changes in reflex gain must be entirely driven by C_B. However, this is misleading. To illustrate, Eq. (10.1) may be rewritten in the form $VOR_{gain} = -s_A m_A - s_B(m_B - gm_A)$. Now, we can see that because $(m_B - gm_A) = 0$ the term associated with s_B drops out of the equation. In fact, changes in synapse s_B have absolutely no effect on reflex performance. The plasticity associated with synapse s_A on cell C_A is entirely responsible for the changes in reflex gain. Thus, from the standpoint of localizing sites for plasticity supporting reflex gain changes, it is in fact the synaptic changes on cell C_A that are *apparently* most relevant, despite the failure to observe any change in this cell's modulation following motor learning. Does this now mean that C_A plays the more important "role" in reflex adaptation? Clearly, such a conclusion is also inappropriate. Both cells make an important contribution to VOR adaptation, although one cell does so because of the changes in the strength of its synaptic input whereas the other cell is important in conveying to the eye the changes in modulation associated with this plasticity.

In Figures 10.2D and 10.2E, the synapse m_B was instead assigned a strength $m_B = m_A/4$. Now, because $m_B - gm_A \neq 0$, changes in synapse s_B do indeed make a contribution to changes in reflex performance. Even if plasticity only occurs there, it is possible to achieve the desired changes in reflex gain (Fig. 10.2D). In addition, because of the interconnectivity of the cells, modulations in activity with gain state are now apparent on both neuron types. However, the modulation in the activity of cell C_B appears to be in an "inappropriate" direction to support the changes in behavioral reflex gain (i.e., cell modulation decreases for an increase in gain). Based on a simple correlation of the responses of these cells with reflex performance, one might be tempted to conclude that cell C_A, with changes in the "appropriate" direction, is mainly responsible for the VOR gain adaptation. Yet, in this simple example, only C_B constituted a site for plasticity. In Figure 10.2E, both cells are sites for plasticity such that $s_{Aa} = s_{Ba}$. Nevertheless, as in Figure 10.2C, changes in modulation associated with changes in VOR gain are only observed on cell C_B. Given the imbalance in the strength of motor projections m_A and m_B, the changes apparent on C_B are now disproportionately large relative to the behavioral gain change. Notice that the simulated changes in the modulation of each cell with VOR gain state are identical to those previously illustrated in Figure 10.1D. However, in this case, sites for plasticity exist on *both* cells.

Although the examples here were clearly oversimplified and incorporate specific assumptions with respect to the relationship between model parameters and cell interconnectivity, they illustrate an important point. In any situation where there is some degree of interconnectivity between cell types, there need no longer exist a direct correlation between sites for plasticity and observed response changes following motor learning. Although, as reviewed above, it has often been acknowledged that the observation of a change in modulation depth associated with reflex adaptation need not indicate a site for plasticity on a given cell (i.e., the change could be conveyed from a modified site elsewhere, as in Fig. 10.2D), the converse has rarely been considered. In fact, a site for plasticity may exist on a cell where there is no observed change in response (e.g., Figs. 10.2C and 10.2E; Green and Galiana 1996; Galiana and Green 1998; Green 2000).

This result may be considered in the context of recent experimental findings. Given the simple assumption of parallel feedforward pathways in the VOR (i.e., as in Fig. 10.1A), the failure to observe significant changes in the average activity of PVP cells following motor learning has led to the suggestion that these cells are unlikely to constitute an important site for plasticity in the VOR. In contrast, because FTN cells exhibit large changes in activity that are correlated with changes in gain state, these cells are currently thought to be the key neurons involved in VOR adaptation. Despite its simplicity, the example here raises the possibility that, depending on the PVP cell's placement within the premotor circuitry, primary afferent synapses onto PVP cells could indeed represent a site for plasticity, yet evidence of this plasticity may simply not be observed. Clearly, therefore, the ability to localize potential sites for plasticity in the VOR depends significantly on our knowledge of the underlying connectivity of the cells that participate in driving the reflex. The same arguments are valid whether the interconnections form part of a feedforward system (as in Fig. 10.2), involve more complex feedback interconnections between the VN and other brain areas involved in eye movements, such as the prepositus hypoglossi and the cerebellum (see below), or involve interconnections between the two sides of the brain (Galiana 1986; Green 2000). For example, similar issues are relevant when evaluating observed changes in the behavior of GVP cells associated with VOR adaptation; these cells demonstrate net changes in modulation in an appropriate direction to support VOR adaptation but *apparently* inappropriately directed changes in sensitivity to head velocity (see the review above).

2.2. Dynamic Models of the VOR

Long-term adaptation of the VOR is typically achieved using magnifying or miniaturizing lenses to train the reflex over the long term during normal behavioral tasks. Under such circumstances, restoration of visual stability requires changes in reflex gain over a broad frequency range, and such broadband changes in reflex performance are indeed observed (e.g., Miles and Eighmy 1980; 0.1–1 Hz). As noted above, investigations of the neural correlates for motor learning in the VOR have focused mainly on identifying those central sites at which there is evidence for a change in sensitivity to *head velocity*, as required to drive high-frequency changes in reflex gain.

Alternatively, visual–vestibular mismatch stimuli may be used to adapt the VOR over the short term at a single temporal frequency. This training approach gives rise to frequency-selective changes in reflex gain that are optimal at the training frequency (Godaux et al. 1983; Lisberger et al. 1983; Powell et al. 1991; Raymond and Lisberger 1996). To date, only a few cells have been recorded *during* such a short-term frequency-specific training protocol (e.g., Partsalis et al. 1995a). The fact that the extent of behavioral adaptation at frequencies well above (or below) the training frequency is typically limited or negligible and accompanied by prominent phase shifts relative to the ideal compensatory phase, however, suggests that plasticity in neural sensitivities to "head velocity" alone is insufficient to account for the changes in reflex behavior. This observation thus raises questions as to whether the plasticity responsible for frequency-specific versus more generalized broadband reflex adaptation occurs at the same sites. Furthermore, different sites may be involved to varying extents following a short-term versus long-term training procedure.

Because the neural correlates for motor learning have been investigated mainly using long-term broadband training stimuli, the majority of models employed to date to aid in the process of localizing sites for plasticity in the VOR have focused on identifying those sites associated with high-frequency changes in reflex performance (i.e., changes in head velocity sensitivity). As a result, these models often do not incorporate (or include only to a limited extent) the lower-frequency dynamic characteristics of the VOR (i.e., information related to "eye position" is often not included; e.g., Lisberger 1994). Clearly, however, to localize the neural correlates for reflex adaptation over a broad frequency range requires that potential sites for plasticity be examined in a model structure that more completely incorporates the full dynamic characteristics of the reflex. Indeed, several studies have proposed models for achieving frequency-specific or VOR phase adaptation and have demonstrated the requirement for sites for plasticity in sensorimotor pathways other than the shortest-latency direct vestibular ones to explain experimentally observed behavior (i.e., changes in sensitivity to head velocity alone are not sufficient; Lisberger et al. 1983; Quinn et al. 1992a, 1992b; Tiliket et al. 1994; Kramer et al. 1995; Powell et al. 1996).

In the following section, we will investigate two realizations for the central *dynamic* processing of semicircular canal signals in the VOR to explore several key issues: (1) How does the presumed realization for the neural circuitry of the VOR influence the locations and number of central sites required to achieve reflex adaptation for different training conditions? (2) Is a central change in "head velocity sensitivity" always required to achieve observed changes in reflex performance? (3) What other brain sites should be investigated for plastic changes using neural recording techniques?

2.2.1. Feedforward Realization of the VOR

The schematic in Figure 10.3A illustrates the classical feedforward realization of the VOR first proposed by Skavenski and Robinson (1973) in which sensory signals are conveyed to the extraocular motoneurons via a set of parallel pathways from the semicircular canals (modeled as high-pass filters of head velocity $C(s) = T_c s/(T_c s + 1)$, where $T_c = 6$ s) to the extraocular motoneurons (MN). Direct pathways associated with weights a and b convey vestibular signals relatively directly via vestibular neurons to the motor nuclei to drive the eye plant (modeled as a first-order low-pass filter, $P(s) = 1/(T_p s + 1)$, where $T_p = 0.25$ s), whereas in the "indirect" pathway (associated with weights c and d) sensory signals are prefiltered by a "neural integrator" or low-pass filter with a very long time constant, T_I ($T_I = 20$ s in Fig. 10.3A). The function of this latter pathway is to provide a neural integration of sensory signals in the VOR so that head velocity signals originating from the semicircular canals are transformed into an internal estimate of desired eye position ($E^*(s)$; above ~0.03 Hz for $T_I = 20$ s, $T_c = 6$ s). A weighted balance of signal projections from the "direct" and "indirect" pathways ensures compensation for the dynamic properties of the eye plant (see below) so that the reflex exhibits a robust response with compensatory phase over a broad frequency range extending from 0.01–0.03 Hz to well above 10 Hz (Raphan et al. 1979; Minor et al. 1999) in the absence of visual feedback (see the solid curves in the Bode plots of Figs. 10.3B through 10.3E).[4] For example, in the system of Figure 10.3A, the ocular response to head rotation is given by the equation

---

[4] It has been demonstrated experimentally that an estimate of the head velocity sensed by the semicircular canals is stored centrally so that the dynamic range of the VOR extends below that expected based on the high-pass characteristics of the semicircular canals (Raphan et al. 1979). However, for simplicity, this effect, known as "velocity storage," has not been incorporated in the models here because neural correlates for motor learning in the VOR are typically investigated at frequencies well above those where velocity storage plays an important role.

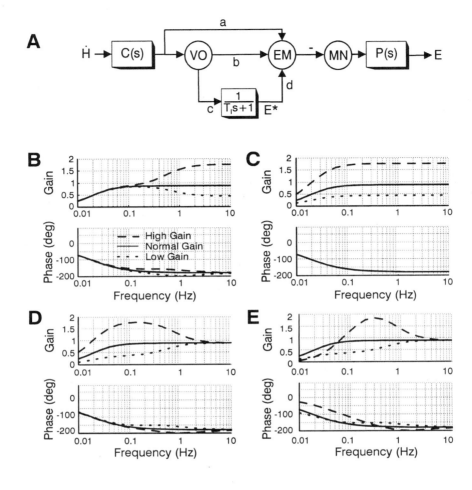

$$\dot{E}(s) = -\frac{G_v s(T_v s + 1)}{T_I s + 1} \frac{1}{T_p s + 1} \frac{T_c s}{T_c s + 1} \dot{H}(s), \tag{10.4}$$

where

$$G_v = a + b + cd \quad \text{and} \quad T_v = \frac{(a+b)T_I}{a+b+cd}.$$

Under circumstances where the weights in the "direct" and "indirect" pathways are adjusted to force $T_v = T_p$, the numerator term $(T_v s + 1)$ in Eq.

◄─────────────────────────────────

FIGURE 10.3. Feedforward dynamic model for the VOR. (**A**) Schematic of a feedforward realization for the VOR adapted from the original proposal of Skavenski and Robinson (1973). Circles in the schematic are summing junctions used to represent particular cell populations, and boxes are dynamic elements that represent either a sensor, motor plant, or neural filtering process (see below). Parameters associated with signal projections are gain elements that denote the strength or weight of the projection. Angular head velocity, $\dot{H}$, sensed by the semicircular canals, $C(s) = T_c s / T_c s + 1$, is conveyed to motor neurons that drive the eye plant, $P(s) = 1/(T_p s + 1)$, via two parallel pathways: (1) a "direct" pathway via vestibular neurons (VO and EM); (2) an "indirect" pathway via a low-pass filter, $1/(T_I s + 1)$, with a very large time constant, $T_I = 20$ s, such that the filter approximates a "neural integrator." A weighted balance of the signal projection strengths in the "direct" and "indirect" pathways ensures appropriate dynamic processing of sensory signals so that the reflex elicits compensatory deviations in eye position, E, over a broad frequency range (see the text). Vestibular neurons include vestibular-only cells (VO), as in the original proposal of Skavenski and Robinson (1973), and eye-movement-sensitive neurons (EM), which are now known to be the key interneurons in the shortest-latency disynaptic pathways. EM cells are shown here to sum both sensory head velocity signal inputs and an internal estimate of eye position, E^*, provided at the output of the "neural integrator." Under normal gain conditions (i.e., prior to any simulated adaptation), baseline model parameters are: $a = 0.12$; $b = 0.1$; $c = 17.38$; $d = 1$; $T_I = 20$; $T_p = 0.25$; $T_c = 6$. Plots B–E illustrate VOR gain and phase as a function of frequency for the normal gain condition (solid curves; gain of 0.88) and for simulated adapted gain states when different assumptions are made with respect to sites for plasticity. (**B**) The sensitivity of cell EM to primary afferent inputs (i.e., weight a associated with the direct pathway) is modified to achieve either a doubling ($a = 0.34$) or halving ($a = 0.01$) of the high-frequency VOR gain. (**C**) Modifications of both a and c to simulate plasticity in both the direct and indirect pathways give rise to a doubling ($a = 0.34$, $c = 34.76$) or halving ($a = 0.01$, $c = 8.69$) of the VOR gain over a broad frequency range. (**D**) Modifications in weight c only (indirect pathway) give rise to frequency-selective increases ($c = 36.6$) or decreases ($c = 7.27$) in reflex gain centered at approximately 0.15 Hz. (**E**) Concurrent changes in the efficacy of the neural integrator and projection weight c result in a shift in the frequency at which adaptation is optimal ($T_I = 1.23$, $c = 2.48$ for high gain; $T_I = 47$, $c = 17.4$ for low gain).

(10.4) cancels the eye plant denominator term $(T_p s + 1)$ (i.e., zero-pole cancellation). The VOR gain is then equal to G_v/T_I at frequencies above both the vestibular and neural integrator poles (e.g., at frequencies above $1/(2\pi T_I)$ and $1/(2\pi T_c)$ Hz).

The schematic in Figure 10.3 has been extended slightly from the original proposition of Skavenski and Robinson (1973; also see Robinson 1981), which explicitly represented a single vestibular-only cell type (VO) in the VN. Here, an eye-movement-sensitive vestibular neuron population (EM) has also been included (i.e., the EM cell group sums both direct sensory head velocity signal inputs and eye position signals from the output of the "neural integrator") because it is in fact a subpopulation of these EM cells (e.g., PVP and FTN/EHV cells), which are currently thought to represent the key interneurons in the most direct VOR pathways and to be involved in reflex adaptation (e.g., Scudder and Fuchs 1992). To date, there is no evidence that reflex adaptation is associated with a change in the sensitivity of VO cells (Lisberger and Miles 1980). The neural integrator is proposed to be located at least partially in the nucleus prepositus hypoglossi (NPH), in keeping with the observation that many neurons in this area code for eye position (Lopez-Barneo et al. 1981; Escudero et al. 1992; McFarland and Fuchs 1992), and lesions of this area (as well as parts of the VN and cerebellum) result in a deterioration in integrator function (Cannon and Robinson 1987; Cheron and Godaux 1987; Kaneko 1997).

After motor learning in the feedforward realization of Figure 10.3A, changes in the head velocity sensitivity of the EM cell types would be associated with modifications in the strengths of weights a and/or b. For simplicity, we will therefore begin our investigation by assuming changes only in the strength of monosynaptic head velocity inputs onto the EM cell type (i.e., weight a) and examine the impact of assuming this site for plasticity on behavioral responses (similar results are obtained by distributing changes between weights a and b). The solid curve in the frequency response plot of Figure 10.3B illustrates the normal gain condition in which parameters were adjusted for a high-frequency gain in the dark of 0.88 and the relative strengths of the "direct" and "indirect" pathways were adjusted appropriately to compensate for the dynamic characteristics of the eye plant in order to achieve compensatory phase over a broad frequency range (i.e., ocular response 180° out of phase with head velocity; see caption of Fig. 10.3 for model parameters). When either a high-gain or low-gain state following motor learning is simulated simply by adjusting a appropriately to either double (dashed curve) or halve (dotted curve) the high-frequency gain, however, this *also* results in a change in the overall dynamics of the behavioral response. Specifically, there is significantly less adaptation at midrange frequencies (e.g., 0.2–1 Hz) and almost none at lower frequencies (<0.2 Hz). Furthermore, changes in gain are accompanied by either a phase lead (high-gain case) or phase lag (low-gain case) relative to the ideal compensatory phase at midrange frequencies. This is not surprising because selective changes in the "direct" pathway upset the balance between T_v and T_p so that the eye plant pole is no longer canceled (Eq. 10.4).

Because adaptation with telescopic lenses typically gives rise to broadband changes in reflex gain (e.g., Miles and Eighmy 1980; 0.1–1 Hz), it is clear that the simple assumption of plasticity only in the "direct" pathways of the feedforward model for the VOR is not sufficient to explain behavioral observations. Rather, this structure implies the requirement for a second site for plasticity in the "indirect" pathway. This case is illustrated in the Bode plot of Figure 10.3C where, in addition to postulating changes in strength of head velocity inputs onto the EM cell (i.e., weight a), the strength of the projection from the VO cell onto "neural integrator" neurons was also modified (weight c). Specifically, weight c was changed in tandem with weight a to ensure that the numerator time constant (T_v) remained appropriate to cancel the dominant eye plant pole. Now we can see that under simulated adapted high- and low-gain conditions, the VOR exhibits the same dynamic characteristics as those prior to motor learning with changes in gain and compensatory phase over a broad frequency range. Hence, if a parallel-pathway, feedforward model structure is assumed for the VOR, then broadband reflex adaptation implies that plasticity must exist at multiple sites. Note, however, that although a second site for plasticity must exist in this model, it need not exist on a different cell type. For example, the changes could occur at multiple sites on the EM cell; one site could exist on synapses conveying head velocity information from the canals (i.e., weights a and/or b), whereas a second site could be on synapses associated with inputs from the neural integrator (i.e., associated with weight d). With the current model structure, however, if a single "neural integrator" subserves multiple eye-movement systems (e.g., both slow eye movements and saccades), as traditionally assumed (Robinson 1981),[5] then changes in weight d would imply changes in the sensitivity to eye position of at least some EM cells during all types of movements.

In contrast, short-term adaptation of the VOR gain is typically achieved using visual–vestibular mismatch stimuli at a specific frequency. Under such circumstances, frequency-specific adaptation of the VOR has been observed. Changes in reflex gain are most robust at the adapting frequency and accompanied by changes in phase above and below this frequency (Godaux et al. 1983; Lisberger et al. 1983; Powell et al. 1991; Raymond and Lisberger 1996). To reproduce this behavior in the simple structure of Figure 10.3A, we may begin by considering only the impact of changes in the strength of vestibular inputs to the "neural integrator" (i.e., weight c). In this case (Fig. 10.3D), the changes in VOR gain are now centered at a frequency of approximately 0.15 Hz but are negligible at very high fre-

---

[5] Recent studies have reported that there may be multiple oculomotor integrator systems for horizontal eye movements (Kaneko 1997, 1999).

quencies (i.e., >3–4 Hz), as observed experimentally (e.g., see Lisberger et al. 1983 for frequency training at 0.2 Hz). If modifications are instead presumed in *both* the strength of vestibular inputs to the integrator and in its internal efficacy (i.e., changes in both weight c and the neural integrator time constant T_I), then the peak of the VOR gain curve can also be shifted. In Figure 10.3E, simulated frequency-specific adaptation at approximately 0.3 Hz is illustrated. Notice that the gain changes are now also accompanied by significant phase shifts relative to the normal condition below and above the preferred frequency. For the high-gain case illustrated here, there are phase leads relative to the ideal compensatory phase below the simulated adapting frequency and phase lags above this frequency, as observed experimentally (Lisberger et al. 1983; Raymond et al. 1996).

It has been suggested previously that such frequency-specific adaptation can be modeled by altering the relative gains in a series of parallel channels in the VOR with different dynamic characteristics (Lisberger et al. 1983). Each channel was postulated to consist of a feedforward parallel-pathway arrangement essentially equivalent to the one illustrated in Figure 10.3A. Channels differed from one another in that different relative weights were assumed in the "direct" versus "integrator" pathways of each one. The approach illustrated here is therefore a lumped version of the same concept with one key difference. In the parallel channel model, the strengths of head velocity inputs in the "direct" pathways were modified by virtue of reweighing the contribution of individual channels to the total motor response. As a result, the model predicted changes in gain at high frequencies that were not observed experimentally. Notice here, however, that the general characteristics of frequency-specific adaptation have been reproduced without postulating any change in the "direct" pathway. Hence, it is clear that the sites for plasticity may depend significantly on the training paradigm used to drive motor learning. Although changes in the "head velocity sensitivity" of vestibular neurons in the direct pathway are required to achieve high-frequency changes in reflex gain, this need not be the case if training occurs at a specific midrange frequency.

In summary, the requirement for a site of plasticity in the integrator pathway has been discussed extensively and proposed in previous models of frequency-specific adaptation or VOR phase adaptation (Lisberger et al. 1983; Tiliket et al. 1994; Kramer et al. 1995; Powell et al. 1996). However, it would appear that if indeed a feedforward parallel-pathway structure for the VOR is appropriate, then a site for plasticity in the "integrator pathway" is also required to achieve the broadband changes in gain observed following long-term adaptation with telescopic spectacles (also see Quinn et al. 1992a, 1992b). Nevertheless, a direct investigation of potential sites for VOR plasticity in the integrator pathway (e.g., in the NPH) of primates has not yet been performed and represents an area for future experimental exploration.

2.2.2. Feedback Realization of the VOR

The predictions of the feedforward form for the VOR may be contrasted with those of an alternative feedback realization of the VOR first proposed by Galiana and Outerbridge (1984) and illustrated in Figure 10.4A. As in the feedforward realization, the cell labeled EM represents a lumped premotor eye-movement-sensitive vestibular cell. In this case, however, cell EM is interconnected in a positive feedback loop with a neural filter, $F(s)$, that represents an internal model of the eye plant ($F(s) = P(s)$). The output of $F(s)$ provides an internal estimate (efference copy) of eye position, E^*, that, as in the feedforward realization, can represent eye-position-sensitive neurons in the nucleus prepositus hypoglossi (NPH). However, in this realization, the neural integration relies on distributed positive feedback loops around a neural filter with a relatively small time constant (i.e., close to the dominant eye plant time constant of ~0.25 s), as opposed to a localized integrator with a large time constant. The feedback realization is more physiologically realistic in its representation of the highly interconnected nature of the premotor circuitry (see McCrea et al. 1981, 1987). Furthermore, such a structure is supported by lesion studies demonstrating that the integrative properties of the premotor network involve several brain areas, including the vestibular nuclei, the NPH, and the cerebellum (Robinson 1974; Cannon and Robinson 1987; Cheron and Godaux 1987; Kaneko 1997). The VOR response properties of the feedback realization are equivalent to the feedforward parallel-pathway model of Robinson (1981; Fig. 10.3A) in terms of overall motor response. However, they can have different implications in terms of localizing sites for plasticity associated with motor learning. Specifically, in the feedback realization presented here, ocular responses to head rotation are described by the equation

$$\dot{E}(s) = -\frac{G_v s(T_f s + 1)}{T_{fb} s + 1} \frac{1}{T_p s + 1} \frac{T_c s}{T_c s + 1} \dot{H}(s) \qquad (10.5)$$

where

$$G_v = \frac{am_p}{1 - m_f b} \quad \text{and} \quad T_{fb} = \frac{T_f}{1 - m_f b}.$$

Notice that if $T_f = T_p$, the zero in Eq. (10.5) cancels the eye plant pole, as in the feedforward realization above. Furthermore, by appropriate adjustment of the strengths of the weights in the feedback loop (i.e., weights m_f and b), the time constant T_{fb} can be set to a large value of 20 s so that it is equivalent to the time constant T_I in the feedforward realization (Fig. 10.3A). Under these conditions, Eq. (10.5) produces the same result as Eq. (10.4), as can be seen by comparison of the normal gain curves (i.e., solid curves) in the frequency response plots of Figures 10.3B and 10.4B.

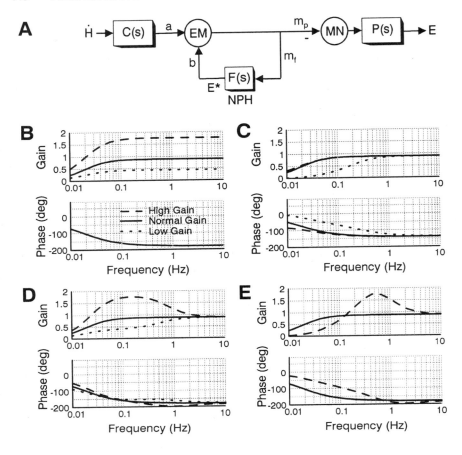

To examine the predictions of this model concerning sites for plasticity, we may again begin by presuming that the plasticity associated with reflex adaptation occurs only at the site of head velocity inputs to the EM cell (weight a). Notice that, in contrast to the feedforward realization, here changes at this single site produce broadband changes in reflex gain like those observed following long-term adaptation with magnifying or minimizing lenses (i.e., compare traces in Fig. 10.4B with those in Figs. 10.3B,C). This occurs in the feedback model because the EM cell both projects to and receives feedback from the neural filter, $F(s)$, whose output provides an internal estimate of current eye position. As a result of the interconnectivity in the circuit, the efference copy signal at the output of neural filter $F(s)$ is automatically updated appropriately when the head velocity inputs to cell EM are modified. Hence, the conclusions of the feedback realization differ from those of the feedforward VOR model. Whereas multiple sites for plasticity are required in the feedforward case to achieve broadband changes in reflex gain, only a *single* site is required in the feedback model. Clearly, therefore, the presumed structure of the premotor circuitry influences conclusions with respect to the number of sites required to achieve observed reflex performance following motor learning.

◄───

FIGURE 10.4. Feedback dynamic model for the VOR. (**A**) Schematic of a feedback realization of the VOR originally proposed by Galiana and Outerbridge (1984). Eye-movement-sensitive cell population, EM, is interconnected in a positive feedback loop with neural filter $F(s) = 1/(T_f s + 1)$ presumed to be located in the nucleus prepositus hypoglossi (NPH). When $F(s)$ represents an internal model of the eye plant, $P(s) = 1/(T_p s + 1)$ (i.e., $T_f = T_p$), the overall dynamic processing of sensory information from the semicircular canals, $C(s) = T_c s/(T_c s + 1)$, is equivalent to that in the feedforward VOR realization illustrated in Figure 10.3A (see the text). In addition, if the EM cell population makes a collateral projection to $F(s)$ equal in strength to its motor projection (i.e., $m_f = m_p$), then the output of $F(s)$ provides an ideal internal estimate of eye position, E^*. Under normal gain conditions (i.e., prior to any simulated adaptation), baseline model parameters are: $a = 0.22$; $b = 0.9875$; $m_f = 1$; $m_p = 1$; $T_f = 0.25$; $T_p = 0.25$; $T_c = 6$. Plots B–E illustrate VOR gain and phase as a function of frequency for the normal gain condition (solid curves; gain of 0.88) and for simulated adapted gain states when different assumptions are made with respect to sites for plasticity. (**B**) Modification in the sensitivity of cell population EM to primary afferent inputs (i.e., weight a) gives rise to broadband increases ($a = 0.44$) or decreases ($a = 0.11$) in reflex gain. (**C**) Impact of increasing or decreasing the strength of projection weight m_f under contraints of maintaining system stability. Dotted curve: $m_f = 1.012$; dashed curve: $m_f = 0.608$. (**D**) Modifications in neural filter time constant T_f give rise to frequency-selective changes in reflex gain centered at a preferred frequency of about 0.15 Hz (high gain: $T_f = 0.12$; low gain: $T_f = 0.54$). (**E**) Concurrent changes in both T_f and m_f make possible a shift in the optimal frequency of adaptation to approximately 0.5 Hz ($T_f = 0.081$, $m_f = 0.82$).

Now, we can consider the impact of postulating other sites for plasticity in this feedback VOR model. For example, in Figure 10.4C, the effect of changes in the strength of synaptic inputs to the neural filter $F(s)$ is illustrated (i.e., weight m_f). Changes at this site can be used effectively to lower the VOR gain at low to midrange frequencies, but it is not a suitable site for increasing the gain. Specifically, large changes in m_f alone only have an effect on the system pole; lowering this parameter can give rise to decreases in gain accompanied by a deterioration in the integrative properties of the premotor network, whereas increases in m_f alone rapidly cause the system to become unstable without giving rise to a significant increase in gain. This is in contrast to the case of the feedforward model, where a change in the strength of inputs to the neural integrator allowed the simulation of both increases and decreases in gain that were frequency-specific. Hence, modifications at different sites can give rise to dynamic effects on the VOR that depend on the postulated structure of the neural circuitry.

If instead changes within the neural filter itself are presumed, such that the time constant, T_f, of the filter is modified, then we observe changes in VOR gain that are largest at a frequency of approximately 0.15 Hz accompanied by negligible changes at high frequencies (Fig. 10.4D). Hence, it appears that with the feedback VOR realization it is also possible to achieve adaptive changes in reflex performance that are frequency-selective. In fact, the simulated responses appear almost identical to those illustrated in Figure 10.3D for the feedforward model. By assuming simultaneous changes in weights m_f and time constant T_f, the preferred frequency for adaptive changes can also be shifted (e.g., Fig. 10.4E; concurrent decreases in weights m_f and T_f shift the peak of the high-gain curve to approximately 0.5 Hz). Again, in the high-gain case, the changes are accompanied by phase leads above and phase lags below the preferred adaptation frequency, as observed experimentally (e.g., Lisberger et al. 1983; Raymond and Lisberger 1996). As in the feedforward structure of Figure 10.3, these changes were achieved without the assumption of any change in the strength of head velocity inputs onto the EM cell.

2.2.3. Summary

Both feedforward and feedback dynamic VOR models can reproduce similar types of adapted behavioral responses. However, the locations and number of required plastic sites depend on the circuit realization. Broadband adaptive changes in VOR gain in the feedforward model imply the requirement for plasticity at multiple sites; such changes can be achieved in the feedback model assuming modifications only in the synaptic strengths of sensory vestibular inputs. Hence, if the feedback model represents a more realistic picture of the premotor VOR circuitry, then the types of experiments that have been performed to date to isolate plastic sites associated with broadband reflex adaptation may indeed be sufficient. However, as illustrated in the previous section, the identification of such sites may nevertheless be very difficult because simple correlations of changes in cell sensitivities with motor behavior are not necessarily sufficient to localize these sites uniquely when interconnections exist between individual cell types (see Section 2.3). On the other hand, if in fact the feedforward realization of the VOR is valid, as is commonly assumed, then clearly future investigations should look for evidence of plasticity in the "indirect" pathway. Perhaps most importantly, it is time to address what realization for the VOR is most appropriate because the interpretation of experimental observations will depend on the premotor structure that is assumed. In all cases above, the simple examples illustrate that the locations for plasticity underlying VOR adaptation may also depend significantly on the conditions under which the reflex was trained. There is a need to consider sites for plasticity other than the synapses associated with sensory vestibular inputs, particularly under conditions in which the reflex was adapted at a specific frequency.

2.3. Dynamic VOR Model with Multiple Cell Types

So far, simple models have been used to illustrate that conclusions with respect to potential sites for plasticity in the VOR can be significantly influenced by the presumed interconnectivity of individual cell types. By considering two different realizations for the dynamic processing in the VOR pathways, it has also been demonstrated that the presumed locations of plastic sites may depend on the adaptation paradigm employed (e.g., training to achieve broadband gain changes versus frequency-specific adaptation). Furthermore, both the locations of these sites and number of sites required depend on the presumed model structure (i.e., feedforward versus feedback realization; see also Galiana and Green 1998 for other examples). In this section, we will extend a dynamic model of the VOR to incorporate multiple cell types and use this structure to illustrate the further complexities that arise when the dynamic characteristics of individual cell responses are taken into account. Typical approaches for interpreting the responses of individual cells will also be examined to address whether they indeed are appropriate for isolating potential sites for plasticity.

2.3.1. Model Network

The model schematic illustrated in Figure 10.5 represents an extension of the feedback realization of the VOR in which the single lumped EM cell of Figure 10.4A has been replaced by two distinct eye-movement-sensitive cell types (EM_A and EM_B). Each cell is interconnected in a positive feedback loop with the neural filter $F(s)$. Because of the shared feedback pathways, interconnectivity between the two cells automatically exists. In addition, however, cell EM_B is postulated to make a direct projection onto

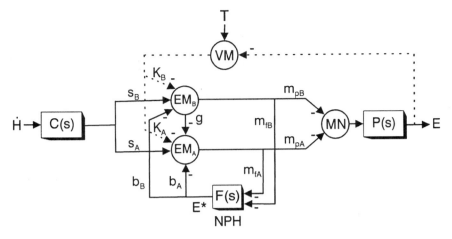

FIGURE 10.5. Extension of the feedback realization for the VOR in Figure 10.4A to incorporate two distinct eye-movement-sensitive premotor vestibular cell types, EM_A and EM_B. Dashed pathways are presumed to be activated by visual feedback and represent a highly simplified lumped pursuit system sufficient to explore model predictions under visual–vestibular interaction conditions. T represents target position. All other labels are as described in the legend of Figure 10.4. Model parameters: $s_A = 0.88$; $s_B = 0.44$; $b_A = 3.95$; $b_B = 1.975$; $m_{pA} = 0.4$; $m_{fA} = 0.4$; $m_{pB} = 0.1$; $m_{fB} = 0.1$; $g = 1$; $K_A = 19.1$; $K_B = 19.1$; $T_f = 0.25$; $T_p = 0.25$; $T_c = 6$.

EM_A (associated with weight g), as was previously illustrated in the simple static example of Figure 10.2A. As described above, potential sites for plasticity are typically identified by evaluating cell responses under different visual–vestibular interaction conditions. Hence, to address such experimental approaches, dashed pathways associated with visuomotor areas (VM) have also been included. These pathways are assumed to be activated by the presence of a visual target and are used here to simulate a simplified lumped pursuit system. Although clearly a highly oversimplified visual–vestibular system model, the arguments to be addressed below using this structure are general and are not restricted by the simplistic nature of the circuit.

2.3.2. VOR in the Dark at Default Gain

First, it is worth briefly considering the performance of this model under normal gain conditions (i.e., prior to any adaptation; solid curves in Figs. 10.6, 10.7, and 10.8). The model parameters chosen to simulate the normal gain condition are indicated in the legend of Figure 10.5. The goal here is to investigate potential problems in *localizing* sites for plasticity in the VOR rather than to reproduce the specific characteristics of known individual cell types. Hence, limited criteria were used to set the model parameters. In particular, model parameters were adjusted, as in previous examples, to achieve a "neural integrator" time constant of 20s and a high-frequency VOR gain of 0.88 in the dark (see Fig. 10.6A, solid traces). For simplicity, both cell types are assumed to exhibit identical dynamic responses during rotation in the dark, with high-frequency sensitivities to head rotation of ~0.44 spikes/s/deg/s (Figs. 10.6B and 10.6C, solid traces). To account for the fact

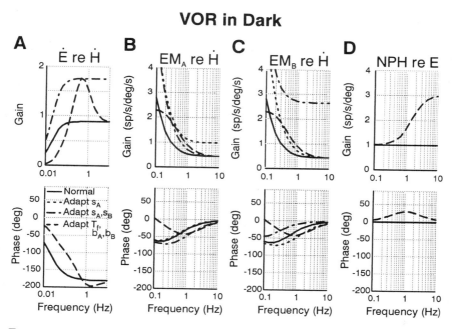

FIGURE 10.6. Frequency response predictions of the model in Figure 10.5 during head rotation in the dark for the normal gain condition and following simulated reflex adaptation assuming different sites for plasticity. (**A**) Behavioral VOR gain ($\dot{E}/\dot{H}$). (**B**) Response of cell EM_A relative to $\dot{H}$. (**C**) Response of cell EM_B relative to $\dot{H}$. (**D**) Response of NPH cells at the output of $F(s)$ relative to E. Solid curves: normal gain condition. Dotted curves (case 1): s_A modified to achieve broadband doubling of the VOR gain ($s_A = 1.43$). Dash-dot curves (case 2): concurrent changes in s_A and s_B to achieve broadband doubling of the VOR gain ($s_A = 3.08$, $s_B = 2.64$). Dashed curves (case 3): changes in T_f, b_A, and b_B to achieve frequency-specific doubling of the VOR gain at 0.5 Hz ($T_f = 0.0825$, $b_B = 1.64$, $b_A = 3.28$). In A, behavioral gains are identical for adaptation cases 1 and 2 (i.e., dash-dot and dashed curves superimpose) and phases for these cases are identical to the normal gain condition (solid, dotted, and dash-dot curves superimpose). NPH output neurons provide a perfect internal estimate of eye position under normal gain conditions and for adaptation cases 1 and 2 (i.e., solid, dotted, and dash-dot curves superimpose in D).

454   A.M. Green et al.

that individual cells may contribute to motor behavior to varying extents (e.g., see Fig. 10.1), the motor projection from cell EM_B was set to one quarter the strength of that from EM_A (i.e., $m_{pB} = 1/4m_{pA}$). Finally, each cell's motor projection was set to be equal in strength to its collateral projection to $F(s)$ (i.e., $m_{pA} = m_{fA}$ and $m_{pB} = m_{fB}$), and $F(s)$ was presumed to represent an internal model of $P(s)$ (i.e., $T_p = T_f$). Hence, the output of $F(s)$, E^*, represents an internal estimate of eye position across all frequencies, as illustrated in Figure 10.6D (solid trace).

2.3.3. Pursuit and VOR Cancellation at Default VOR Gain

Predicted behavioral and cell responses of the model during head-stationary pursuit of a visual target are illustrated in Figure 10.7 when the visual feedback pathways (i.e., dashed pathways in Fig. 10.5) are assumed

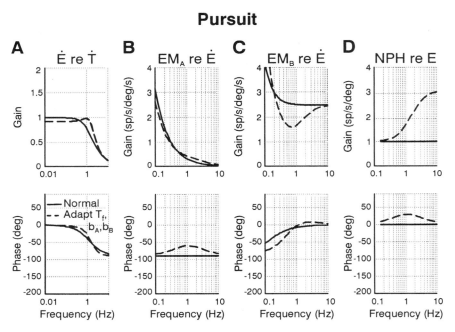

FIGURE 10.7. Frequency response predictions of the model in Figure 10.5 during head-stationary smooth pursuit. (**A**) Behavioral response ($\dot{E}/\dot{T}$). (**B**) Response of cell EM_A relative to $\dot{E}$. (**C**) Response of cell EM_B relative to $\dot{E}$. (**D**) Response of NPH cells at the output of $F(s)$ relative to E. For adaptation cases 1 and 2 (see legend of Fig. 10.6), pursuit performance is identical to that for the normal VOR gain condition (solid curves). Dashed curves denote responses during smooth pursuit for adaptation case 3 (T_f, b_B, b_A modified).

to be active. For simplicity, model parameters were adjusted such that K_A = K_B = K, where K was set to provide close to ideal target tracking performance at low frequencies with a pursuit bandwidth of approximately 1.2 Hz, as observed in primates (e.g., Lisberger et al. 1981; Barnes 1993; see Fig. 10.7A, solid trace). Although both vestibular cell types receive visuomotor projections of identical strength (i.e., K_A = K_B) and exhibit similar response properties during head rotation in the dark (Figs. 10.6B and 10.6C), the cells exhibit quite distinct responses during pursuit. Cell EM_A (Fig. 10.7B, solid traces) modulates closely in phase with contralaterally directed eye position across all frequencies (i.e., lags contralaterally directed eye velocity by 90°). In contrast, cell EM_B (Fig. 10.7C, solid traces) exhibits a much larger modulation that is closely in phase with eye velocity at midrange to high frequencies. Despite the appearance that cell EM_B codes closely for eye velocity during pursuit at higher frequencies, however, the cell does not actually receive any direct "eye velocity" projection per se. In fact, visual tracking in the model relies only on retinal position error feedback. Clearly, the pursuit system implemented in the model here is highly oversimplified.[6] It is nevertheless sufficient to illustrate the potential to observe neural responses that appear (at least over some particular frequency range) to be well correlated with a variable of interest, despite the fact that this variable is not explicitly represented anywhere in the network (i.e., there is no internal estimate of eye velocity in the model). Furthermore, despite the fact that both neurons receive visuomotor input projections of the same strength, it has been illustrated that a distinct behavior for each cell type emerges from its unique placement within the premotor circuitry (i.e., differences in the interconnectivities of the cells and the strengths of the feedback loops in which they are placed).

Figure 10.8 illustrates the predicted responses of the model when the subject stabilizes a target that moves in tandem with the head (head-fixed target) so that the VOR is "cancelled" at frequencies below the pursuit system bandwidth (Fig. 10.8A, solid traces). Notice again that each cell exhibits quite distinct response properties. Cell EM_A exhibits a reduced modulation at low to midrange frequencies under normal gain conditions that is nearly in phase with or slightly lags ipsilaterally directed head velocity (Fig. 10.8B, solid traces). This might be expected because the output, E^*, of filter $F(s)$ provides an internal estimate of eye position (Fig. 10.8D, solid traces) and eye movement is "suppressed" at lower frequencies in this protocol. The projection from $F(s)$ therefore conveys little deviation in "eye

---

[6] The model uses feedback of retinal position error to simulate a simple visual tracking system, sufficient for illustration purposes here. A more realistic pursuit model would include feedback of retinal slip (velocity error) and possibly retinal acceleration error in addition to retinal position error (Rashbass 1961; Tychsen and Lisberger 1986).

VOR Cancellation

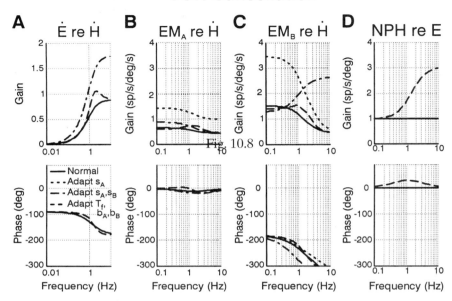

FIGURE 10.8. Frequency response predictions of the model in Figure 10.5 during rotation while stabilizing a head-fixed target (VOR cancellation). (**A**) Behavioral response ($\dot{E}/\dot{H}$). (**B**) Response of cell EM_A relative to $\dot{H}$. (**C**) Response of cell EM_B relative to $\dot{H}$. (**D**) Response of NPH cells at the output of $F(s)$ relative to E. See legend of Figure 10.6 for a description of the various cases. In A, behavioral gain responses are identical for adaptation cases 1 and 2 (i.e., dotted and dash-dot curves superimpose) and phases for these cases are identical to the normal gain condition (solid, dotted, and dash-dot curves superimpose). Response phases for the normal gain condition and adaptation case 1 superimpose in B. NPH output neurons provide a perfect internal estimate of eye position under normal gain conditions and for adaptation cases 1 and 2 (i.e., solid, dotted, and dash-dot curves in D superimpose).

position" to cell EM_A. The VOR cancellation protocol is thus commonly thought to unmask the component of a cell's activity that is related to head velocity (given a near null eye movement contribution and assuming that the cell's activity can be described as the sum of head velocity and efference copy of eye movement signals). Notice that the response of EM_A during VOR cancellation is indeed a reasonably close estimate of the cell's actual net sensitivity to ipsilaterally directed head velocity of 0.44 spikes/s/deg/s (i.e., $s_A - gs_B = 0.44$). This is not, however, the case for cell EM_B. The sensory vestibular projection to EM_B associated with weight s_B

conveys an *ipsilaterally* directed head velocity signal (again with weight $s_B = 0.44$ spikes/s/deg/s), yet this model neuron in fact modulates closely in phase with, or lags, *contralaterally* directed head velocity during VOR cancellation.

The responses of cells EM_A and EM_B illustrated here may be loosely compared with experimentally recorded PVP and FTN cell types, respectively. Specifically, PVP cells are characterized by slow phase sensitivities to eye movement during pursuit that are *oppositely* directed to their sensitivities to head movement during VOR cancellation (e.g., a contralaterally directed sensitivity to eye movement during pursuit and an ipsilaterally directed sensitivity to head velocity during VOR cancellation), whereas FTN cells exhibit eye and head movement sensitivities during pursuit and VOR cancellation, respectively, that are in the *same* direction (e.g., contralaterally directed sensitivities to both eye and head movements). At least a subset of the FTN cells that code for contralaterally directed eye and head movements (i.e., eye-contra FTN cells) are known to receive inputs from ipsilateral semicircular canal afferents (Broussard and Lisberger 1992). Nevertheless, these cells demonstrate contralaterally directed sensitivities to head rotation during VOR cancellation. If it is presumed that the VOR cancellation paradigm provides an accurate means of "unmasking" a cell's sensitivity to head velocity, then their responses imply that eye-contra FTN cells receive a contralaterally directed head velocity input that is stronger than the inputs that they receive directly from the ipsilateral canal. Previously, it has been suggested that this head velocity signal is conveyed via an inhibitory input from the flocculus (Scudder and Fuchs 1992; Lisberger et al. 1994a). However, in the example model here, cell EM_B receives no additional indirect head velocity signal per se. Its modulation in phase with contralaterally directed head movement under head-fixed target viewing conditions is not in this case indicative of an indirect contralaterally directed head velocity input onto the cell; instead, failure to account for the fact that the cell also receives other signals produces that impression. The illustrations here thus call into question the assumption that the VOR cancellation paradigm *always* provides a valid means of estimating a cell's net sensitivity to head velocity. By extension, as will be illustrated below, this paradigm may not always provide an accurate means of estimating changes in a cell's sensitivity to head velocity after motor learning.

2.3.4. VOR in the Dark after Adaptation

With the normal VOR gain state as a reference, we may now examine the effects of introducing parametric changes in the model at different sites to simulate adapted gain conditions. In particular, we will focus on model predictions at 0.5 Hz when the VOR is adapted to a high gain of 1.76. Figure 10.6 illustrates behavioral and cell frequency response predictions during head rotation in the dark. In two cases described below, the reflex was assumed to be trained over the long term using magnifying lenses to achieve broadband changes in reflex performance. As illustrated previously, such broadband adaptation can be simulated in a feedback realization of the VOR by assuming sites for plasticity that are restricted to synaptic inputs from primary canal afferents (see Fig. 10.4). Here, we consider two specific examples: (1) changes in the strength of primary afferent projections onto cell EM_A only (i.e., changes in weight s_A; dotted curves in Fig. 10.6); (2) changes in the strengths of primary afferent projections onto both EM_A and EM_B (i.e., changes in weights s_A and s_B; dash-dot curves in Fig. 10.6). Finally, in a third case, it is assumed that the reflex was trained under visual–vestibular mismatch conditions at 0.5 Hz to achieve frequency-specific changes in reflex performance (dashed curves in Fig. 10.6). As previously illustrated, reflex adaptation localized to a preferred frequency of ~0.5 Hz can be simulated in a feedback realization of the VOR by postulating changes in the time constant, T_f, of the neural filter, $F(s)$, accompanied by modifications in the strength of projection weights in the feedback loops. In this particular case, simulated adaptation to a gain of 1.76 at approximately 0.5 Hz was achieved by simultaneously reducing weight T_f by 67% and each of weights b_A and b_B by 17%.

Cases 1 and 2 assume modifications only in the strength of primary projections, as previously explored in the simple static examples of Figure 10.2, to examine predicted cell responses at high frequencies. We can see that the high-frequency predictions of the dynamic model here (Fig. 10.6) are in agreement with the previous static model; when only s_A is modified, changes in sensitivity following motor learning are evident at high frequencies on cell EM_A but not on EM_B. The observed changes in central responses are thus directly correlated with the site for plasticity. However, when it is instead assumed that both s_A and s_B represent sites for plasticity, then, as before, observed changes in cell responses need not be directly correlated with the locations at which the adaptation took place. Because of the interconnectivity between the cells (i.e., weight g), it remains possible to observe changes at high frequencies only on cell EM_B despite the fact that a site for plasticity also exists on EM_A. Hence, as discussed above, it becomes difficult to localize sites for plasticity without knowledge of the actual interconnectivity in the premotor circuitry.

In this dynamic model, however, additional issues become evident that can also affect the ability to localize adaptation sites. First, notice that the adaptation-related changes observed in the *global modulation* of a given cell depend significantly on the frequency of testing. At low to midrange frequencies (e.g., 0.5 Hz), changes in modulation depth during rotation are always observed on both cell types, regardless of whether a given cell is a site of plasticity. Other predictions relate to the VOR phase. When primary afferent synapses onto vestibular neurons constitute the only sites for plasticity, changes in VOR gain at the behavioral level are accompanied by no change in phase. Central neurons, on the other hand, exhibit changes in *both* the response gain and phase of their modulations to varying extents, depending on their locations within the premotor circuitry and the postulated sites for plasticity. Their global activities reflect more complex changes because they combine both sensory information and activity related to the motor command (i.e., dynamically processed sensory information).

As discussed previously, the activity of each model cell in the dark can be described as the sum of a head velocity signal component and a feedback signal from the output of neural filter $F(s)$. Under normal gain conditions, this feedback signal represents an internal estimate of eye position E^* that remains accurate after VOR adaptation provided the only sites for plasticity are primary afferent synapses onto EM_A and/or EM_B (e.g., see Figs. 10.6D through 10.8D). Following high-gain adaptation, a given head rotation is now associated with a larger deviation in eye position. Unsurprisingly, therefore, because both cells receive efference copy signals, they demonstrate frequency-dependent changes in their global modulations regardless of whether a given cell itself represents a site for plasticity. Clearly, to localize the neural correlates for VOR adaptation in this case, it is necessary to somehow disassociate the component of an observed change in global modulation that is due to efference copy feedback from that representing a true change in a cell's sensitivity to sensory vestibular signals.

Before considering this problem further, we may examine test case 3, in which parameters T_f, b_A, and b_B (see Fig. 10.5) were changed to achieve frequency-specific tuning of the VOR at 0.5 Hz (Fig. 10.6A, dashed line). Both cells exhibit identical changes with increased modulations at low to midrange frequencies and negligible changes only at very high frequencies (>3 Hz; Figs. 10.6B and 10.6C, dashed curves). Again, we are faced with the dilemma of determining whether observed changes at a given frequency were due to local changes in sensitivity or associated with feedback signals. Notice that if we were simply to observe the responses of these cells at a single midrange frequency of 0.5 Hz, the changes in their activities would appear quite similar to those of case 1, in which the adapted gain state was associated with a change in the head velocity sensitivity of cell EM_A (e.g., compare dotted and dashed curves in Figs. 10.6B and 10.6C). How could we then correctly identify that in case 3 there was in fact no change in the head velocity sensitivity of either cell but rather that the plasticity occurred on NPH cells and in their synaptic feedback inputs onto EM_A and EM_B?

2.3.5. Evaluating the "Vestibular" Component of Premotor VOR Cells

When the dynamic characteristics of the premotor VOR circuitry are taken into account, adaptation of the reflex will be accompanied by patterns of frequency-dependent changes in global cell modulations that may reflect both local plasticity as well as changes conveyed indirectly via either feedforward or feedback pathways. To date, emphasis has been placed on localizing central sites at which there is evidence for changes in sensitivity to sensory vestibular signals. Clearly, however, to isolate such sites requires that the head velocity signal content of cell responses be disassociated from those response components associated with eye movement or motor-related signals, whether these are conveyed via feedback pathways, as illustrated in the model here, or alternatively via a feedforward projection as in the classical Robinsonian-style model illustrated in Figure 10.3. Several approaches have been used to extract the sensory vestibular component of cell activities. These include: (1) transient perturbation of the head in the dark to see whether there is evidence of changes at short latency before the eyes start to move; (2) estimation of the sensitivity of the cell to eye movement and removal of this component from the overall modulation to unmask the "vestibular" component; (3) examination of cell responses during the VOR cancellation paradigm in which the head moves but the eyes do not.

The first approach requires the fewest assumptions and has been used successfully to identify at least one potential brain stem site for plasticity on FTN cells. As reviewed above, following motor learning, the shortest-latency ocular responses to a rapid head movement were observed by Lisberger (1984) after 14 ms; however, evidence for reflex gain adaptation became apparent only after an average of 19 ms. Using similar stimuli, the median latency for FTN cell responses to head movement was found to be 11 ms, from which it was estimated that their activities would influence eye movements after about 19 ms, consistent with the latency for observed adaptive changes in motor behavior (Lisberger et al. 1994b). Hence, in this case, transient stimuli could indeed be used to identify short-latency changes in average FTN cell responses that were disassociated from the adaptive change in eye movement.[7] GVP cells, however, responded to ramps in head velocity at a significantly longer latency of 23 ms, well after the eye had started moving and an adaptive change was evident (Lisberger et al. 1994c; du Lac et al. 1995). Therefore, in the case of this cell type, in particular, isolation of a potential change in its head velocity sensitivity requires that some approach be used to disassociate eye movement feedback from head velocity information (e.g., approaches 2 and 3 above). The

---

[7] Although changes in FTN cell responses appeared at short enough latency to be disassociated from adaptation-related increases or decreases in eye movement responses, they could not necessarily be completely disassociated from changes in internal efference copy signals because these may be available at a significantly shorter latency than the actual eye movement. The burst of burst-tonic type NPH cells, for example, precedes the saccadic eye movement (McFarland and Fuchs 1992).

goal here will be to investigate whether typically employed approaches for isolating the sensory vestibular component of a cell's activity always provide accurate estimates of its true sensitivity to head velocity and thus whether such approaches can reliably be used to identify potential sites for plasticity.

Under dark conditions, vestibular neuron responses can be described as a sum of head velocity and efference copy signals. For example, in the model of Figure 10.5, the VN cell types may be described by the following equations:

$$EM_A(s) = (s_A - gs_B)\dot{H}(s) - (b_A - gb_B)E^*(s) \tag{10.6}$$

$$EM_B(s) = s_B\dot{H}(s) - b_B E^*(s) \tag{10.7}$$

Clearly, if the output of filter $F(s)$, E^*, provides an accurate internal estimate of eye position *across all frequencies*, we should be able to estimate the sensitivity of each cell to deviations in eye position and simply remove that component to unmask the cell's sensitivity to head velocity. One approach to estimating this "eye position sensitivity" is to examine cell responses during static fixations assuming that the efference copy signal E^* is appropriately updated during and following saccades to new positions (i.e., assuming that the efference copy center is shared by multiple eye-movement systems such that it provides the same accurate estimate of eye position under all conditions). The result of employing this strategy is illustrated in Figures 10.9A and 10.9B. When sites for plasticity are presumed only in the strengths of synapses associated with sensory vestibular inputs (i.e., cases 1 and 2, where s_A or both s_A and s_B were modified, respectively), we can see that the component of cell activities associated with eye movement feedback is appropriately removed. The same gain with respect to head velocity is now observed across all frequencies (i.e., equal to the high frequency sensitivities of the cells in Fig. 10.6). For example, when only s_A is a site for plasticity, the approach appropriately unmasks the fact that changes in sensitivity to head velocity occur on EM_A but not on EM_B. Note, however, that when both s_A and s_B change, we encounter the same problem in site localization that we observed previously. Because of the interconnectivity between the cells, changes in vestibular sensitivity are only apparent on EM_B, despite the fact that synaptic changes also occur on EM_A. Despite errors in localizing *all* sites for vestibular plasticity, however, the approach does allow us to disassociate global changes in a head velocity pathway from those associated with eye movement feedback, provided that the signal content of the cell can indeed always be described as the sum of head velocity and efference copy signals. Notably, these conditions are not satisfied for case 3, where weights b_A, b_B, and T_f were all modified to simulate frequency-specific adaptation at 0.5 Hz. In this case, as a result of the modification in T_f, an accurate efference copy signal is no longer available across all frequencies. Although E^* always provides an accurate estimate

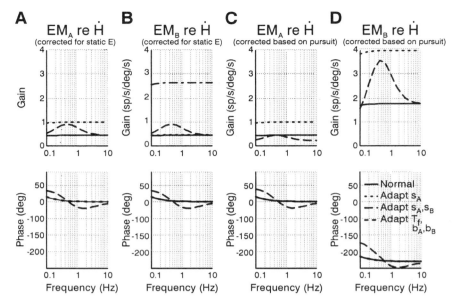

FIGURE 10.9. Vestibular neuron responses during rotation in the dark after subtraction of the portion of their activities estimated to be related to eye movement feedback. (**A, B**) The responses of cells EM_A and EM_B, respectively, were corrected based on their eye position sensitivities estimated from static fixations. (**C, D**) Neural responses were corrected based on their eye position and eye velocity sensitivities estimated from their responses during smooth pursuit (see the text). Response phases for the normal gain condition and cases 1 and 2 superimpose in A–D. Response gains for adaptation cases 1 and 2 superimpose in A. Response gains for adaptation case 1 and the normal gain condition superimpose in B. Response gains for case 2 and the normal gain condition superimpose in C and D.

of eye position at low frequencies (appropriate, for example, to ensure that saccades land on target), this estimate is no longer accurate at higher frequencies (see Figs. 10.6D through 10.8D). Therefore, if cell responses were examined at 0.5 Hz but corrected based on sensitivities to eye position estimated from *static* fixations both prior to and following motor learning, it would *appear* as if there was a change in their underlying sensitivities to head velocity (Figs. 10.9A and 10.9B, dashed curves). Because the approach did not in fact appropriately unmask the vestibular signal content of the cells following motor learning, the result would be an erroneous identification of the vestibular synapses onto these cells as potential sites for plasticity.

In general, it is not known *a priori* whether a given cell's response in the dark can simply be represented as the sum of an efference copy of eye position and a head velocity signal. Furthermore, as illustrated for case 3 above, although such an assumption may provide a reasonably accurate description of a cell's signal content at particular frequencies, it may not at others (i.e., at higher frequencies following motor learning in case 3). Based on observations of cell responses during smooth pursuit, the majority of cells sensitive to eye movement modulate with a phase between that of eye velocity and eye position (e.g., cell EM_B, Fig. 10.7C). The cells thus at least *appear* to also carry signals related to feedback of eye velocity. To account for this observation, a more typical approach to unmasking the head velocity component of cell activities during rotation is to correct for their sensitivities to eye movement estimated from their responses during smooth pursuit. Alternatively, cell responses are often examined under head-fixed target stabilization conditions when head movement is accompanied by negligible ocular deviations.

To simulate the first approach, sensitivities to eye position and eye velocity (prior to and following motor learning) were estimated in the model based on the gains and phases of neural activities during pursuit at 0.5 Hz both prior to and following simulated reflex adaptation. In other words, the response of each cell during pursuit was modeled by the transfer function $FR(s)/E(s) = G(T_{pur}s + 1)$, where $FR(s)$ represents the neural firing rate. Gain, G, then represents the cell's estimated sensitivity to eye position and its sensitivity to eye velocity, $sE(s)$, is GT_{pur} (see Scudder and Fuchs 1992). The results of using these estimated sensitivities to correct for the eye-movement-related components of each cell's response during head rotation in the dark are illustrated in Figures 10.9C and 10.9D. We will focus on examining residual activities at 0.5 Hz because corrections for eye position and eye velocity were based on sensitivities estimated from 0.5 Hz pursuit responses.

In the case of cell EM_A, after pursuit-based correction for eye-movement-related activity, the portion of the neuron's activity related to head velocity at 0.5 Hz is appropriately unmasked in all cases. As before, we would appropriately conclude that there was a modification in the cell's head velocity sensitivity in case 1 (only s_A modified, dotted trace in Fig. 10.9C) and that there was no *net* change in its head velocity sensitivity in case 2 (both s_A and s_B modified, dash-dot trace in Fig. 10.9C) even though, as discussed previously, in this latter case, the weight of the synaptic head velocity input to the cell was actually modified (a *local* partial change). In addition, we also correctly conclude that there was no change in the cell's head velocity sensitivity in case 3 (T_f, b_A, b_B modified, dashed trace in Fig. 10.9C). Notice, however, that in this case postadaptation changes in the cell's pursuit response are apparent (Fig. 10.7), correctly suggesting plasticity in pathways other than those strictly associated with head movement that could be local to the cell and associated with its other signal inputs (as is indeed the case here).[8]

Similar conclusions would be reached if the presence of adaptation-related changes in the head velocity sensitivity of cell EM_A were evaluated based on the cell's response during head-fixed target stabilization (Fig. 10.8B). The modulation depth of cell EM_A during cancellation following motor learning is approximately double that under normal gain conditions in case 1, appropriately reflecting the modification in the strength of synapse s_A (dotted curve). In case 3, (T_f, b_A, b_B modified) the change in the activity of EM_A as compared with the normal gain condition is negligible at 0.5 Hz; thus, we would correctly conclude that there was no change in the head velocity sensitivity of the cell. When both s_A and s_B represent sites for plasticity (case 2), although there is no *net* change in the head velocity sensitivity of cell EM_A (despite the fact that s_A constitutes a site for plasticity), a small increase in the firing rate modulation of the cell is nevertheless observed. This is because VOR cancellation performance in this case is not as good as in the normal gain condition such that there are larger residual deviations in eye position following motor learning (e.g., compare dash-dot and solid curves in Fig. 10.8A). If a correction was applied to account for this deterioration in cancellation performance (not illustrated here), the small increase in the cell's modulation would be eliminated so that no change in the net head velocity sensitivity of EM_A would be apparent, in agreement with previous examples. In general, therefore, both correction for the eye-movement-related components of the cell's activity during rotation in the dark and examination of the activity of the cell under head-fixed target stabilization conditions yield reasonably accurate estimates of the head-velocity-related activity of cell EM_A. Although an actual site for plasticity associated with motor learning was only localized in case 1 (s_A modified), the approaches employed were successful in isolating the instances in which simulated reflex adaptation was associated with *net* changes in the head velocity sensitivity of cell EM_A.

In contrast, when these same approaches are applied to cell EM_B, the cell's actual sensitivity to head velocity signals is never appropriately unmasked. For example, under normal gain conditions (Fig. 10.9D, solid curve), the use of pursuit-based estimated sensitivities to eye movement to "correct" the cell's firing rate during head rotation in the dark yields an

---

[8] In case 3, the sites for plasticity are within the feedback loops of the system, resulting not only in changes in the dynamic characteristics of VOR responses but also in those of pursuit responses (see Fig. 10.7). Notice, however, that whereas the changes in the behavioral pursuit response at 0.5 Hz are relatively small, the changes in the response phase of cell EM_A and in the gain of cell EM_B during pursuit are significant, resulting in new estimates of the eye position and eye velocity sensitivities of these cells follwing motor learning. These new sensitivities were taken into account when correcting for the eye movement component of cell responses to isolate any changes in sensitivity to head velocity.

apparent *contralaterally* directed sensitivity of head velocity of about 1.7 spikes/s/deg/s. Examination of the cell's response under normal gain conditions during VOR cancellation at 0.5 Hz yields an estimated contralaterally directed head velocity sensitivity of close to 1.5 spikes/s/deg/s (Fig. 10.8C, solid trace). Clearly, neither approach provides an accurate estimate of the cell's true ipsilaterally directed sensitivity to head velocity of 0.44 spikes/s/deg/s. Similarly, the head velocity component of the cell's activity is not appropriately isolated following motor learning. As a result, incorrect conclusions are reached concerning sites for plasticity. For example, when only synapse s_A onto cell EM_A constitutes a site for plasticity (case 1), it nevertheless appears that there was a change in the head velocity sensitivity of cell EM_B; there is an apparent increase in the cell's sensitivity to contralaterally directed head velocity, and thus the change appears to be in the wrong direction to support the adapted gain state (Fig. 10.9D; dotted curve). This conclusion is supported by examining the cell's response during VOR cancellation (Fig. 10.8C), where again there is an apparent increase in the cell's sensitivity to contralaterally directed head velocity. When both s_A and s_B constitute sites for plasticity (case 2), after subtraction of the estimated component of the cell's activity related to eye movement, there is no apparent change in the sensitivity of cell EM_B to head velocity, leading to the false conclusion that the cell does not constitute a site for plasticity. The same erroneous conclusion in this case follows from the cell's response during VOR cancellation at 0.5 Hz, where only a tiny increase in the cell's modulation (compared with the normal gain condition) is apparent. Finally, in case 3 (T_f, b_A, b_B modified), comparison of eye movement "corrected" responses prior to and following motor learning leads to the false impression of an increase in its sensitivity to contralaterally directed head velocity. However, examination of the cell's response during VOR cancellation at 0.5 Hz yields a different result. There is negligible change in the cell's activity associated with the adaptation, leading in this case to the correct conclusion that there was no change in the cell's head velocity sensitivity. Nevertheless, the head velocity sensitivity that is estimated during the VOR cancellation paradigm is not an accurate estimate of the cell's true sensitivity to head velocity inputs.

Classical techniques employed to isolate the head velocity sensitivity component of cell responses provided reasonably accurate estimates in the case of cell EM_A but not for EM_B. The head velocity component of the EM_B cell's activity was generally incorrectly estimated because the techniques failed to correctly model all inputs to the cell. It was assumed that because the cell's response during 0.5 Hz pursuit could be approximated (incorrectly) as the sum of eye position and eye velocity signals, that during rotation in the dark it could be modeled as the sum of eye position, eye velocity, and head velocity signals. However, as previously discussed, there is in fact no explicit eye velocity signal input to the cell. In fact, as is clear from Eq. (10.7), during head rotation in the dark, the cell's response reflects the sum of an ipsilaterally directed head velocity input and an efference copy signal E^*. In all cases where an accurate efference copy is maintained following motor learning (i.e., cases 1 and 2), the cell's response in the dark thus reflects a sum of head velocity and estimated eye position signals (i.e., no eye velocity component). On the other hand, in case 3, at higher frequencies where an accurate efference copy is no longer available, the cell's response is indeed more closely modeled as a combination of head velocity, eye position, and eye velocity signals. The apparent eye velocity component, however, is frequency-dependent and need not be the same eye velocity sensitivity that would be measured during pursuit. Another way to state this is simply to say that sensitivities estimated under visual feedback conditions cannot always be assumed to apply to dark conditions because in general the activation of visual feedback loops gives rise to overall changes in the dynamic properties of the system and introduces additional inputs. Although the same issues apply to the case of cell EM_A, due to slightly different interconnectivity, its signal inputs combine in a different proportion such that the net contribution of retinal position error to the cell's response is negligible. As a result, the cell's modulation during pursuit indeed provides an accurate estimate of its sensitivity to eye position in the dark.

2.3.6. Summary

A very simple model was used to illustrate problems that arise in the interpretation of experimental observations when individual cell responses are examined in a highly interconnected circuit processing several sensory signals. In general, all cells may exhibit adaptation-related changes in modulation that vary in extent depending on the frequency of testing, regardless of whether they actually represent sites for plasticity. To isolate a site for plasticity on a specific cell population requires that localized changes in sensitivity be distinguished from changes in sensitivity conveyed from other locations either via feedback or feedforward projections. Given incomplete knowledge, several types of assumptions are typically made to simplify the problem of identifying the neural correlates for motor learning in the VOR. These include assumptions with respect to (1) the structural characteristics of the system; (2) the signal content of individual neurons; (3) the sites at which plasticity is expected to occur. At least two key problems with such assumptions have been illustrated here with the goal of stimulating further investigation.

Specifically, we began with a very simple model structure and assumed that we had explicit knowledge of this structure and thus of the signal content of individual neural populations (i.e., Eqs. (10.6) and (10.7)). It was shown that we could indeed correctly isolate actual changes in neural sensitivities to head velocity from efference copy feedback signals *when the only plasticity responsible for reflex adaptation occurred in cell sensitivities to head velocity inputs* (cases 1 and 2). Nevertheless, because of the interconnectivity of the model vestibular neurons, a site for plasticity could not always be localized appropriately on both cells (e.g., case 2). In other words, an accurate evaluation of changes in modulation is not necessarily correlated with a *local* change in cell sensitivity. A larger problem became evident, however, when adaptation was achieved at 0.5 Hz by changing the efficacy of the neural integrator (i.e., a presumed site for plasticity in the NPH; case 3). Corrections for the eye position sensitivities of the cells were made under the false assumption that the only sites for plasticity were in central sensitivities to head velocity inputs. Specifically, it was presumed that the efference copy signal and central sensitivities to this signal were unaffected by VOR adaptation. As a result, it was incorrectly concluded that there were changes in the head velocity sensitivities of both cells when in fact adaptation at 0.5 Hz was achieved without any central changes in sensitivity to head velocity. Hence, despite complete knowledge of the model structure, a preconceived notion with respect to sites for plasticity biased the interpretation of the experimental results.

Because in fact we do not have *a priori* knowledge of the circuit structure, the problem of isolating sites for plasticity was reinvestigated, this time relying on observed cell responses under different visual–vestibular stimulus conditions to provide an estimate of the signal content of each neuron. Specifically, cell responses during smooth pursuit were used to estimate cell sensitivities to eye movement, whereas the VOR cancellation paradigm was used to estimate cell sensitivities to head movement in the absence of eye movement. Problems in interpreting experimental observations nevertheless arose because estimates of the signal content of each neuron were assumed to be invariant to the stimulus condition and inputs to the cells were incorrectly modeled. During pursuit at 0.5 Hz, for example, each cell was modeled by a weighted sum of eye position and eye velocity signals. It was then assumed that each cell receives a combination of dedicated efference copy of eye position and eye velocity signal inputs, both of which were accounted for in interpreting cell responses during rotation in the dark. Yet, in fact no explicit central estimate of eye velocity exists in the model. Correction of cell responses in the dark for an assumed efference copy of eye velocity signal that did not in fact exist biased the interpretation of the experimental observations. Here, assumptions with respect to the signal content of the neurons both prior to and following adaptation therefore led to inappropriate conclusions with respect to sites for plasticity in many cases.

On the basis of these simple examples, there would appear to be a need to develop techniques for interpreting neural responses that rely on fewer structural or functional assumptions that can bias data interpretation. In particular, it is inadequate to investigate simply the head velocity sensitivity of central activities to probe for plastic sites. The parameters in a postulated model should allow for more global dynamic changes over a broad frequency range. Errors in evaluations of the signal content of individual neurons can be minimized by accounting for results over a broad range of frequencies and by attempting to account for *whole* cell responses in a variety of alternate model forms that incorporate the full range of VOR response dynamics (i.e., not just head/eye velocity). In the following section, a system identification approach will be presented that can be used to provide less biased estimates of plastic sites *if appropriate protocols are used to investigate neural responses prior to and following motor learning and adequate model structures are used to evaluate experimentally observed responses.*

3. Estimating Parametric Changes in Dynamic VOR Models: System Identification Approach

As illustrated above, investigation of neural systems with multiple recursive feedback loops presents a significant problem in terms of data analysis and interpretation. This is especially true when the goal is to identify changes in a specific signal component of a cell's response (e.g., head velocity) or to distinguish an actual change in sensitivity from changes that are conveyed from another location. So far, we have examined an analytic approach to localizing potential sites for plasticity in which the relationship between dynamic cellular and behavioral responses to sensory stimuli is examined in anatomically plausible assumed model structures. We illustrated several caveats in exposing sites of plasticity associated with incomplete knowledge of neural signal content and interconnectivity as well as with *a priori* assumptions about potential adaptive sites.

In the following section, we illustrate a global system identification approach to investigating reflex adaptation in which the system is divided into subsystems related to potential adaptive sites without imposing a particular structure at the single neuron level. Although plastic sites are not necessarily identified in terms of specific anatomical location, using this method it is possible to make tentative conclusions with respect to *processes* or *signaling pathways* in which adaptation takes place based on recordings from a single cell type. The approach addresses several of the caveats outlined above because *a priori* assumptions about potential sites for plasticity are not made and issues of relevant structure exist at a more global level (i.e., between subsystems as opposed to local cell interconnectivity). Once adaptive subprocesses are identified at this more global level, possible anatomical substrates for motor learning can be examined within a neuronal structure by recording from cells postulated to exist within a given subprocess (see below and Hirata and Highstein 2001).

3.1. System Identification in Partitioned VOR Models

The first step in system identification is to postulate a valid model structure for the questions at hand. A structure in this case has two meanings. One is a global structure: how subsystems or processes are connected. Another is a local structure: the structure or order of each subsystem. The structures must be based on known anatomy and physiology if the model is used to localize the site of any adaptation. To localize adaptive sites in the VOR circuitry, for example, we need to decompose the circuit into subsystems each of which contains a possible locus of adaptation. Then, we identify characteristics of each subsystem before and after the adaptation. Subsystems that changed their characteristics after the adaptation are the ones containing an adaptive element or elements.

To identify any model characteristics, we need input and output relationships acquired in well-designed experiments. Inputs should contain rich information in terms of amplitude and frequency (e.g., Gaussian white noise) that covers the dynamic range of the system because the model characteristics can be identified only within the bandwidth of the input signal. If a band-limited input is used, then the system identification is restricted to the narrower stimulus range.

The description of the model can be either parametric or nonparametric. Generally, if the system is linear, autoregression (AR), moving average (MA), autoregressive moving average (ARMA), or other types of regression models are applied for a parametric model and the impulse response (or, equivalently, transfer function) is estimated for a nonparametric model. If it is nonlinear, nonlinear versions of AR, MA, and ARMA models (Nikias and Raghuveer 1987; Toda and Usui 1991) are applied for a parametric model and Volterra or Wiener kernels (Marmarelis and Marmarelis 1978; Schetzen 1980) are estimated for a nonparametric model. In many cases of biological system identification, however, applying a rich input controllable from outside to a neuronal structure (or a subsystem) and recording its output is not always possible. In such a case, we must either limit the input range or configure the system so that a rich input can be applied to each subsystem.

The second step in system identification is to estimate model parameters. The model parameters are estimated so that the output of the model and the experimental data in response to the same input coincide. Experimental data are always contaminated by various kinds of noise. Hence, regression techniques are used to optimally estimate the parameters that minimize errors (usually squared sum of residual) in the fit. One must be careful to restrict the number of parameters to that justified by the noise level. If the model structure is too complex (contains more parameters than required), the model output tends to fit the noise and estimated parameters can be seriously biased. Thus, an important issue is determining a valid model complexity (number of free parameters). Information criteria such as AIC (Akaike's information criterion; Akaike 1973), BIC (Bayesian information criterion; Schwarz 1978) and MDL (Minimum description length; Rissanen 1986) are used for that purpose.

The model validity is examined in the third step, which can be an iterative process in fixing the model complexity. In general, the validity of the selected model is demonstrated by its accuracy in predicting the system responses to new stimuli that were not used in the parameter estimation step. This approach will be illustrated in an evaluation of adaptation in the vertical VOR.

3.2. Application to Vertical VOR Adaptation

3.2.1. Determination of the Model Structure

The known anatomy relevant for the vertical VOR is illustrated in Figure 10.10A. As reviewed in the Introduction, the flocculus is essential in motor learning. Its only output elements, Purkinje cells, show small but consistent changes in firing modulation during VOR in darkness after acute VOR adaptation induced by visual–vestibular mismatch training at a particular frequency for several hours (Watanabe 1984; Hirata and Highstein 2001). However, this does not necessarily mean that the flocculus is the site responsible for the adaptation. As illustrated in Figure 10.10A, the flocculus receives multimodal input: vestibular (from FPNs), visual (from

⟶

FIGURE 10.10. Vertical VOR and OKR neuronal circuit (**A**) and model (**B**). In A, FPN and FTN are the floccular projecting neurons and the floccular target neurons, respectively, in the superior vestibular nuclei (SVN); Y is the dorsal y group; Int. represents the interneuron in the SVN that projects to the y group; P is a floccular Purkinje cell; g is a floccular granular cell; and PVP represents position-vestibular-pause neurons in the medial vestibular nuclei (MVN). MT, MST, and DLPN are the middle temporal visual area, the medial superior temporal area, and the dorsolateral pontine nuclei, respectively. LTN indicates the lateral terminal nucleus of the accessory optic system. MN represents extraocular motor neurons PMT represents the paramedian tracts. In B, $G_{\text{pre-FL\&FL}}^{\text{ecopy}}(s)$, $G_{\text{pre-FL\&FL}}^{\text{vestib}}(s)$, and $G_{\text{pre-FL\&FL}}^{\text{visual}}(s)$ are prefloccular/floccular subsystems represented by transfer functions associated with efference copy, vestibular pathways, and visual pathways, respectively. The three components are added in the flocculus and form the Purkinje cell simple spike (SS) output. $G_{\text{post-FL}}(s)$ represents the transfer function of the postfloccular pathway that transfers the Purkinje cell simple spike (SS) activity into one portion of the net motor command. $G_{\text{non-FL}}^{\text{visual}}(s)$ and $G_{\text{non-FL}}^{\text{vestib}}(s)$ represent transfer functions of the nonfloccular visual and vestibular pathways, respectively. The neuronal circuit corresponding to $G_{\text{non-FL}}^{\text{visual}}(s)$ is not shown in A. The block on the left-hand side of $G_{\text{pre-FL\&FL}}^{\text{visual}}(s)$ represents a nonlinear transformation of the retinal slip to linearize its relationship to Purkinje cell activity. $h(t)$, $d(t)$, $r(t)$, $f(t)$, ecopy(t), and $x(t)$ are head movement, optokinetic stimulus movement, retinal slip movement, floccular Purkinje cell SS activity, efference copy signal, and eye movement, respectively.

A

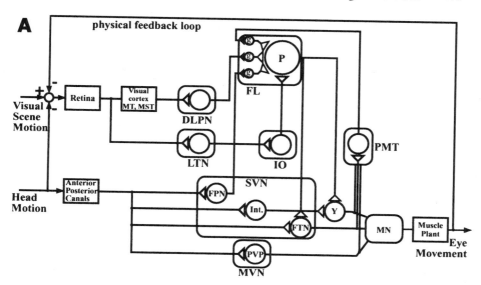

B

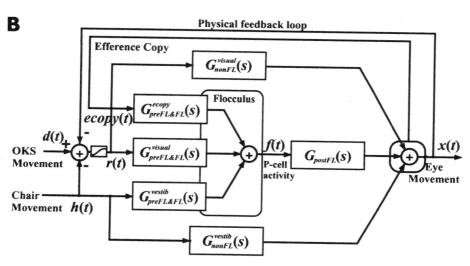

DLPN), and efference copy (from PMT). The visual pathway does not function in darkness and hence cannot account for the plastic changes observed in the absence of visual feedback. Yet, it is not obvious which signal (vestibular, efference copy, or both) accounts for the changes in floccular cell responses observed following adaptation. A second question is which other sites (i.e., outside the flocculus) might contribute to the adaptation. To address these questions based simply on neural responses recorded from floccular neurons, we began by partitioning the vertical VOR system into subprocesses containing potential adaptive elements. First, we differentiated the flocculus pathway from other nonflocculus pathways and divided the former into subsystems representing each of the three modalities. We also divided the nonflocculus pathway into those processing visual and vestibular signals based on the anatomy. Figure 10.10B illustrates a possible model structure used for the system identification approach. The prefloccular and floccular (pre-FL&FL) system includes signal processing executed in prefloccular pathways and in the flocculus itself, and its output is indicated as floccular Purkinje cell simple spike (SS) activity. In the current study, only SS firing was analyzed. For this reason, Figure 10.10B does not include the pathway from the inferior olive (10) to the flocculus, which is illustrated in Figure 10.10A. The three kinds of mossy fiber input to the flocculus illustrated in Figure 10.10A are explicitly described as separate pathways in the pre-FL&FL system in Figure 10.10B. In each pathway, the sensory or efference copy signal is processed by a subsystem $G_{\text{pre-FL\&FL}}^{\text{vestib}}(s)$, $G_{\text{pre-FL\&FL}}^{\text{visual}}(s)$, or $G_{\text{pre-FL\&FL}}^{\text{ecopy}}(s)$ for vestibular, visual, or efference copy signals, respectively, and is converted into floccular Purkinje cell SS activity. The post-FL system $G_{\text{post-FL}}(s)$ and subsequent oculomotor muscle plant transfer Purkinje cell activity into part of the overall oculomotor response. The non-FL system consists of two pathways processing vestibular or visual sensory signals and converting them into another portion of the net eye movement. Signal processing executed in the non-FL visual and vestibular pathways is described by the transfer functions $G_{\text{non-FL}}^{\text{visual}}(s)$ and $G_{\text{non-FL}}^{\text{vestib}}(s)$, respectively.

Presently, only slow-phase eye movements and corresponding Purkinje cell SS activities have been studied. Also, it was assumed that each subsystem is time-invariant during one recording set that consists of about five minutes of several visual–vestibular mismatch paradigms. Finally, the technique below presumed linear visual–vestibular operation (Robinson 1977) and explored parametric estimates for responses during 0.5 Hz protocols at various levels of VOR adaptation.

3.2.2. Parameter Estimation

1. Identifying Prefloccular and floccular systems: $G_{\text{pre-FL\&FL}}^{\text{visual}}(s)$, $G_{\text{pre-FL\&FL}}^{\text{ecopy}}(s)$, $G_{\text{pre-FL\&FL}}^{\text{vestib}}(s)$.
The pre-FL&FL systems form a multiple-input–single-output system that is described in the Laplace transform domain as

$$F(s) = H(s)G_v(s)G_{\text{pre-FL\&FL}}^{\text{vestib}}(s) + R(s)G_r(s)G_{\text{pre-FL\&FL}}^{\text{visual}}(s) + E(s)G_{\text{pre-FL\&FL}}^{\text{ecopy}}(s),$$

where s denotes the Laplace operator and $F(s)$, $H(s)$, $R(s)$, and $E(s)$ denote Laplace transforms of floccular Purkinje cell SS firing rate, $f(t)$, angular head position, $h(t)$, retinal slip, $r(t)$, and efference copy, ecopy(t), respectively. $G_v(s)$ and $G_r(s)$ represent transfer functions for the vestibular and retinal sensory processes, respectively. They are presumed to be unaffected by the gain of the VOR (e.g., Miles et al. 1980b) and thus are incorporated into their downstream processes $G_{\text{pre-FL\&FL}}^{\text{vestib}}(s)$ and $G_{\text{pre-FL\&FL}}^{\text{visual}}(s)$ hereafter. Therefore, the equation above becomes

$$F(s) = H(s)G_{\text{pre-FL\&FL}}^{\text{vestib}}(s) + R(s)G_{\text{pre-FL\&FL}}^{\text{visual}}(s) + E(s)G_{\text{pre-FL\&FL}}^{\text{ecopy}}(s). \quad (10.8)$$

This defines the relative roles of the subcomponents but does not indicate the nature (complexity) of each element—one must now determine the order of each of these based on physiological evidence. It has been shown that (1) head velocity and acceleration signals (Goldberg and Fernández 1971) and (2) retinal slip velocity and acceleration signals (Kobayashi et al. 1998) contribute to Purkinje cell firing modulation and that (3) efference copy signals convey mainly eye velocity information (Miles et al. 1980a; Stone and Lisberger 1990). If we assume that each of these components interacts linearly, the system equation in the time domain can be expressed as the multiple linear regression model

$$f(t) = \alpha_h \text{acc}_h(t - \tau_h) + \beta_h \text{vel}_h(t - \tau_h) + \alpha_r \text{acc}_r(t - \tau_r)$$
$$+ \beta_r \text{vel}_r(t - \tau_r) + \beta_e \text{ecopy}(t - \tau_e) + \delta + \varepsilon(t), \quad (10.9)$$

where α_h (spikes/s per deg/s^2), β_h (spikes/s per deg/s), α_r (spikes/s per deg/s^2), β_r (spikes/s per deg/s), β_e (spikes/s per deg/s), and δ (spikes) are regression coefficients that denote sensitivities of Purkinje cell firing to head acceleration, $\text{acc}_h(t)$, head velocity, $\text{vel}_h(t)$, retinal slip acceleration, $\text{acc}_r(t)$, retinal slip velocity, $\text{vel}_r(t)$, efference copy signals, ecopy(t), and the dc component of Purkinje cell firing, δ, respectively. τ_h, τ_r, and τ_e denote delays between head movement and Purkinje cell activity, retinal slip and Purkinje cell activity, and efference copy signals and Purkinje cell activity, respectively. $\varepsilon(t)$ is the model error term whose mean is 0. The transfer functions of pre-FL&FL subsystems are defined from estimated regression coefficients as follows:

$$G_{\text{pre-FL\&FL}}^{\text{visual}}(s) = (\alpha_r s^2 + \beta_r s)e^{-\tau_r s},$$
$$G_{\text{pre-FL\&FL}}^{\text{vestib}}(s) = (\alpha_h s^2 + \beta_h s)e^{-\tau_h s},$$
$$G_{\text{pre-FL\&FL}}^{\text{ecopy}}(s) = \beta_e e^{-\tau_e s}.$$

The least squares method is usually employed to estimate regression coefficients. In general, if we use an experimental paradigm that activates all of the subsystems in the model (e.g., visual–vestibular interaction paradigms), then we can determine all parameters at the same time. In the case of the VOR system, however, head velocity $vel_h(t)$ and eye velocity $ecopy(t)$ usually are highly correlated because that is the function of the VOR. By assuming that subsystems are invariant in the different protocols, at a given VOR gain level, piecewise (sequential) regression was used as an alternative. The visual following (VF) paradigm was first used to estimate the coefficients of $G^{\text{visual}}_{\text{pre-FL\&FL}}(s)$ and $G^{\text{ecopy}}_{\text{pre-FL\&FL}}(s)$, when head movement is zero. Equation (10.9) can then be simplified to

$$f_{\text{VF}}(t) = \alpha_r \text{acc}_r(t - \tau_r) + \beta_r \text{vel}_r(t - \tau_r)$$
$$+ \beta_e \text{ecopy}(t - \tau_e) + \delta_{\text{VF}} + \varepsilon_{\text{VF}}(t). \quad (10.10)$$

The regression coefficients α_r, β_r, β_e, and δ_{VF} were determined by solving the least squares normal equations derived from Eq. (10.10) to minimize the square sum of $\varepsilon_{\text{VF}}(t)$ while τ_e and τ_r were globally searched between -0.015 and $0.035\,\text{s}$ in $0.001\,\text{s}$ steps and between 0.025 and $0.075\,\text{s}$ in $0.001\,\text{s}$ steps, respectively. The latency between Purkinje cell activity and eye movement was referred to as τ_e because the eye velocity trace was substituted for a measured efference copy signal. Similarly, the VORd (in the dark) paradigm was used next to define $G^{\text{vestib}}_{\text{pre-FL\&FL}}(s)$ by finding the regression coefficients related to head movement (no retinal slip) in

$$f_{\text{VORd}}(t) = \alpha_h \text{acc}_h(t - \tau_h) + \beta_h \text{vel}_h(t - \tau_h)$$
$$+ \beta_e \text{ecopy}(t - \tau_e) + \delta_{\text{VORd}} + \varepsilon_{\text{VORd}}(t). \quad (10.11)$$

Because β_e and τ_e are presumed known from VF data fits, the least squares fit was executed to minimize the squared sum of $\varepsilon_{\text{VORd}}(t)$ in $\{f_{\text{VORd}}(t) - \beta_e \text{ecopy}(t - \tau_e)\}$, and τ_h was globally searched between -0.015 and $0.035\,\text{s}$ (see Hirata and Highstein 2001 for rationale). Thus, the computation of prefloccular and floccular transfer functions depends critically on the assumption of identical Purkinje cell dependence on efference copy signals in both VF and VORd paradigms. The plausibility of this assumption will be examined when evaluating the validity of the model.

2. *Identifying Postfloccular and Nonfloccular subsystems:* $G_{\text{post-FL}}(s)$, $G^{\text{visual}}_{\text{non-FL}}(s)$, $G^{\text{vestib}}_{\text{non-FL}}(s)$.

The post-FL and non-FL systems form a multiple-input–single-output system relating eye position (behavior) to floccular and direct (nonfloccular) inputs. The system equation is described in the Laplace domain as

$$X(s) = H(s)G_v(s)G^{\text{vestib}}_{\text{non-FL}}(s)G_m(s)$$
$$+ R(s)G_r(s)G^{\text{visual}}_{\text{non-FL}}(s)G_m(s) + F(s)G_{\text{post-FL}}(s)G_m(s),$$

where $G_m(s)$ and $X(s)$ denote the Laplace transform of the ocular muscle plant and eye position, respectively. Because it is again assumed that the characteristics of $G_m(s)$ as well as $G_v(s)$ and $G_r(s)$ do not change with VOR gain change, they are incorporated into $G_{\text{non-FL}}^{\text{vestib}}(s)$, $G_{\text{non-FL}}^{\text{visual}}(s)$, or $G_{\text{post-FL}}(s)$ so that

$$X(s) = H(s)G_{\text{non-FL}}^{\text{vestib}}(s) + R(s)G_{\text{non-FL}}^{\text{visual}}(s) + F(s)G_{\text{post-FL}}(s). \qquad (10.12)$$

This can be rewritten with Purkinje cell firing rate as the measured response:

$$F(s) = X(s)/G_{\text{post-FL}}(s) - H(s)G_{\text{non-FL}}^{\text{vestib}}(s)/G_{\text{post-FL}}(s)$$
$$- R(s)G_{\text{non-FL}}^{\text{visual}}(s)/G_{\text{post-FL}}(s).$$

To obtain a regression formulation, one must again make some assumptions on the order of each transfer function. It has been shown that Purkinje cell firing encodes a part of eye movement inverse dynamics as a linear combination of eye acceleration, velocity, and position (Shidara et al. 1993). This means that the post-FL pathway can be described by a second-order linear system. If we assume that non-FL subsystems can also be characterized by a second-order linear system within the input range currently employed, the equation above can be expressed in the time domain as

$$f(t) = a_x \text{acc}_x(t + \tau_x) + b_x \text{vel}_x(t + \tau_x) + c_x \text{pos}_x(t + \tau_x) - \alpha_r \text{acc}_r(t - \tau_r)$$
$$- b_r \text{vel}_r(t - \tau_r) - a_h \text{acc}_h(t - \tau_h) - b_h \text{vel}_h(t - \tau_h) + K + \zeta(t), \qquad (10.13)$$

where a_x (spikes/s per deg/s^2), b_x (spikes/s per deg/s), and c_x (spikes/s per deg) denote sensitivities of Purkinje cell firing rate to eye acceleration, $\text{acc}_x(t)$, velocity, $\text{vel}_x(t)$, and position, $\text{pos}_x(t)$, respectively, and τ_x denotes a delay time between Purkinje cell activity and eye movement. K is the dc term and $\zeta(t)$ is the error term whose mean is 0. a_r, b_r, a_h, b_h, τ_r, and τ_h together with a_x, b_x, c_x, and τ_x determine the second-order transfer functions of the non-FL subsystems and post-FL subsystem as follows:

$$G_{\text{post-FL}}(s) = e^{-\tau_x s}/(a_x s^2 + b_x s + c_x),$$
$$G_{\text{non-FL}}^{\text{visual}}(s) = (a_r s^2 + b_r s)e^{-(\tau_r + \tau_x)s}/(a_x s^2 + b_x s + c_x),$$
$$G_{\text{non-FL}}^{\text{vestib}}(s) = (a_h s^2 + b_h s)e^{-(\tau_h + \tau_x)s}/(a_x s^2 + b_x s + c_x).$$

However, the available visual–vestibular protocols again would lead to highly correlated data in Eq. (10.13). In the same manner as in the previous sections, the issue was resolved (with similar caveats) by first finding a_x, b_x, c_x, a_r, b_r, τ_x, and τ_r from the VF paradigm, thereby defining $G_{\text{post-FL}}(s)$ and $G_{\text{non-FL}}^{\text{visual}}(s)$. The coefficients thus obtained were then substituted back into Eq. (10.13) for the VORd paradigm (no retinal inputs) to obtain $G_{\text{non-FL}}^{\text{vestib}}(s)$.

3.2.3. Evaluation of the Model Validity

The model adequacy was checked by examining the residuals of predicted versus measured Purkinje cell firing rates in response to protocols that were not used to estimate the parameters, such as enhancement of the VOR (VORe), reversal of the VOR (VORr), and suppression of the VOR (VORs) (model cross validation). The amplitude distribution and auto-correlation function of the residuals were compared with those of the Purkinje cell firing rate during no external stimulation for this examination. The same residual checks were performed for the (fitted) VORd and VF protocols after the parameter estimation. The idea is that if the amplitude distribution and autocorrelation functions of the residual match those of the Purkinje cell firing during the no-stimulus condition, then the information encoded in the Purkinje cell firing by the applied external stimuli has been successfully extracted or predicted. Both types of verification supported the accuracy of information extraction in the postulated model format. Therefore, the VOR model structure was found to be valid within the stimulus range currently employed. Refer to Hirata and Highstein (2001) for details on Purkinje cell predictive fits. Here we will focus on the results related to potential sites for parametric change during continuous tracking of VOR adaptive changes.

3.3. *Estimated Sites for Parametric Change*

By using 46 Purkinje cells recorded at 69 VOR gains altered acutely by visual–vestibular mismatch paradigms (VORs, VORr, or VORe) at 0.5 Hz, changes in the characteristics of the model subsystems during VOR adaptation were evaluated. Figure 10.11 plots VOR gain versus system characteristics (gains and phases) at 0.5 Hz for the pre-FL&FL ($G_{\text{pre-FL\&FL}}^{\text{vestib}}$, $G_{\text{pre-FL\&FL}}^{\text{visual}}$, $G_{\text{pre-FL\&FL}}^{\text{ecopy}}$), post-FL ($G_{\text{post-FL}}$), and non-FL ($G_{\text{non-FL}}^{\text{visual}}$, $G_{\text{non-FL}}^{\text{vestib}}$) subsystems in a 3D polar representation. Each circle represents an estimate from each Purkinje cell. The projection of the circle atop each thin vertical line onto the ordinate represents the VOR gain at which each Purkinje cell was recorded. The intersection of the base of each thin vertical line with the polar plane indicates the estimated gain (radius) and phase (angle) characteristics of the system. In each figure, the slanted heavy black line is a regression line (the first principal component) indicating the average characteristics of the system in relation to VOR gain. The line is in the plane (dotted line) perpendicular to the polar plane. It was found that system characteristics of the pre-FL&FL vestibular pathway, $G_{\text{pre-FL\&FL}}^{\text{vestib}}$, and non-FL vestibular pathway, $G_{\text{non-FL}}^{\text{vestib}}$, changed in parallel with VOR gain (slopes of the regression lines are significantly different from 0: $P < 0.0075$ ($F_0 = 7.8635$), $P < 0.000011$ ($F_0 = 24.5431$), respectively), whereas the post-FL pathway, $G_{\text{post-FL}}$, pre-FL&FL efference copy pathway, $G_{\text{pre-FL\&FL}}^{\text{ecopy}}$, and non-FL visual pathway, $G_{\text{non-FL}}^{\text{visual}}$, showed minimal changes ($P > 0.804$ ($F_0 = 0.0621$), $P > 0.4380$ ($F_0 = 0.6125$), $P > 0.273$ ($F_0 = 1.2290$), respectively). The slope of changes in the pre-FL&FL retinal slip pathway, $G_{\text{pre-FL\&FL}}^{\text{visual}}$, also showed no significant difference from 0 ($P > 0.2026$ ($F_0 = 1.6723$)).

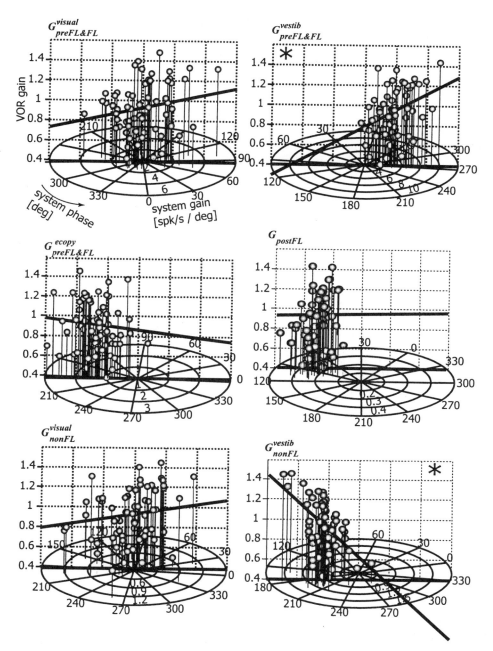

FIGURE 10.11. 3D polar representation of VOR gain versus estimated system characteristics at 0.5 Hz for each subsystem. Each circle represents an estimate from a single Purkinje cell. The length of the thin black line under each circle represents the gain of the VOR at which the cell was recorded. The intersection of the base of each thin vertical line with the polar plane indicates the estimated gain (radius) and phase (angle) of the system. The heavy black line is a regression line (First principal component) that is in the plane (dotted lines) perpendicular to the polar plane. Asterisks indicate that the slope of the regression line is statistically different from 0 ($P < 0.0075$).

3.4. Summary

The system identification approach employed here revealed that there are multiple neuronal sites responsible for acute VOR adaptation at 0.5 Hz: at least one in the flocculus (FL) or upstream from the FL ($G_{\text{pre-FL\&FL}}^{\text{vestib}}(s)$) and another in the non-FL pathway ($G_{\text{non-FL}}^{\text{vestib}}(s)$). These conclusions are consistent with prior experimental work, as discussed more fully in Hirata and Highstein (2001). Notably, however, these conclusions were reached based on the responses of a single cell type, and the approach was not biased in that we did not make any presumption with respect to the adaptability of each subsystem in the model. All of the subsystem parameters were free to change in accordance with experimental data. Several assumptions, however, were made in obtaining these parameter estimates. We used sequential estimation of subsystems from two light–dark protocols, assuming that related subsystems are identical across these protocols and that the dependence of central activities on efference copy is the same. Neither of these assumptions has yet been verified experimentally, although, as outlined in previous sections, they could seriously bias experimental conclusions if proven wrong. The fact that the model could predict various system behaviors accurately with the parameters estimated under these assumptions assures that the assumptions are plausible within the stimulus range (both frequency and amplitude) currently employed. This time, only a single frequency was used for training and testing stimuli to directly compare the results with those demonstrated in previous work; however, more rich input, especially in terms of frequency, can and should be applied to this system identification approach to provide more robust estimation of broadband dynamics in the presence of noise, to explore VOR adaptation in multiple frequency channels or to investigate the influence of training at a single frequency on other frequencies. (Collewijn and Grootendorst 1979; Godaux et al. 1983; Lisberger et al. 1983; Powell et al. 1991; Raymond and Lisberger 1996; Hirata et al. 2002, 2003).

4. Discussion and Conclusions

Much progress has been made in identifying the neural correlates for motor learning in the vestibuloocular reflex. However, the simple examples presented in this chapter have illustrated several caveats in the interpretation of experimental observations. These point to the requirement for further experimental investigation and the use of innovative analysis techniques that are less sensitive to *a priori* assumptions concerning both network structure and the locations within a given structure that represent candidate sites for plasticity in the VOR. In the following sections, the key implications of this chapter will be summarized with the goal of proposing directions for future study.

4.1. Implications of Assumptions with Respect to Model Structure

4.1.1. Local Interconnectivity

The majority of the identified caveats in the interpretation of experimental data are related to incomplete knowledge of circuit structure. In particular, we have shown that both the number and location of presumed plastic sites in the VOR are strongly influenced by our conception of the premotor topology. In our first examples, simple static models were used to illustrate that a direct correlation between sites for plasticity and observed changes in neural activity, associated with adaptive changes in behavior, need not exist whenever there is some degree of interconnectivity between cell types. Changes in neural sensitivity of a given cell population need not be associated with *local* plastic sites. Similarly, neurons that are involved in plastic changes need not *express* any change in global activity. Furthermore, attempts to assign a particular "role" (or lack thereof) in reflex adaptation to particular cell populations based on simple correlations of their activities with changes in behavior can result in misleading conclusions. Because of the highly interconnected nature of premotor circuits underlying the VOR, such possibilities must be taken into consideration during data interpretation. Unfortunately, however, although alternative possibilities are easy to illustrate theoretically, the requirement for explicit knowledge of the anatomical interconnections between individual cell types is a formidable problem to address experimentally.

In the absence of this information, one solution may be to group cell populations with similar physiological properties and/or gross anatomical interconnectivity into subsystems (as was done to illustrate the system identification approach) and to focus globally on identifying those subsystems involved in motor adaptation, with the goal of gradually breaking these subsystems down. Again, however, the conclusions reached using such an approach will depend significantly on how those subsystems are separated out and what types of interconnections between subsystems are assumed. Furthermore, as will be discussed in more detail below, it may be difficult to decide to which subsystem an individual cell population belongs. Does this then mean that little progress can be made in further understanding the neural substrate for motor learning in the VOR without more detailed anatomical information?

Simplified static models for the VOR at high frequencies illustrated that a lack of anatomical knowledge may limit our ability to conclusively localize sites for plasticity on particular synapses of individual cell populations or to define specific "roles" for populations of cells in motor learning. For example, at present, we may not conclusively determine whether PVP cells do or do not play some role in VOR adaptation. However, as addressed in our subsequent examples and as we will argue below, further consideration, both experimentally and theoretically, of the *dynamic* characteristics of the reflex offers the potential for new insight. The picture is not necessarily bleak: examinations of alternate sites for plasticity in candidate models can provide new clues with regard to premotor structure, the global strategies and signal processing pathways involved in reflex adaptation, and the optimal protocols to test alternate models.

4.1.2. Dynamic Substrate for the VOR

The importance of considering the low-frequency dynamic characteristics of the VOR was illustrated in our second set of examples, in which different potential sites for plasticity were examined in both feedforward and feedback realizations for the dynamic processing of canal signals in the VOR. We showed that, depending on network realization, there may be untested sites for plasticity in the VOR. Specifically, to date, investigation of the neural correlates for motor learning has focused on identifying sites at which there are changes in sensitivity to head velocity responsible for high-frequency reflex gain changes. However, the use of optics to train the VOR typically gives rise to broadband changes in reflex performance. In the parallel-pathway model, this implies that, in addition to a site for plasticity in the direct or "head velocity" pathway, there must be a second site somewhere in the "indirect" or integrator pathway. Alternatively, in the feedback model, changes in the head velocity sensitivities of eye-movement-sensitive cells alone are sufficient to produce broadband changes in reflex gain.

To achieve frequency-specific adaptation, on the other hand, apparently requires adaptive changes at multiple sites in either feedback or feedforward structural realizations for the neural integrator, as has previously been proposed (e.g., Lisberger et al. 1983; Tiliket et al. 1994). Notably, however, gain adaptation at a specific low to midrange frequency may not rely predominantly on changes in central sensitivities to head velocity inputs but rather may be achieved *mainly* by changes at alternative sites that affect the efficacy of neural integrator function. Support for this possibility is provided by the observation that for frequency-specific training at a midrange frequency of 0.2 Hz, for example, large changes in reflex gain close to the training frequency are not accompanied by the significant changes in performance at high frequencies that might be expected if in fact plasticity in central sensitivities to head velocity played a prominent role (Lisberger et al. 1983; Raymond and Lisberger 1996). Furthermore, the observation of changes in eccentric gaze-holding ability following frequency-specific VOR gain training or adaptation of VOR phase point clearly to the importance of modifications in integrator performance (Tiliket et al. 1994; Kramer et al. 1995, 1998).

The results of these examples thus have several key implications for future study. First, it is clear that one should evaluate experimental observations in the context of *dynamic* models. Second, the time has clearly come to address what brain stem network is responsible for neural integration. Although anatomy clearly supports a distributed integrator, a feedforward parallel-pathway type structure nevertheless continues to be assumed in many instances, and each has different implications for plastic sites. Thus, new directions for future experimental research should involve an examination of sites for plasticity other than those strictly related to changes in sensitivity to sensory head velocity signals. Indeed, as will be discussed further below, failure to do so can significantly bias the interpretation of experimental observations.

4.1.3. Identification of Neural Signal Components Within a Dynamic Network

A key problem in localizing sites for VOR plasticity to particular subpopulations of neurons is that of distinguishing local changes in sensitivity from *apparent* changes due to plasticity at other sites. Because the key interneurons in the premotor VOR network carry both sensory head velocity information and signals correlated with the motor response (i.e., signals related to eye movement), all such cells will exhibit changes in their whole-cell modulations following a change in behavioral reflex gain. To date, the primary focus has been on identifying those central sites at which there is evidence for changes in sensitivity to head velocity responsible for high-frequency changes in reflex performance. This requires that the vestibular component of central responses be isolated from other signal components. An inherent assumption in the most commonly employed techniques for extracting the vestibular component is the idea that there are areas of the brain that provide dedicated internal estimates of kinematic eye movement parameters (i.e., eye velocity and eye position) and that the populations of cells of interest receive scaled copies of these signals related to eye movement. Thus, if it is possible to estimate a cell's sensitivity to eye movement under conditions where the head does not move, then it is possible to eliminate the eye movement component under conditions where both the head and the eyes move to isolate the vestibular component of a cell's response. Alternatively, the vestibular component could be isolated more directly under conditions where stabilization of a target during head movement does not require any eye movement (i.e., VOR cancellation protocol).

In our third set of examples, a very simple dynamic model that included multiple vestibular neuron types was used as a "strawman" to illustrate the problems that can arise in attempts to isolate the vestibular component of cell activities. Specifically, biases may be introduced by the assumption that the appearance of eye position or eye velocity signals on a cell's modulation is tied to direct projections of such efference copies. Because in certain example cases this was an invalid assumption, erroneous combinations of signals were postulated to model a cell's response, and hence the wrong conclusions were reached. For example, with *a priori* knowledge that the model neurons reflected a summation of head velocity and efference copy of eye position signals under normal gain conditions, we examined a case in which doubling of the reflex gain at 0.5 Hz was achieved by modifying feedback loop parameters, including the dynamic characteristics of the neural filter. No change was applied to the synaptic strengths of sensory head velocity inputs. Correction of neural responses on the basis of their static eye position sensitivities both prior to and following simulated motor learning nevertheless led to the false conclusion that there was indeed a plasticity in neural sensitivities to sensory head velocity signals. The problem arose in this case because of the dynamic changes in the neural filter, which under normal gain conditions represented an internal model of the eye plant but was no longer a perfect internal model following motor learning. As a result, an accurate efference copy preadaptation and postadaptation at low frequencies was not associated with identical efference copy contributions to cell responses at the testing frequency of 0.5 Hz.

Is such an example realistic? One might be tempted to argue that in practice the eye movement sensitivity of a given cell would not be evaluated simply during static fixations in the dark but also during head-stationary pursuit or optokinetic tracking of a visual target at the appropriate testing frequency. Yet, here again, assuming dedicated efference copy signals can be an issue. Using our example structure, we illustrated that cell responses during 0.5 Hz pursuit could be modeled as a sum of eye position and eye velocity components. This observation was assumed to indicate an eye velocity signal input onto each cell that was then taken into account in the assessment of cell responses during rotation in the dark. However, because there was in fact no dedicated efference copy of eye velocity signal input to the cells in the model, inappropriate conclusions with respect to sites for plasticity were reached in many instances.

In view of these problems, we make several suggestions for future work. First, central responses should be examined over a broad frequency range *both prior to and following* motor learning. Given potential problems associated with applying observations under visual feedback conditions to the dark condition, emphasis should be placed on examining central responses at multiple frequencies *in the dark* when the goal is to evaluate adaptive changes in reflex pathways that are unrelated to visual feedback. Second, caution must be exercised in attempting to isolate changes in a particular component of a cell's response by subtracting off or correcting for other signal components, as this approach imbeds assumptions about the existence of dedicated signals in the network. Specifically, as illustrated here, problems may arise when attempts are made to correlate cell responses with both sensory inputs and motor outputs simultaneously because in fact we cannot always be sure that accurate internal estimates of these motor parameters exist at all frequencies or that they exist at the same frequencies following reflex adaptation. A better approach is to evaluate *whole*-cell slow-phase responses to known sensory inputs only (i.e., whole-cell responses to head rotation in the dark), making corrections only for changes in average eye position associated with saccadic eye movements where necessary and when this is clearly appropriate. Several studies have illustrated such an analysis approach in which the dynamics of behavioral and total-cell responses are expressed simply in terms of known stimuli, and the effects of sudden changes in eye position at the beginning of each slow phase are accounted for using transient analysis techniques (variable initial conditions) (Rey and Galiana 1993; Mettens et al. 1994; Kukreja et al. 1999). Experimentally observed changes in the gain and phase of whole-cell responses to head rotation can then be considered in the context of different potential model structures in which multiple potential adaptation sites are considered *simultaneously*.

4.1.4. Global Localization of Plastic Sites in a Network: System Identification Approach

An alternative approach based on systems identification analysis was presented to investigate the neuronal substrate for motor learning in the VOR. Rather than postulating a particular network model for the VOR, incorporating specific assumptions with respect to the interconnectivity and signal content of individual cells, sensory motor processes potentially involved in motor learning in the VOR were broken into major subcomponents. Those processes or subsystems involved in adaptive changes were identified based on behavioral responses and neural recordings from a single cell type. Because this approach relied on larger subsystems, questions of local interconnectivity between cell types and the particular brain stem implementation of the neural integrator were not an issue. Furthermore, the approach did not make assumptions about potential sites for plasticity and hence allowed exploration of changes in all parameters, regardless of their location in any subprocess. Hence, it would appear that several of the problems outlined above are alleviated by employing this approach. Plastic sites should be better localized, at least within computational subprocesses.

A limitation of the approach in terms of identifying sites for plasticity, however, is that computational subprocesses cannot necessarily be tied to anatomical locations. At some point, the subprocesses have to be broken down into their substrates, and there may be much overlap. For example, the vestibular nuclei and cerebellum participate in both visual and vestibular reflexes, and a given cell within either structure carries both visually and vestibularly derived signals. Thus, in general, if changes are found in a vestibular process, for example, it may still be almost impossible to localize these changes to specific CNS sites unless the system processes are mapped onto a relevant anatomical network. Furthermore, as argued above, the systems identification approach must also be used with experimental protocols over broad bandwidths to achieve reliable and consistent results. In addition, because the postulated interconnectivity of the subsystems may impact on conclusions, multiple potential model structures should be examined while preserving anatomical relevance. The model used in the current example has a general structure to which many potential variations of connectivity may be assigned without violating known anatomy. The approach presented here thus represents an important first stage in localizing sites for plasticity that must be extended and used in conjunction with more distributed modeling approaches to truly localize sites for plasticity (see Section 4.3 below).

4.2. Exploring Alternate Sites for Plasticity—Experimental Issues

Available experimental and theoretical observations imply that the nature of training protocols is likely to play an important role in the location and number of presumed plastic sites for the VOR. In our simple examples above, postulating changes at multiple sites to different degrees reproduced different experimentally observed behaviors, including broadband adaptation of the VOR (e.g., after long-term reflex training with telescopic spectacles) and what appeared to be frequency-specific tuning (e.g., after visual–vestibular mismatch training at a specific midrange frequency). Experimental investigations have also illustrated the ability to adapt reflex phase, that both frequency-specific and phase training are accompanied by changes in gaze-holding performance and that the degree of alteration in gaze-holding performance depends on training context (Tiliket et al. 1994; Kramer et al. 1995, 1998). Reflex training with brief versus longer-duration visual–vestibular stimulus pulses produces differential effects on the gain and dynamics of the VOR, with longer-duration stimuli inducing larger effects on VOR dynamics (Raymond and Lisberger 1996). These observations suggest multiple sites for plasticity recruited to different extents, depending on the training paradigm. Furthermore, it is clear that at least a subset of such sites substantially impact the overall dynamic characteristics of the premotor circuitry involved in the control of eye movements.

So far, investigations of the neural correlates for motor learning in the VOR have focused on identifying sites for plasticity on cells in the vestibular nuclei or cerebellum mainly following long-term, broadband reflex training. This follows from the emphasis on identifying those cells that demonstrate evidence for at least semipermanant changes in sensitivity to head velocity. The neural correlates for motor learning have thus been investigated to date predominantly for relatively specific training conditions at very specific sites. Given the broad range of behavioral observations associated with different reflex-training paradigms, a clear direction for future research is thus a comparative investigation of differences in the types of changes observed on these cell populations for different training conditions (e.g., frequency-specific versus broadband reflex training, chronic versus short-term adaptation, interleaving of different protocols to change training context).

In addition, we have emphasized here the requirement for evaluating potential changes in signal components other than sensory head velocity signals. The requirement for a site of plasticity at some point in the integrator pathway has been discussed extensively and proposed in previous models of frequency-specific adaptation (Lisberger et al. 1983; Tiliket et al. 1994) or VOR phase adaptation (Kramer et al. 1995, 1998). However, it would appear that such a site is also required in a feedforward parallel-pathway realization of the VOR circuitry to achieve the broadband changes in gain observed following long-term adaptation with telescopic spectacles (see Quinn et al. 1992a). Yet, although the majority of researchers support the parallel-pathway model for the VOR, there has been no direct attempt to test for neural correlates of plasticity in the "indirect" integrator pathway. A prime candidate brain area to examine in this regard is the NPH. Adaptive changes might occur on integrator neurons themselves (i.e., on NPH neurons as the proposed site of the neural integrator) or on subpopulations of eye-movement-sensitive cells in the vestibular nuclei and flocculus at the site of synapses associated with inputs from the neural integrator. Furthermore, the locations of sites for plasticity are likely to depend on the training paradigm and whether reflex adaptation is accompanied by a change in integrator function or simply by a change in synaptic strengths within the "integrator pathway."

To achieve broadband reflex adaptation in a feedforward model for the VOR, for example, requires no change in the efficacy of neural integration per se. Rather, adaptive changes must occur somewhere in the integrator pathway either at the site of head velocity inputs onto neurons in the NPH (e.g., on eye-ipsi EHV-type cells in the NPH that are known to be sensitive to head velocity in addition to eye movement and thus might be cells that participate at an early stage in a gradual integration process) or at the synapses associated with output projections from the neural integrator onto eye-movement-sensitive neurons in the VN. Alternatively, as previously discussed, if a feedback realization of the VOR is more appropriate, adaptation of the strength of sensory head velocity inputs onto vestibular (and cerebellar) neurons is sufficient to achieve the broadband adaptive gain changes; hence, in this case one might not expect to find evidence for plasticity in other signaling pathways.

The large changes in phase that accompany frequency-specific reflex training or phase training of the reflex, on the other hand, point to changes in the integrator function itself. Indeed, following VOR phase training, there is direct evidence for changes in gaze-holding ability, implying some change in the efficacy of the neural integrator. In both feedback and feedforward VOR realizations, one might expect to see evidence of such changes on the populations of NPH cells that code mainly for eye position under normal

VOR gain conditions; the actual locations of sites for plasticity may be more distributed in a feedback VOR realization, however, occurring anywhere within a feedback pathway. Hence, we suggest that directions for future work should include not only an examination of the responses of neural populations in areas such as the NPH that have not yet been explored in association with VOR adaptation but also that evidence for potential changes in the strengths of signal components other than head velocity be examined more closely on the cell populations in the VN and cerebellum that have previously been investigated. However, as previously discussed, *a priori* assumptions concerning potential sites for plasticity or the signal content of neurons should be avoided by considering the possibility of multiple sites *simultaneously* in different model structures. Such investigations may not only shed light on strategies for VOR adaptation under different training conditions but may also assist in deciding between feedforward and feedback realizations for the VOR.

4.3. Proposed Quantitative Approach for Interpreting Experimental Data

The ultimate goal of investigating neural systems in the brain is to understand the basic principles underlying the organization of these systems and how they function to generate appropriate behavior. This does not necessarily require an understanding of the role of every individual neuron, nor is this likely possible in systems that are very complex in terms of the number of neurons or neural populations involved. Nevertheless, study of the brain in sufficient detail to understand the basic principles of system function does require an exploration of the responses of individual neurons. With regard to the current topic, understanding the mechanisms of motor learning ultimately requires identification of the sites of plasticity underlying adaptive changes in reflex performance. We have seen that a lack of explicit knowledge of connectivity at the individual neuron level currently limits our ability to localize such plastic sites. Nevertheless, there remains much to explore at the conceptual or systems level in terms of the basic strategies and signaling pathways involved in motor learning under different conditions. Indeed, we suggest that at the current time a better understanding of these strategies at a systems level is necessary before the localization of sites for plasticity at the individual neuron level will be possible.

In particular, we advocate an approach to investigating the neural correlates for motor learning in the VOR that combines the techniques of system identification with a mapping of results onto more anatomically relevant model structures. A first step is to develop a process or subsystem-style model and to use system identification techniques to aid in conceptualizing the signaling pathways involved in adaptation and in quantifying parametric changes without *a priori* restrictions. For this approach to be effective, however, it must be extended from the example presented here to include data from multiple frequencies (i.e., responses to richer dynamic stimuli).

The second stage involves expanding larger subsystems into more realistic anatomical components. Experimental data should be interpreted in multiple possible structures. By evaluating the results across candidate circuit topologies and evaluating both differences and consistencies, it should be possible to more clearly identify those subsystems or signal pathways involved in motor learning for different training paradigms. However, this may not be a trivial task because individual cells need not easily fall into one subsystem or another. For example, to illustrate the system identification approach above, postfloccular and nonfloccular subsystems were isolated. However, FTN cells are known to receive projections from floccular Purkinje cells and as such at least a portion of this cell's response is represented in the "postfloccular" subsystem. On the other hand, FTN cells also receive direct sensory vestibular projections and can be considered part of the "nonfloccular" subsystem. Nevertheless, although the FTN cell does not fit clearly into a specific subsystem component, the information gained from the system identification approach can provide useful constraints in a more anatomically realistic model structure. For example, because adaptation-related changes were evident in the nonfloccular subsystem but not in the postfloccular subsystem, this information could be used to postulate that those synapses associated with sensory vestibular inputs to FTN cells are more likely to be sites for plasticity than are the synapses associated with their inputs from floccular neurons. Similarly, if use of the system identification approach for a particular experimental paradigm were to reveal changes in signal components involved in neural integration, this would point to the need to consider synapses associated with the feedback of efference copy signals and/or the participation of cells in areas such as the NPH as potential sites for plasticity.

4.4. Conclusions

The identification of potential sites for plasticity in the VOR has thus far been limited mainly to the investigation of very specific signal components that are modified following a particular reflex-training paradigm (i.e., broadband reflex training). Yet, the richness of different behavioral observations associated with different training paradigms points to the existence of multiple potential adaptation sites within the VOR pathways and the use of different adaptation strategies that thus far remain virtually unexplored. Although the ability to explicitly localize sites for plasticity to individual cells is currently limited by incomplete knowledge of network structure, the use of innovative analysis and modeling approaches that are less sensitive to *a priori* assumptions can aid in conceptualizing such strategies and in identifying additional sites for plasticity. At the present time, therefore, an apparent step back toward more process-oriented models and interpretation may be required to make further progress in identifying sites for plasticity at the level of individual neurons. Hence, despite much progress in identifying the neural correlates for motor learning in the VOR, the story is far from complete. The VOR system remains an excellent model system for the investigation of the neural correlates for motor learning and in particular for investigating learning strategies that are context-dependent.

References

Akaike H (1973) Information theory and extension of the maximum likelihood principle. In: Petrov BN, Csaki F (eds) 2nd Intl Symposium on Information Theory. Budapest: Akademiai Kiado, pp. 267–281.

Albus JS (1971) A theory of cerebellar function. Math Biosci 10:25–61.

Angelaki DE, Hess BJM (1998) Visually induced adaptation in three dimensional organization of primate vestibuloocular reflex. J Neurophysiol 79:791–807.

Artola A, Singer W (1993) Long-term depression of excitatory synaptic transmission and its relationship to long-term potentiation. Trends Neurosci 11:480–487.

Baker JF, Harrison REW, Isu N, Wickland C, Peterson B (1986) Dynamics of adaptive change in vestibuloocular reflex. II. Sagittal plane rotations. Brain Res 371: 166–170.

Baker JF, Wickland C, Peterson BW (1987) Dependence of cat vestibuloocular reflex direction adaptation on animal orientation during adaptation and rotation in darkness. Brain Res 408:339–343.

Baker R, Precht W, Llinas R (1972) Cerebellar modulatory action on the vestibulo-trochlear pathway in the cat. Exp Brain Res 15:364–385.

Barnes GR (1993) Visual–vestibular interaction in the control of head and eye movement: the role of visual feedback and predictive mechanisms. Prog Neurobiol 41:435–472.

Bear MF, Malenka RC (1994) Synaptic plasticity: LTP and LTD. Curr Opin Neurobiol 4:389–399.

Bello S, Paige GD, Highstein SM (1991) The squirrel monkey vestibuloocular reflex and adaptive plasticity in yaw, pitch and roll. Exp Brain Res 87:57–66.

Bliss TV, Collingridge GL (1993) A synaptic model of memory: long-term potentiation in the hippocampus. Nature 361:31–39.

Brindley CS (1964) The use made by the cerebellum of the information that it receives from sense organs. Int Brain Res Org Bull 3:80.

Brontë-Stewart HM, Lisberger SG (1994) Physiological properties of vestibular primary afferents that mediate motor learning and normal performance of the vestibulo-ocular reflex in monkeys. J Neurosci 14:1290–1308.

Broussard DM, Lisberger SG (1992) Vestibular inputs to brain stem neurons that participate in motor learning in the primate vestibuloocular reflex. J Neurophysiol 68:1906–1909.

Broussard DM, Bronte-Stewart HM, Lisberger SG (1992) Expression of motor learning in the response of the primate vestibuloocular reflex pathway to electrical stimulation. J Neurophysiol 67:1493–1508.

Bures J, Fenton AA, Kaminsky Y, Zinyuk L (1997) Place cells and place navigation. Proc Natl Acad Sci U.S.A 94:343–350.

Cannon SC, Robinson DA (1987) Loss of the neural integrator of the oculomotor system from brain stem lesions in monkey. J Neurophysiol 57:1383–1409.

Cheron G, Godaux E (1987) Disabling of the oculomotor neural integrator by kainic acid injections in the prepositus-vestibular complex of the cat. J Physiol (Lond) 394:267–290.

Clendaniel RA, Lasker DM, Minor LB (2001) Horizontal vestibuloocular reflex evoked by high-acceleration rotations in the squirrel monkey. IV. Responses after spectacle-induced adaptation. J Neurophysiol 86:1594–1611.

Collewijn H, Grootendorst AF (1979) Adaptation of optokinetic and vestibuloocular reflexes to modified visual input in the rabbit. Prog Brain Res 50:772–781.

Collewijn H, Martins AJ, Steinman RM (1983) Compensatory eye movements during active and passive head movements: fast adaptation to changes in visual magnification. J Physiol (Lond) 340:259–286.

Cullen KE, McCrea RA (1993) Firing behavior of brain stem neurons during voluntary cancellation of the horizontal vestibuloocular reflex. I. Secondary vestibular neurons. J Neurophysiol 70:828–843.

Cullen KE, Chen-Huang C, McCrea RA (1993) Firing behavior of brainstem neurons during voluntary cancellation of the horizontal vestibuloocular reflex. II. Eye movement related neurons. J Neurophysiol 70:844–856.

Curthoys IS, Halmagyi GM (1995) Vestibular compensation: a review of the oculomotor, neural, and clinical consequences of unilateral vestibular loss. J Vestib Res 5:67–107.

Daniel H, Levenes C, Crepel F (1998) Cellular mechanisms of cerebellar LTD. Trends Neurosci 21:401–407.

Dieringer N (1995) "Vestibular compensation": neural plasticity and its relations to functional recovery after labyrinthine lesions in frogs and other vertebrates. Prog Neurobiol 46:97–129.

Dufosse M, Ito M, Jastreboff PJ, Miyashita Y (1978) A neuronal correlate in rabbit's cerebellum to adaptive modification of the vestibuloocular reflex. Brain Res 150:611–616.

du Lac S, Raymond JL, Sejnowski TJ, Lisberger SG (1995) Learning and memory in the vestibuloocular reflex. Annu Rev Neurosci 18:409–441.

Escudero M, De La Cruz RR, Delgado-Garcia JM (1992) A physiological study of vestibular and prepositus hypoglossi neurones projecting to the abducens nucleus in the alert cat. J Physiol (Lond) 458:539–560.

Fujita M (1982) Adaptive filter model of the cerebellum. Biol Cybern 45:195–206.

Galiana HL (1986) A new approach to understanding adaptive visual–vestibular interactions in the central nervous system. J Neurophysiol 55:349–374.

Galiana HL, Green AM (1998) Vestibular adaptation: how models can affect data interpretations. Otolaryngol Head Neck Surg 119:231–243.

Galiana HL, Outerbridge JS (1984) A bilateral model for central neural pathways in the vestibulo-ocular reflex. J Neurophysiol 51:210–241.

Gauthier GM, Robinson DA (1975) Adaptation of the human vestibuloocular reflex to magnifying lenses. Brain Res 92:331–335.

Godaux E, Halleux J, Gobert C (1983) Adaptive change of the vestibuloocular reflex in the cat: the effects of a longterm frequency selective procedure. Exp Brain Res 49:28–34.

Goldberg JM, Fernández C (1971) Physiology of peripheral neurons innervating semicircular canals of the squirrel monkey. I. Resting discharge and response to constant angular accelerations. J Neurophysiol 34:635–660.

Gomi H, Kawato M (1992) Adaptive feedback control models of the vestibulo-cerebellum and spinocerebellum. Biol Cybern 68:105–114.

Gonshor A, Melvill Jones G (1971) Plasticity in the adult human vestibuloocular reflex arc. Proc Can Fed Biol Soc 14:11.

Gonshor A, Melvill Jones G (1976a) Shortterm adaptive changes in the human vestibuloocular reflex arc. J Physiol (Lond) 256:361–379.

Gonshor A, Melvill Jones G (1976b) Extreme vestibuloocular adaptation induced by prolonged optical reversal of vision. J Physiol (Lond) 256:381–414.

Green AM (2000) Visual–vestibular interaction in a bilateral model of the rotational and translational vestibulo-ocular reflexes: an investigation of viewing-context-dependent reflex performance. Ph.D. Thesis, McGill University, Montreal, Canada.

Green A, Galiana HL (1996) Exploring sites for short-term VOR modulation using a bilateral model. Ann N Y Acad Sci 781:625–628.

Harrison REW, Baker JF, Isu N, Wickland CR, Peterson BW (1986) Dynamics of adaptive change in vestibuloocular reflex direction. I. Rotation in the horizontal plane. Brain Res 371:162–165.

Highstein SM (1973) Synaptic linkage in the vestibulo-ocular and cerebello-vestibular pathways to the VIth nucleus in the rabbit. Exp Brain Res 17:301–314.

Highstein SM (1998) Role of the flocculus of the cerebellum in motor learning of the vestibulo-ocular reflex. Otolaryngol Head Neck Surg 119:212–220.

Highstein SM, Partsalis A, Arikan R (1997) Role of Y-group of the vestibular nuclei and flocculus of the cerebellum in motor learning of the vertical vestibulo-ocular reflex. Prog Brain Res 114:383–397.

Hirata Y, Highstein SM (2001) Acute adaptation of the vestibuloocular reflex: signal processing by floccular and ventral parafloccular Purkinje cells. J Neurophysiol 85:2267–2288.

Hirata Y, Lockard JM, Highstein SM (2002) Capacity of vertical VOR adaptation in squirrel monkey. J Neurophysiol 88:3194–3207.

Hirata Y, Takeuchi I, Highstein SM (2003) A dynamical model for the vertical vestibuloocular reflex and optokinetic response in primate. Neurocomputing 52–54:531–540.

Ito M (1972) Neural design of the cerebellar motor control system. Brain Res 40:81–84.

Ito M (1989) Long-term depression. Annu Rev Neurosci 12:85–102.

Ito M, Shiida T, Yagi N, Yamamoto M (1974) The cerebellar modification of rabbit's horizontal vestibuloocular reflex induced by sustained head rotation combined with visual stimulation. Proc Jpn Acad 50:85–89.

Ito M, Nisimaru N, Yamamoto M (1977) Specific patterns of neuronal connexions involved in the control of the rabbit's vestibulo-ocular reflexes by the cerebellar flocculus. J Physiol (Lond) 265:833–854.

Ito M, Sakurai M, Tongroach P (1981) Evidence for modifiability of parallel fiber-Purkinje cell synapses. In: Szentagothai J, Hamori J, Palkovits M (eds) Advances in Physiological Sciences, Volume 2. Oxford: Pergamon Press, pp. 97–105.

Ito M, Sakurai M, Tongroach P (1982) Climbing fibre induced depression of both mossy fibre responsiveness and glutamate sensitivity of cerebellar Purkinje cells. J Physiol (Lond) 324:113–134.

Kaneko CR (1997) Eye movement deficits after ibotenic acid lesions of the nucleus preositus hypoglossi in monkeys. I. Saccades and fixation. J Neurophysiol 78:1753–1768.

Kaneko CR (1999) Eye movement deficits following ibotenic acid lesions of the nucleus prepositus hypoglossi in monkeys. II. Pursuit, vestibular, and optokinetic responses. J Neurophysiol 81:668–681.

Keller EL, Precht W (1979) Adaptive modification of central vestibular neurons in response to visual stimulation through reversing prisms. J Neurophysiol 42:896–911.

Khater TT, Quinn KJ, Pena J, Baker JF, Peterson BW (1993) The latency of the cat vestibulo-ocular reflex before and after shortterm and longterm adaptation. Exp Brain Res 94:16–32.

Kim JJ, Thompson RF (1997) Cerebellar circuits and synaptic mechanisms involved in classical eyeblink conditioning. Trends Neurosci 20:177–181.

Kobayashi Y, Kawano K, Takemura A, Inoue Y, Kitama T, Gomi H, Kawato M (1998) Temporal firing patterns of Purkinje cells in the cerebellar ventral paraflocular during ocular following responses in monkeys. II. Complex spikes. J Neurophysiol 80:832–848.

Kramer PD, Shelhamer M, Zee DS (1995) Short-term adaptation of the phase of the vestibulo-ocular reflex (VOR) in normal human subjects. Exp Brain Res 106:318–326.

Kramer PD, Shelhamer M, Peng GCY, Zee DS (1998) Context-specific short-term adaptation of the phase of the vestibulo-ocular reflex. Exp Brain Res 120:184–192.

Kukreja S, Galiana HL, Smith HLH, Kearney RE (1999) Parametric identification of non-linear hybrid systems. Proc BMES/IEEE-EMBS Ann Conf 21:991.

Lisberger SG (1984) The latency of pathways containing the site of motor learning in the monkey vestibulo-ocular reflex. Science 225:74–76.

Lisberger SG (1988) The neural basis for learning of simple motor skills. Science 242:728–735.

Lisberger SG (1994) Neural basis for motor learning in the vestibuloocular reflex of primates. III. Computational and behavioral analysis of the sites of learning. J Neurophysiol 72:974–998.

Lisberger SG (1998) Physiologic basis for motor learning in the vestibulo-ocular reflex. Otolaryngol Head Neck Surg 119:43–48.

Lisberger SG, Fuchs AF (1978) Role of primate flocculus during rapid behavioral modification of vestibuloocular reflex. I. Purkinje cell activity during visually guided horizontal smooth-pursuit eye movements and passive head rotation. J Neurophysiol 41:733–763.

Lisberger SG, Miles FA (1980) Role of primate medial vestibular nucleus in long-term adaptive plasticity of vestibuloocular reflex. J Neurophysiol 43:1725–1745.

Lisberger SG, Pavelko TA (1986) Vestibular signals carried by pathways subserving plasticity of the vestibulo-ocular reflex. J Neurosci 6:346–354.

Lisberger SG, Pavelko TA (1988) Brain stem neurons in modified pathways for motor learning in the primate vestibulo-ocular reflex. Science 242:771–773.

Lisberger SG, Sejnowski TJ (1992) Motor learning in a recurrent network model based on the vestibulo-ocular reflex. Nature 360:159–161.

Lisberger S, Evinger C, Johanson G, Fuchs A (1981) Relationship between eye acceleration and retinal image velocity during foveal smooth pursuit in man and monkey. J Neurophysiol 46:229–249.

Lisberger SG, Miles FA, Optican LM (1983) Frequency-selective adaptation: evidence for channels in the vestibulo-ocular reflex? J Neurosci 3:1234–1244.

Lisberger SG, Miles FA, Zee DS (1984) Signals used to compute errors in monkey vestibuloocular reflex: possible role of flocculus. J Neurophysiol 52:1140–1153.

Lisberger SG, Pavelko TA, Broussard DM (1994a) Responses during eye movements of brain stem neurons that receive monosynaptic inhibition from the flocculus and ventral paraflocculus in monkeys. J Neurophysiol 72:909–927.

Lisberger SG, Pavelko TA, Broussard DM (1994b) Neural basis for motor learning in the vestibuloocular reflex of primates. I. Changes in responses of brain stem neurons. J Neurophysiol 72:928–953.

Lisberger SG, Pavelko TA, Brontë-Stewart HM, Stone LS (1994c) Neural basis for motor learning in the vestibuloocular reflex of primates. II. Changes in the responses of horizontal gaze velocity Purkinje cells in the cerebellar flocculus and ventral paraflocculus. J Neurophysiol 72:954–973.

Llinás R, Lang EJ, Welsh JP (1997) The cerebellum, LTD, and memory: alternative views. Learn Mem 3:445–455.

Lopez-Barneo J, Ribas J, Delgado-Garcia JM (1981) Identification of prepositus neurons projecting to the oculomotor nucleus in the cat. Brain Res 214:174–179.

Luebke AE, Robinson DA (1994) Gain changes of the cat's vestibulo-ocular reflex after flocculus deactivation. Exp Brain Res 98:379–390.

Marmarelis PD, Marmarelis VZ (1978) Analysis of physiological systems. New York: Plenum Press.

Marr D (1969) A theory of cerebellar cortex. J Physiol (Lond) 202:437–470.

Mauk MD (1997) Roles of cerebellar cortex and nuclei in motor learning: contradictions or clues? Neuron 18:343–346.

Mauk MD, Garcia KS, Medina JF, Steele PM (1998) Does cerebellar LTD mediate motor learning? Toward a resolution without a smoking gun. Neuron 20:359–362.

McCrea RA, Yoshida K, Evinger C, Berthoz A (1981) The location, axonal arborization and termination sites of eye-movement-related secondary vestibular neurons demonstrated by intra-axonal HRP injection in the alert cat. In: Fuchs A, Becker W (eds) Progress in Oculomotor Research. Amsterdam: Elsevier, pp. 379–386.

McCrea RA, Strassman A, May A, Highstein SM (1987) Anatomical and physiological characteristics of vestibular neurons mediating the horizontal vestibulo-ocular reflex of the squirrel monkey. J Comp Neurol 264:547–570.

McElligott JG, Beeton P, Polk J (1998) Effect of cerebellar inactivation by lidocaine microdialysis on the vestibuloocular reflex in goldfish. J Neurophysiol 79:1286–1294.

McFarland JL, Fuchs AF (1992) Discharge patterns in nucleus prepositus hypoglossi and adjacent medial vestibular nucleus during horizontal eye movement in behaving macaques. J Neurophysiol 68:319–332.

McNaughton BL (1998) The neurophysiology of reminiscence. Neurobiol Learn Mem 70:252–267.

Melvill Jones G, Davies P (1976) Adaptation of cat vestibulo-ocular reflex to 200 days of optically reversed vision. Brain Res 103:551–554.

Mettens P, Godaux E, Cheron G, Galiana HL (1994) Effect of muscimol micro-injections into the prepositus hypoglossi and the medial vestibular nuclei on cat eye movements. J Neurophysiol 72:785–802.

Miles FA, Eighmy BB (1980) Long-term adaptive changes in primate vestibuloocular reflex. I. Behavioral observations. J Neurophysiol 43:1406–1425.

Miles FA, Fuller JH (1974) Adaptive plasticity in the vestibulo-ocular responses of the rhesus monkey. Brain Res 80:512–516.

Miles FA, Lisberger SG (1981) Plasticity in the vestibulo-ocular reflex: a new hypothesis. Annu Rev Neurosci 4:273–299.

Miles FA, Fuller JH, Braitman DJ, Dow BM (1980a) Long-term adaptive changes in primate vestibuloocular reflex. III. Electrophysiological observations in flocculus of normal monkeys. J Neurophysiol 43:1437–1476.

Miles FA, Braitman DJ, Dow BM (1980b) Long-term adaptive changes in primate vestibuloocular reflex. IV. Electrophysiological observations in flocculus of adapted monkeys. J Neurophysiol 43:1477–1493.

Minor LB, Lasker DM, Backous DD, Hullar TE (1999) Horizontal vestibuloocular reflex evoked by high-acceleration rotations in the squirrel monkey. I. Normal responses. J Neurophysiol 82:1254–1270.

Nagao S (1983) Effects of vestibulocerebellar lesions upon dynamic characteristics and adaptation of vestibulo-ocular and optokinetic responses in pigmented rabbits. Exp Brain Res 53:36–46.

Nagao S (1989) Behavior of floccular Purkinje cells correlated with adaptation of vestibulo-ocular reflex in pigmented rabbits. Exp Brain Res 77:531–540.

Neville HJ, Bavelier D (1998) Neural organization and plasticity of language. Curr Opin Neurobiol 8:254–258.

Nikias CL, Raghuveer MR (1987) Bispectrum estimation: a digital signal processing framework. Proc IEEE 75:869–891.

Paige GD (1983a) Vestibuloocular reflex and its interactions with visual following mechanisms in the squirrel monkey. I. Response characteristics in normal animals. J Neurophysiol 49:134–151.

Paige GD (1983b) Vestibuloocular reflex and its interactions with visual following mechanisms in the squirrel monkey. II. Response characteristics and plasticity following unilateral inactivation of horizontal canal. J Neurophysiol 49:152–168.

Paige GD, Sargent EW (1991) Visually-induced adaptive plasticity in the human vestibulo-ocular reflex. Exp Brain Res 84:25–34.

Partsalis AM, Zhang Y, Highstein SM (1995a) Dorsal Y group in squirrel monkey. I. Neuronal responses during rapid and long-term modifications of the vertical VOR. J Neurophysiol 73:615–631.

Partsalis AM, Zhang Y, Highstein SM (1995b) Dorsal Y group in the squirrel monkey. II. Contribution of the cerebellar flocculus to neuronal responses in normal and adapted animals. J Neurophysiol 73:632–650.

Pastor AM, De La Cruz RR, Baker R (1992) Characterization and adaptive modification of the goldfish vestibuloocular reflex by sinusoidal and velocity step vestibular stimulation. J Neurophysiol 68:2003–2015.

Pastor AM, De La Cruz RR, Baker R (1994) Cerebellar role in adaptation of the goldfish vestibuloocular reflex. J Neurophysiol 72:1383–1394.

Pastor AM, De La Cruz RR, Baker R (1997) Characterization of Purkinje cells in the goldfish cerebellum during eye movement and adaptive modification of the vestibulo-ocular reflex. Prog Brain Res 114:359–381.

Powell KD, Quinn KJ, Rude SA, Peterson BW, Baker JF (1991) Frequency dependence of cat vestibulo-ocular reflex direction adaptation: single frequency and multifrequency rotations. Brain Res 550:137–141.

Powell KD, Peterson BW, Baker JF (1996) Phase-shifted direction of adaptation of the vestibulo-ocular reflex in cat. J Vestib Res 6:277–293.

Quinn KJ, Schmajuk N, Baker JF, Peterson BW (1992a) Simulation of adaptive mechanisms in the vestibulo-ocular reflex. Biol Cybern 67:103–112.

Quinn KJ, Schmajuk N, Baker JF, Peterson BW (1992b) Vestibulo-ocular reflex arc analysis using an experimentally constrained neural network. Biol Cybern 67:113–122.

Raphan T, Matsuo V, Cohen B (1979) Velocity storage in the vestibulo-ocular reflex arc (VOR). Exp Brain Res 35:229–248.

Rashbass C (1961) The relationship between saccadic and smooth tracking eye movements. J Physiol (Lond) 159:326–338.

Raymond JL, Lisberger SG (1996) Behavioral analysis of signals that guide learned changes in the amplitude and dynamics of the vestibulo-ocular reflex. J Neurosci 16:7791–7802.

Raymond JL, Lisberger SG, Mauk MD (1996) The cerebellum: a neuronal learning machine? Science 272:1126–1130.

Rey C, Galiana HL (1993) Transient analysis of vestibular nystagmus. Biol Cybern 69:395–405.

Rissanen J (1986) Stochastic complexity and modeling. Annals of statistics 14:1080–1100.

Robinson DA (1974) The effect of cerebellectomy on the cat's vestibulo-ocular integrator. Brain Res 71:195–207.

Robinson DA (1976) Adaptive gain control of vestibulo-ocular reflex by the cerebellum. J Neurophysiol 39:954–969.

Robinson DA (1977) Linear addition of optokinetic and vestibular signals in the vestibular nucleus. Exp Brain Res 30:447–450.

Robinson DA (1981) The use of control systems analysis in the neurophysiology of eye movements. Annu Rev Neurosci 4:463–503.

Sato Y, Kawasaki T (1987) Target neurons of floccular caudal zone inhibition in Y-group nucleus of vestibular nucleus complex. J Neurophysiol 57:460–480.

Sato Y, Kanda K-I, Kawasaki T (1988) Target neurons of floccular middle zone inhibition in medial vestibular nucleus. Brain Res 446:225–235.

Schairer JO, Bennett MVL (1981) Cerebellectomy in goldfish prevents adaptive gain control of the VOR without affecting the optokinetic system. In: Gualtierotti T (ed) The Vestibular System: Function and Morphology. New York: Springer-Verlag, pp. 463–477.

Schetzen M (1980) The Volterra and Wiener theories of nonlinear systems. New York: Wiley.

Schultheis LW, Robinson DA (1981) Directional plasticity of the vestibulo-ocular reflex in the cat. Ann N Y Acad Sci 374:504–512.

Schwarz G (1978) Estimating the dimension of a model. Annals of Statistics 6:461–464.

Scudder CA, Fuchs AF (1992) Physiological and behavioral identification of vestibular nucleus neurons mediating the horizontal vestibuloocular reflex in trained rhesus monkeys. J Neurophysiol 68:244–264.

Seidenberg MS (1997) Language acquisition and use: learning and applying probabilistic constraints. Science 275:1599–1603.

Seidman SH, Paige GD, Tomko DL (1999) Adaptive plasticity in the naso-occipital linear vestibulo-ocular reflex. Exp Brain Res 125:485–494.

Shidara M, Kawano K, Gomi H, Kawato M (1993) Inverse dynamics model eye movement control by Purkinje cells in the cerebellum. Nature 365:50–52.

Silva AJ, Giese KP, Fedorov NB, Frankland PW, Kogan JH (1998) Molecular, cellular, and neuroanatomical substrates of place learning. Neurobiol Learn Mem 70:44–61.

Skavenski AA, Robinson DA (1973) Role of abducens neurons in vestibuloocular reflex. J Neurophysiol 36:724–738.

Snyder LH, King WM (1988) Vertical vestibuloocular reflex in cat: asymmetry and adaptation. J Neurophysiol 59:279–298.

Stone LS, Lisberger SG (1990) Visual response of Purkinje cells in the cerebellar floccular during smooth-pursuit eye movements in monkeys. I. Simple spikes. J Neurophysiol 63:1241–1261.

Tabata H, Yamamoto K, Kawato M (2002) Computational study on monkey VOR adaptation and smooth pursuit based on the parallel control-pathway theory. J Neurophysiol 87:2176–2189.

Thach WT (1998) A role for the cerebellum in learning movement coordination. Neurobiol Learn Mem 70:177–188.

Tiliket C, Shelhamer M, Roberts D, Zee DS (1994) Short-term vestibulo-ocular reflex adaptation in humans. I. Effect on the ocular motor velocity-to-position neural integrator. Exp Brain Res 100:316–327.

Toda N, Usui S (1991) An overview of biological signal processing: non-linear and non-stationary aspects. Front Med Biol Eng 3:125–129.

Tychsen L, Lisberger SG (1986) Visual motion processing for the initiation of smooth-pursuit eye movements in humans. J Neurophysiol 56:953–968.

Watanabe E (1984) Neuronal events correlated with long-term adaptation of the horizontal vestibulo-ocular reflex in the primate flocculus. Brain Res 297:169–174.

Wei M, Angelaki DE (2001) Cross-axis adaptation of the translational vestibulo-ocular reflex. Exp Brain Res 138:304–312.

11
Clinical Applications of Basic Vestibular Research

G. Michael Halmagyi, Ian S. Curthoys, Swee T. Aw, and Joanna C. Jen

1. Introduction

Basic vestibular research is the cornerstone of clinical research on the diagnosis and treatment of patients with disorders of the vestibular system. Although *ad hoc* observations and empirical treatments of vestibular disorders and diseases will be with us until we know everything about the normal and abnormal vestibular systems, science at all levels from molecular biology to mathematical modeling is nevertheless the most promising means by which we might one day achieve that happy state. In this chapter, we describe three selected clinical advances that were derived from advances in basic vestibular research: clinical vestibular function testing, vestibular compensation, and inherited vestibular diseases. The opposite sometimes occurs, too: basic science picks up ideas from the clinic. For example, the observation that the GABA(B) agonist baclofen interferes with central vestibular function was first made in patients with periodic alternating nystagmus (Halmagyi et al. 1980), and this observation led to studies of its effects on brain stem velocity storage in the rhesus monkey (*Macaca mulatta*) (Cohen et al. 1987).

2. Clinical Tests of Vestibular Function

2.1. Impulsive Testing of Semicircular Canal Function

2.1.1. Physiologic Background

In a normal subject, any head rotation, even one restricted to a single plane, will change the activity from at least one pair of semicircular canals (SCCs) so that the brain stem signal that eventually drives the vestibuloocular reflex (VOR) is produced by excitation from one SCC and disfacilitation from the other SCC of the pair in the plane of rotation.

To illustrate this principle, consider the VOR in response to a yaw plane head rotation; this is the *horizontal* VOR and arises mainly from the lateral SCCs. During leftward head rotation, the activity of left lateral SCC primary afferent neurons increases, while at the same time the activity of right lateral SCC primary neurons decreases from the normal resting rate, which is about 90 spikes/s in the squirrel monkey (*Saimiri sciureus*) (Goldberg and Fernández 1971). Therefore, the increase in activity of type I position-vestibular-pause (PVP) secondary vestibular neurons in the left medial vestibular nucleus, the neurons that drive the rightward compensatory eye rotation (i.e., the vestibuloocular reflex—VOR), will be the result of both direct excitation from the left lateral SCC primary neurons and indirect commissural disinhibition from the right lateral SCC primary neurons (Shimazu and Precht 1966; Scudder and Fuchs 1992). In other words, the horizontal VOR normally functions as a push–pull system from the two lateral SCCs (Fig. 11.1A).

Direct excitation and indirect disinhibition are, however, potentially asymmetrical. Although the discharge rate of a vestibular neuron can increase linearly without obvious saturation in response to a rapid yaw head rotation in the excitatory (i.e., "on") direction, it can decrease only to zero in response to a rotation in the disfacilitatory direction (i.e., in the "off" direction). This might be especially true for the nonlinear, velocity-dependent component of the VOR, which seems to derive from the activity of irregularly discharging primary afferents (Clendaniel et al. 2002). In contrast, regularly discharging afferents, which might drive the linear component of the VOR, do not show saturation even at high head velocities (Hullar and Minor 1999). Because most secondary SCC neurons have a lower resting rate and a higher sensitivity to angular accelerations than do primary SCC neurons, they are even more easily silenced by rapid off-direction accelerations than primary neurons (Shimazu and Precht 1965) so that response asymmetry is even more marked at the level of secondary PVP neurons in the vestibular nuclei.

In this regard, anterior and posterior SCC neurons function similarly to lateral SCC neurons. For example, the mixed vertical–torsional VOR, which occurs in response to a forward and CW head rotation (i.e., a head rotation in the plane of the right anterior and left posterior SCC, the so-called RALP plane), is produced by the excitation of right vestibular nucleus secondary neurons, which are themselves directly excited by right anterior SCC primary neurons and indirectly disinhibited by left posterior SCC primary neurons. However, just as in the case of lateral SCC primary neurons, direct excitation and indirect disinhibition are inherently asymmetrical. The discharge rate of primary and secondary neurons from both the anterior and the posterior SCCs can increase linearly without obvious saturation in response to rapid RALP or LARP (left anterior right posterior) head rotations in the on direction, but it can decrease only to zero in response to rotations in the opposite off direction (Reisine and Raphan 1992).

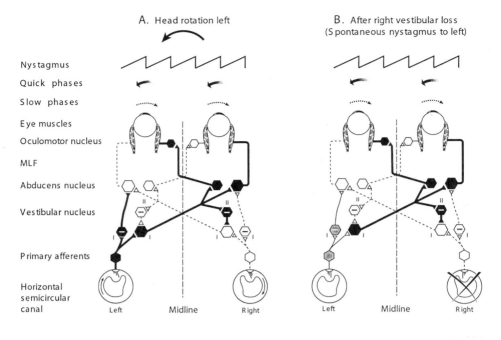

A. Head rotation left

B. After right vestibular loss
(Spontaneous nystagmus to left)

Nystagmus

Quick phases

Slow phases

Eye muscles

Oculomotor nucleus

MLF

Abducens nucleus

Vestibular nucleus

Primary afferents

Horizontal
semicircular
canal

Left   Midline   Right   Left   Midline   Right

⬡ Normal Resting Activity   ⬢ Increased Activity   ◯ Reduced Activity   ⊖ Inhibitory neuron
— Normal Resting Activity   — Increased Activity   --- Reduced Activity

FIGURE 11.1. Cartoon to show similarity between vestibular nucleus neural activity (A) during a leftward head rotation and (B) spontaneously at rest after a right vestibular deafferentation. (A) The sequence of events during a leftward head rotation is as follows. Ampullopetal endolymph flow in the left lateral SCC leads to an increase in the firing rate of primary afferents in the left vestibular nerve above the normal resting rate, which leads to an increased firing rate of the left medial vestibular nucleus type I secondary neurons (PVP neurons). There is also ampullofugal endolymph flow in the right lateral SCC, which leads to decreased firing of primary afferents in the right vestibular nerve, which leads to decreased firing of right medial vestibular nucleus type I secondary neurons, which leads to decreased firing of left vestibular nucleus type II (inhibitory) neurons, which also leads to increased firing of left vestibular nucleus type I neurons. The critical point is that activation of left vestibular nucleus PVP neurons, the neurons that drive motoneurons innervating the right lateral and left medial recti to produce the rightward compensatory eye rotations that comprise the slow phase of the VOR, is produced both by direct ipsilateral excitation and by indirect contralateral disinhibition. (B) After a right vestibular deafferentation, there is decreased activity of right medial vestibular nucleus type I neurons at rest due to two mechanisms. There is not only loss of excitation by right vestibular nerve primary afferents but also decreased activity of right vestibular nucleus type I neurons, which leads to decreased activity of left vestibular nucleus type II (inhibitory) neurons, which then leads to increased activity of left vestibular nucleus type I neurons, which then leads to increased activity of right vestibular nucleus type II neurons, which then also leads to increased inhibition (i.e., decreased spontaneous resting activity of right vestibular nucleus type II neurons), which produces the rightward slow phases of the left-beating spontaneous nystagmus, the hallmark of the right vestibular deafferentation.

2.1.2. Clinical Applications

This inherent asymmetry or nonlinearity of SCC responses is not only normally concealed by the bilateral interaction between the two labyrinths but also artificially concealed by the methods used in most laboratories to test and analyze vestibular function: namely, responses to low-acceleration sinusoidal rotations analyzed by algorithms that ignore threshold cutoff and directional asymmetry and calculate gain only for the excitatory direction stimulus. This is a pity because the on–off asymmetry provides an excellent opportunity for the clinician to test for unilateral impairment of SCC canal function, which is the basis for most complaints of vertigo.

Testing the VOR with head "impulses" in patients who have had a total unilateral vestibular deafferentation (uVD) yields scientifically interesting and clinically important results. Head impulses are rapid, passive, low-amplitude (10–20°), intermediate-velocity (120–180°/s), high-acceleration (3000–4000°/s²), unpredictable rotations of the head with respect to the trunk. They are delivered by an examiner who holds the patient's head firmly and at random rapidly rotates it in the yaw plane either to the left or to the right, or, in the RALP or LARP plane, either forward or backward (Fig. 11.2). The patient's task is to fixate a target at 1 m. To minimize

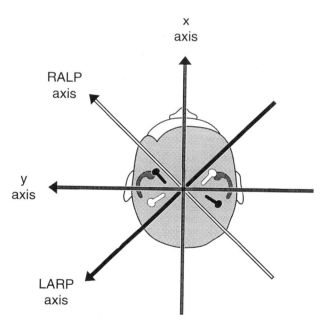

FIGURE 11.2. Axial view of the head from above. The Left Anterior—Right Posterior canal (LARP) axis is within the Right Anterior—Left Posterior canal (RALP) plane; similarly the RALP axis is within the LARP plane.

any contribution from the cervicoocular reflex, the visual pursuit reflex, or the saccadic system, only those compensatory eye movement responses that occur in the first 150 ms after the onset of head acceleration are analyzed. To represent VOR gain, eye velocity can be plotted as a function of head velocity.

In normal subjects, the horizontal VOR in response to yaw plane head impulses has a velocity gain of 0.94 ± 0.08 (SD) at an arbitrary 122°/s head velocity. In contrast, the vertical–torsional VOR in response to RALP and LARP plane head impulses has a gain of only 0.7 to 0.8, probably because the gain of the roll-torsional VOR is lower than the gain of the pitch-vertical VOR in normal subjects.

Following uVD, the VOR in response to ipsilesional yaw plane head impulses, now generated only by disfacilitation from the single functioning lateral SCC, is severely deficient (Halmagyi et al. 1990). Eye velocity gain decreases with increasing head velocity and appears to saturate at about 0.20 (Fig. 11.3). In contrast, the VOR in response to contralesional yaw

Yaw Impulses

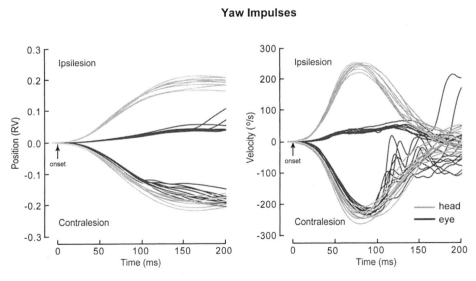

FIGURE 11.3. Horizontal head (gray lines) and eye (black lines) position and velocity during 8 superimposed yaw head impulses with a peak acceleration of about 5000 deg/s/s in a patient 1 year after unilateral vestibular deafferentation during vestibular schwannoma surgery. With contralesional head impulses eye position approximated head position and eye velocity approximated head velocity throughout the impulse; with ipsilesional head impulses at peak head velocity of about 250 deg/s, eye velocity was only about 25 deg/s (gain ~ 0.1).

plane head impulses, generated by excitation from the single functioning lateral SCC, is only mildly deficient, with a maximal velocity gain of 0.92. A high-acceleration rotation saturates the ipsilesional (i.e., off-direction), horizontal VOR in the human and in the guinea pig (Gilchrist et al. 1998), just as it saturates the off-direction discharge rate of lateral SCC afferents in the cat, monkey, rat, and guinea pig (Goldberg and Fernández 1971; Blanks et al. 1975a).

Silencing of irregularly discharging primary lateral SCC neurons in the contralesional (i.e., intact) vestibular nerve, leading to maximal disinhibition of type I secondary lateral SCC neurons in the ipsilesional vestibular nucleus (Goldberg and Fernández 1971), could be the reason why a symmetrical head rotation stimulus produces an asymmetrical eye rotation response in a subject in whom only one SCC is being stimulated (Ewald's second law) and is also the reason why the magnitude of the response asymmetry is a function of the magnitude of the stimulus. Furthermore, in response to high-acceleration stimulation, excitation of a single lateral SCC can by itself produce a near-normal horizontal VOR. This suggests that, in the human as well as in the monkey (Fetter and Zee 1988), disinhibition of ipsilateral type I neurons from the contralateral lateral SCC makes only a small contribution to the horizontal VOR.

Following uVD, the vertical–torsional VOR in response to RALP and LARP plane head impulses behaves similarly to the horizontal VOR (Cremer et al. 1998). In response to head impulses toward the lesioned anterior or posterior SCC—that is, in the off direction of the intact posterior or anterior SCC—the VOR is severely deficient (Fig. 11.4, Fig. 11.5A, Fig. 11.6). With impulsive testing, selective deficits of vertical SCC function can be detected in acute superior vestibular neuritis, which affects only the anterior and lateral SCC nerves, and acute inferior vestibular neuritis, which affects only the posterior SCC nerve (Aw et al. 2001). Impulsive testing can be a useful way to follow progress in patients who have had intratympanic gentamicin therapy for Meniere's disease (Carey et al. 2002).

With practice, it is possible to recognize VOR deficits in response to head impulses clinically. If the VOR is completely normal, then the patient will be able to maintain visual fixation during head impulses in any direction. If the VOR is severely defective, then the patient will not be able to maintain fixation and will need to make one or two refixating saccades, which the clinician can observe (Halmagyi and Curthoys 1988). For example, if, in response to a leftward head impulse, the patient makes a rightward saccade in order to maintain fixation, this indicates that the left lateral SCC is not functioning properly. Similarly, if, in response to a backward and CW head impulse, the patient makes a downward saccade, this indicates that the right posterior SCC is not working properly. In general, the deficit in SCC function needs to be severe in order for the compensatory saccadic eye movements to be large enough to be observed clinically.

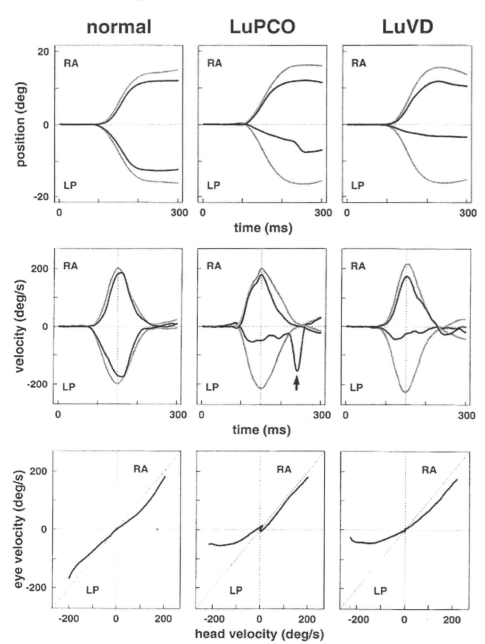

2.1.3. Semicircular Canal Occlusion

Occluding an SCC duct inactivates it by preventing endolymph flow. A procedure dating to Ewald (1892) is now used to treat patients with intractable benign paroxysmal positional vertigo (BPPV) (Pohl 1996; Agrawal and Parnes 2001). The cause of BPPV is movement of displaced otoconia in the duct of the SCC, usually a posterior, causing cupular displacement and hence vertigo and nystagmus. Because neither humans nor animals show any static symptoms such as spontaneous nystagmus after SCC occlusion, as would be expected if the labyrinth had been damaged, primary afferents from the occluded SCC presumably continue to fire at the usual resting rate. In humans (Cremer et al. 1998) as well as in guinea pigs (Gilchrist et al. 2000) and squirrel monkeys (*S. sciureus*) (Lasker et al. 2000), impulsive testing detects the SCC that has been occluded (Figs. 11.3–11.5). The high-acceleration VOR deficit is permanent in humans (Cremer et al. 1998) and in guinea pigs (Gilchrist et al. 2000) but only temporary in squirrel monkeys (*S. sciureus*) and rhesus monkeys (*M. mulatta*) (Hess et al. 2000; Lasker et al. 2000) and absent in the toadfish (*Opsanus tau*) (Rabbitt et al. 1999).

Confirmation that uVD produces a permanent VOR deficit in response to high-acceleration stimulation in primates (Lasker et al. 1999) as well as in humans (Tabak et al. 1997a, 1997b) has implications for the diagnosis and treatment of vestibular diseases (Halmagyi 1994; Reid et al. 1996) and also raises questions about the extent of the dynamic compensation that really occurs when accelerations within the range produced by natural head movements are encountered. The diagnostic significance of these observations is that it is possible to demonstrate at the bedside the severe permanent deficit in the ipsilesional VOR produced by uVD. For example, a yaw head impulse toward the lesioned side will produce a clinically obvious compensatory saccade or series of saccades toward the intact side

FIGURE 11.4. Head impulses are shown from a normal subject (left column), a subject whose left posterior SCC had been surgically occluded (LuPCO) (center column), and from a subject whose left vestibular nerve had been cut (LuVD) (right column). Head position and head velocity are in gray; inverted eye position and eye velocity are in black. Top row: head and eye position during a single head impulse in the direction of the right anterior SCC (RA) and another in the direction of the left posterior SCC (LP). Middle row: head and eye velocity corresponding to the head position in the top row. The vertical broken lines indicate the point of maximum head velocity; data are taken from the onset of the head impulse to maximum head velocity. The sharp peak in eye velocity in the uPCO patient's data is a catch-up saccade. Bottom row: eye velocity as a function of head velocity for each of the head impulses shown above. A perfect VOR would yield a plot superimposed on the diagonal line. In both patients, the VOR is deficient for head impulses only in the left posterior SCC direction and not in the right anterior SCC direction. (From Cremer et al. 1998, with permission.)

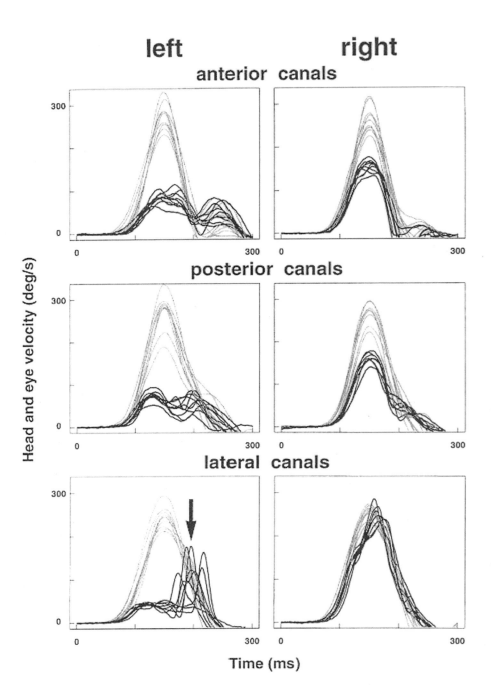

FIGURE 11.5. (**A**) Head velocity (gray) and inverted eye velocity (black) from a subject following a left vestibular neurectomy. A total of 54 head impulses are shown, comprising nine toward each of the six SCCs. For head impulses directed toward the left anterior, left posterior, and left lateral SCCs, the VOR is markedly deficient. The sharp peaks in the eye velocity traces are catch-up saccades. (From Cremer et al. 1998, with permission.)

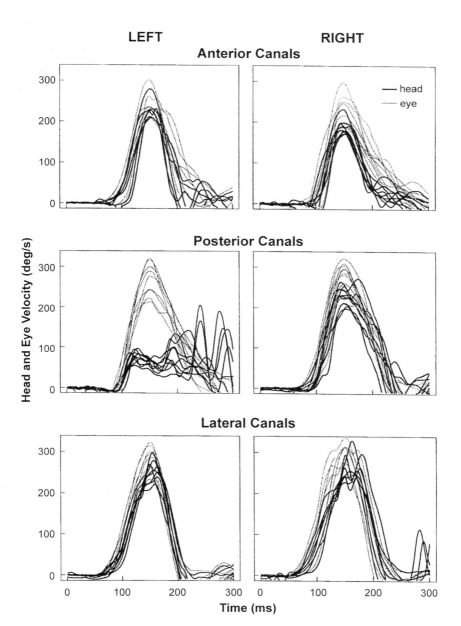

FIGURE 11.5. (**B**) Head velocity (gray) and inverted eye velocity (black) in a subject following a left posterior SCC occlusion. A total of 54 head impulses are shown, comprising nine head impulses toward each of the six SCCs. In response to head impulses directed toward the inactivated left posterior SCC, the VOR is markedly deficient, whereas head impulses directed toward any of the five intact SCCs elicit a normal VOR. The sharp peaks in the eye velocity traces during head impulses directed toward the left posterior SCC are catch-up saccades, which partially compensate for the deficient vestibular response. (From Cremer et al. 1998, with permission.)

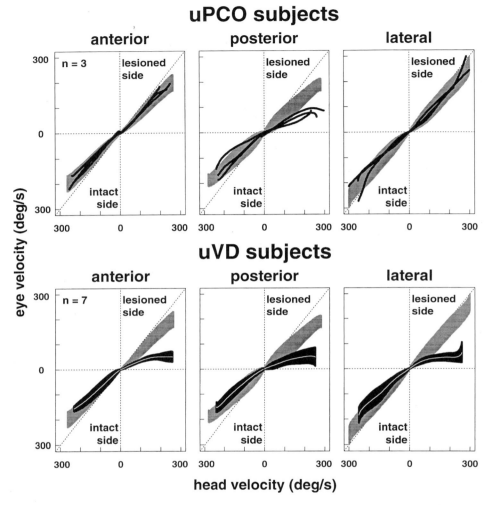

FIGURE 11.6. Eye velocity as a function of averaged head velocity for each of three unilateral posterior SCC occlusion subjects (top row) and averaged eye velocity (95% confidence intervals) for seven uVD subjects (bottom row). The VOR during head impulses directed toward the anterior SCC is shown in the left column, toward the posterior SCC in the middle column, and toward the lateral SCC in the right column. The normal range (average 95% confidence intervals) is shown as a gray band in each plot. The VOR during head impulses toward the lesioned side is shown on the right-hand side of each graph, and for head impulses toward the intact side on the left-hand side of each graph. The eye velocity is limited to <50°/s during head impulses in the direction of any lesioned SCC. (From Cremer et al. 1998, with permission.)

(Halmagyi and Curthoys 1988). Low-frequency, low-acceleration sinusoidal testing of the type routinely used for testing the horizontal VOR in humans and in experimental animals is artificially restricted to the narrow linear response range of the horizontal SCC and is therefore unsuitable for detecting or monitoring unilateral vestibular lesions. Only when high-acceleration passive head rotations are used does the expected deficit in compensatory eye movements become apparent (Della Santina et al. 2002).

However, some patients with unilateral impairment of the lateral SCC on caloric testing can have normal impulsive tests and conversely some patients with unilateral impairment of the lateral SCC on impulsive testing have normal caloric tests (Schmid-Priscoveanu et al. 2001). These findings suggest that the SCC pathology, just like cochlear pathology, can produce frequency-specific functional deficits: whereas impulsive testing represents a 4–5 Hz stimulus, the dominant frequency of the caloric test is around 0.025 Hz. Because impulsive and caloric testing examine different aspects of SCC function, there is a need to be able to measure anterior and posterior SCC function not just with impulsive tests but also with caloric tests.

2.2. Caloric Testing of Vertical Semicircular Canal Function

2.2.1. Physiologic Background

The vestibular system is stimulated when air or water, above or below body temperature, is instilled into the external auditory canal. The response is nystagmus that is largely horizontal: the fast phases beat toward the stimulated ear when the water is warm and the subject is supine and also when the water is cool and the subject is prone. This much has been known since Robert Bárány, winner of the Nobel Prize for Medicine, in 1917 first proposed that the mechanism of this "caloric" nystagmus was cupular displacement as a result of the hydrostatic forces produced by thermal convective flow within the lateral SCC duct being transmitted to the cupula (Baloh 2002).

Hydrostatic stimulation of an individual SCC (Ewald 1892) or electrical stimulation of an individual SCC nerve (Cohen and Suzuki 1963; Cohen et al. 1964) produces nystagmus with the rotation axis of the nystagmus orthogonal to the SCC plane (i.e., parallel to the SCC axis). Because caloric nystagmus is predominantly horizontal, it is assumed to result predominantly from stimulation of the lateral SCC. The thermal gradient created by caloric stimulation also reaches the vertical SCCs and, as long as one vertical SCC plane is gravitationally vertical, it will stimulate that vertical SCC, producing a vertical component of the nystagmus. This is the theoretical basis for trying to measure vertical SCC function in response to caloric stimulation. However, in order to do so, it is necessary to measure the vertical and torsional components as well as the horizontal components of the caloric nystagmus and to analyze the data in such a way as to build a mathematically correct reconstruction of the three-dimensional rotation axis of the nystagmus. Recent technological advances in eye movement measurement as well as improvements in the understanding of the mathematics of three-dimensional rotations (at least as far as vestibular researchers are concerned) have now brought measurement of vertical SCC caloric responses close to a practical diagnostic possibility. Although at present the scleral magnetic search coil method is still the only one with sufficiently high spatial and temporal resolution and sufficiently low cross talk for accurate three-dimensional measurement of nystagmus, video techniques are improving (Yagi et al. 1992; MacDougall et al. 2002). However, three-axis measurement does not automatically mean three-dimensional analysis; that is, simply summing the horizontal, vertical, and torsional position components of nystagmus one after the other does not yield a mathematically correct description of the eye position. Because rotations are noncommutative, one must calculate the vector sum using some appropriate description, such as Euler angles, rotation vectors, or quaternions (for a review, see Haslwanter 1995).

In intact rhesus monkeys (*M. mulatta*) (Böhmer et al. 1996) and humans (Aw et al. 1998; Fetter et al. 1998), a caloric stimulus activates all three SCCs. In humans, the lateral SCC gives the largest response; the maximal anterior SCC response is about 30% of the lateral SCC response, and the maximal posterior SCC response is about 10% of the lateral SCC response. Caloric responses can be elicited from individual vertical SCCs by first irrigating the ear to set up a thermal gradient with the head positioned so that the plane of the vertical SCC to be tested is gravitationally horizontal (i.e., in the position for minimal convective activation of that SCC). In this position, the plane of the other vertical SCC must be gravitationally vertical (i.e., in the position for maximal convective activation of that SCC). Recording the torsional and vertical components of the nystagmus elicited in this position should reflect the function of the gravitationally vertical, vertical SCC. Reorienting the head by turning it 90° shifts the vertical SCC that was in the gravitationally horizontal null position into the gravitationally vertical maximal position and vice versa.

2.2.2. Lateral SCC Reorientation

Consider a subject sitting upright with the head pitched 30° forward. In this position, the lateral SCCs are gravitationally horizontal and the anterior and posterior SCCs are gravitationally vertical (Fig. 11.7). Warm caloric stimulation of the right ear will produce little or no nystagmus response. There is no response from the right lateral SCC because it is gravitationally horizontal. There is no response from the vertical SCCs probably because the center of the heat source is below the lowest point of both the anterior and the posterior SCC and therefore the convective pressure in

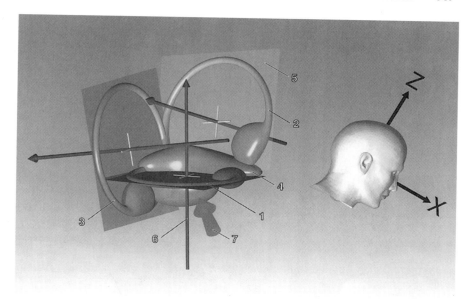

FIGURE 11.7. Three-dimensional rendering of the right labyrinth from the original data of Blanks et al. (1975b). 1 = plane of the lateral SCC; 2 = duct of the anterior SCC; 3 = duct of the posterior SCC; 4 = vestibule; 5 = plane of the anterior SCC; 6 = rotation axis of the lateral SCC; 7 = deduced position and direction of the heat source during caloric stimulation.

the two limbs of each SCC is equal. If the subject is now pitched 90° backward into the supine position, so that the lateral SCC is gravitationally vertical and the vertical SCCs are at 45° to the gravitational vertical, a vigorous nystagmus with leftward, CCW, and upward slow phases appears (Fig. 11.8, left column). When this nystagmus is resolved in SCC plane coordinates (i.e., into SCC vectors), it is clear that there has been not only excitation of the right lateral SCC (due to ampullopetal endolymph flow) but also excitation of both the right anterior SCC and, to a lesser extent, the right posterior SCC (due to ampullofugal endolymph flow).

2.2.3. Anterior SCC Reorientation

Consider now the same subject in the supine position as before but now with the head turned 45° to the right (Fig. 11.8, center column). In this position, the right anterior SCC plane will be gravitationally horizontal and the right posterior and right lateral SCC planes will be gravitationally vertical. Warm caloric stimulation of the right ear produces a nystagmus with leftward horizontal and CCW torsional components. When this nystagmus is resolved into SCC vectors, it indicates activation predominantly of the right lateral SCC (due to ampullopetal endolymph flow) but also slight activa-

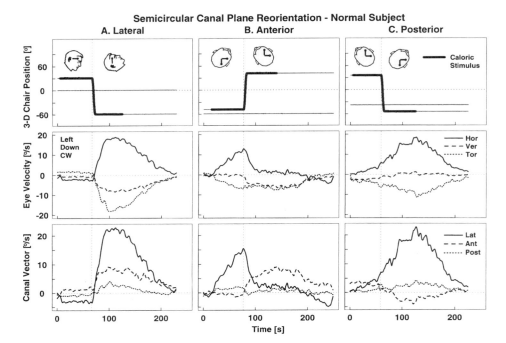

FIGURE 11.8. Representative examples of the nystagmus responses to (**A**) lateral SCC, (**B**) anterior SCC, and (**C**) posterior SCC plane reorientation from gravitational horizontal to gravitational vertical in a normal subject. The top row shows the subject's head position obtained from the chair position with reorientation of the head about the interaural axis in (**A**) and about the rostrocaudal axis in (**B**) and (**C**). The middle row shows the time series of the subject's horizontal (solid line), vertical (dashed line), and torsional (dotted line) slow-phase eye velocities during the SCC plane reorientations after warm caloric stimulation of the right ear. The bottom row shows the derived lateral SCC (solid line), anterior SCC (dashed line), and posterior SCC (dotted line) vectors before, during, and after SCC reorientations. The vertical dotted line shows the onset of the SCC reorientation.

tion of the right posterior SCC (due to ampullofugal endolymph flow). If the subject's head is now turned to the left by 90° into the 45° left position, the right anterior SCC plane will be gravitationally vertical (i.e., it will be in the maximal position), and the right posterior SCC plane will be gravitationally horizontal (i.e., it will now be in the null position). Vector analysis of the resulting nystagmus shows activation of the right anterior SCC and disactivation of the right lateral and posterior SCCs.

It is also interesting to consider the right lateral SCC responses in this particular experiment. When the right anterior SCC is horizontal, the right lateral SCC cupula is positioned at about 60° to the right (as viewed from the top of the head) within the gravitationally vertical lateral SCC plane and is above the heat source. Caloric stimulation in this position produces vigorous leftward slow phases. When the head is repositioned so that the right anterior SCC is now gravitationally vertical (the cupula of the lateral SCC is now at the same elevation as the heat source), the horizontal component of the nystagmus rapidly declines to about 20% of its previous value (Fig. 11.2B, bottom panel). This result shows the importance of the elevation of the cupula with respect to the heat source, and of the inclination of the SCC duct segment between the heat source and the cupula, to the magnitude of the response.

2.2.4. Posterior SCC Reorientation

The subject is again supine, but this time the head is turned 45° to the left. In this position, the right posterior SCC is gravitationally horizontal and the right anterior and right lateral SCCs are gravitationally vertical. Warm caloric stimulation of the right labyrinth produces a nystagmus with leftward, CCW, and upward slow phases. On vector analysis, the nystagmus indicates approximately equal activation of the right lateral and right anterior SCCs (Fig. 11.8, right-hand column). The head is then turned into the 45° right position so that the right posterior SCC becomes gravitationally vertical and the right anterior SCC becomes gravitationally horizontal. Theoretically, this should disactivate the right anterior SCC and activate the right posterior SCC. Vector analysis shows that the anterior SCC is not just inactivated but disfacilitated—the right anterior SCC vector actually reverses. In contrast, right posterior SCC activation is small—only about one-tenth of the maximal lateral SCC activation and only one-third of the maximal anterior SCC activation.

2.2.5. Determinants of SCC Caloric Responses

The magnitude of caloric nystagmus from an individual SCC in an intact animal or human is a function of the hydrostatic pressure across its cupula. From the data reviewed here and from the work of Coats and Smith (1967), O'Neill (1987), and Gentine et al. (1990), one can list the major physical factors that determine the temperature gradient across the two limbs of the SCC and therefore the hydrostatic pressure across its cupula:

1. The size of the SCC. This means the cross-sectional area and radius of curvature of the SCC.
2. The acceleration of gravity.
3. The inclination of the SCC plane with respect to gravity. This is the cosine of the angle between the SCC plane and the gravitational horizontal plane.
4. The position of the heat source relative to the SCC plane. The caloric response should be minimal if the assumed point heat source is orthogonal to the SCC plane. It will also be minimal if the point heat source is in the

SCC plane but is at its lowest or highest elevation and will be maximal when it is between these two points.

    5. The elevation of the cupula above the heat source.

Within these conditions, the caloric response should be, as Bárány proposed, maximal when the SCC duct segment between the heat source and the cupula is gravitationally vertical. However, recent data from rhesus monkeys (*M. mulatta*) with all six SCCs occluded suggests that mechanisms other than convective flow can produce a nystagmus response, at least to a cold caloric stimulus. These mechanisms include inhibition of activity in ampullary nerves, contraction of endolymph in the stimulated canals, and orientation of eye velocity to gravity through velocity storage (Arai et al. 2002). The relevance of these data to caloric testing in humans is not yet clear; for example, we have observed absent ipsilesional caloric (as well as impulsive) VOR responses in three patients who had lateral SCC canal occlusion for intractable positional vertigo. There appear to be fundamental differences between humans and animals in the effects of SCC occlusion on the vestibular function as tested either by high-acceleration impulses or by caloric stimulation. These functional differences might reflect differences in the exact site and nature of the occlusion produced by human and animal surgery.

2.3. Evoked Potential (VEMP) Testing of Saccular Function

2.3.1. Physiologic Background

Brief (0.1 ms), loud (>95 dB above normal hearing level, NHL), monaural clicks (Colebatch and Halmagyi 1992; Colebatch et al. 1994) and short tonebursts (Murofushi et al. 1999; Welgampola and Colebatch 2001) produce a large (60–300 μV), short-latency (8 ms) inhibitory potential in the tonically contracting ipsilateral sternocleidomastoid muscle. The initial positive–negative potential, which has peaks at 13 ms (p13) and 23 ms (n23), is abolished by selective vestibular neurectomy (Fig. 11.9A) but not by profound sensorineural hearing loss (Fig. 11.9B). In other words, even if the patient cannot hear the clicks, there can nonetheless be normal p13–n23 responses. Later components of the evoked response do not share the properties of the p13–n23 potential and probably do not depend on vestibular afferents. Failure to distinguish between these early and late components could explain why earlier work along similar lines was inconclusive.

    For the reasons above, we called the p13–n23 response the vestibular evoked myogenic potential, or VEMP. Unlike a neural evoked potential such as the brain stem auditory evoked potential, which is generated by the synchronous discharge of nerve cells, the VEMP is generated by synchronous discharges of muscle cells or, rather, motor units. Being a

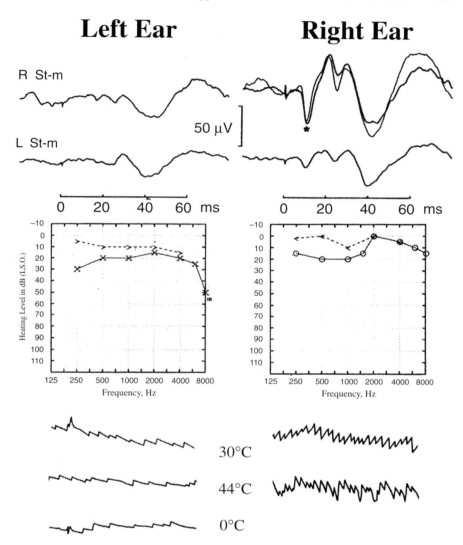

FIGURE 11.9. (**A**) Effect of unilateral vestibular deafferentation on the VEMP in a patient who had previously undergone a selective left vestibular nerve section for intractable vertigo. The left part of the figure refers to results for the left ear, the right part to the right. The audiograms confirm that hearing was well-preserved, and the caloric tests (bottom row) show only spontaneous right-beating nystagmus and no response to caloric stimulation of the left ear, consistent with previous vestibular deafferentation. Clicks of 100 dB intensity delivered to the right ear generate a normal p13(*)–n23 response in the ipsilateral right sternomastoid muscle with a weak crossed response in the left sternomastoid (top row). In contrast, clicks of the same intensity applied to the left ear generated no p13–n23 response in either sternomastoid muscle, although later potentials were still apparent. (From Colebatch et al. 1994, with permission.)

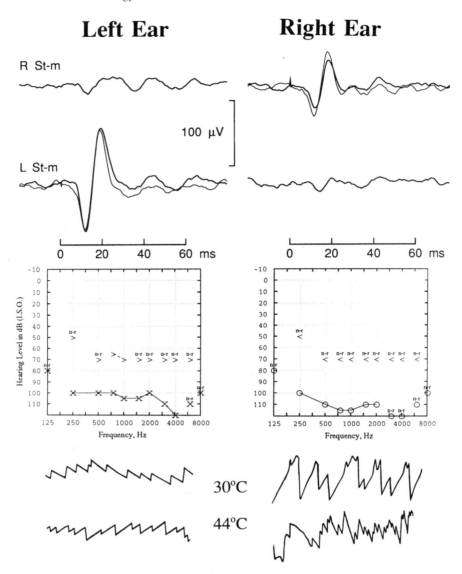

FIGURE 11.9. (**B**) The lack of effect of severe cochlear loss on the VEMP in a patient with intact lateral SCC function, as indicated by preserved horizontal nystagmus in response to caloric stimulation (bottom panel). Clicks of 100dB intensity to the left and right ears (top row) each generated normal p13–n23 responses in the ipsilateral sternocleidomastoid muscles, although the patient could not hear them. (From Colebatch et al. 1994, with permission.)

myogenic potential, the VEMP can be 500–1000 times larger than a brain stem potential (Todd 2001), 200 µV versus less than 1 µV. Single motor unit recordings in the tonically contracting sternocleidomastoid muscle show a decreased firing rate synchronous with the surface VEMP (see Halmagyi et al. 1994a, Fig. 8).

The amplitude of the VEMP is linearly related to the intensity of the click and to the intensity of sternomastoid activation during the period of averaging, as measured by the mean rectified electromyogram (EMG) (Colebatch et al. 1994). Inadequate sternomastoid contraction produces spurious results by reducing the amplitude of the VEMP (see, e.g., Ferber-Viart et al. 1999). A conductive hearing loss abolishes the response by attenuating the intensity of the stimulus (see Halmagyi et al. 1994a, Fig. 9). In such cases, the VEMP can be elicited by a tap to the forehead (Halmagyi et al. 1995) or by a bone vibrator (Sheykholeslami et al. 2000, 2001; Welgampola et al. 2003) or a dc current applied to the mastoid bone (Watson et al. 1998).

There are two main reasons to suppose that the VEMP arises from stimulation of the saccule. First, the saccule is the most sound-sensitive of the vestibular end organs (Young et al. 1977; Didier and Cazals 1989), possibly because it lies just under the stapes footplate (Anson and Donaldson 1973; Backous et al. 1999), in an ideal position to receive the full impact of a loud click delivered to the tympanic membrane. Second, not only do click-sensitive neurons in the vestibular nerve respond to tilts (Murofushi et al. 1995; Murofushi and Curthoys 1997) but most originate in the saccular macula (McCue and Guinan 1997; Murofushi and Curthoys 1997) and project to the lateral and descending vestibular nuclei as well as to other structures (Kevetter and Perachio 1986; Murofushi et al. 1996a). The VEMP measures vestibular function through what appears to be a disynaptic vestibulocollic reflex originating in the saccule and transmitted via the ipsilateral medial vestibulospinal tract to sternomastoid motoneurons (Kushiro et al. 1999).

2.3.2. Method

Any equipment suitable for recording brain stem auditory potentials will also record VEMPs. Because the amplitude of the VEMP is linearly related both to the intensity of the click and to the intensity of sternomastoid activation during the period of averaging, it is essential to ensure that the sound source is correctly calibrated and that the background level of rectified sternomastoid EMG activation is measured. Two reasons why the VEMPs could be absent or less than 50 µV in amplitude are a conductive hearing loss and inadequate contraction of the sternomastoid muscles.

For clinical testing, three superimposed runs of 128 averages for each ear in response to clicks of 100 dB intensity are usually sufficient. The test cannot be done on uncooperative or unconscious patients. The patient lies down and activates the sternomastoid muscles for the averaging period by keeping her or his head raised from a pillow. An alternative method useful, for example, in patients with painful neck problems is to ask the patient to turn the head, which continues to rest on the pillow, to one side. It is then possible to measure the VEMP in the sternomastoid muscle on the side opposite the rotation.

The peak-to-peak amplitude of the p13–n23 potential from each side can be expressed relative to the level of background mean rectified EMG to create a ratio that largely removes the effect of differences in muscle activity. A more accurate but more time-consuming correction can be made by making repeated observations with differing levels of tonic activation (Colebatch et al. 1994). One ear is best evaluated by comparing the amplitude of its VEMP with the amplitude of the VEMP from the other ear. We take asymmetry ratios of 2.5 to 1 to be the upper limit of normal—a value similar to that obtained by others (Brantberg and Fransson 2001). Minor left–right differences in latency commonly occur and might reflect differences in electrode placement over the muscle or differing muscle anatomy.

2.3.3. Clinical Applications

2.3.3.1. Superior Semicircular Dehiscence

A third window into the bony labyrinth allows sound to activate the vestibular system in animals (Tullio 1929; Dohlman and Money 1963) and in humans (Minor et al. 1998; Watson et al. 2000; Brantberg et al. 2001). Patients with a bony opening or dehiscence from the superior SCC to the middle cranial fossa (Fig. 11.10a) not only have sound- and pressure-induced vestibular nystagmus but also have abnormally large, low-threshold VEMPs (Colebatch et al. 1998; Watson et al. 2000; Brantberg et al. 2001; Streubel et al. 2001). In normal subjects, the VEMP, just like the acoustic reflex, has a threshold, usually 90–95 dB NHL. In patients with the superior SCC dehiscence, the VEMP threshold is about 20 dB lower than in normal subjects (Fig. 11.10b), and the VEMP amplitude at the usual 100–105 dB stimulus level can be abnormally large (>300 μV). If a VEMP can be consistently elicited at 70 dB NHL, this indicates that the patient has a superior SCC dehiscence.

Superior semicircular canal dehiscence also produces interesting changes in hearing. Patients notice that they are super-sensitive to bone conducted sounds. For example they can hear their own eyes move and their hearts beat (pulsatile tinnitus). Their own chewing sounds so loud to them that they cannot eat and listen at the same time. They can hear a tuning-fork placed at a remote bony prominence such as the ankle (Watson et al. 2000). Audiograms at a low-frequencies show that air conduction thresholds are raised while bone-conduction thresholds are lowered. This pattern can be mistaken for ossicular fixation due to otosclerosis and some of these

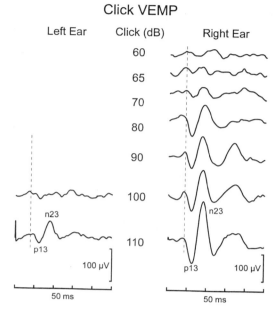

FIGURE 11.10. (**A**) Large-amplitude, low-threshold VEMPs from the symptomatic right ear in a patient with bilateral superior semicircular dehiscence (see Fig. 11.10B). There was no response from the left (or right) sternomastoid muscle in response to a 70 dB click in the left ear, but there was a VEMP (p13–n23) of about 40 μV amplitude from the right sternomastoid muscle in response to the same 70 dB click in the right ear. From the left ear, the VEMP appeared at a threshold within the normal range: 110 dB. At 110 dB, the VEMP from the symptomatic right ear was about twice the amplitude of that from the asymptomatic left ear.

patients have an unnecessary stapedectomy (Minor et al. 2003; Halmagyi et al. 2003b).

2.3.3.2. Meniere's Disease

VEMPs can be either too small (de Waele et al. 1999) or too large (Young et al. 2002a) in Meniere's disease as well as in delayed endolymphatic hydrops (Young et al. 2002b). In some cases, glycerol dehydration can reduce the size of VEMPs that are too large and increase the size of VEMPs that are too small (Murofushi et al. 2001a). VEMPs can be used to monitor intratympanic gentamicin therapy in patients with Meniere's disease (de Waele et al. 2002).

2.3.3.3. Vestibular Neurolabyrinthitis and BPPV

After an attack of vestibular neuritis, about one patient in three will develop posterior SCC benign paroxysmal positioning vertigo (BPPV), usually

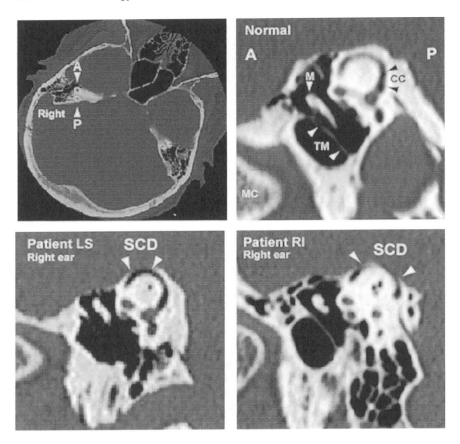

FIGURE 11.10. (**B**) High-resolution CT scan of the temporal bones reconstructed in the plane of the superior SCCs in a patient with bilateral superior SCC dehiscence (SCD) into the middle cranial fossa. Although there was a bilateral SCD, the patient only had sound- and pressure-induced nystagmus from the right. A normal CT is shown for comparison. M = head of the malleus; TM = tympanic membrane; CC = crus communis; A = anterior, P = posterior, MC = mandibular condyle. (From Halmagyi et al. 2003a, with permission.)

within 3 months (Murofushi et al. 1996b). The patients who develop BPPV after vestibular neuritis have intact VEMPs, whereas those who do not have absent VEMPs. In other words, an intact VEMP seems to be a prerequisite for the development of postvestibular neuritis BPPV. The reason for this could be that in those patients who develop postvestibular neuritis BPPV, only the superior vestibular nerve, which innervates the anterior SCC, lateral SCC, and the utricle, is involved. Because the inferior vestibular nerve innervates the posterior SCC and the saccule, the presence of posterior canal BPPV and the preservation of the VEMP imply that the inferior vestibular nerve must have been spared. Support for such an explanation comes from data that show preservation of posterior SCC impulsive VOR in some patients with vestibular neuritis—patients who presumably have only involvement of the superior vestibular nerve (Fetter and Dichgans 1996; Aw et al. 2001). The VEMPs evoked by galvanic current are generally abolished in those vestibular neuritis patients in whom the click-evoked VEMPs are abolished, indicating that the site of lesion is truly in the vestibular nerve rather than, or as well as, in the labyrinth (Murofushi et al. 2002).

2.3.3.4. Acoustic Neuroma

Although most patients with acoustic neuromas (vestibular schwannomas) present with unilateral hearing loss, some present with vestibular ataxia. This is not entirely surprising because most "acoustic" neuromas in fact arise not from the acoustic nerve but from one of the vestibular nerves, usually the inferior vestibular nerve (Komatsuzaki and Tsunoda 2001). The VEMP, which is transmitted via the inferior vestibular nerve, is abnormal—of low amplitude or absent—in perhaps four out of five patients with acoustic neuromas (Murofushi et al. 1998, 2001b; Tsutsumi et al. 2000). Because the VEMP does not depend on cochlear or lateral SCC functions, it can be diagnostically valuable in a patient suspected of having an acoustic neuroma because the VEMP can be abnormal even if brain stem auditory-evoked potentials cannot be measured because the patient is too deaf and even if the caloric test of lateral SCC function is normal.

Multiple sclerosis. VEMPs can also be abnormal in diseases affecting central vestibular pathways, especially white matter diseases such as multiple sclerosis (Shimizu et al. 2000; Versino et al. 2002), which affect the medial vestibulospinal tract, the continuation of the medial longitudinal fasciculus, a site commonly involved by demyelination.

3. Unilateral Vestibular Deafferentation and Vestibular Compensation

3.1. Behavioral Observations

Both scientists and clinicians need to understand the neural changes that occur after unilateral vestibular deafferentation (uVD) (Curthoys and Halmagyi 1995, 1998; Dieringer 1995; Vidal et al. 1998a), and our current understanding of the consequences of and recovery from uVD in humans—vestibular compensation—is a good example of the contribution of science to the clinic.

Sudden complete loss of the function of one intact labyrinth causes immediate stereotyped behavioral changes that are virtually the same in humans as in animals. These include intense spontaneous nystagmus (Fig.

11.1B), sensations of rotation (vertigo), and disturbances of stance and gait. Because such symptoms are present even when the patient or animal is at rest, they are called *static* symptoms (Fig. 11.1B). The *dynamic* symptoms of uVD are the changes in vestibuloocular reflexes. Remarkably, within about a week, these static symptoms largely disappear. What then are the neural mechanisms responsible for the invariable appearance and rapid disappearance of these static symptoms?

Early direct neurophysiological studies of neurons in the vestibular nuclei (Precht and Shimazu 1965; Shimazu and Precht 1965, 1966; Precht et al. 1966) showed the major types of horizontal SCC-driven neurons and demonstrated functionally inhibitory interactions between the two sides (Fig. 11.1A).

These findings confirmed older ideas that had emphasized the importance of neural interconnections between the vestibular nuclei (e.g., Spiegel and Demetriades 1925) and were of profound importance to our present understanding of clinical vestibular disorders after uVD. They led to the prediction that immediately after uVD neurons in the ipsilesional vestibular nucleus should have reduced resting discharge and neurons in the contralesional vestibular nucleus should have increased resting discharge. Many studies have confirmed this prediction and have shown the functional equivalence of such an imbalance between that of the vestibular nucleus and the imbalance in discharge that occurs during a long-duration unidirectional angular acceleration. The vertigo and the intense spontaneous nystagmus that uVD patients experience are readily understandable as the appropriate perceptual and behavioral responses to a large maintained vestibular stimulus. Although the time course of behavioral recovery does not exactly correspond to that of the return of activity in the vestibular nuclei, single neuron recordings in alert guinea pigs before and after uVD have confirmed the major results from anesthetized animals (Ris et al. 1995, 1997; Ris and Godaux 1998). A similar functionally inhibitory bilateral commissural interaction that exists between neurons in the otolith system (Kushiro et al. 1999) helps explain compensation of the static otolith-induced symptoms of uVD such as the ocular torsion, the shift of the visual horizontal, the head tilt, and the skew deviation (Halmagyi et al. 1979; Halmagyi and Curthoys 1999).

The rapid resolution of the static symptoms is in contrast with the incomplete recovery of the dynamic deficits following uVD. The head impulse test shows that there is little recovery of SCC function even years after uVD (Halmagyi et al. 1990; Aw et al. 1995, 1996a, 1996b; Cremer et al. 1998; Tabak et al. 1997a, 1997b; Della Santina et al. 2002). Guinea pigs (Gilchrist et al. 1998) and squirrel monkeys (*S. sciureus*) (Lasker et al. 1999) also show a severe permanent deficit in the ipsilesional horizontal VOR gain after uVD.

During ipsilesional head impulses in humans, there are permanent deficits not only in VOR velocity but also in the VOR rotation axis (Aw et al. 1996a, 1996b; Cremer et al. 1998) (Figs. 11.3–11.5). For the retinal image to remain stable during a head rotation, eye velocity must not only match head velocity but the axis of eye rotation must match the axis of head rotation. In the yaw-horizontal VOR of normal subjects, both eye velocity and the axis of eye rotation are appropriate, but in uVD patients, the eye not only rotates at an inadequate velocity but rotates around an incorrect axis, which changes during the head rotation. All of these errors will produce retinal image smear during head rotation. The axis misalignment after uVD highlights the challenge facing clinicians attempting vestibular rehabilitation. The assumption that the goal of vestibular rehabilitation should be to boost horizontal VOR gain is naive in light of the axis shift of uVDs—those shifts show that uVD patients require more complex changes in eye movement response for image stabilization than just the boosting of horizontal VOR gain.

3.2. Neural Mechanisms of Recovery after Unilateral Vestibular Deafferentation

A major assumption is that the imbalance in resting activity between the ipsilesional and contralesional vestibular nuclei after uVD is responsible for the acute symptoms after unilateral loss, and as the imbalance in neural activity is reduced, the acute symptoms decrease. But how is that neural imbalance reduced? In particular, what is the mechanism that initiates neural recovery within the first day? We consider evidence below, but whatever that mechanism is, the functionally inhibitory interconnections between the vestibular nuclei that cause the neural imbalance after uVD will act to assist the restoration of balanced neural activity between the vestibular nuclei during the recovery. As the cells in the ipsilesional vestibular nucleus start to fire again, they will exert inhibition on cells in the contralesional vestibular nucleus (Ris and Godaux 1998) via those functionally inhibitory commissural interconnections between the vestibular nuclei.

Some of the changes in the vestibular nuclei may not be caused by the vestibular loss itself but indirectly by the very behavioral effects that are produced by the loss. For example, head tilt will cause changes in the spinal afferent input to vestibular nuclei, which in turn will affect the resting discharge of vestibular nucleus neurons (Dieringer 1995). But the role of these indirect changes in the mechanism of the return of balanced activity between the vestibular nuclei is still not clear. In somewhat parallel fashion, it may be that cerebellar input triggered by the uVD may also act to restore the balanced activity between the bilateral vestibular nuclei.

Although long-lasting changes in the brain stem, probably predominantly in the vestibular nuclei, are likely to be responsible for maintaining the recovery of static symptoms (Guyot et al. 1995; de Waele et al. 1996; Gacek and Schoonmaker 1997; Gacek et al. 1988, 1989, 1991), it is clear that different neural processes must be responsible for the initiation of compensation. The very earliest phase of vestibular compensation—just hours after the vestibular loss—is too early for some neuronal mechanisms (such as axonal sprouting) to contribute (see, most recently, Aldrich and Peusner 2002), although there are clear glial changes even just a few hours after uVD (de Waele et al. 1996). Strong evidence for different processes underlying initiation and maintenance of compensation comes from studies where a second uVD on the remaining functional labyrinth is carried out at varying times after the first. If the remaining labyrinth is removed a few days or weeks after the first uVD, the animal shows a near-complete pattern of static symptoms, just as if this second uVD on the compensated animal were the first uVD on a normal animal. So, after this second uVD, there is spontaneous nystagmus, roll head tilt, and static eye deviation, all toward the most recently operated side. This behavioral pattern after the second uVD is called the Bechterew phenomenon (Bechterew 1883; Zee et al. 1982). It is interpreted as showing that neural rebalancing in both oculomotor and postural control systems must have taken place in the interval between the two labyrinthectomies. The rebalanced system is then again "unbalanced" by the second uVD so that although there are no vestibular sensory inputs present at all after the second uVD, the animal still displays symptoms, such as nystagmus, just as if all vestibular afferent inputs were intact and were subject to a strong maintained vestibular stimulus (Fig. 11.11).

How long is this "initiation" period? The answer seems to be about 3 days. If the second uVD is carried out within 3 days of the first, the consequences of it are mild and there is no Bechterew phenomenon. However, if the second uVD is carried out 3 or more days after the first uVD, the Bechterew phenomenon is present and becomes more pronounced at progressively later intervals. It is argued that the absence of the Bechterew phenomenon very early in compensation shows that the process responsible for the early disappearance of spontaneous nystagmus must be different from the processes responsible for the absence of nystagmus later after uVD.

Although there are many neuronal mechanisms that could have a role in the long-term restoration of neural activity and the maintenance of compensation, the number of neuronal mechanisms that can initiate the changes in the first few days after uVD is limited. De Waele et al. (1996) have identified the glial changes that commence soon (1 day) after uVD. These glial changes are likely to have a major role in the maintenance of vestibular compensation, and they occur so soon after uVD that they may even have a role in the initiation of vestibular compensation. To identify other mechanisms requires detailed investigation of membrane and synaptic processes using physiological and pharmacological procedures conducted on isolated slices of brains from animals at various stages of compensation. Understanding the synaptic and membrane changes occurring during vestibular compensation, especially those processes responsible for the initiation of

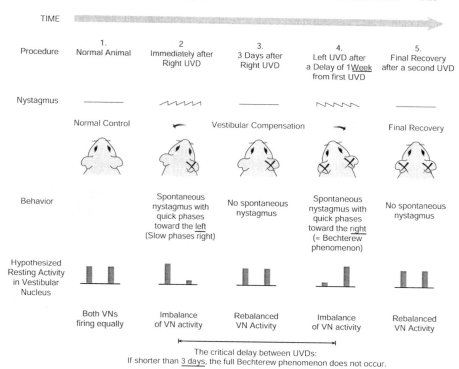

FIGURE 11.11. Cartoon to show the manifestations and proposed mechanisms of the Bechterew phenomenon, which is the critical dependence of the appearance of a second uVD syndrome on the time elapsed since the first. Note that if the second uVD is carried out less than one week after the first (that is, before vestibular compensation has occurred), only a partial uVD syndrome will occur; if the second uVD is carried out within a few hours of the first, not only will a second uVD syndrome not occur but the first will be terminated.

compensation, holds out the possibility of pharmacological treatments that may accelerate vestibular compensation in human patients.

3.3. Brain Slice and Isolated Whole Brain Studies

The development of brain slice preparations of the vestibular nuclei has promoted active research on neurotransmitters in the vestibular and ocu-lomotor systems (Serafin et al. 1991a, 1991b; de Waele et al. 1995; Cameron and Dutia 1997). The isolated whole-brain preparation from the guinea pig allows even more comprehensive physiological and pharmacological studies (Babalian et al. 1997; Vibert et al. 1997). The major problem in relat-ing this work to vestibular compensation in living animals is that the very process of removing the brain for preparing the slice or isolated whole brain requires that both vestibular nerves be cut (i.e., bilateral deafferentation of the vestibular nuclei). How can data after such bilateral deafferentation be related to the processes occurring in whole animals after unilateral deafferentation? Work from Vibert et al. (1999) might solve this problem.

After a uVD, the guinea pig is allowed a few days to recover before the brain is removed for a slice or isolated whole-brain preparation. Cutting the sole remaining vestibular nerve during the surgery for the removal of the brain to prepare the slice, Vidal et al. argue, generates a Bechterew phenomenon (Fig. 11.11) and therefore is equivalent to the first uVD in an intact animal. Both slice and isolated whole brain do show the asymmetries of neural activity between the two medial vestibular nuclei, which are similar to those that have been recorded *in vivo* both in anesthetized and in awake guinea pigs post-uVD. It is the side of the second uVD, effected at the time the brain is removed, that shows the lower resting discharge (Vibert et al. 1999).

There is now good evidence concerning some of the alterations at the synaptic and membrane levels of neurons in the vestibular nuclei that act to restore the balance of resting activity (Vibert et al. 1999, 2000; Tighilet and Lacour 2001). Some studies have directly measured the neurochemical changes in the vestibular nuclei that accompany vestibular compensation (Flohr et al. 1985; Henley and Igarashi 1991, 1993; de Waele et al. 1994, 1995; Cirelli et al. 1996; Cransac et al. 1996; Darlington and Smith 1996; Li et al. 1996; Duflo et al. 1999; Saxon et al. 2001). Dutia's group has shown changes at the synaptic and membrane level of neurons in the ipsilesional vestibular nucleus that are in the correct direction to restore the balance in resting activity between the two vestibular nuclei. They have shown a decrease of the effectiveness of inhibitory GABA receptors in neurons in the ipsilesional medial vestibular nucleus neurons, whereas the intrinsic membrane excitability of these same neurons increases (Yamanaka et al. 2000; Graham and Dutia 2001; Him and Dutia 2001). These basic physiological processes appear to be slowed in older animals (Him et al. 2001), perhaps corresponding to the anecdotal observation that older human patients tend to show slower vestibular compensation.

Most neurochemical studies have tested substances that affect the time course of vestibular compensation—usually the disappearance of spontaneous nystagmus or the change in posture. But any substance that alters, however indirectly, the delicate balance of neural activity between the vestibular nuclei will result in behavioral manifestation of vestibular symptoms such as nystagmus, vertigo, and ataxia and thus appear to affect compensation. The vestibular nuclei contain neurons with cholinergic, glutaminergic, dopaminergic, and GABAergic receptors so that substances that affect any of these transmitter systems (and probably many others) will appear to affect vestibular compensation whether those transmitter systems are directly involved in the recovery process or not. Glutamate is clearly an important transmitter in the vestibular system, and NMDA receptors have a role in the transmission of vestibular information and in vestibular compensation (de Waele et al. 1995). Nonetheless, it is likely that the different neural systems for oculomotor and postural effects use different neurotransmitters. Also, in light of the evidence about the difference between the initiation and maintenance of compensation, it must be recognized that the role of different transmitters may change during the course of compensation.

3.4. Mathematical Models of Vestibular Compensation

Some of the anatomical and physiological evidence has been incorporated into neural network models of the VOR that have sought to account for the static and dynamic behavioral symptoms after uVD and vestibular compensation (e.g., Galiana et al. 1984; Anastasio 1992; Weissenstein et al. 1996; Cartwright et al. 2003). Recently, "realistic" neural network models constructed to be consistent with established anatomical and physiological results, and trained on actual (guinea pig) eye movement responses to high-acceleration test stimuli before and after uVD, have produced results that account for the behavioral changes after uVD (Cartwright and Curthoys 1996; Cartwright et al. 2003; Gilchrist et al. 2003). In such models, it is possible to identify which neurons in the neural circuit show the greatest changes during compensation and therefore which neurons are of greatest importance for the process of compensation. It seems that the type I neurons on the contralesional (intact) side show the most change in gain. Data such as these could be the clue to understanding the mechanisms of compensation.

Of particular significance is that the Cartwright–Curthoys models are trained only on dynamic eye movement responses to head rotations before and after uVD. However, the neural changes that take place in the neural network to generate this asymmetry of dynamic response also produce an imbalance between the two abducens nuclei corresponding to the static symptom of horizontal spontaneous nystagmus, suggesting that the static and dynamic symptoms of uVD might be more closely related than has been believed.

4. Inherited Vestibular Diseases

In contrast with our increasing knowledge of the genetics of inherited hearing loss (Martini et al. 1997), our knowledge of the genetics of inherited vestibular loss is scanty. There are several reasons why this is so.

1. In patients with inherited vestibulopathies, the vestibular loss is usually bilateral and symmetrical so that they experience only mild to mod-

erate symptoms, namely chronic gait ataxia and oscillopsia, rather than the much more distressing symptoms of recurrent acute spontaneous vertigo. This is especially so in patients with inherited cochleovestibulopathies—the imbalance is considered to be only a minor inconvenience compared with the obvious disability produced by the deafness.

2. Vestibular function is more difficult to test than hearing so that patients with vestibulopathy are often not identified or tested.

3. There have been few large families informative enough genetically to facilitate mapping the disease loci, screening and identifying candidate genes.

4. Compared with many mouse models for genetic deafness, there are few animal models with genetic vestibulopathy, again perhaps a reflection of difficulty in assessing vestibular function even in a mouse.

In this section, we focus on patients with inherited vestibular loss without hearing loss, as at this stage so little is known about the vestibular function of patients with genetically defined nonsyndromic hearing loss (Morell et al. 1998).

4.1. Isolated Hereditary Peripheral Vestibulopathy

Baloh et al. (1994) described three families with a dominantly inherited bilateral peripheral vestibulopathy. All of those affected had suffered recurrent attacks of acute spontaneous vertigo and had as a result lost all or at least most SCC function bilaterally and eventually symmetrically. Hearing was consistently normal, but migraine was prominent in both affected and unaffected family members. The episodes of vertigo were commonly triggered by stress and fatigue. Some patients responded to acetazolamide with decreased attacks of vertigo. The mechanism of action of acetazolamide, a carbonic anhydrase inhibitor, is thought to be mediated by changes in extracellular pH and potassium ion concentration (Bain et al. 1992). Verhagen et al. (1987) reported a similar family with three affected siblings. The genetic abnormality responsible for this rare inherited condition has not yet been found, and it is also possible that some cases of sporadic vestibular failure are caused by a germline mutation of the same gene that is responsible for the hereditary form of the disease.

Hereditary vestibulopathy shares clinical features with diseases such as periodic paralyses and episodic ataxias, disorders known to be caused by defects in ion channel genes, channelopathies—the paroxysmal, recurrent nature of the symptoms, the development of progressive baseline deterioration, and clinical response to acetazolamide. The prominence of migraine in hereditary vestibulopathy is intriguing, as migraine could also be a channelopathy (Ptacek 1998). In fact, hemiplegic migraine with episodic ataxia is in some patients caused by mutations in CACNA1A encoding a neuronal voltage-gated calcium channel subunit (Ophoff et al. 1996; Ducros et al.

2001). We screened for mutations in CACNA1A in the three families reported by Baloh et al. (1994) as well as in an Australian family with six affected members in three generations but have so far found none. Ion channel genes expressed in both the brain and the inner ear are therefore likely candidate genes for hereditary vestibulopathy with migraine. Another candidate gene codes for a slowly activating potassium channel (isK) (Swanson et al. 1993). This channel is concentrated in the inner ear and could be responsible for generating the high potassium concentration in endolymph (Vetter et al. 1996); cloned isK knockout mice have severe receptor hair cell degeneration in both the cochlea and in the SCCs, with relative preservation of otolith hair cells.

4.2. Hereditary Peripheral Vestibular Loss with Hearing Loss: Usher Syndrome Type 1B

An abnormal gene codes for myosin type VIIa in the human disorder Usher syndrome type 1B, as well as in the deaf mouse mutant, *Shaker-1* (Gibson et al. 1995; Weil et al. 1995). Usher syndrome type 1B is an autosomal recessive disease characterized by severe congenital hearing and vestibular loss and retinitis pigmentosa (Kimberling et al. 1989). Myosin VIIa has been found in stereocilia (Corey et al. 1996), but its function is not yet clear. It might be involved in the "actin–myosin motor" that tensions tip links, the glycoprotein strands connecting adjacent receptor hair cells (Garcia-Añoveros and Corey 1997). Tip links have a role in opening and closing the transduction ion channels during deflection of the stereocilia, thereby producing cell depolarization or hyperpolarization (Pickles et al. 1984).

4.3. Hereditary Meniere's Disease

Although most patients with genetically defined nonsyndromic deafness are not known to have vestibular involvement, patients with DNFA9 mutations and dominantly inherited nonsyndromic deafness experience variable vestibular dysfunction, with both cochlear and vestibular symptoms reminiscent of Meniere's disease (Manolis et al. 1996). Meniere's disease is characterized by episodic vertigo, fluctuating low-frequency hearing loss, and tinnitus or aural fullness. DFNA9 patients generally have high-frequency hearing loss with onset in the third or fourth decade and deafness by age 40–50 years. Many DFNA9 patients also have vertigo and instability in the dark, with diminished or absent vestibular function on caloric testing. Histopathological studies on postmortem temporal bone from DFNA9 patients showed prominent acidophilic deposition in the inner ear (Khetarpal et al. 1991; Khetarpal 1993, 2000). Mutations causing DFNA9 were subsequently identified in COCH, a novel gene within the candidate region of DFNA9 preferentially expressed in high levels in the inner ear

(Robertson et al. 1998, 2001). The function of the COCH gene product, cochlin, is not known, but it is thought to play a structural role in the extracellular matrix affecting fluid homeostasis or afferent nerve function.

Patients with genuine Meniere's disease usually do not have a positive family history. Where there is a family history, it is almost always associated with migraine (Martini 1982; Oliveira et al. 1997; Neuhauser et al. 2001; Radtke et al. 2002). Patients with "vestibular Meniere's," which might be the same as "benign recurrent vertigo" (Slater 1979; Moretti et al. 1980; Kentala and Pyykko 1997), suffer from recurrent attacks of vertigo but have no tinnitus or hearing loss. They often also have migraine and a family history of migraine, vertigo, or both (Baloh and Andrews 1999). As with headache in migraine, the vertigo in benign recurrent vertigo can be triggered by stress, sleep deprivation, and exercise. Familial benign recurrent vertigo appears to be an autosomal dominant migraine syndrome with decreased penetrance in men. Identifying and documenting large families with benign recurrent vertigo is the first step in the genetic characterization of this condition (Kim et al. 1998; Oh et al. 2001).

"Benign paroxysmal vertigo of childhood" (Basser 1964) occurs in otherwise normal children with recurring attacks of staggering, pallor, vomiting, sweating, and crying, with complete resolution of symptoms in minutes. Some children report a true spinning sensation. The attacks eventually disappear, and many of these children develop migraine in adult life (Lanzi et al. 1994). There is usually no family history of vertigo.

4.4. Vestibular Dysfunction in Hereditary Spinocerebellar Ataxia (SCA)

The clinical characterization and identification of large, informative families has made possible the mapping and gene identification of several autosomal dominant spinocerebellar ataxias (SCAs) numbered in a chronological order of disease loci mapping to specific chromosomal regions. So far, at least 21 different chromosomal loci have been cataloged (Vuillaume et al. 2002), with disease-causing mutations identified in several genes, including SCA1 (ataxin 1), SCA2 (ataxin 2), SCA3 (ataxin 3), and SCA6. SCA1, SCA2, SCA3, and SCA6 are caused by glutamine-encoding CAG repeat expansions. These conditions are associated with distinct oculomotor phenotypes, which reflect *central* vestibular dysfunction with different disease mechanisms (Buttner et al. 1998; Bürk et al. 1999; Durig et al. 2002). Horizontal gaze-evoked nystagmus and impaired smooth pursuit were common in all SCAs. Slowing of saccades, indicating pontine involvement, was moderate in SCA1 but severe in SCA2. Low VOR gain, such as in the family reported by Philcox et al. (1975) and now known to have SCA1 and in SCA3, could indicate vestibular nerve or vestibular nucleus involvement. In SCA6 patients, there is downbeat nystagmus with abnormal OKN and VOR suppression but normal saccades and VOR gain, suggesting involvement of the cerebellar vermis (Gomez et al. 1997).

Although the functions of the genes causing SCA1, SCA2, and SCA3 are not known, it is known that the gene underlying SCA6 (CACNA1A) encodes a neuronal calcium channel subunit. CACNA1A mutations can also cause two other diseases: episodic ataxia type 2 (EA2) and familial hemiplegic migraine (Ophoff et al. 1996; Zhuchenko et al. 1997). Episodic ataxia type 2 is characterized by attacks of vertigo and ataxia lasting hours to days, with interictal gaze-evoked and rebound nystagmus. The episodes of vertigo and ataxia could be triggered by stress and fatigue, and the attacks can be prevented with regular acetazolamide (Griggs et al. 1978). Some SCA6 patients, like EA2 patients, experience vertigo attacks that also respond to acetazolamide (Jen et al. 1998).

The gene CACNA1A codes for the alpha1 subunit of a P/Q-type voltage-gated calcium channel. Mutations causing EA2 have been nonsense, splice site, or frame-shift mutations that disrupt the open reading frame, which may lead to truncated mutant protein products that are hypothesized to be nonfunctional. It is noteworthy that several missense mutations that altered single highly conserved amino acid residues have also been found to cause EA2 and, in one case, severe progressive ataxia (Yue et al. 1997; Guida et al. 2001; Jen et al. 2001).

About half of the families with hemiplegic migraine tested appeared to be linked to 19p13, where the gene CACNA1A resides (Joutel et al. 1993, 1994; Ophoff et al. 1994). Of those familial hemiplegic migraine families with mutations in CACNA1A who have been examined in detail, most have associated episodic or progressive cerebellar features (Ducros et al. 2001). At least two other loci, both on chromosome 1, have been reported in familial hemiplegic migraine without cerebellar features, but so far no disease-causing gene has been identified in the candidate region (Ducros et al. 1997, 2001; K. Gardner et al. 1997). Small expansions of CAG repeats in the last exon of CACNA1A from the normal range of 4–18 to 22–28 cause SCA6. The genetic findings are interesting but also raise some important unanswered questions. For example, by what mechanism does a CACNA1A mutation alter calcium channel function, and how do abnormal calcium channels lead to disease? Unlike in SCA1, SCA2, and SCA3 (Orr et al. 1993; Kawaguchi et al. 1994; Imbert et al. 1996), in SCA6 the number of CAG repeats does not correlate with disease severity or age of onset (Gomez et al. 1997), and how do defects in the same gene (the genotype) cause different diseases (the phenotype) such as familial hemiplegic migraine (Ophoff et al. 1996), central positional vertigo (Jen et al. 1998), EA2 (Ophoff et al. 1996; Jen et al. 1998), and a more aggressive form of cerebellar ataxia (Yue et al. 1997)? Variations in the phenotype might be due to the modifying effects of other genes (including the normal allele), somatic mosaicism (differences between the concentration of the abnormal gene in different tissues), or environmental influences.

The identification of missense CACNA1A mutations in patients with familial hemiplegic migraine raises questions regarding a possible role of CACNA1A in more common forms of migraine. Although there are some linkage data suggestive of a role of CACNA1A in migraine with or without aura (May et al. 1995), linkage has not been established in other studies of migraine patients (Noble-Topham et al. 2002). Furthermore, to date, no polymorphism or mutation in CACNA1A has been identified in any migraineur without recurrent hemiplegia or ataxia (Brugnoni et al. 2002). Whether CACNA1A is involved in common forms of migraine remains controversial.

4.5. Mitochondrial Mutations and Predisposition to Aminoglycoside Otoxicity

Despite their potential toxic effects on cochlear as well as vestibular hair cells, aminoglycoside antibiotics are commonly used throughout the world. The reason is that they are not only effective in life-threatening gram-negative bacterial infections as well as in tuberculosis but they are also inexpensive. In humans, gentamicin, tobramycin, and streptomycin are mainly vestibulotoxic, whereas neomycin, kanamycin, and amikacin are mainly cochleotoxic (Ballantyne 1984). It is of interest that some individuals appeared to be predisposed, possibly on a genetic basis, to aminoglycoside ototoxicity. Given enough gentamicin, anyone's vestibular system can be wiped out. In fact, systemic gentamicin has been used to destroy vestibular function in patients with intractable vertigo due to bilateral Meniere's disease (Schuknecht 1957). However, in some patients, just three or four standard doses of gentamicin will produce severe permanent vestibulotoxicity, despite normal renal function and despite "nontoxic" blood levels (Halmagyi et al. 1994b). Although at this stage there are no data about a genetic predisposition to gentamicin vestibulotoxicity, there are data about a familial predisposition to streptomycin cochleotoxicity. Abnormalities of (maternally inherited) mitochondrial DNA have been demonstrated both in familial and sporadic cases of streptomycin-induced deafness. These mtDNA mutations might confer selective vulnerability to aminoglycoside ototoxicity, possibly by inactivation of enzymes responsible for the metabolism of aminoglycosides (Hu et al. 1991; Prezant et al. 1993; J. Gardner et al. 1997). In the future, genetic screening might guide clinicians to avoid aminoglycosides in patients who are predisposed to toxicity. Mitochondrial DNA mutations have also been shown to cause severe bilateral hearing loss without clinically obvious involvement of any other systems or tissues (Sue et al. 1998). For example, the A3243G mutation in humans produces the MELAS syndrome (mitochondrial encephalomyopathy with lactic acidosis and strokelike episodes). Cochlear deafness can be part of MELAS or can occur without any of the other features of MELAS in some of the affected kindred. In some patients, the deafness is only mild; in others, it is severe enough to require cochlear implantation. At this stage, little is known about vestibular abnormalities in MELAS, but the possibility of mitochondrial inheritance of vestibular disorders is an area that will be receiving attention in the future.

4.6. Hereditary Vestibular Nerve Schwannomas in Neurofibromatosis Type 2

Although these tumors are usually called acoustic neuromas, they are neither acoustic nor neuromas. They are schwannomas and usually arise at the junction between the peripheral and central myelin covering the inferior vestibular nerve (Jackler 1994; Komatsuzaki and Tsunoda 2001). They can occur sporadically as unilateral tumors (95% of cases) or bilaterally (5% of cases), in which case the patient has neurofibromatosis type 2 (NF2). NF2 is characterized by bilateral vestibular schwannomas as well as other nervous system tumors such as meningiomas, gliomas, ependymomas, and neurofibromas. Autosomal dominant inheritance is observed in half of NF2 cases, while sporadic cases, many with demonstrable *de novo* mutations, account for the other half (Evans et al. 1992, 2000; Parry et al. 1994).

The abnormal gene is on chromosome 22, and its product schwannomin (also known as *merlin*), is a tumor-suppressor protein that shares homology with Protein 4.1 molecules, a family of proteins that link the actin cytoskeleton to cell surface glycoproteins (Rouleau et al. 1993; Trofatter et al. 1993). Inactivation of both copies of the NF2 gene is necessary for the development of both sporadic and inherited vestibular schwannomas (Moffat and Irving 1995). Patients with NF2 inherit one abnormal allele and acquire a second abnormal allele during their lifetime, allowing the tumor to develop (Knudson 1971).

Many different germline and somatic mutations have been found in the first 15 exons of the 17-exon NF2 gene in two-thirds of patients with NF2. Most mutations—nonsense, splice site, and frame shift—result in barely detectable truncated and nonfunctional proteins in all NF2-related tumors, whereas the rarer nontruncating missense and small in-frame deletion mutations produce mutant proteins with altered functional domains. Nonsense and frame-shift mutations appear to cause early onset of symptoms and a large number of tumors at the time of diagnosis (Parry et al. 1996; Evans et al. 1998). Splice site mutations are associated with a high degree of phenotypic variability (Kluwe et al. 1998). Missense mutations and small in-frame deletions are associated with mild to severe phenotypes (Bourn et al. 1994; Scoles et al. 1996; Welling 1998). In patients with late-onset bilateral vestibular schwannomas without nonvestibular tumors, mutations are often not detected in the NF2 gene (Parry et al. 1996).

Because in some patients with typical NF2 no mutation has been detected, a negative result from mutation screening does not exclude the diagnosis. Molecular genetics might become important in screening at-risk individuals and perhaps in prenatal diagnosis in families with known mutations (Kluwe et al. 2002). Intrafamilial phenotypic variability may reflect differences in the timing of the "second hit" but also emphasizes the importance of modifier genes.

5. Summary

For progress in the diagnosis and treatment of vestibular diseases and disorders, clinicians depend on progress in scientific ideas, methods, and techniques. In reverse, scientists can, without being forced into "mission-orientated" research, pick up useful new ideas by exposure to the experiments of nature found in the Dizzy Clinic. Although some of the clinical advances that we have described here, such as in caloric testing and in vestibular compensation, link directly to vestibular science, others such as the high-resolution CT required to diagnose superior semicircular canal dehiscence and the genetics of migraine vestibulopathy depend on a broader science and technology. For continued progress, not only do vestibular scientists need to talk to and work with vestibular clinicians but they also need to be familiar with the impact that advances in fields such as genetics and imaging could be making in vestibular research.

Acknowledgments. This work was supported by the National Health and Medical Research Council (Australia), by the National Institutes for Deafness and Communication Disorders (USA), and by the Garnett Passe and Rodney Williams Memorial Foundation. Dr. P.D. Cremer helped prepare the first draft of the manuscript.

References

Agrawal SK, Parnes LS (2001) Human experience with canal plugging. Ann N Y Acad Sci 942:300–305.

Aldrich EM, Peusner KD (2002) Vestibular compensation after ganglionectomy: ultrastructural study of the tangential vestibular nucleus and behavioral study of the hatchling chick. J Neurosci Res 67:122–138.

Anastasio TJ (1992) Simulating vestibular compensation using recurrent back-propagation. Biol Cybern 66:389–397.

Anson BJ, Donaldson JA (1973) Surgical Anatomy of the Temporal Bone and Ear. Philadelphia: Saunders, p. 285.

Arai Y, Yakushin SB, Cohen B, Suzuki J, Raphan T (2002) Spatial orientation of caloric nystagmus in semicircular canal-plugged monkeys. J Neurophysiol 88:914–928.

Aw ST, Halmagyi GM, Pohl DV, Curthoys IS, Yavor RA, Todd MJ (1995) Compensation of the human vertical vestibulo-ocular reflex following occlusion of one vertical semicircular canal is incomplete. Exp Brain Res 103:471–475.

Aw ST, Haslwanter T, Halmagyi GM, Curthoys IS, Yavor RA, Todd MJ (1996a) Three-dimensional vector analysis of the human vestibuloocular reflex in response to high-acceleration head rotations. I. Responses in normal subjects. J Neurophysiol 76:4009–4020.

Aw ST, Halmagyi GM, Haslwanter T, Curthoys IS, Yavor RA, Todd MJ (1996b) Three-dimensional vector analysis of the human vestibuloocular reflex in response to high-acceleration head rotations. II. Responses in subjects with unilateral vestibular loss and selective semicircular canal occlusion. J Neurophysiol 76:4021–4030.

Aw ST, Haslwanter T, Fetter M, Heimberger J, Todd MJ (1998) Contribution of the vertical semicircular canals to the caloric nystagmus. Acta Otolaryngol (Stockh) 118:618–627.

Aw ST, Fetter M, Cremer PD, Karlberg M, Halmagyi GM (2001) Individual semicircular canal function in superior and inferior vestibular neuritis. Neurology 57:768–774.

Babalian A, Vibert N, Assie G, Serafin M, Muhlethaler M, Vidal PP (1997) Central vestibular networks in the guinea-pig—functional characterization in the isolated whole brain in vitro. J Neurosci 81:405–426.

Backous DD, Minor LB, Aboujaoude ES, Nager GT (1999) Relationship of the utriculus and sacculus to the stapes footplate: anatomic implications for sound- and/or pressure-induced otolith activation. Ann Otol Rhinol Laryngol 108:548–553.

Bain PG, O'Brien MD, Keevil SF, Porter DA (1992) Familial periodic ataxia: a problem of cerebellar intracellular pH homeostasis. Ann Neurol 31:147–154.

Ballantyne J (1984) Ototoxicity. In: Oosterveld WJ (ed) Otoneurology. New York: Wiley, pp. 41–51.

Baloh RW (2002) Robert Barany and the controversy surrounding his discovery of the caloric reaction. Neurology 58:1094–1099.

Baloh RW, Andrews JC (1999) Migraine and Meniere's disease. In: Harris JP (ed) Meniere's Disease. The Hague, The Netherlands: Kugler Publications, pp. 281–289.

Baloh RW, Jacobson K, Fife T (1994) Familial vestibulopathy: a new dominantly inherited syndrome. Neurology 44:20–25.

Basser LS (1964) Benign paroxysmal vertigo of childhood: a variety of vestibular neuritis. Brain 87:141–152.

Bechterew W (1883) Ergebnisse der Durchschneidung des N. Acusticus, nebst Erorterung der Bedeutung der semicircularen Kanale fur das Korpergleichwicht. Pflügers Arch Gesamte Physiol Menschen Tiere 30:312–347 (as quoted by Zee et al. Bechterew's phenomenon in a human patient. Ann Neurol 1982;12:495–496).

Blanks RHI, Estes MS, Markham CH (1975a) Physiologic characteristics of first-order canal neurons in the cat. II. Response to constant accelerations. J Neurophysiol 38:1250–1268.

Blanks RHI, Curthoys IS, Markham CH (1975b) Planar relationships of the semicircular canals in man. Acta Otolaryngol (Stockh) 80:185–196.

Böhmer A, Straumann D, Suzuki JI, Hess BJM, Henn V (1996) Contributions of single semicircular canals to caloric nystagmus as revealed by canal plugging in rhesus monkey. Acta Otolaryngol (Stockh) 116:513–520.

Bourn D, Carter SA, Evans DG, Goodship J, Coakham H, Strachan T (1994) A mutation in the neurofibromatosis type 2 tumor-suppressor gene, giving rise to widely different clinical phenotypes in two unrelated individuals. Am J Hum Genet 55:69–73.

Brantberg K, Fransson PA (2001) Symmetry measures of vestibular evoked myogenic potentials using objective detection criteria. Scand Audiol 30:189–196.

Brantberg K, Bergenius J, Mendel L, Witt H, Tribukait A, Ygge J (2001) Symptoms, findings and treatment in patients with dehiscence of the superior semicircular canal. Acta Otolaryngol (Stockh) 121:68–75.

Brugnoni R, Leone M, Rigamonti A, Moranduzzo E, Cornelio F, Mantegazza R, Bussone G (2002) Is the CACNA1A gene involved in familial migraine with aura? Neurol Sci 23:1–5.

Bürk K, Fetter M, Abele M, Laccone F, Brice A, Dichgans J, Klockgether T (1999) Autosomal dominant cerebellar ataxia type I: oculomotor abnormalities in families with SCA1, SCA2 and SCA3. J Neurol 246:789–797.

Buttner N, Geschwind D, Jen J, Perlman S, Pulst SM, Baloh RW (1998) Oculomotor phenotypes in autosomal dominant ataxias. Arch Neurol 55:1353–1357.

Cameron SA, Dutia MB (1997) Cellular basis of vestibular compensation: changes in intrinsic excitability of MVN neurones. Neuroreport 8:2595–2599.

Carey JP, Minor LB, Peng GC, Della Santina CC, Cremer PD, Haslwanter T (2002) Changes in the three-dimensional angular vestibulo-ocular reflex following intratympanic gentamicin for Meniere's disease. J Assoc Res Otolaryngol 3:430–443.

Cartwright AD, Curthoys IS (1996) A neural network simulation of the vestibular system: implications on the role of intervestibular nuclear coupling during vestibular compensation. Biol Cybern 75:485–493.

Cartwright AD, Gilchrist DPD, Burgess AM, Curthoys IS (2003) A realistic neural network simulation of both slow and quick phase components of the guinea pig VOR. Exp Brain Res 149:299–311.

Cirelli C, Pompeiano M, D'Ascanio P, Arrighi P, Pompeiano O (1996) c-fos expression in the rat brain after unilateral labyrinthectomy and its relation to the uncompensated and compensated stages. Neuroscience 70:515–546.

Clendaniel RA, Lasker DM, Minor LB (2002) Differential adaptation of the linear and nonlinear components of the horizontal vestibuloocular reflex in squirrel monkeys. J Neurophysiol 88:3534–3540.

Coats AC, Smith SY (1967) Body position and the intensity of caloric nystagmus. Acta Otolaryngol (Stockh) 63:515–532.

Cohen B, Suzuki JI (1963) Eye movements induced by ampullary nerve stimulation. Am J Physiol 204:347–351.

Cohen B, Suzuki JI, Bender MB (1964) Eye movements from semicircular canal nerve stimulation in the cat. Ann Otol Rhinol Laryngol 73:153–169.

Cohen B, Helwig D, Raphan T (1987) Baclofen and velocity storage: a model of the effects of the drug on the vestibulo-ocular reflex in the rhesus monkey. J Physiol (Lond) 393:703–725.

Colebatch JG, Halmagyi GM (1992) Vestibular evoked potentials in human neck muscles before and after unilateral vestibular deafferentation. Neurology 42:1635–1636.

Colebatch JG, Halmagyi GM, Skuse NF (1994) Myogenic potentials generated by a click-evoked vestibulocollic reflex. J Neurol Neurosurg Psychiatry 57:190–197.

Colebatch JG, Day BL, Bronstein AM, Davies RA, Gresty MA, Luxon LM, Rothwell JC (1998) Vestibular hypersensitivity to clicks is characteristic of the Tullio phenomenon. J Neurol Neurosurg Psychiatry 65:670–678.

Corey DP, Hasson T, Chen Z-Y, Garcia J, et al. (1996) Location and function of myosins in auditory hair cells. Cold Spring Harb Symp Quant Biol 61:80.

Cransac H, Peyrin L, Farhat F, Cottet-Emard JM, Pequignot JM, Reber A (1996) Effect of hemilabyrinthectomy on monoamine metabolism in the medial vestibular nucleus, locus coeruleus, and other brainstem nuclei of albino and pigmented rats. J Vestib Res 6:243–253.

Cremer PD, Halmagyi GM, Aw ST, Curthoys IS, McGarvie LA, Todd MJ, Black RA, Hannigan IP (1998) Semicircular canal plane head impulses detect absent function of individual semicircular canals. Brain 121:699–716.

Curthoys IS, Halmagyi GM (1995) Vestibular compensation. A review of the oculomotor, neural and clinical consequences of unilateral vestibular loss. J Vestib Res 5:67–107.

Curthoys IS, Halmagyi GM (1998) Vestibular compensation. In: Büttner U (ed) Vestibular dysfunction and its therapy. Advances in Otorhinolaryngology, Volume 55. Basel: Karger, pp. 82–110.

Darlington CL, Smith PF (1996) What neurotransmitters are important in the vestibular system? In: Baloh RW, Halmagyi GM (eds) Disorders of the Vestibular System. New York: Oxford University Press, pp. 140–144.

de Waele C, Abitbol M, Chat M, Menini C, Mallet J, Vidal PP (1994) Distribution of glutamatergic receptors and GAD mRNA-containing neurons in the vestibular nuclei of normal and hemilabyrinthectomized rats. Eur J Neurosci 6:565–576.

de Waele C, Muhlethaler M, Vidal PP (1995) Neurochemistry of central vestibular pathways: a review. Brain Res Rev 20:24–46.

de Waele C, Torres AC, Josset P, Vidal PP (1996) Evidence for reactive astrocytes in rat vestibular and cochlear nuclei following unilateral inner ear lesion. Eur J Neurosci 8:2006–2018.

de Waele C, Huy PT, Diard JP, Freyss G, Vidal PP (1999) Saccular dysfunction in Meniere's disease. Am J Otol 20:223–232.

de Waele C, Meguenni R, Freyss G, Zamith F, Bellalimat N, Vidal PP, Tran Ba Huy P (2002) Intratympanic gentamicin injections for Meniere disease: vestibular hair cell impairment and regeneration. Neurology 59:1442–1444.

Della Santina CC, Cremer PD, Carey JP, Minor LB (2002) Comparison of head thrust test with head autorotation test reveals that the vestibulo-ocular reflex is enhanced during voluntary head movements. Arch Otolaryngol Head Neck Surg 128:1044–1054.

Didier A, Cazals Y (1989) Acoustic responses recorded from the saccular bundle on the eighth nerve of the guinea pig. Hear Res 37:123–128.

Dieringer N (1995) "Vestibular compensation": neural plasticity and its relations to functional recovery after labyrinthine lesions in frogs and other vertebrates. Prog Neurobiol 46:97–129.

Dohlman G, Money K (1963) Experiments on the Tullio vestibular fistula reaction. Acta Otolaryngol (Stockh) 56:271–278.

Ducros A, Joutel A, Vahedi K, Cecillon M, Ferreira A, Bernard E, Verier A, Echenne B, Lopez de Munain A, Bousser MG, Tournier-Lasserve E (1997) Mapping of a second locus for familial hemiplegic migraine to 1q21–q23 and evidence of further heterogeneity. Ann Neurol 42:885–890.

Ducros A, Denier C, Joutel A, Cecillon M, Lescoat C, Vahedi K, Darcel F, Vicaut E, Bousser MG, Tournier-Lasserve E (2001) The clinical spectrum of familial hemiplegic migraine associated with mutations in a neuronal calcium channel. N Engl J Med 345:17–24.

Duflo SGD, Gestreau C, Tighilet B, Lacour M (1999) Fos expression in the cat brainstem after unilateral vestibular neurectomy. Brain Res 824:1–17.

Durig JS, Jen JC, Demer JL (2002) Ocular motility in genetically defined autosomal dominant cerebellar ataxia. Am J Ophthalmol 133:718–721.

Evans DG, Huson SM, Donnai D, Neary W, Blair V, Teare D, Newton V, Strachan T, Ramsden R, Harris R (1992) A genetic study of type 2 neurofibromatosis in the United Kingdom. I. Prevalence, mutation rate, fitness, and confirmation of maternal transmission effect on severity. J Med Genet 29:841–846.

Evans DG, Trueman L, Wallace A, Collins S, Strachan T (1998) Genotype/phenotype correlations in type 2 neurofibromatosis (NF2): evidence for more severe disease associated with truncating mutations. J Med Genet 35:450–455.

Evans DG, Sainio M, Baser ME (2000) Neurofibromatosis type 2. J Med Genet 37:897–904.

Ewald EJR (1892) Physiologische Untersuchungen uber das Endorgans des Nervus Octavus. Wiesbaden: Bergmann.

Ferber-Viart C, Dubreuil C, Duclaux R (1999) Vestibular evoked myogenic potentials in humans: a review. Acta Otolaryngol (Stockh) 119:6–15.

Fetter M, Dichgans J (1996) Vestibular neuritis spares the inferior division of the vestibular nerve. Brain 119:755–763.

Fetter M, Zee DS (1988) Recovery from unilateral labyrinthectomy in rhesus monkey. J Neurophysiol 59:370–393.

Fetter M, Aw S, Haslwanter T, Hemiberger J, Dichgans J (1998) Three-dimensional eye movement analysis during caloric stimulation used to test vertical semicircular canal function. Am J Otol 19:180–187.

Flohr H, Abeln W, Luneburg U (1985) Neurotransmitter and neuromodulator systems involved in vestibular compensation. Rev Oculomot Res 1:269–277.

Gacek RR, Schoonmaker JE (1997) Morphologic changes in the vestibular nerves and nuclei after labyrinthectomy in the cat: a case for the neurotrophin hypothesis in vestibular compensation. Acta Otolaryngol (Stockh) 117:244–249.

Gacek RR, Lyon MJ, Schoonmaker J (1988) Ultrastructural changes in vestibuloocular neurons following vestibular neurectomy in the cat. Ann Otol Rhinol Laryngol 97:42–51.

Gacek RR, Lyon MJ, Schoonmaker JE (1989) Morphologic correlates of vestibular compensation in the cat. Acta Otolaryngol Suppl (Stockh) 462:1–16.

Gacek RR, Lyon MJ, Schoonmaker J (1991) Ultrastructural changes in contralateral vestibulo-ocular neurons following vestibular neurectomy in the cat. Acta Otolaryngol Suppl (Stockh) 477:1–14.

Galiana HL, Flohr H, Melvill Jones G (1984) A reevaluation of intervestibular nuclear coupling: its role in vestibular compensation. J Neurophysiol 51:242–259.

Garcia-Añoveros J, Corey DP (1997) The molecules of mechanosensation. Annu Rev Neurosci 20:567–594.

Gardner JC, Goliath R, Viljoen D, Sellars S, Cortopassi G, Hutchin T, Greenberg J, Beighton P (1997) Familial streptomycin ototoxicity in a South African family: a mitochondrial disorder. J Med Genet 34:904–906.

Gardner K, Barmada MM, Ptacek LJ, Hoffman EP (1997) A new locus for hemiplegic migraine maps to chromosome 1q31. Neurology 49:1231–1238.

Gentine A, Eichhorn J-L, Kopp C, Conraux C (1990) Modelling the action of caloric stimulation of the vestibule. I. The hydrostatic model. Acta Otolaryngol (Stockh) 110:328–333.

Gibson F, Walsh J, Mburu P, Varela A, Brown KA, Antonio M, Beisel KW, Steel KP, Brown SD (1995) A type VII myosin encoded by the mouse deafness gene Shaker-1. Nature 374:62–64.

Gilchrist DP, Curthoys IS, Burgess AM, Cartwright AD, Jinnouchi K, MacDougall HG, Halmagyi GM (2000) Semicircular canal occlusion causes permanent VOR changes. Neuroreport 11:2527–2531.

Gilchrist DPD, Curthoys IS, Cartwright AD, Burgess AM, Topple AN, Halmagyi GM (1998) High acceleration impulsive rotations reveal severe long-term deficits of the horizontal vestibulo-ocular reflex in the guinea pig. Exp Brain Res 123:242–254.

Gilchrist DPD, Cartwright AD, Burgess AM, Curthoys IS (2003) Behavioural characteristics of the quick phase of vestibular nystagmus before and after unilateral labyrinthectomy in guinea pig. Exp Brain Res 149:289–298.

Goldberg JM, Fernández C (1971) Physiology of peripheral neurons innervating semicircular canals of the squirrel monkey. I. Resting discharge and response to constant angular accelerations. J Neurophysiol 34:635–660.

Gomez CM, Thompson RM, Gammack JT, Perlman SL, Dobyns WB, Truwit CL, Zee DS, Clark HB, Anderson JH (1997) Spinocerebellar ataxia type 6: gaze-evoked and vertical nystagmus, Purkinje cell degeneration, and variable age of onset. Ann Neurol 42:933–950.

Graham BP, Dutia MB (2001) Cellular basis of vestibular compensation: analysis and modelling of the role of the commissural inhibitory system. Exp Brain Res 137:387–396.

Griggs RC, Moxley RT III, Lafrance RA, McQuillen J (1978) Hereditary paroxysmal ataxia: response to acetazolamide. Neurology 28:1259–1264.

Guida S, Trettel F, Pagnutti S, Mantuano E, Tottene A, Veneziano L, Fellin T, Spadaro M, Stauderman K, Williams M, Volsen S, Ophoff R, Frants R, Jodice C, Frontali M, Pietrobon D (2001) Complete loss of P/Q calcium channel activity caused by a CACNA1A missense mutation carried by patients with episodic ataxia type 2. Am J Hum Genet 68:759–764.

Guyot JP, Lyon MJ, Gacek RR, Magnin C (1995) Ultrastructural changes in superior vestibular commissural neurons following vestibular neurectomy in the cat. Ann Otol Rhinol Laryngol 104:381–387.

Halmagyi GM (1994) Vestibular insufficiency following unilateral vestibular deafferentation. Aust J Otolaryngol 1:510–512.

Halmagyi GM, Curthoys IS (1988) A clinical sign of canal paresis. Arch Neurol 45:737–739.

Halmagyi GM, Curthoys IS (1999) Clinical testing of otolith function. Ann N Y Acad Sci 871:195–204.

Halmagyi GM, Gresty MA, Gibson WP (1979) Ocular tilt reaction with peripheral vestibular lesion. Ann Neurol 6:80–83.

Halmagyi GM, Rudge P, Gresty MA, Leigh RJ, Zee DS (1980) Treatment of periodic alternating nystagmus. Ann Neurol 8:609–611.

Halmagyi GM, Curthoys IS, Cremer PD, Henderson CJ, Todd MJ, Staples MJ, D'Cruz DM (1990) The human horizontal vestibulo-ocular reflex in response to high-acceleration stimulation before and after unilateral vestibular neurectomy. Exp Brain Res 81:479–490.

Halmagyi GM, Curthoys IS, Colebatch JG (1994a) New tests of vestibular function. Baillieres Clin Neurol 3:485–500.

Halmagyi GM, Fattore CM, Curthoys IS, Wade S (1994b) Gentamicin vestibulo-toxicity. Otolaryngol Head Neck Surg 111:571–574.

Halmagyi GM, Yavor RA, Colebatch JG (1995) Tapping the head activates the vestibular system: a new use for the clinical reflex hammer. Neurology 45:1927–1929.

Halmagyi GM, McGarvie LA, Aw ST, Yavor RA, Todd MJ (2003a). The click-evoked vestibulo-ocular reflex in superior semicircular canal dehiscence. Neurology 60:1172–1175.

Halmagyi GM, Aw ST, McGarvie LA, Todd MJ, Bradshaw A, Yavor RA, Fagan PA (2003b) Superior semicircular canal dehiscence simulating otosclerosis. J Laryngol Otol 117:553–557.

Haslwanter T (1995) Mathematics of three-dimensional eye rotations. Vision Res 35:1727–1739.

Henley C, Igarashi M (1993) Polyamines in the lateral vestibular nuclei of the squirrel monkey and their potential role in vestibular compensation. Acta Otolaryngol (Stockh) 113:235–238.

Henley CM, Igarashi M (1991) Amino acid assay of vestibular nuclei 10 months after unilateral labyrinthectomy in squirrel monkeys. Acta Otolaryngol Suppl (Stockh) 481:407–410.

Hess BJ, Lysakowski A, Minor LB, Angelaki DE (2000) Central versus peripheral origin of vestibuloocular reflex recovery following semicircular canal plugging in rhesus monkeys. J Neurophsiol 84:3078–3082.

Him A, Dutia MB (2001) Intrinsic excitability changes in vestibular nucleus neurons after unilateral deafferentation. Brain Res 908:58–66.

Him A, Johnston AR, Yau JLW, Seckl J, Dutia MB (2001) Tonic activity and GABA responsiveness of medial vestibular nucleus neurons in aged rats. Neuroreport 12:3965–3968.

Hu DN, Qiu WQ, Wu BT, Fang LZ, Zhou F, Gu YP, Zhang QH, Yan JH, Ding YQ, Wong H (1991) Genetic aspects of antibiotic induced deafness: mitochondrial inheritance. J Med Genet 28:79–83.

Hullar TE, Minor LB (1999) High-frequency dynamics of regularly discharging canal afferents provide a linear signal for angular vestibuloocular reflexes. J Neurophysiol 82:2000–2005.

Imbert G, Saudou F, Yvert G, Devys D, Trottier Y, Garnier JM, Weber C, Mandel JL, Cancel G, Abbas N, Durr A, Didierjean O, Stevanin G, Agid Y, Brice A (1996) Cloning of the gene for spinocerebellar ataxia 2 reveals a locus with high sensitivity to expanded CAG/glutamine repeats. Nat Genet 14:285–291.

Jackler RJ (1994) Acoustic neuroma (vestibular schwannoma). In: Jackler RJ, Brackmann DE (eds) Neurotology. St Louis: Mosby, pp. 729–786.

Jen J, Wan J, Graves M, Yu H, Mock AF, Coulin CJ, Kim G, Yue Q, Papazian DM, Baloh RW (2001) Loss-of-function EA2 mutations are associated with impaired neuromuscular transmission. Neurology 57:1843—1848.

Jen JC, Yue Q, Karrim J, Nelson SF, Baloh RW (1998) Spinocerebellar ataxia type 6 with positional vertigo and acetazolamide responsive episodic ataxia. J Neurol Neurosurg Psychiatry 65:565–568.

Joutel A, Bousser MG, Biousse V, Labauge P, et al. (1993) A gene for familial hemiplegic migraine maps to chromosome 19. Nat Genet 5:40–45.

Joutel A, Ducros A, Vahedi K, Labauge P, et al. (1994) Genetic heterogeneity of familial hemiplegic migraine. Am J Hum Genet 55:1166–1172.

Kawaguchi Y, Okamoto T, Taniwaki M, Aizawa M, Inoue M, Katayama S, Kawakami H, Nakamura S, Nishimura M, Akiguchi I, Kimura J, Narumiya S, Kakizuka A (1994) CAG expansions in a novel gene for Machado–Joseph disease at chromosome 14q32.1. Nat Genet 8:221–228.

Kentala E, Pyykko I (1997) Benign recurrent vertigo—true or artificial diagnosis. Acta Otolaryngol Suppl (Stockh) 529:101–103.

Kevetter GA, Perachio AA (1986) Distribution of vestibular afferents that innervate the sacculus and posterior canal in the gerbil. J Comp Neurol 254:410–424.

Khetarpal U (1993) Autosomal dominant sensorineural hearing loss. Further temporal bone findings. Arch Otolaryngol Head Neck Surg 119:106–108.

Khetarpal U (2000) DFNA9 is a progressive audiovestibular dysfunction with a microfibrillar deposit in the inner ear. Laryngoscope 110:1379–1384.

Khetarpal U, Schuknecht HF, Gacek RR, Holmes LB (1991) Autosomal dominant sensorineural hearing loss. Pedigrees, audiologic findings, and temporal bone findings in two kindreds. Arch Otolaryngol Head Neck Surg 117:1032–1042.

Kim JS, Yue Q, Jen JC, Nelson SF, Baloh RW (1998) Familial migraine with vertigo: no mutations found in CACNA1A. Am J Med Genet 79:148–151.

Kimberling WJ, Moller CG, Davenport SLH, Lund G, Grissom TJ, Priluck I, White V, Weston MD, Biscone-Halterman K, Brookhouser PE (1989) Usher syndrome: clinical findings and gene localization studies. Laryngoscope 99:66–72.

Kluwe L, MacCollin M, Tatagiba M, Thomas S, Hazim W, Haase W, Mautner VF (1998) Phenotypic variability associated with 14 splice-site mutations in the NF2 gene. Am J Med Genet 77:228–233.

Kluwe L, Friedrich RE, Tatagiba M, Mautner VF (2002) Presymptomatic diagnosis for children of sporadic neurofibromatosis 2 patients: a method based on tumor analysis. Genet Med 4:27–30.

Knudson AG (1971) Mutation and cancer: a statistical study. Proc Natl Acad Sci U S A 68:820–823.

Komatsuzaki A, Tsunoda A (2001) Nerve origin of the acoustic neuroma. J Laryngol Otol 115:376–379.

Kushiro K, Zakir M, Ogawa Y, Sato H, Uchino Y (1999) Saccular and utricular inputs to sternocleidomastoid motoneurons of decerebrate cats. Exp Brain Res 126:410–416.

Lanzi G, Ballotin U, Fazzi E, Tagliasacchi M, Manfrin M, Mira E (1994) Benign paroxysmal vertigo of childhood: a long term follow-up. Cephalalgia 14:458–460.

Lasker DM, Backous DD, Lysakowski A, Davis GL, Minor LB (1999) Horizontal vestibuloocular reflex evoked by high-acceleration rotations in the squirrel monkey. II. Responses after canal plugging. J Neurophysiol 82:1271–1285.

Lasker DM, Hullar TE, Minor LB (2000) Horizontal vestibuloocular reflex evoked by high-acceleration rotations in the squirrel monkey. III. Responses after labyrinthectomy. J Neurophysiol 83:2482–2496.

Li H, Godfrey TG, Godfrey DA, Rubin AM (1996) Quantitative changes of amino acid distributions in the rat vestibular nuclear complex after unilateral vestibular ganglionectomy. J Neurochem 66:1550–1564.

MacDougall HG, Brizuela AE, Burgess AM, Curthoys IS (2002) Between-subject variability and within-subject reliability of the human eye-movement response to bilateral galvanic (DC) vestibular stimulation. Exp Brain Res 144:69–78.

Manolis EN, Yandavi N, Nadol JB Jr, Eavey RD, McKenna M, Rosenbaum S, Khetarpal U, Halpin C, Merchant SN, Duyk GM, MacRae C, Seidman CE, Seidman JG (1996) A gene for non-syndromic autosomal dominant progressive postlingual sensorineural hearing loss maps to chromosome 14q12–13. Hum Mol Genet 5:1047–1050.

Martini A (1982) Hereditary Meniere's disease: report of two families. Am J Otolaryngol 3:163–167.

Martini A, Mazzoli M, Kimberling W (1997) An introduction to the genetics of normal and defective hearing. Ann N Y Acad Sci 830:361–374.

May A, Ophoff RA, Terwindt GM, Urban C, van Eijk R, Haan J, Diener HC, Lindhout D, Frants RR, Sandkuijl LA, Ferrari MD (1995) Familial hemiplegic migraine locus on 19p13 is involved in the common forms of migraine with and without aura. Hum Genet 96:604–608.

McCue MP, Guinan JJ (1997) Sound-evoked activity in primary afferent neurons of the mammalian vestibular system. Am J Otol 18:355–360.

Minor LB, Solomon D, Zinreich JS, Zee DS (1998) Tullio's phenomenon due to deshiscence of the superior semicircular canal. Arch Otolaryngol Head Neck Surg 124:249–258.

Minor LB, Carey JP, Cremer PD, Lustig LR, Streubel SO, Ruckenstein MJ (2003) Dehiscence of bone overlying the superior canal as a cause of apparent conductive hearing loss. Otol Neurotol 24:270–278.

Moffat DA, Irving R (1995) The molecular genetics of vestibular schwannoma. J Laryngol Otol 109:381–384.

Morell RJ, Hung JK, Hood LJ, Goforth L, Friderici K, Fisher R, Van Camp G, Berlin Cl, Oddoux C, Ostrer H, Keats B, Friedman TB (1998) Mutations in the connexin 26 gene (GJB2) among ashkenazi Jews with nonsyndromic recessive deafness. N Engl J Med 339:1500–1505.

Moretti G, Manzoni GC, Caffarra P, Parma M (1980) "Benign recurrent vertigo" and its connection with migraine. Headache 20:344–346.

Murofushi T, Curthoys IS (1997) Physiological and anatomical study of click-sensitive primary vestibular afferents in the guinea-pig. Acta Otolaryngol (Stockh) 117:66–72.

Murofushi T, Curthoys IS, Topple AN, Colebatch JG, Halmagyi GM (1995) Responses of guinea pig primary vestibular neurons to clicks. Exp Brain Res 103:174–178.

Murofushi T, Curthoys IS, Gilchrist DP (1996a) Response of guinea pig vestibular nucleus neurons to clicks. Exp Brain Res 111:149–152.

Murofushi T, Halmagyi GM, Yavor RA, Colebatch JG (1996b) Vestibular evoked myogenic potentials in vestibular neuritis: an indicator of inferior vestibular nerve involvement. Arch Otolaryngol Head Neck Surg 122:845–848.

Murofushi T, Matsuzaki M, Mizuno M (1998) Vestibular evoked myogenic potentials in patients with acoustic neuromas. Arch Otolaryngol Head Neck Surg 124:509–512.

Murofushi T, Matsuzaki M, Wu CH (1999) Short tone burst-evoked myogenic potentials on the sternocleidomastoid muscle: Are these potentials also of vestibular origin? Arch Otolaryngol Head Neck Surg 125:660–664.

Murofushi T, Matsuzaki M, Takegoshi H (2001a) Glycerol affects vestibular evoked myogenic potentials in Meniere's disease. Auris Nasus Larynx 28:205–208.

Murofushi T, Shimizu K, Takegoshi H, Cheng PW (2001b) Diagnostic value of prolonged latencies in the vestibular evoked myogenic potential. Arch Otolaryngol Head Neck Surg 127:1069–1072.

Murofushi T, Takegoshi H, Ohki M, Ozeki H (2002) Galvanic-evoked myogenic responses in patients with an absence of click-evoked vestibulo-collic reflexes. Clin Neurophysiol 113:305–309.

Neuhauser H, Leopold M, von Brevern M, Arnold G, Lempert T (2001) The interrelations of migraine, vertigo, and migrainous vertigo. Neurology 56:436–441.

Noble-Topham SE, Dyment DA, Cader MZ, Ganapathy R, Brown JD, Rice GP, Ebers GC (2002) Migraine with aura is not linked to the FHM gene CACNA1A or the chromosomal region, 19p13. Neurology 59:1099–1101.

Oh AK, Jacobson KM, Jen JC, Baloh RW (2001) Slowing of voluntary and involuntary saccades: an early sign in spinocerebellar ataxia type 7. Ann Neurol 49:801–804.

Oliveira CA, Bezerra RL, Araujo MF, Almeida VF, Messias CI (1997) Meniere's syndrome and migraine: incidence in one family. Ann Otol Rhinol Laryngol 106:823–829.

O'Neill G (1987) The caloric stimulus—temperature generation within the temporal bone. Acta Otolaryngol (Stockh) 103:266–272.

Ophoff RA, van Eijk R, Sandkuijl LA, Terwindt GM, Grubben CP, Haan J, Lindhout D, Ferrari MD, Frants RR (1994) Genetic heterogeneity of familial hemiplegic migraine. Genomics 22:21–26.

Ophoff RA, Terwindt GM, Vergouwe MN, van Eijk R, Oefner PJ, Hoffman SMG, Lamerdin JE, Mohrenweiser HW, Bulman DE, Ferrari M, Haan J, Lindhout D, van Ommen G-JB, Hofker MH, Ferrari MD, Frants RR (1996) Familial hemiplegic migraine and episodic ataxia type-2 are caused by mutations in the Ca^{2+} channel gene CACNL1A4. Cell 87:543–552.

Orr HT, Chung M-Y, Banfi S, Kwiatkowski TJ Jr, Servadio A, Beaudet AL, McCall AE, Duvick LA, Ranum LP, Zoghbi HY (1993) Expansion of an unstable trinucleotide CAG repeat in spinocerebellar ataxia type 1. Nat Genet 4:211–226.

Parry DM, Eldridge R, Kaiser-Kupfer MI, Bouzas EA, Pikus A, Patronas N (1994) Neurofibromatosis 2 (NF2): clinical characteristics of 63 affected individuals and clinical evidence for heterogeneity. Am J Med Genet 52:450–461.

Parry DM, MacCollin MM, Kaiser-Kupfer MI, Pulaski K, Nicholson HS, Bolesta M, Eldridge R, Gusella JF (1996) Germ-line mutations in the neurofibromatosis 2 gene: correlations with disease severity and retinal abnormalities. Am J Hum Genet 59:529–539.

Philcox DV, Sellars SL, Pamplett R, Beighton P (1975) Vestibular dysfunction in hereditary ataxia. Brain 98:309–316.

Pickles JO, Comis SD, Osborne MP (1984) Cross-links between stereocilia in the guinea pig organ of Corti, and their possible relation to sensory transduction. Hear Res 15:103–112.

Pohl DV (1996) Surgical procedures for benign postitional vertigo. In: Baloh RW, Halmagyi GM (eds) Disorders of the Vestibular System. New York: Oxford University Press, pp. 563–574.

Precht W, Shimazu H (1965) Functional connections of tonic and kinetic vestibular neurons with primary vestibular afferents. J Neurophysiol 28:1014–1028.

Precht W, Shimazu H, Markham CH (1966) A mechanism of central compensation of vestibular function following hemilabyrinthectomy. J Neurophysiol 29:996–1010.

Prezant TR, Agapian JV, Bohlman CM, Bu X, Oztas S, Qiu WQ, Arnos KS, Cortopassi G, Jaber L, Rotter JI, Shohat M, Fischel-Ghodsian N (1993) Mitochondrial ribosomal RNA mutation associated with both antibiotic-induced and non-syndromic deafness. Nat Genet 4:289–294.

Ptacek LJ (1998) The place of migraine as a channelopathy. Curr Opin Neurol 11:217–226.

Rabbitt RD, Boyle R, Highstein SM (1999) Influence of surgical plugging on horizontal semicircular canal mechanics and afferent response dynamics. J Neurophysiol 82:1033–1053.

Radtke A, Lempert T, Gresty MA, Brookes GB, Bronstein AM, Neuhauser H (2002) Migraine and Meniere's disease: Is there a link? Neurology 59:1700–1704.

Reid CB, Eisenberg R, Halmagyi GM, Fagan PA (1996) The outcome of vestibular nerve section for intractable vertigo: the patients' point of view. Laryngoscope 106:1553–1556.

Reisine H, Raphan T (1992) Neural basis for eye velocity generation in the vestibular nuclei of alert monkeys during off-vertical axis rotation. Exp Brain Res 92:209–226.

Ris L, Godaux E (1998) Neuronal activity in the vestibular nuclei after contralateral or bilateral labyrinthectomy in the alert guinea pig. J Neurophysiol 80:2352–2367.

Ris L, de Waele C, Serafin M, Vidal PP, Godaux E (1995) Neuronal activity in the ipsilateral vestibular nucleus following unilateral labyrinthectomy in the alert guinea pig. J Neurophysiol 74:2087–2099.

Ris L, Capron B, de Waele C, Vidal PP, Godaux E (1997) Dissociations between behavioural recovery and restoration of vestibular activity in the unilabyrinthectomized guinea-pig. J Physiol (Lond) 500:509–522.

Robertson NG, Lu L, Heller S, Merchant SN, Eavey RD, McKenna M, Nadol JB Jr, Miyamoto RT, Linthicum FH Jr, Lubianca Neto JF, Hudspeth AJ, Seidman CE, Morton CC, Seidman JG (1998) Mutations in a novel cochlear gene cause DFNA9, a human nonsyndromic deafness with vestibular dysfunction. Nat Genet 20:299–303.

Robertson NG, Resendes BL, Lin JS, Lee C, Aster JC, Adams JC, Morton CC (2001) Inner ear localization of mRNA and protein products of COCH mutated in the sensorineural deafness and vestibular disorder DFNA9. Hum Mol Genet 10:2493–2500.

Rouleau GA, Merel P, Lutchman M, Sanson M, Zucman J, Marineau C, Hoang-Xuan K, Demczuk S, Desmaze C, Plougastel B, Pulst S, Lenoir G, Bijlsma E, Fashold K, Dumanski J, de Jong P, Parry D, Eldridge R, Aurias A, Delattre O, Thomas G (1993) Alteration in a new gene encoding a putative membrane-organising protein causes neurofibromatosis type 2. Nature 363:515–521.

Saxon DW, Anderson JH, Beitz AJ (2001) Transtympanic tetrodotoxin alters the VOR and Fos labeling in the vestibular complex. Neuroreport 12:3051–3055.

Schmid-Priscoveanu A, Bohmer A, Obzina H, Straumann D (2001) Caloric and search-coil head-impulse testing in patients after vestibular neuritis. J Assoc Res Otolaryngol 2:72–78.

Schuknecht H (1957) Ablation therapy in the management of Meniere's disease. Acta Otolaryngol Suppl (Stockh) 132:1–42.

Scoles DR, Baser ME, Pulst SM (1996) A missense mutation in the neurofibromatosis 2 gene occurs in patients with mild and severe phenotypes. Neurology 47:544–546.

Scudder CA, Fuchs AF (1992) Physiological and behavioural identification of vestibular nucleus neurons mediating the horizontal vestibuloocular reflex in trained rhesus monkeys. J Neurophysiol 68:244–264.

Serafin M, de Waele C, Khateb A, Vidal PP, Muhlethaler M (1991a) Medial vestibular nucleus in the guinea-pig. I. Intrinsic membrane properties in brainstem slices. Exp Brain Res 84:417–425.

Serafin M, de Waele C, Khateb A, Vidal PP, Muhlethaler M (1991b) Medial vestibular nucleus in the guinea-pig. II. Ionic basis of the intrinsic membrane properties in brainstem slices. Exp Brain Res 84:426–433.

Sheykholeslami K, Murofushi T, Kermany MH, Kaga K (2000) Bone-conducted evoked myogenic potentials from the sternocleidomastoid muscle. Acta Otolaryngol (Stockh) 120:731–734.

Sheykholeslami K, Habiby Kermany M, Kaga K (2001) Bone-conducted vestibular evoked myogenic potentials in patients with congenital atresia of the external auditory canal. Int J Pediatr Otorhinolaryngol 57:25–29.

Shimazu H, Precht W (1965) Tonic and kinetic responses of cat's vestibular neurons to horizontal angular acceleration. J Neurophysiol 28:991–1013.

Shimazu H, Precht W (1966) Inhibition of central vestibular neurons from the contralateral labyrinth and its mediating pathway. J Neurophysiol 29:467–492.

Shimizu K, Murofushi T, Sakurai M, Halmagyi M (2000) Vestibular evoked myogenic potentials in multiple sclerosis. J Neurol Neurosurg Psychiatry 69:276–277.

Slater R (1979) Benign recurrent vertigo. J Neurol Neurosurg Psychiatry 42:363–367.

Spiegel EA, Demetriades TD (1925) Die zentrale Kompensation des Labyrinthverlustes. Pflügers Arch Gesamte Physiol Menschen Tiere 210:215–222.

Streubel S-O, Cremer PD, Carey JP, Weg N, Minor LB (2001) Vestibular-evoked myogenic potentials in the diagnosis of superior semicircular canal dehiscence. Acta Otolaryngol Suppl (Stockh) 545:41–49.

Sue CM, Lipsett LJ, Crimmins DS, Tsang CS, Boyages SC, Presgrave CM, Gibson WPR, Byrne E, Morris JGL (1998) Cochlear origin of hearing loss in MELAS syndrome. Ann Neurol 43:350–359.

Swanson R, Hice RE, Folander K, Sanguinetti MC (1993) The IsK protein, a slowly activating voltage-dependent K^+ channel. Semin Neurosci 5:117–124.

Tabak S, Collewijn H, Boumans LJJM, Vandersteen J (1997a) Gain and delay of human vestibulo-ocular reflexes to oscillation and steps of the head by a reactive torque helmet. I. Normal subjects. Acta Otolaryngol (Stockh) 117:785–787.

Tabak S, Collewijn H, Boumans LJJM, Vandersteen J (1997b) Gain and delay of human vestibulo-ocular reflexes to oscillation and steps of the head by a reactive torque helmet. II. Vestibular-deficient subjects. Acta Otolaryngol (Stockh) 117:796–809.

Tighilet B, Lacour M (2001) Gamma amino butyric acid (GABA) immunoreactivity in the vestibular nuclei of normal and unilateral vestibular neurectomized cats. Eur J Neurosci 13:2255–2267.

Todd N (2001) Evidence for a behavioral significance of saccular acoustic sensitivity in humans. J Acoust Soc Am 110:380–390.

Trofatter JA, MacCollin MM, Rutter JL, Murrell JR, Duyao MP, Parry DM, Eldridge R, Kley N, Menon AG, Pulaski K, Haase V, Ambrose C, Munroe D, Bove C, Haines J, Martuza R, MacDonald M, Seizinger B, Short MP, Buckler A, Gusella J (1993) A novel moesin-, ezrin-, radixin-like gene is a candidate for the neurofibromatosis 2 tumor suppressor. Cell 72:791–800.

Tsutsumi T, Tsunoda A, Noguchi Y, Komatsuzaki A (2000) Prediction of the nerves of origin of vestibular schwannomas with vestibular evoked myogenic potentials. Am J Otol 21:712–715.

Tullio P (1929) Das Ohr und die Entstehung der Sprache und Schrift. Berlin: Urban and Schwarzenberg.

Verhagen WI, Huygen PL, Horstink MW (1987) Familial congenital vestibular areflexia. J Neurol Neurosurg Psychiatry 50:933–935.

Versino M, Colnaghi S, Callieco R, Bergamaschi R, Romani A, Cosi V (2002) Vestibular evoked myogenic potentials in multiple sclerosis patients. Clin Neurophysiol 113:1464–1469.

Vetter DE, Mann JR, Wangemann P, Jianzhong L, McLaughlin KJ, Lesage F, Marcus DC, Lazdunski M, Heinemann SF, Barhanin J (1996) Inner ear defects induced by null mutation of the isk gene. Neuron 17:1251–1264.

Vibert N, de Waele C, Serafin M, Babalian A, Muhlethaler M, Vidal PP (1997) The vestibular system as a model of sensorimotor transformations. A combined in vivo and in vitro approach to study the cellular mechanisms of gaze and posture stabilization in mammals. Prog Neurobiol 51:243–286.

Vibert N, Babalian A, Serafin M, Gasc JP, Muhlethaler M, Vidal PP (1999) Plastic changes underlying vestibular compensation in the guinea-pig persist in isolated, in vitro whole brain preparations. Neuroscience 93:413–432.

Vibert N, Beraneck M, Bantikyan A, Vidal PP (2000) Vestibular compensation modifies the sensitivity of vestibular neurons to inhibitory amino acids. Neuroreport 11:1921–1927.

Vidal PP, de Waele C, Vibert N, Muhlethaler M (1998a) Vestibular compensation revisited. Otolaryngol Head Neck Surg 119:34–42.

Vuillaume I, Devos D, Schraen-Maschke S, Dina C, Lemainque A, Vasseur F, Bocquillon G, Devos P, Kocinski C, Marzys C, Destee A, Sablonniere B (2002) A new locus for spinocerebellar ataxia (SCA21) maps to chromosome 7p21.3–p15.1. Ann Neurol 52:666–670.

Watson SR, Fagan P, Colebatch JG (1998) Galvanic stimulation evokes short-latency EMG responses in sternocleidomastoid which are abolished by selective vestibular nerve section. Electroencephalogr Clin Neurophysiol 109:471–474.

Watson SR, Halmagyi GM, Colebatch JG (2000) Vestibular hypersensitivity to sound (Tullio phenomenon): structural and functional assessment. Neurology 54:722–728.

Weil D, Blanchard S, Kaplan J, Guilford P, Gibson F, Walsh J, Mburu P, Varela A, Levilliers J, Weston MD, Kelley PM, Kimberling WJ, Wagenaar M, Levi-Acobas F, Larget-Piet D, Munnich A, Steel KP, Brown SDM, Petit C (1995) Defective myosin VII gene responsible for Usher syndrome type 1B. Nature 374:60–61.

Weissenstein L, Ratnam R, Anastasio TJ (1996) Vestibular compensation in the horizontal vestibulo-ocular reflex of the goldfish. Behav Brain Res 75:127–137.

Welgampola MS, Colebatch JG (2001) Characteristics of tone burst-evoked myogenic potentials in the sternocleidomastoid muscles. Otol Neurotol 22:796–802.

Welgampola MS, Rosengren SM, Halmagyi GM, Colebatch JG (2003) Vestibular activation by bone-conducted sound. J Neurol Neurosurg Psychiatry 74:771–778.

Welling DB (1998) Clinical manifestations of mutations in the neurofibromatosis type 2 gene in vestibular schwannomas (acoustic neuromas). Laryngoscope 108:178–189.

Yagi T, Kurosaki S, Yamanobe S, Morizono T (1992) Three-component analysis of caloric nystagmus in humans. Arch Otolaryngol Head Neck Surg 118:1077–1080.

Yamanaka T, Him A, Cameron SA, Dutia MB (2000) Rapid compensatory changes in GABA receptor efficacy in rat vestibular neurons after unilateral labyrinthectomy. J Physiol (Lond) 523:413–424.

Young ED, Fernández C, Goldberg JM (1977) Responses of squirrel monkey vestibular neurons to audio-frequency sound and head vibration. Acta Otolaryngol (Stockh) 84:352–360.

Young YH, Wu CC, Wu CH (2002a) Augmentation of vestibular evoked myogenic potentials: an indication for distended saccular hydrops. Laryngoscope 112:509–512.

Young YH, Huang TW, Cheng PW (2002b) Vestibular evoked myogenic potentials in delayed endolymphatic hydrops. Laryngoscope 112:1623–1626.

Yue Q, Jen JC, Nelson SF, Baloh RW (1997) Progressive ataxia due to a missense mutation in a calcium-channel gene. Am J Hum Genet 61:1078–1087.

Zee DS, Preziosi TJ, Proctor LR (1982) Bechterew's phenomenon in a human patient. Ann Neurol 12:495–496.

Zhuchenko O, Bailey J, Bonnen P, Ashizawa T, Stockton DW, Amos C, Dobyns WB, Subramony SH, Zoghbi HY, Lee CC (1997) Autosomal dominant cerebellar ataxia (SCA6) associated with small polyglutamine expansions in alpha-1-A-voltage-dependent calcium channel. Nat Genet 15:62–69.

Index